Binge Eating

Guido K.W. Frank • Laura A. Berner
Editors

Binge Eating

A Transdiagnostic Psychopathology

 Springer

Editors
Guido K.W. Frank
Department of Psychiatry,
UCSD Eating Disorder Center
for Treatment and Research
University of California San Diego
San Diego, CA, USA

Laura A. Berner
Department of Psychiatry, Center of
Excellence in Eating and Weight
Disorders
Icahn School of Medicine at Mount Sinai
New York, NY, USA

ISBN 978-3-030-43564-6   ISBN 978-3-030-43562-2   (eBook)
https://doi.org/10.1007/978-3-030-43562-2

This Springer imprint is published by the registered company Springer Nature Switzerland AG.
The registered company address is: Gewerbestrasse 11, 6330 Cham, Switzerland

Preface

Binge eating, or feeling out of control while consuming a large amount of food in a discrete time period, is characteristic of several eating disorders. It is sometimes followed by compensatory behaviors like self-induced vomiting and is often highly distressing and resistant to treatment.

Binge eating as a specific symptom has been understudied. In this book, written for both clinicians and researchers, experts in the eating disorders field review our current knowledge of binge eating. Research on the binge eating phenotype is discussed across cultures and psychosocial factors and integrated into theoretical models of binge eating development and maintenance. This is followed by basic science and human research on reward and control brain circuits, hormonal influences, sex, and genotype. Next, current psychosocial and pharmacological interventions for binge eating are reviewed. Final chapters propose a research agenda and discuss new directions for treatment development.

Overall, this is the most up-to-date, integrative, and comprehensive synthesis of research on binge eating.

San Diego, CA, USA
New York, NY, USA

Guido K.W. Frank
Laura A. Berner

Contents

About the Editors

Guido K.W. Frank Dr. Frank is board certified in adult as well as child and adolescent psychiatry. He earned his medical degree at the Ludwig Maximilian University in Munich, Germany. He trained in psychosomatics at the Center for Behavioral Health Klinik Roseneck, Prien, Germany, and then received clinical and research training at the Western Psychiatric Institute and Clinic, University of Pittsburgh, and the University of California San Diego, USA. He holds an appointment as professor in the Department of Psychiatry at the University of California San Diego.

Dr. Frank has done extensive postgraduate work including receiving research training in the neurobiology of psychiatric disorders. He has also received extensive certified training in cognitive behavioral and other psychotherapies. Dr. Frank is a fellow of the Academy for Eating Disorders and the American College of Neuropsychopharmacology. He is an expert consultant to local and national law firms. He has received multiple awards, including an NIH Minority Access to Research Career Program (NIMH) Mentor Recognition Award and the first Eating Disorder Foundation Greg Hueni Memorial Award for excellence in research.

Dr. Frank has been funded through the National Institute of Mental Health and multiple private foundation grants for the past fifteen years to study the biological domains that underlie eating disorder behavior in youth and adults. His research work has introduced computational modeling to the eating disorder field, and his overarching goal is to develop translational research designs that bridge clinical presentation with neuroscience to develop more effective treatments.

Laura A. Berner Dr. Berner is a clinical psychologist interested in how cognitive neuroscience can help us better understand and treat eating disorders. She earned her PhD from Drexel University and completed her postdoctoral fellowship at the University of California San Diego Eating Disorders Center for Treatment and Research. She is currently an Assistant Professor of Psychiatry at the Icahn School of Medicine at Mount Sinai.

Her research aims to understand how altered self-control may promote cycles of binge eating, purging, and restricted eating. To this end, her work combines innovative behavioral tasks with brain imaging and self-report and laboratory-based symptom measures. In addition, Dr. Berner is a licensed clinician with extensive experience and specialized training in the treatment of eating disorders and co-occurring conditions. Her ultimate goals are to build new explanatory models of eating disorders and to translate research findings into tools for clinical decision-making and novel interventions.

Dr. Berner's research has been funded by the National Institute of Mental Health, the Hilda and Preston Davis Foundation, the American Psychological Association, and the Academy for Eating Disorders. She has received early career awards from the American College of Neuropsychopharmacology, the Society of Biological Psychiatry, the Academy for Eating Disorders, the Society for the Study of Ingestive Behavior, and the Eating Disorders Research Society.

Epidemiology of Binge Eating

Madeline R. Wick, Elizabeth H. Fitzgerald, and Pamela K. Keel

Abstract

This chapter reviews information on the epidemiology of binge eating and disorders defined by the presence of recurrent binge-eating episodes, bulimia nervosa (BN), and binge-eating disorder (BED). Prevalence estimates indicate that binge eating affects 4.9% of females and 4% of males over their lifetimes. BN occurs in about 1.9% of females and 0.6% of males over their lifetimes, and 1.5% of females and 0.1% of males currently have BN. BED is more common than BN: about 2.8% of females and 1% of males will be diagnosed with BED in their lifetimes, and 2.3% of females and 0.3% of males currently have BED. Both BN and BED are more common in women than men, and binge eating and BN are more common in young adulthood compared to middle and late adulthood. Mixed findings exist for gender differences in prevalence of any binge eating, and the prevalence of BED appears similar throughout young and middle adulthood. Prevalence estimates for binge eating are higher among racial/ethnic minorities and prevalence estimates for BN are higher among Latinos and African-Americans compared to non-Latino White Americans. In contrast, prevalence estimates for BED are largely similar among racial/ethnic minorities and non-Latino White Americans. Incidence rates for binge eating and BED have not been examined for changes over time. For BN, studies support a significant increase in incidence during the latter half of the twentieth century, with rates either remaining stable or decreasing during the transition to the twenty-first century. The chapter ends with a discussion of factors relevant to interpreting the epidemiological data.

Keywords

Epidemiology · Point prevalence · Twelve-month prevalence · Lifetime prevalence · Incidence · Binge eating · Bulimia nervosa · Binge-eating disorder

Learning Objectives

In this chapter, you will:

- Learn how to distinguish different kinds of prevalence estimates from one another, as well as recognize how prevalence differs from incidence.
- Be able to identify how many people experience binge eating and disorders characterized by binge eating, and who is at increased risk.
- Understand if rates of binge eating and disorders characterized by binge eating are increasing, decreasing, or remaining constant over time.

M. R. Wick · E. H. Fitzgerald · P. K. Keel (✉)
Florida State University, Tallahassee, FL, USA
e-mail: wick@psy.fsu.edu; fitzgerald@psy.fsu.edu; keel@psy.fsu.edu

© Springer Nature Switzerland AG 2020
G. K.W. Frank, L. A. Berner (eds.), *Binge Eating*, https://doi.org/10.1007/978-3-030-43562-2_1

1 Introduction

Binge eating is a core feature of both bulimia nervosa (BN) and binge-eating disorder (BED) and their other-specified counterparts, as well as a possible feature of the binge/purge subtype of anorexia nervosa (AN) in the *Diagnostic and Statistical Manual of Mental Disorders*, 5th Edition (DSM-5, American Psychiatric Association 2013) and *International Classification of Disease*, Eleventh Revision (ICD-11, World Health Organization 2018). The present chapter provides epidemiological data for binge eating, BN, and BED, including prevalence estimates and incidence rates. This information allows us to understand how many individuals have these disorders, if there are demographic risk factors for these disorders, and if rates have changed over time.

2 What Is Binge Eating?

In the DSM-5 (American Psychiatric Association 2013), binge eating refers to the act of eating a large amount of food in a discrete period of time (e.g., 2 h) while experiencing a loss of control. Most recently, the 11th revision of the *International Classification of Diseases* (ICD-11; World Health Organization 2018) introduced a broader definition of binge eating, defined as "a distinct period of time during which the individual experiences a subjective loss of control over eating, eating notably more or differently than usual, and feels unable to stop eating or limit the type or amount of food eaten." With this definition, binge eating requires a loss of control but no longer requires consumption of a large amount of food. While the DSM dominates eating disorders research in the United States as well as several other nations, the ICD impacts epidemiological findings from other parts of the world.

3 What Eating Disorders Are Defined by Binge Eating?

Binge eating is a core feature of two eating disorders in the DSM-5 and ICD-11: BN and BED. Individuals with AN can also have binge-eating episodes, but it is not a core defining feature of AN (i.e., individuals can meet diagnostic criteria for AN but never engage in binge eating). The DSM-5 diagnostic criteria for BN require individuals to engage in binge eating, as well as inappropriate compensatory behaviors, at least once a week for 3 months, whereas the diagnostic criteria for BN in the ICD-11 require individuals to engage in binge eating at least once a week for at least 1 month. Additionally, the ICD-11 specifies no minimum frequency for inappropriate compensatory behaviors, other than indicating that these behaviors must be "repeated." For a diagnosis of DSM-5 BED, binge eating must occur at least once a week for 3 months, whereas for a diagnosis of ICD-11 BED, binge eating must occur at least once a week over a period of several months. As such, the ICD-11 implies that BED should have a longer minimum duration than BN, while BED and BN share the same minimum duration in the DSM-5.

4 What Are Prevalence and Incidence?

Prevalence refers to an estimate of how many individuals in a population have a given disorder. Prevalence estimates for eating disorders are commonly measured with one of three metrics: lifetime prevalence, 12-month prevalence, or point prevalence. Lifetime prevalence estimates how many individuals in the population have ever met diagnostic criteria for a given disorder in their lifetime. Twelve-month prevalence estimates how many individuals in the population had a given disorder at any point in the last year, regardless of whether or not they currently suffer from that disorder or when they first developed the disorder. Finally, point prevalence estimates how many individuals in the population have a given disorder at the time of measurement. Because prevalence does not capture the time of onset, it is important to refer to it as an estimate rather than a rate. Prevalence estimates answer several important questions, including how many individuals have a given problem or disorder and whether gender, age group, or racial/

ethnic groups differ in likelihood of having a given problem or disorder.

An incidence rate identifies how many new cases of a disorder emerge over a specified time period per a unit of population. For eating disorders, the incidence is commonly measured as the number of new cases of a disorder per 100,000 individuals in the population in a year, or, stated differently, per 100,000 person-years (Smink et al. 2012). Unlike prevalence estimates, incidence over time reflects a true rate of change and provides information about whether a problem or disorder has become more or less common over time.

5 How Many Individuals Experience Binge Eating?

Data from the U.S. National Comorbidity Survey-Replication (NCS-R) reported that the lifetime prevalence of binge eating was 4.5% and that the 12-month prevalence for binge eating was 2.1% (Hudson et al. 2007). This means that over one in 25 individuals will engage in binge eating at some point in their lives, and that in a single year, about one in 50 individuals will engage in binge eating. Of note, the NCS-R defined binge eating as eating an objectively large amount of food at least twice a week for 3 months but did not assess for loss of control over eating. Instead, authors used alternative criteria to evaluate distress as a means to differentiate episodes from simple overeating. Therefore, estimates from the NCS-R may be comparable or slightly higher than what would have been observed had a loss of control been required. Very elevated point prevalence estimates of binge eating occur in certain populations: for example, for bariatric surgery patients, an estimated 61.3%, or about three in five patients, experienced recurrent binge-eating episodes (Saunders et al. 1998). To our knowledge, only limited epidemiological data exist on binge eating from non-U.S. populations. In a sample of 50 Fijian women, the estimated point prevalence of binge-eating episodes was 10% (Becker et al. 2003). More research is needed to determine the prevalence of binge eating in non-U.S. populations.

6 Does Risk for Binge Eating Differ Across Demographic Groups: Gender, Age Group, and Race/Ethnicity?

Data from the NCS-R indicated that almost one in 20 women (4.9%) and one out of 25 men (4%) had engaged in recurrent episodes of overeating at some point in their lives. Furthermore, 12-month prevalence estimates of binge eating were 2.5% among women and 1.7% among men (Hudson et al. 2007). This means that in a given year, between one and two out of 50 women and about one in 50 men have repeated episodes in which they eat a large amount of food. Although both estimates suggest that binge eating may be most common in women, neither reflected a statistically significant difference between genders. Using this same definition for binge eating, point prevalence estimates for binge eating were significantly higher in female (5.1%) compared to male (0.4%) college students, meaning that about one in 20 female college students compared to one in 200 male college students engaged in twice weekly binge eating (Keel et al. 2006). When no minimum frequency of binge eating was considered, lifetime prevalence was 28.7% for females and 8.1% for males, and point prevalence of binge eating was 14.8% for female students and 3.8% for male students. This means that about three in 10 female students and two in 25 male students had engaged in binge eating in their lifetime.

One reason that estimates for college students may have been higher is because they fall in a high-risk age group. In the NCS-R, binge eating was more common in younger compared to older age groups (Hudson et al. 2007). Furthermore, binge eating is more common among ethnic minorities than it is among non-Latino whites (Marques et al. 2011). Specifically, African-Americans, Latinos, and Asians living in the United States were at least twice as likely as non-Latino whites to engage in any binge eating over the course of a year (Marques et al. 2011). Additionally, research has indicated that lifetime prevalence of binge eating among Latinos is 5.61%, meaning that between 5 and 6 out of 100 Latinos will engage in some sort of binge eating during their lifetime (Alegria et al. 2007). Similarly, the lifetime prevalence for binge eating among African-Americans was 5.08% (Taylor et al. 2007). Of note, these data come from the United States, and little is known about the prevalence of binge eating among different demographic groups outside of the United States.

7   Have Rates of Binge Eating Changed in Recent Years?

Few studies have examined incidence rates of binge eating. In a longitudinal study of a large group of US adolescents, the incidence rate for binge eating was 10.1 per 1000 person-years among females and 6.6 per 1000 person-years among males (Field et al. 2008). Findings supported significantly higher rates in females compared to males. However, the study did not examine incidence by year. Future research is needed to determine whether rates of binge eating, in the United States or globally, have changed over time.

8   How Many Individuals Have BN?

According to studies of nationally representative samples of US adults, lifetime prevalence estimates of BN range from about 0.3% (Udo and Grilo 2018) to 1.5% (Hudson et al. 2007). This suggests that between one and five out of 300 American adults will meet diagnostic criteria for BN in their lifetimes. In these same samples, 12-month prevalence estimates for BN were 0.14% (Udo and Grilo 2018) and 0.3% (Hudson et al. 2007). This means that in a given year, between seven and 15 out of 500 adults living in the United States will meet criteria for BN. Given that Hudson et al. (2007) did not include loss of control in their definition of binge eating, it is

possible that the lower estimates reported by Udo and Grilo (2018) better reflect lifetime and 12-month prevalence of BN using DSM-based criteria.

A large epidemiological study of adolescents in Australia suggested a point prevalence of 0.6% for BN, suggesting that three out of 500 adolescents living in Australia met diagnostic criteria for BN at the time of measurement (Allen et al. 2009). Point prevalence estimates for BN among Australian adults range from 0.33% to 0.66%, depending on the diagnostic criteria employed (Hay et al. 2015). According to a recent literature review of epidemiological data from various countries in the twenty-first century, the average point prevalence of BN was 1.5% for females and 0.1% for males (Galmiche et al. 2019).

9   Does Risk for BN Differ Across Demographic Groups: Gender, Age Group, and Race/Ethnicity?

Data from large epidemiological studies indicate that the prevalence of BN differs by gender. For both lifetime and 12-month prevalence estimates, BN is consistently more common in females than in males. Specifically, results from the NCS-R suggest that lifetime prevalence of BN is three times higher in women than in men, and that 12-month prevalence is five times higher (Hudson et al. 2007). Epidemiological data collected between 2012 and 2013 suggested an even greater gender split in terms of lifetime prevalence: lifetime prevalence of BN in women was almost six times greater than that of men (Udo and Grilo 2018). Point prevalence estimates reveal an even larger discrepancy between the two genders: point prevalence of BN in females is 15 times greater than point prevalence of BN in males (Galmiche et al. 2019).

Risk for BN has been shown to differ by age. Lifetime prevalence estimates of BN differed significantly across age groups: 0.40% for ages 18–44, 0.21% for ages 45–50, and 0.1% for 60 years and older (Udo and Grilo 2018). Findings on just the adult population suggest that 12-month prevalence estimates for BN are identical from ages 18–29 and 30–44 (0.23%), but then sharply decrease at age 45 years and beyond (Udo and Grilo 2018). On average, studies including adolescents (e.g., Beato-Fernández et al. 2004; Swanson et al. 2011) produce lower point prevalence estimates than those studying young adults (e.g., Ghaderi and Scott 1999; Santonastaso et al. 2006). Studies including middle-aged or older adults (e.g., Mangweth-Matzek et al. 2006) produce comparable point prevalence estimates to those studying adolescents.

Lifetime prevalence of BN is significantly greater among Latino and African-Americans compared to non-Latino Whites (Marques et al. 2011). Specifically, lifetime prevalence for BN among non-Latino Whites is 0.51%, compared to 2.03% among Latinos and 1.31% among African-Americans (Marques et al. 2011). Lifetime prevalence for BN among Asians was 1.50%, which was not statistically significantly greater than the lifetime prevalence of BN among non-Latino Whites (Marques et al. 2011). These findings echo results from the comparison of binge-eating prevalence among racial/ethnic groups and contradict popular stereotypes that racial and ethnic minorities are protected from the development of eating disorders. Future research should examine factors that may account for elevated risk in racial/ethnic minority groups in the U.S. Research on the prevalence of BN in different racial/ethnic minority groups outside of the United States is very limited.

10 Have Rates of BN Changed in Recent Years?

A meta-analysis of studies examining incidence rates of BN from Western nations found a significant increase in the incidence of BN during the latter half of the twentieth century, which was associated with a statistically significant and large effect size (Keel and Klump 2003). In accordance with the meta-analysis's finding, a 2005 study found that incidence of BN increased overall from 1988 to 1996 in the United Kingdom for women ages 10–39 years but that incidence rates

declined significantly by 38.9% from 1996 to 2000 (Currin et al. 2005). The incidence rate's steep rise and then decline has several possible explanations, including that BN was newly coined as a disorder in 1980. Thus, an unmet need for treatment may have led to greater numbers of those affected to seek care, as well as increased practitioners' awareness of BN. However, once the backlog of undiagnosed and untreated patients had entered the treatment system, incidence may have more accurately captured new onset cases rather than just newly seeking treatment cases (Currin et al. 2005).

Echoing findings from the United Kingdom, a study in the Netherlands found that incidence of BN remained constant from 1985–1989 to 1995–1999, with a nonsignificant rate of decrease from 8.6 to 6.1 per 100,000 person-years (van Son et al. 2006). A subsequent study extending follow-up to 2005–2009 found significantly decreasing rates of BN in this population, with a rate of 3.2 per 100,000 person-years from 2005 to 2009 (Smink et al. 2014). Other studies in the twenty-first century have found stable rates of BN: A Danish study examining rates of BN for individuals 10–49 years old from 1995 to 2010 found stable rates, with a nonsignificant increase of 6.3–7.2 per 100,000 person-years (Steinhausen and Jensen 2015). Additionally, a study in northwestern Spain from 2005 to 2009 found stable rates of BN at 4.4 per 100,000 person-years for individuals 15 years and older (Larrañaga et al. 2012). Taken together, studies in the twenty-first century suggest that rates of BN are either stable or decreasing. Given the influence of culture on BN (Keel and Klump 2003), rates of BN could be decreasing in some cultures while remaining stable or even increasing in others.

11 How Many Individuals Have BED?

Between 0.85% and 2.8% of adults living in the United States will meet diagnostic criteria for BED in their lifetime, according to results from large epidemiological studies (Cossrow et al. 2016; Hudson et al. 2007; Udo and Grilo 2018).

This means that throughout their lifetime, between one and three out of 100 adults will meet diagnostic criteria for BED, on average. The lifetime prevalence for BED when using DSM-IV criteria was 2.03%, compared to 1.64% when using DSM-5 criteria. Additionally, data from two large epidemiological studies in the United States reported 12-month prevalence estimates for BED of 0.44% (Udo and Grilo 2018) and 1.2% (Hudson et al. 2007). This means that within the United States, between 44 and 120 adults out of 1000 have BED in a year. As mentioned previously, because Hudson et al. (2007) did not include loss of control in their definition of binge eating, their study likely overestimated the true prevalence of BED.

Point prevalence estimates for BED range from 0.5% (Flament et al. 2014) to 3.6% (Solmi et al. 2014). Outside of the United States, the mean point prevalence estimate for BED in Latin American countries is 3.53%, indicating that between 3 and 4 out of 100 individuals living in Latin America meet diagnostic criteria for BED (Kolar et al. 2016). In Australia, the point prevalence of BED is 5.58%, indicating that between 5 and 6 out of 100 Australian individuals meet diagnostic criteria for BED at the time of measurement. Among those seeking weight loss, BED point prevalence ranged from 14% to 35% depending on diagnostic criteria and assessment method (Çelik Erden et al. 2016; Dymek-Valentine et al. 2004), indicating that between 14 and 35 out of 100 individuals seeking bariatric surgery currently meet diagnostic criteria for BED.

12 Does Risk for BED Differ Across Demographic Groups: Gender, Age Group, and Race/Ethnicity?

Epidemiological data suggest that BED is more common in women than in men. Specifically, one epidemiological study reported lifetime prevalence of BED in the United States to be three times greater in women than in men (Udo and

Grilo 2018), while another reported lifetime prevalence of BED to be only about one and a half times greater in women than in men (Hudson et al. 2007). Comparing 12-month prevalence estimates for BED in the United States using both DSM-IV and DSM-5 criteria indicates that BED is about twice as common in women compared to men using DSM-IV criteria, compared to less than twice as common in women compared to men using DSM-5 criteria (Cossrow et al. 2016). Overall, epidemiological data suggest that both 12-month and lifetime prevalence estimates for BED are about three times greater in females compared to males (Galmiche et al. 2019). Point prevalence estimates for BED are also about three times greater in women compared to men (Kinzl et al. 1999a, b), as indicated by Austrian epidemiological data. Overall, these data demonstrate that BED is the most common eating disorder in males, and the ratio of males to females with BED is greater than the ratio of males to females with BN.

Epidemiological data in the United States indicated that lifetime prevalence of BED remains largely consistent from ages 18 to 59 years (about 1%), but significantly declines at age 60 years and beyond to 0.54% (Udo and Grilo 2018). Twelve-month prevalence estimates for BED based on age group demonstrated a similar pattern: between ages 18 and 59 years, about 0.5% of individuals had BED within a 12-month period, whereas at age 60 years and beyond, 0.33% had BED (Udo and Grilo 2018).

Within the United States, racial and ethnic groups do not differ in either lifetime or 12-month prevalence of BED (Marques et al. 2011). Specifically, lifetime prevalence was 1.41% among non-Hispanic whites, compared to 1.24–2.11% for minority groups, and 12-month prevalence estimates for BED in non-Hispanic whites was 0.55%, compared to 0.68–1.11% for minority groups (Marques et al. 2011). Given that lifetime prevalence includes individuals who had the disorder at any point in their lifetime, whereas 12-month prevalence includes individuals who had the disorder at any point in the last year, it makes sense that we would see greater lifetime prevalence estimates for BED compared to 12-month prevalence.

13 Have Rates of BED Changed in Recent Years?

Compared to BN, much less is known about incidence rates of BED, given that the disorder only obtained a distinct diagnostic code in the DSM-5 in 2013. Reflecting this, a 2012 article reported that, to their knowledge, no studies examining the incidence of BED existed at this time (Smink et al. 2012). One study used a cohort of over 2800 Finnish twins born from 1975 to 1979 and found the incidence rate for BED of 35 per 100,000 person-years for women ages 10–24 years (Mustelin et al. 2015). A study of 496 girls aged 12–15 years in the United States from 1998 to 2005 reported the incidence of BED of 343 per 100,000 person-years (Stice et al. 2013). Although these data could be interpreted as suggesting an increase in BED incidence given the larger prevalence rate found by Stice et al. (2013), methodological differences constrain comparisons. In a smaller sample, a diagnosis in one or two individuals can dramatically alter incidence as it gets multiplied to reflect what would be expected in a larger population. For example, an incidence of 343 per 100,000 person-years in 496 girls translates into almost two girls being diagnosed with BED each year. Given this information, it would not be accurate to conclude from Stice et al.'s (2013) limited sample size that rates of BED are increasing. Studies examining BED incidence over successive years are needed to determine whether rates of BED are changing.

14 Summary and Discussion

Binge eating and the disorders characterized by binge eating affect a considerable portion of the US population and populations of other countries. Most data indicate that women are more likely than men to binge eat, and the greatest gender differences are observed for BN. Both binge eating and BN tend to predominantly affect young adults, but BED affects a similar proportion of individuals during both young and middle adulthood. Binge eating and BN are more common among racial and ethnic minorities than in whites; however, both 12-month and lifetime prevalence estimates for BED are similar among all racial and ethnic groups, including non-Latino whites.

Several factors impact interpretation of prevalence estimates and incidence rates. One is sociocultural factors, including access to highly palatable food, urbanization, and internalization of thin ideals. Individuals typically consume highly palatable foods during their binge-eating episodes (Novelle and Diéguez 2018). Both industrialization and transition to a consumer-based economy dramatically increased access to readily edible and highly palatable foods over the course of the twentieth century, potentially accounting for increased incidence of BN. In addition, highly processed foods that are high in sugar and fat tend to be less expensive (Ramirez 1990), with high concentrations of these foods in poor urban neighborhoods in which a larger proportion of ethnic and racial minorities live (Drewnowski 2012; James et al. 2014). Thus, economic disparities may contribute to differential access to highly palatable foods and contribute to ethnic/racial differences in binge eating and BN. Moreover, recent data suggest a connection between food insecurity and risk for binge eating (Rasmusson et al. 2019). Future research might aim to examine whether racial and ethnic minorities experience more food insecurity in the United States, which might mediate an increased risk for binge eating. Additionally, urbanization increased over the twentieth century, and urbanization may increase anonymity in accessing food and have detrimental effects on mental health, in part due to increased stress levels (Berry 2007). Studies have noted increased incidence of BN in urban as compared to nonurban areas of a given population (Hoek et al. 1995; Nadaoka et al. 1996). Finally, over the twentieth century, internalization of a thin ideal became increasingly common, especially among women (Klimek et al. 2018). Given that overevaluation of weight and shape is a core component of BN (American Psychiatric Association 2013), this could account for increased prevalence of BN in women compared to men, as well as increased BN incidence during the latter half of the twentieth century (Keel and Klump 2003).

Biological factors may also contribute to epidemiological patterns. Data from twin studies and studies of gonadal hormones suggest that ovarian hormones may play a role in the onset of binge eating (Fowler et al. 2019; Klump et al. 2007, 2011, 2017). Gender differences may reflect biologically-based sex differences, and the increased risk for eating disorders characterized by binge eating during adolescence could reflect the activating effects of ovarian hormones on genetic risk for these disorders (Klump et al. 2011). Body mass index (BMI) represents another biological factor that may contribute to time trends as well as differences observed across racial/ethnic groups. There is a well-documented positive association between BMI and binge eating (Mason and Lewis 2014). Although it is impossible to determine a causal relationship from correlational data, some findings support that overeating and weight gain may alter brain responses to food, making it more rewarding (Stice and Burger 2019). If true, then population-based changes in BMI may reflect increased neural vulnerability for binge eating over time.

Finally, estimates of binge eating, BN, and BED depend on how these terms are defined. Diagnostic criteria for each of these terms differ between the DSM-5 and the ICD-11. Criteria have also changed over time, directly affecting who gets counted. Although such changes would not explain differences within a study that applied the same diagnostic criteria to all participants, it complicates comparisons between studies that use different definitions, making it difficult to compare epidemiological data from different countries. In addition, changing definitions could dramatically impact incidence based on patient registers. Given dramatic changes in definitions over the last 50 years, it is likely that definitions will continue to evolve and impact understanding of the epidemiology of binge eating and disorders characterized by binge eating episodes.

References

Alegria M, Woo M, Cao Z, Torres M, Meng XL, Striegel-Moore R (2007) Prevalence and correlates of eating disorders in Latinos in the United States. Int J Eat Disord 40:S15–S21

Allen KL, Byrne SM, Forbes D, Oddy WH (2009) Risk factors for full- and partial-syndrome early adolescent eating disorders: a population-based pregnancy cohort study. J Am Acad Child Adolesc Psychiatry 48:800–809

American Psychiatric Association (2013) Diagnostic and statistical manual of mental disorders, 5th edn. Author, Arlington, VA

Beato-Fernández L, Rodriguez-Cano T, Belmonte-Llario A, Martinez-Delgado C (2004) Risk factors for eating disorders in adolescents: a Spanish community-based longitudinal study. Eur Child Adolesc Psychiatry 13:287–294

Becker AE, Burwell RA, Navara K, Gilman SE (2003) Binge eating and binge eating disorder in a small-scale, indigenous society: the view from Fiji. Int J Eat Disord 34:423–431

Berry H (2007) Crowded suburbs and killer cities: a brief review of the relationship between urban environments and mental health. NSW Public Health Bull 18:222–227

Çelik Erden S, Seyit H, Yazisiz V, Türkyilmaz Uyar E, Önem Akçakaya R, Besirli A et al (2016) Binge-eating disorder prevalence in bariatric surgery patients: evaluation of presurgery and postsurgery quality of life, anxiety, and depression levels. Bariatr Surg Pract Patient Care 11:61–66

Cossrow N, Pawaskar M, Witt EA, Ming EE, Victor TW, Herman BK et al (2016) Estimating the prevalence of binge-eating disorder in a community sample from the United States: comparing DSM-IV-TR and DSM-5 criteria. J Clin Psychiatry 77:e968–e974

Currin L, Schmidt U, Treasure J, Jick H (2005) Time trends in eating disorder incidence. Br J Psychiatry 186:132–135

Drewnowski A (2012) The economics of food choice behavior: why poverty and obesity are linked. Nestle Nutr Inst Workshop Ser 73:95–112

Dymek-Valentine M, Rienecke-Hoste R, Alverdy J (2004) Assessment of binge eating disorder in morbidly obese patients evaluated for gastric bypass: SCID versus QEWP-R. Eat Weight Disord 9:211–216

Field AE, Javaras KM, Aneja P, Kitos N, Camargo CA, Taylor CB, Laird NM (2008) Family, peer, and media predictors of becoming eating disordered. Arch Pediatr Adolesc Med 162:574

Flament MF, Buchholz A, Henderson K, Obeid N, Maras D, Schubert N et al (2014) Comparative distribution and validity of DSM-IV and DSM-5 diagnoses of eating disorders in adolescents from the community. Eur Eat Disord Rev 23:100–110

Fowler N, Keel PK, Burt SA, Neale M, Boker S, Sisk CL, Klump KL (2019) Associations between ovarian hormones and emotional eating across the menstrual cycle: do ovulatory shifts in hormones matter? Int J Eat Disord 52:195–199

Galmiche M, Déchelotte P, Lambert G, Tavolacci MP (2019) Prevalence of eating disorders over the 2000–2018 period: a systematic literature review. Am J Clin Nutr 109:1402–1413

Ghaderi A, Scott B (1999) Prevalence and psychological correlates of eating disorders among females aged 18–30 years in the general population. Acta Psychiatr Scand 99:261–266

Hay P, Girosi F, Mond J (2015) Prevalence and sociodemographic correlates of DSM-5 eating disorders in the Australian population. J Eat Disord 3:1–7

Hoek HW, Bartelds AIM, Bosveld JJF, van der Graaf Y, Limpens VEL, Maiwald M, Spaaij CJK (1995) Impact of urbanization on detection rates of eating disorders. Am J Psychiatr 152:1272–1278

Hudson JI, Hiripi E, Pope HG Jr, Kessler RC (2007) The prevalence and correlates of eating disorders in the national comorbidity survey replication. Biol Psychiatry 61:348–358

James P, Arcaya MC, Parker DM, Tucker-Seeley RD (2014) Do minority and poor neighborhoods have higher access to fast-food restaurants in the United States? Health Place 29:10–17

Keel PK, Klump KL (2003) Are eating disorders culture-bound syndromes? Implications for conceptualizing their etiology. Psychol Bull 129:747–769

Keel PK, Heatherton TF, Dorer DJ, Joiner TE, Zalta AK (2006) Point prevalence of bulimia nervosa in 1982, 1992, and 2002. Psychol Med 36:119–127

Kinzl JF, Traweger C, Trefalt E, Mangweth B, Biebl W (1999a) Binge-eating disorder in females: a population-based investigation. Int J Eat Disord 25:287–292

Kinzl JF, Traweger C, Trefalt E, Mangweth B, Biebl W (1999b) Binge-eating disorder in males: a population-based investigation. Eat Weight Disord 4:169–174

Klimek P, Murray SB, Brown T, Gonzales M, Blashill AJ (2018) Thinness and muscularity internalization: associations with disordered eating and muscle dysmorphia in men. Int J Eat Disord 51:352–357

Klump KL, Perkins PS, Burt SA, Mcgue M, Iacono WG (2007) Puberty moderates genetic influences on disordered eating. Psychol Med 37(05):627

Klump KL, Culbert KM, Slane JD, Burt SA, Sisk CL, Nigg JT (2011) The effects of puberty on genetic risk for disordered eating: evidence for a sex difference. Psychol Med 42(3):627–637

Klump KL, Culbert KM, Sisk CL (2017) Sex differences in binge eating: gonadal hormone effects across development. Annu Rev Clin Psychol 13:183–207

Kolar DR, Mejía Rodriguez DL, Mebarak Chams M, Hoek HW (2016) Epidemiology of eating disorders in Latin America: a systematic review and meta-analysis. Curr Opin Psychiatry 29:363–371

Larrañaga A, Docet MF, García-Mayor RV (2012) High prevalence of eating disorders not otherwise specified in northwestern Spain: population-based study. Soc Psychiatry Psychiatr Epidemiol 47:1669–1673

Mangweth-Matzek B, Rupp CI, Hausmann A, Assmayr K, Mariacher E, Kemmier G et al (2006) Never too old for eating disorders or body dissatisfaction: a community study of elderly women. Int J Eat Disord 39:583–586

Marques L, Alegria M, Becker AE, Chen C, Fang A, Chosak A, Belo Diniz J (2011) Comparative prevalence, correlates of impairments, and service utilization for eating disorders across U.S. ethnic groups: implications for reducing ethnic disparities in health care access for eating disorders. Int J Eat Disord 44:412–420

Mason TB, Lewis RJ (2014) Profiles of binge eating: the interaction of symptoms, eating styles, and body mass index. Eat Disord 22:450–461

Mustelin L, Raevuori A, Hoek HW, Kaprio J, Keski-Rahkonen A (2015) Incidence and weight trajectories of binge eating disorder among young women in the community. Int J Eat Disord 48:1106–1112

Nadaoka T, Oiji A, Takahashi S, Morioka Y, Kashiwakura M, Totsuka S (1996) An epidemiological study of eating disorders in a northern area of Japan. Acta Psychiatr Scand 93:305–310

Novelle MG, Diéguez C (2018) Food addiction and binge eating: lessons learned from animal models. Nutrients 10:71

Ramirez I (1990) What do we mean when we say "palatable food"? Appetite 14:159–161

Rasmusson G, Lydecker JA, Coffino JA, White MA, Grilo CM (2019) Household food insecurity is associated with binge-eating disorder and obesity. Int J Eat Disord 52:28–35

Santonastaso P, Scicluna D, Colombo G, Zaneti T, Favaro A (2006) Eating disorders and attitudes in Maltese and Italian female students. Psychopathology 39:153–157

Saunders R, Johnson L, Teschner J (1998) Prevalence of eating disorders among bariatric surgery patients. Eat Disord 4:309–317

Smink FR, van Hoeken D, Hoek HW (2012) Epidemiology of eating disorders: incidence, prevalence and mortality rates. Curr Psychiatry Rep 14(4):406–414

Smink FR, van Hoeken D, Oldehinkel AJ, Hoek HW (2014) Prevalence and severity of DSM-5 eating disorders in a community cohort of adolescents. Int J Eat Disord 47:610–619

Solmi F, Hatch SL, Hotopf M, Treasure J, Micali N (2014) Prevalence and correlates of disordered eating in a general population sample: the south east London community health (SELCoH) study. Soc Psychiatry Psychiatr Epidemiol 49:1335–1346

Steinhausen H, Jensen CM (2015) Time trends in lifetime incidence rates of first-time diagnosed anorexia nervosa and bulimia nervosa across 16 years in a Danish nationwide psychiatric registry study. Int J Eat Disord 48:845–850

Stice E, Burger K (2019) Neural vulnerability factors for obesity. Clin Psychol Rev 68:38–53

Stice E, Marti CN, Rohde P (2013) Prevalence, incidence, impairment, and course of the proposed DSM-5 eating disorder diagnoses in an 8-year prospective community study of young women. J Abnorm Psychol 122:445–457

Swanson SA, Crow SJ, Le Grange D, Swendsen J, Merikangas KR (2011) Prevalence and correlates of eating disorders in adolescents: results from the national comorbidity survey replication adolescent supplement. Arch Gen Psychiatry 68:714–723

Taylor JY, Howard Caldwell CH, Baser RE, Faison N, Jackson JS (2007) Prevalence of eating disorders among blacks in the national survey of American life. Int J Eat Disord 40:S10–S14

Udo T, Grilo CM (2018) Prevalence and correlates of DSM-5-defined eating disorders in a nationally representative sample of U.S. adults. Biol Psychiatry 84:345–354

van Son GE, van Hoeken D, Bartelds AI, Furth EF, Hoek HW (2006) Time trends in the incidence of eating disorders: a primary care study in The Netherlands. Int J Eat Disord 39:565–569

World Health Organization (2018) International statistical classification of diseases and related health problems (11th revision). https://icd.who.int/browse11/l-m/en

Binge Eating Assessment

Deena Peyser, Mia Campbell, and Robyn Sysko

Abstract

This chapter provides an overview of binge eating assessments that are commonly used for diagnosis, case conceptualization, treatment planning, ongoing assessment, and treatment outcome. Included is a review of well-validated clinical interviews, self-report measures, and laboratory eating paradigms that are available for the assessment of binge-eating disorder (BED) in clinical and research contexts. Assessments with adequate psychometric properties are described, and benefits, limitations, and criteria assessed to aid in the selection of an appropriate measure for each context and question. This chapter describes disorders characterized by binge eating based on DSM-5 criteria and relevant updates from ICD-11. A review of clinical interviews including the Eating Disorder Examination (EDE), Eating Disorder Assessment for DSM-5 (EDA-5), and the Structured Clinical Interview for DSM-5 (SCID-5) are provided along with evaluations of self-report measures including the Eating Disorder Examination-Questionnaire (EDE-Q), the Eating Disorder Diagnostic Scale (EDDS), the Eating Pathology Symptoms Inventory (EPSI), Questionnaire on Eating and Weight Patterns (QWEP), and self-monitoring completed within the context of cognitive-behavioral therapy. Finally, laboratory eating paradigms are discussed as a useful and objective real-time assessment of eating behavior. In sum, this chapter provides information that may assist clinicians and researchers in understanding and selecting appropriate measures to evaluate binge eating.

Keywords

Binge-eating disorder · Bulimia nervosa · Eating disorders · Feeding disorders · Assessment, diagnosis, and classification · DSM-5

Learning Objectives

Readers will be able to:

1. Identify and select an assessment of binge eating that is a best fit in clinical and research contexts.
2. Distinguish assessment tools based on their characteristics, limitations, and benefits.

D. Peyser · M. Campbell · R. Sysko (✉)
Eating and Weight Disorders Program, Department of Psychiatry, Icahn School of Medicine at Mount Sinai, New York, NY, USA
e-mail: Deena.peyser@mssm.edu; Mia.campbell@mssm.edu; Robyn.sysko@mssm.edu

© Springer Nature Switzerland AG 2020
G. K.W. Frank, L. A. Berner (eds.), *Binge Eating*, https://doi.org/10.1007/978-3-030-43562-2_2

1 Introduction

Binge eating episodes, as defined by the Diagnostic and Statistical Manual of Mental Disorders (DSM-5; APA 2013), are characterized by consuming a large amount of food in a discrete period of time (given the context) and experiencing a sense of loss of control, or not being able to resist or stop eating once started. Binge eating is a behavior required for two of the eating disorders (bulimia nervosa, BN, and binge-eating disorder, BED) and can be observed in a third (individuals with the binge eating/purging subtype of anorexia nervosa, AN) as well as in residual categories of Other Specified Feeding or Eating Disorder and some presentations of Unspecified Feeding or Eating Disorder. To be diagnosed with BED in DSM-5, binge eating also must be accompanied by marked distress about the episode, and three of five additional indicators: eating more rapidly than usual, eating until uncomfortably full, eating large amounts of food when not physically hungry, eating alone because of embarrassment about what or how much one is eating, and feeling disgusted, depressed, or guilty after eating. In the 11th edition of the *International Classification of Diseases*, binge eating will have an even broader definition, in which loss of control is the only required element, and size of the eating episode will be irrelevant in the diagnosis of eating disorders.

There are numerous assessments used to measure binge eating, including clinical interviews, self-report measures, and laboratory eating paradigms. For example, a systematic review by Burton et al. (2016) of self-report measures identified 29 that assess binge eating symptoms. However, the two measures with the strongest psychometric data (Bulimic Investigatory Test–Edinburgh, BITE, Ricca et al. 2000; Bulimia Test–Revised, BULIT-R, Thelen et al. 1991) assess related symptoms of BN and BED, rather than the frequency of binge eating to allow for a diagnosis of disorders characterized by binge eating. In this chapter, we review the assessment measures with adequate psychometrics and report psychometric ratings and criteria assessed by each measure.

2 Clinical Interviews

2.1 Eating Disorder Examination (EDE)

The Eating Disorder Examination (EDE, current version 17.0D; Fairburn et al. 2014), is widely viewed as the gold standard assessment of eating disorders (Wilson et al. 1993; Berg et al. 2012). The EDE is a semi-structured, investigator-based interview, which takes 45–75 min to administer. The EDE provides frequency data on key features of eating disorders, such as number of episodes and, in some cases, number of days on which the behavior occurred; and it also assesses severity. This measure assigns DSM-5 eating disorder diagnoses and it is publically available at http://www.cred-oxford.com/pdfs/EDE_17.0D.pdf.

The EDE assesses a broad range of psychopathology commonly associated with AN, BN, and BED and has been adapted to allow for the diagnosis of eating disorders based on the DSM-5 criteria. It includes four subscales (Restraint, Eating Concern, Shape Concern, and Weight Concern) as well as a global score. This measure was created primarily as a research tool for assessing the severity of symptoms, which has led to frequent usage in studies of treatment response, and captures behaviors during the past 28 days and up to 3 months for diagnostic items related to BED.

Assessors using the EDE rate the amounts of food included in eating episodes and accompanying feelings of loss of control to categorize four different types of eating. This categorization is critical for diagnosing BN and BED. The EDE assesses four different types of eating viewed by the individual as excessive:

(a) objective bulimic episodes (OBEs), defined as the consumption of an objectively large amount of food while experiencing a sense of loss of control; (b) subjective bulimic episodes (SBEs), defined as experiencing loss of control while consuming smaller amounts of food; (c) objective overeating, defined as eating an objectively large amount of food without loss of control; and (d) subjective overeating, defined as eating a small amount of food without a sense of loss of control, which the individual believes is excessive (Fairburn and Cooper 1993). The EDE questions related to these four constructs probe details about the types and amount of food, the social context in which they occur, and feelings of loss of control during the episode. While the trained interviewer determines whether the food consumed constitutes an "objectively large" amount of food, the EDE includes an appendix, with standardized amounts of food that are considered objectively large. For example, the consumption of three main courses (e.g., three Big Macs) or more than one pint of ice cream would be considered large when rating OBEs.

The EDE also includes a module to evaluate the extra specifiers for BED if at least 12 OBEs have been present over the prior 3 months. Recurrent episodes of OBEs should be associated with three or more of the following: (1) eating more rapidly than usual, (2) eating until uncomfortably full, (3) consuming large amounts of food when not physically hungry, (4) eating alone because of embarrassment regarding amount of food, or (5) disgust with self, depressed, or guilty after binge episode. In addition, the level of distress regarding binge eating is assessed, and confirmation is obtained that binge episodes occur at least once per week for 3 months and that these episodes are not associated with compensatory behavior, as in BN, and do not occur exclusively during episodes of AN.

While the EDE has advantages, as the interviewer can assist in defining complicated eating disorder concepts and can probe the participant for additional clinical information, it also has disadvantages, as it requires extensive interviewer training to ensure competency (Fairburn et al. 2014). To achieve reliable and valid administration of this assessment, interviewers must demonstrate familiarity with the interview format, co-rating of interviews, and supervision from a previously EDE trained individual. The training requirements are a significant investment (20–30 h) in addition to the time it takes to administer the measure to each individual (~1–2 h), and in combination, may prove prohibitive in settings that are primarily involved in the provision of clinical care (Sysko and Alavi 2018). However, despite these limitations, the EDE (versions 12–16) shows evidence of strong psychometric properties, including norms on large samples with nonclinical populations, excellent inter-rater reliability (primarily kappas ≥ 0.85), use of multiple groups of judges to develop the instrument with quantitative ratings, and a preponderance of evidence supporting the use of the EDE in different demographic groups (e.g., age, gender, and ethnicity) and across multiple contexts (e.g., community and inpatient setting).

2.2 Structured Clinical Interview for DSM-5 (SCID-5)

The Structured Clinical Interview for DSM-5 (SCID-5; First et al. 2015) allows for the assessment and diagnosis of binge-eating disorders based on DSM-5 criteria. This assessment is available exclusively for purchase through the American Psychiatric Publishing. The SCID has been commonly used in research settings, particularly in earlier iterations for DSM-IV, though it has several limitations for the measurement of binge eating. First, it does not assess the overall frequency of binge eating as a stand-alone factor, only the presence or absence of this feature, and it does not provide guidance for quantifying whether a binge episode is large. Therefore, when using the SCID, clinicians must rely on their judgment to determine what constitutes a large amount of food. Additionally, data cannot be collected to quantify subjective binge eating using the SCID. The SCID, like the EDE, requires significant training for interviewers (~20–30 h) primarily because of the breadth of information and symptoms collected across DSM diagnoses, which requires familiarity with a wide range of psychopathology. Furthermore, it is time consuming to administer and includes complicated skip logic that may lead researchers and clinicians to miss opportunities to capture important diagnostic information (Thomas et al. 2016). Thus, while the SCID is a useful measure for the diagnosis of eating and other psychiatric disorders, it cannot be used as an indicator of changes in symptom presentation. The prohibitive cost of the SCID is a limitation for many clinicians and researchers who need to use a free or inexpensive tool to collect clinically relevant information on patients. Psychometric data are not currently available for eating disorders based on the current version of the SCID. Inter-rater reliability of DSM-IV eating disorder diagnoses are good (kappa = 0.77), however, the test–retest reliability estimates for DSM-IV eating disorder diagnoses (correlation = 0.64) are not consistent with a rating of acceptable (minimum correlation over several days or weeks = 0.70; Zanarini et al. 2000).

2.3 Eating Disorder Assessment for the DSM-5 (EDA-5)

The Eating Disorder Assessment for the DSM-5 (EDA-5; Sysko et al. 2015) is an adaptive semistructured interview that was developed to assess all DSM-5 feeding and eating disorders or related conditions in adults. The questions assess the current problem within the last 3 months. The measure requires limited training to administer and it is portable, accessible, and brief (~15 min; Sysko et al. 2015), which reduces the burden on the interviewer and the participant. The EDA-5 also is delivered in an electronic format that only elicits the information necessary to assign a diagnosis and therefore may offer additional utility in settings with more limited resources for assessment as compared to lengthier and more burdensome tools (Kornstein et al. 2016). Nonetheless, clinical judgment must be used when administering this measure.

The EDA-5 includes a section entitled "Binge Eating and Compensatory Behaviors." This section includes questions assessing loss of control while eating, specific types of foods eaten during this loss of control in order to distinguish between OBEs and SBEs, and the frequency of binge episodes. If the individual does not meet the criteria for AN, BN, or Avoidant/Restrictive Food Intake Disorder (ARFID), the interviewer then assesses for BED. This section, like the EDE BED module, includes yes/no questions about the following: rapid eating, eating until uncomfortably full, avoiding eating around others due to shame or embarrassment, negative affect associated with the episode, and marked distress regarding binge episode.

The EDA-5 has several advantages, particularly in comparison to several of the other available assessment tools discussed. Studies that compared the EDA-5 to the EDE and to unstructured clinical interviews indicate preliminary evidence of the validity (Sysko et al. 2015) and test–retest reliability of the EDA-5 (Sysko et al. 2015). The EDE-5 also includes an informative appendix entitled "Is it a binge?" that describes discriminating features of binge episodes in further detail and provides specific guidelines and examples for assessing lack of control and for discriminating between an OBE and an SBE. For example, an objectively large amount of food could be 2 pints of ice cream or 10 apples or 1 family-size bag of chips. In contrast, a subjectively large amount of food maybe 2 bowls of cereal or 3 slices of pizza, or 2 Big Macs. Despite its strengths, the EDA-5 has several limitations including a lack of data on the assessment of feeding disorders, limited data on efficiently distinguishing between case and non-case status, and minimal dimensional data (Sysko et al. 2015).

3 Self-Report Measures

3.1 Eating Disorder Examination-Questionnaire (EDE-Q)

The Eating Disorder Examination-Questionnaire (EDE-Q; Fairburn and Beglin 1994, 2008) is a 38-item self-report measure based on the EDE interview (Cooper and Fairburn 1987; Fairburn et al. 2014) and it is designed to be completed in 15 min (see Table 1 for more characteristics). This measure assesses symptoms over the past 28 days, including symptom frequency and severity. Diagnostic criteria assessed by the EDE-Q are provided in Table 1. The EDE includes questions such as, "Over the past 28 days, how many times have you eaten what other people would regard as an unusually large amount of food (given the circumstances)?" and "On how many of these times did you have a sense of having lost control over your eating (at the time you were eating)?" The ease of administration of the EDE-Q supports its use to measure the course and outcomes of treatment for eating disorders.

The EDE-Q, similar to the EDE, contains four subscales: restraint, eating concern, shape concern, and weight concern, as well as a global score. The EDE-Q has been shown to discriminate successfully between individuals with and without eating disorders and the EDE-Q subscale scores positively correlate with EDE subscale scores (Forbush and Berg 2016). The EDE-Q has fewer items than the EDE and is by design less comprehensive, with a focus on identifying symptoms rather than collecting all of the information needed for a diagnosis. In the current version of the EDE-Q, information about large eating episodes with loss of control is collected to allow for the measurement of OBEs. Prior versions also evaluated SBEs.

The EDE-Q can be used in both research and clinical settings. Data exploring the differences between the EDE and EDE-Q have suggested differences in patient reports of complex behaviors, such as binge eating (Black and Wilson 1996; Carter et al. 2001; Fairburn and Beglin 1994; Wilfley et al. 1997). Other studies have demonstrated good agreement between the EDE and the EDE-Q for OBEs, but not for SBEs or overeating (Grilo et al. 2001a, b). As such, studies aiming to assess binge eating should consider utilizing the same measure (EDE or EDE-Q) if reliability in the measurement of binge eating over time is an important outcome of the research. When completed by the same individual across time points, the EDE-Q was found to be responsive enough to indicate if an individual has improved, recovered, deteriorated, or remained stable over time (Dingemans and vanFurth 2017). The EDE-Q can also be completed with or without a set of instructions, which provides specifications for determining OBEs. This enhanced version of the EDE-Q achieves similar results to the EDE interview. In contrast, the EDE-Q without instructions lacks correlation with the EDE, which further suggests a greater degree of variance in how respondents interpret the definition of a binge (Goldfein et al. 2005).

3.2 Eating Disorder Diagnostic Scale

The Eating Disorder Diagnostic Scale (EDDS; Stice et al. 2000) is a brief 22-item self-report scale to assess eating disorder symptoms over the prior 3 months. This assessment can be used to generate a diagnosis of BED, AN, BN, and an overall composite score for eating disorder symptoms. This measure was developed to diagnose eating disorders for etiological research, ongoing assessments in research, and for diagnosing individuals with eating disorders in clinical practice in both psychological and medical (primary care) settings (Stice et al. 2000).

The EDDS describes binge eating behaviorally rather than using the term "binge," in order to address problems stemming from the subjectivity in assessing binge eating by self-report (Peterson and Mitchell 2005). The most updated EDDS, based on the DSM-5, prompts the interviewee about eating unusually large amounts of food, loss of control, episode frequency, marked distress, and the five behavioral components of BED: rapid eating, eating until uncomfortably full, eating large amounts when not physically hungry, eating alone due to embarrassment, and feelings of disgust, depression, or guilt after eating. Diagnostic criteria assessed by the EDDS are provided in Table 1. The measure uses a diverse format of questions including questions rated on a Likert scale, dichotomous response questions, questions regarding symptom frequency (e.g., average number of times in the

Table 1 Characteristics and diagnostic criteria assessed by instruments used for screening or diagnosis

Instrument	Time to administer	Time frame assessed	OBE	SBE	Loss of control	Rapid eating	Eating until uncomfortably full	Consuming large amounts of food when not physically hungry	Eating alone due to shame and being embarrassed	Negative affect	Marked distress	Quantity of food	Frequency
Clinical interviews													
EDE	50–90	28 days	x	x	x	x	x	x	x	x	x	x	x
EDA-5	~15	3 mos.	x	x	x	x	x	x	x	x	x	x	x
Self-report measures													
EDE-Q	~15	28 days	x		x				x	x			x
EDDS	~15	3 mos.	x		x	x	x	x	x	x	x		x
EPSI	~20	28 days	x			x	x	x					
QEWP-R	~15	6 mos.	x		x	x	x	x	x	x	x		x
QEWP-5	~15	6 mos.	x	x	x	x	x	x	x		x	x	

Note: *EDE* Eating Disorder Examination, *EDA-5* Eating Disorder Assessment for the DSM-5, *SCID* Structured Clinical Interview for DSM-5, *EDE-Q* Eating Disorder Examination-Questionnaire, *EDDS* Eating Disorder Diagnostic Scale, *EPSI* Eating Pathology Symptoms Inventory, *QEWP-R* Questionnaire on Eating and Weight Patterns—Revised, *QEWP-5* Questionnaire on Eating and Weight Patterns-5, *OBE* Objective Binge Eating, *SBE* Subjective Binge Eating past 3 months that one ate an unusually large amount of food and felt a loss of control), and open-ended questions. To date, data regarding the psychometric properties of the most updated version of the EDDS, based on the DSM-5, are not yet available, but prior versions of the EDDS have examined the reliability and validity of the EDDS scores for DSM-IV diagnoses (internal consistency, alpha = 0.89; test–retest reliability, $r = 0.87$; Stice et al. 2000, 2004) and the measure is well suited for clinical practice because it is brief, straightforward to score, and freely available.

3.3 Eating Pathology Symptoms Inventory

The Eating Pathology Symptoms Inventory (EPSI; Forbush et al. 2013) is a 45-item self-report questionnaire that assesses eating disorder dimensions relevant to treatment outcome. This assessment measures eight subscale measurements including "Binge Eating," based on items assessing eating large amounts of food, eating until uncomfortably full, inability to resist eating food once offered, rapid eating, and accompanying cognitive symptoms (Forbush and Berg 2016). Diagnostic criteria assessed by the EPSI are provided in Table 1. While this measure assesses a construct labeled binge eating, it does not explicitly state these words in any of the questions, and some items may be better understood as mindless eating. Questions related to the "Binge Eating" subscale include, "I stuffed myself with food to the point of being sick," "I did not notice how much I ate until after I had finished eating," "I snacked throughout the evening without realizing it," and "I ate as if I was on autopilot."

Several studies have looked at the validity, reliability, and stability of the EPSI and found strong psychometric properties (Forbush et al. 2013, 2014). Studies have examined the reliability and validity of the EPSI scores for discriminating symptoms of DSM-IV diagnostic groups (Forbush et al. 2013, 2014) and the measure is appropriate for clinical practice.

3.4 Questionnaire on Eating and Weight Patterns

The Questionnaire on Eating and Weight Patterns-Revised (QEWP-R) is a 20-item self-report questionnaire that assesses symptoms of eating and weight disorders, including binge eating (Yanovski et al. 1993). The prompts inquire about eating unusually large amounts of food in a short period of time, loss of control, marked distress, frequency of binge eating, and the five behavioral components of BED: rapid eating, eating until uncomfortably full, eating large amounts when not physically hungry, eating alone due to embarrassment, and feelings of disgust, depression, or guilt after eating. Studies have supported the measure's psychometric properties and its ability to identify individuals who binge eat (Barnes et al. 2011; Elder et al. 2006).

While the measure was originally developed based on the DSM-IV criteria, it has since been adapted based on the DSM-5 criteria (QEWP-5, Yanovski et al. 2015). The QEWP-5 is a 26-item self-report measure that assesses the frequency and severity of binge eating and compensatory behaviors and assesses for a possible diagnosis of binge-eating disorder. Unlike many of the other assessments, it also asks respondents to "list everything you ate and drank during the [binge] episode." In contrast to the QEWP-R, this measure includes questions that assess feelings of loss of control over eating, even in the absence of consuming an objectively large amount of food. As such, the key difference between the QEWP-R and QEWP-5 is the additional questions focusing on SBEs and the feelings and behavioral symptoms surrounding the episode. Specifically, this measure asks "During the past 3 months, how often did you have...the feeling that your eating was out of control, but you did not consume what most people would think was an unusually large amount of food." Notably, while the QEWP-5 can be used as a screening instrument in both the research and clinical settings, it should not be used to make a diagnosis in the absence of a more comprehensive clinical interview as it is sensitive, but not specific for a BED diagnosis (Yanovski et al. 2015).

3.5 Self-Monitoring

Self-monitoring is a common tool used in cognitive-behavioral therapy to assess eating behaviors, whereby patients record all food intake and related information in real time. Self-monitoring records completed by patients and in the context of a cognitive-behavioral treatment include the time of day one eats or drinks, exactly what one ate or drank, where and when the food or drink was consumed, meals or snacks that felt excessive, use of vomiting, laxatives and/or diuretics, and anything that seems to be influencing eating (Fairburn et al. 2014). Self-monitoring can provide important information about binge eating episodes and can lead to changes in eating behaviors (Fairburn et al. 2014). This assessment tool may reduce inaccuracies related to retrospective recall of binge eating that can arise when using other assessment measures (e.g., EDE; Wilson and Vitousek 1999). Self-monitoring is most commonly used in the clinical setting and is useful for case conceptualization, treatment planning, and ongoing assessment of eating behaviors (Fairburn et al. 2003), but could also be employed in research settings to capture daily data on binge eating.

Self-monitoring has been shown to be effective for reducing binge eating episodes in cognitive behavior therapy-guided self-help (CBT-GSH; Hildebrandt and Latner 2006; Latner and Wilson 2002); however, traditional self-monitoring can be time consuming and inconvenient to integrate and adhere to on a daily basis outside of the therapy context. While self-monitoring has historically been administered using paper and pencil format, in recent years there has been the development and implementation of smartphone-based technology to approximate traditional methods (Fairburn and Rothwell 2015). Ecological Momentary Assessment (EMA; Engel et al. 2016; Farchaus and Corte 2003; Smyth et al. 2001) allows for recording eating behavior and/or binge eating in real time using smartphone-based technology (Fairburn

and Rothwell 2015; Farchaus and Corte 2003). For example, Noom Monitor is a smartphone application developed to facilitate guided self-help treatments by simplifying and digitizing self-monitoring records (Hildebrandt et al. 2017), which may increase accessibility and adherence, and reduce treatment burden.

4　Laboratory Eating

Binge eating can also be assessed in vivo through laboratory studies of eating behavior, which offer a clinically useful alternative to self-report or other subjective assessments (see also Sysko et al. 2018). Feeding laboratory paradigms allow for an objective real-time assessment of eating behavior. They also may enhance the scientific rigor and reproducibility of the data by minimizing recall errors or bias and by quantifying behaviors that may be difficult to capture accurately in questionnaires or interviews, such as the size of the episode or the degree of loss of control experienced during a binge episode.

In a laboratory eating assessment, participants are provided with a single food item (e.g., yogurt shake) or an array of foods (multi-item meals) in a standardized way (e.g., amount of food, type of food, instructions), and behavior is monitored (see Sysko et al. 2018 for further details). Although individuals with BED and BN are instructed to eat in a manner consistent with a binge episode, binge eating may not occur naturally in this setting. Several experimental adjustments are made to capture behavior that is more consistent with an episode of loss of control eating. For example, in a multi-item meal, foods typical of binge eating episodes (e.g., ice cream, cookies, and chips) are provided in amounts that are larger than with a normal meal (e.g., 11,342 kcal, Sysko et al. 2013), and participants are asked to not eat anything for several hours before the meal to approximate the dietary restriction that often precedes binge episodes. In addition, participants may complete multiple meals in the laboratory to increase comfort with the experience, or be provided with access to a bathroom to ensure that an inability to self-induce vomiting is not a barrier to participation. Following the meal, participants are asked how typical the laboratory meal was in comparison to binge eating episodes that occur outside the laboratory, with only those considered to be "moderately," "very," or "extremely" typical classified as a binge episode. Common outcome measures from this assessment include total kilocalorie and macronutrient composition of food consumed, as well as meal timing and the types and order of foods consumed.

Studies have shown that individuals diagnosed with BED and BN consistently engage in objectively distinct and abnormal eating behavior in the laboratory setting. Several characteristics are particularly notable, including total energy consumed, macronutrient patterns, and rate of eating. Individuals with BED consume more than overweight or obese individuals without BED. For example, in a laboratory binge meal, BED patients consumed 943.15 + 271.44 g, obese controls consumed 552.06 + 252.16 g, and normal-weight controls consumed 475.83 + 161.04 g (Sysko et al. 2007b). Binge size correlates significantly and positively with body mass index (BMI) among patients with BED (Guss et al. 2002), and in-laboratory binge meals among patients with BED are characterized by consuming a higher proportion of energy from fat and a lower proportion from protein compared with those of controls (Yanovski et al. 1992). Similarly, when asked to binge eat, individuals with BN consume far greater amounts of food and a smaller fraction of energy derived from protein than do healthy individuals who are instructed to "let themselves go" (Van der Ster et al. 1994; Walsh et al. 1989). Individuals with BN tend to initiate in-laboratory meals with dessert and snack foods, in contrast to healthy controls, who typically start with fish and meat (Hadigan et al. 1989). Additionally, individuals with BN demonstrate an accelerated rate of food consumption during binge episodes, particularly when provided with a single-item liquid meal (Kissileff et al. 1986; Walsh et al. 1989). Interestingly, while individuals with BED consume significantly more food during their binge meals than comparable controls, this appears primarily due to meals lasting significantly longer than those of controls, rather than to a faster rate of eating (Walsh and Boudreau 2003).

Measures of eating behavior in a laboratory setting have been shown to be sensitive and reproducible. For example, effect sizes for energy (in kcal) consumed in a binge meal as compared with eating by healthy individuals were large both in a study of BN ($d = 1.4$, Sysko et al. 2017) and of BED ($d = 1.5$, Sysko et al. 2007a), even with modest sample sizes.

Other subjective and objective outcome measures can also be used in combination with meals administered in this setting to provide a more comprehensive assessment of eating among individuals with BED and BN. For example, visual analog scales (VAS) capture subjective responses to eating such as hunger, fullness, sickness, and loss of control. Among individuals with BED, hunger and fullness ratings were generally similar to those of healthy controls (with some exceptions, e.g., Guss et al. 2002), but unlike controls, participants with BED do not use these signals to terminate the meal (Samuels et al. 2009; Sysko et al. 2007b).

Along with intake and subjective measures, a wide range of physiological measures can be assessed during laboratory meals, including the sympathetic/parasympathetic state, physiological markers of stress, and gastrointestinal humoral factors (e.g., CCK, ghrelin, GLP1). Measurements of appetitive hormones and gastrointestinal function during laboratory meals have elucidated the psychobiology of maladaptive eating and identified potential markers of abnormal eating and recovery. For example, abnormalities in the development of satiety during a single-item meal have been documented among patients with BN across multiple domains, including release of the hormone cholecystokinin (CCK; Devlin et al. 1997; Geracioti and Liddle 1988; Keel et al. 2007), gastric emptying (Cuellar et al. 1988; Devlin et al. 1997; Geliebter et al. 1982; Inui et al. 1995), gastric capacity (Geliebter et al. 1982), and gastric relaxation (Walsh et al. 2003). These biological variables also can be assessed longitudinally and may provide a useful metric for detecting improvement among patients with eating disorders.

While laboratory meals provide a controlled setting for the momentary assessment of eating behaviors in a reliable manner, this approach has several limitations. First, in-laboratory binge eating episodes lack ecological validity, as they are intrinsically artificial and occur outside of the natural environment. Additionally, there are few laboratory eating studies with adolescent samples (Tanofsky-Kraff et al. 2011), and it is therefore unknown whether the results of research in adults with eating disorders generalize across different stages of development. Despite large effects, prior research samples are generally small and homogenous, with primarily female and Caucasian participants. Finally, numerous challenges in conducting laboratory eating behavior studies (e.g., cost of setting up a lab, time to execute lab-based meals, training for staff to ensure standardization) limit the broad use of this type of assessment (Sysko and Alavi 2018).

5   Conclusion

Future research may help to refine existing measures to increase accessibility and usability for clinical and research purposes. Further, studies to better understand the limits of existing assessment tools (e.g., examining the psychometrics of self-reported binge eating on the basis of gender or other demographic characteristics; Hildebrandt and Craigen 2015) would help inform the selection of a measure of binge eating in a way that is not currently possible. Additionally, adaptive assessments (e.g., Gibbons et al. 2016) that can focus on measures that quickly and accurately assess binge eating could reduce the burden placed on the patient and the provider and ensure the availability of standardized assessments in a wider range of care settings. In conclusion, a number of structured clinical interviews, self-report instruments, and objective measures, such as laboratory eating paradigms, are useful in the assessment of binge eating. Although a wide range of assessment tools is useful, as described above, important differences and pros and cons exist between the measures. The assessments measure different aspects of eating behaviors (e.g., OBEs/SBEs) and vary in

the number of questions that assess binge eating, other diagnostic symptoms, frequency, and severity (Table 1). Furthermore, many of the measures require extensive training and time to administer, which may impact their feasibility for routine use. Therefore, clinicians and researchers must be thoughtful in selecting an appropriate measure for diagnosis (e.g., of BED), case conceptualization, ongoing assessment (e.g., changes in frequency/severity of binge episodes), treatment planning, or evaluation of treatment outcome.

References

American Psychiatric Association (2013). Diagnostic and statistical manual of mental disorders (DSM-5®). American Psychiatric Pub

Barnes RD, Masheb RM, White MA, Grilo CM (2011) Comparison of methods for identifying and assessing obese patients with binge eating disorder in primary care settings. Int J Eat Disord 44(2):157–163

Berg KC, Peterson CB, Frazier P, Crow SJ (2012) Psychometric evaluation of the eating disorder examination and eating disorder examination-questionnaire: a systematic review of the literature. Int J Eat Disord 45 (3):428–438

Black CMD, Wilson GT (1996) Assessment of eating disorders: interview versus questionnaire. Int J Eat Disord 20(1):43–50

Burton AL, Abbott MJ, Modini M, Touyz S (2016) Psychometric evaluation of self-report measures of binge eating symptoms and related psychopathology: a systematic review of the literature. Int J Eat Disord 49 (2):123–140

Carter JC, Steward DA, Fairburn CG (2001) Eating disorder examination questionnaire: norms for young adolescent girls. Behav Res Ther 39(5):625–632

Cooper PJ, Fairburn CG (1987) The eating disorder examination: a semi-structured interview for the assessment of the specific psychopathology of eating disorders. Int J Eat Disord 6(1):1–8

Cuellar RE, Kaye WH, Hsu LKG, Van Thiel DH (1988) Upper gastrointestinal tract dysfunction in bulimia. Dig Dis Sci 33(12):1549–1553

Devlin MJ, Walsh BT, Guss JL et al (1997) Postprandial cholecystokinin release and gastric emptying in patients with bulimia nervosa. Am J Clin Nutr 65 (1):114–120

Dingemans AE, vanFurth EF (2017) Measuring changes during the treatment of eating disorders: a comparison of two types of questionnaires. Tijdschr Psychiatr 59 (5):278–285

Elder KA, Grilo CM, Masheb RM et al (2006) Comparison of two self-report instruments for assessing binge eating in bariatric surgery candidates. Behav Res Ther 44(4):545–560

Engel SG, Crosby RD, Thomas G et al (2016) Ecological momentary assessment in eating disorder and obesity research: a review of the recent literature. Curr Psychiatry Rep 18(4):37

Fairburn CG, Beglin SJ (1994) Assessment of eating disorders: interview or self-report questionnaire? Int J Eat Disord 16(4):363–370

Fairburn CG, Beglin SJ (2008) Eating questionnaire. Available http://www.credo-oxford.com/pdfs/EDE-Q_6.0.pdf. Accessed 8 July 2019

Fairburn CG, Cooper Z (1993) The eating disorder examination. In: Fairburn CG, Wilson GT (eds) Binge eating: nature, assessment, and treatment, 12th edn. Guilford Press, New York, pp 317–360

Fairburn CG, Cooper Z, Shafran R (2003) Cognitive behavior therapy for eating disorders: a "transdiagnostic" theory and treatment. Behav Res Ther 41(5):509–528

Fairburn CG, Cooper Z, O'Connor M (2014) Eating disorder examination. Available http://www.credo-oxford.com/pdfs/EDE_17.0D.pdf. Accessed 15 July 2019

Fairburn CG, Rothwell ER (2015) Apps and eating disorders: a systematic clinical appraisal. Int J Eat Disord 48(7):1038–1046

Farchaus SK, Corte CM (2003) Ecologic momentary assessment of eating-disordered behaviors. Int J Eat Disord 34:349–360

First MB, Williams JBW, Karg RS, Spitzer RL (2015) Structured clinical interview for DSM-5 disorders, clinician version (SCID-5-CV). American Psychiatric Association, Arlington, VA

Forbush KT, Berg KC (2016) Self-report assessments of eating pathology. In: Walsh BT, Attia E, Glasofer DR, Sysko R (eds) Handbook of assessment and treatment of eating disorders. American Psychiatric Association, Arlington, pp 157–174

Forbush KT, Wildes JE, Pollack LO et al (2013) Development and validation of the eating pathology symptoms inventory (EPSI). Psychol Assess 25:859–878

Forbush KT, Wildes JE, Hunt TK (2014) Gender norms, psychometric properties, and validity for the eating pathology symptoms inventory. Int J Eat Disord 47(1):85–91

Geliebter A, Melton PM, McCray RS et al (1982) Gastric capacity, gastric emptying, and test-meal intake in normal and bulimic women. Am J Clin Nutr 56(4):656–661

Geracioti TD, Liddle RA (1988) Impaired cholecystokinin secretion in bulimia nervosa. N Engl J Med 319(11):683–688

Gibbons RD, Weiss DJ, Frank E, Kupfer D (2016) Computerized adaptive diagnosis and testing of mental health disorders. Annu Rev Clin Psychol 12:83–104

Goldfein JA, Devlin MJ, Kamenetz C (2005) Eating disorder examination-questionnaire with and without instruction to assess binge eating in patients with binge eating disorder. Int J Eat Disord 37(2):107–111

Grilo CM, Masheb RM, Wilson GT (2001a) A comparison of different methods for assessing the features of eating disorders in patients with binge eating disorder. J Consult Clin Psychol 69(2):317–322

Grilo CM, Masheb RM, Wilson GT (2001b) Different methods for assessing the features of eating disorders in patients with binge eating disorder: a replication. Obes Res 9(1):418–422

Guss JL, Kissileff HR, Devlin MJ et al (2002) Binge size increases with body mass index in women with binge-eating disorder. Obes Res 10(10):1021–1029

Hadigan CM, Kissileff HR, Walsh BT (1989) Patterns of food selection during meals in women with bulimia. Am J Clin Nutr 50(4):759–766

Hildebrandt T, Craigen K (2015) Eating-related pathology in men and boys. In: Walsh BT, Attia E, Glasofer DR, Sysko R (eds) Handbook of assessment and treatment of eating disorders. American Psychiatric Association, Arlington, pp 105–118

Hildebrandt T, Latner J (2006) Effect of self-monitoring on binge eating: treatment response of 'binge drift'? Eur Eat Disord Rev 14(1):17–22

Hildebrandt T, Michaelides A, Mackinnon D et al (2017) Randomized controlled trial comparing smartphone assisted versus traditional guided self-help for adults with binge eating. Int J Eat Disord 50(11):1313–1322

Inui A, Okano H, Miyamoto M et al (1995) Delayed gastric emptying in bulimic patients. Lancet 346(8984):1240

Keel PK, Wolfe BE, Liddle RA et al (2007) Clinical features and physiological response to a test meal in purging disorder and bulimia nervosa. Arch Gen Psychiatry 64(9):1058–1066

Kissileff HR, Walsh BT, Kral JG, Cassidy SM (1986) Laboratory studies of eating behavior in women with bulimia. Physiol Behav 38(4):563–570

Kornstein SG, Kunovac JL, Herman BK, Culpepper L (2016) Recognizing binge-eating disorder in the clinical setting: a review of the literature. Prim Care Companion CNS Disord 18(3)

Latner JD, Wilson GT (2002) Self-monitoring and the assessment of binge eating. Behav Ther 33(3):467–477

Peterson CB, Mitchell JE (2005) Self-report measures. In: Mitchell JE, Peterson CB (eds) Assessment of eating disorders. Pergamon, New York, pp 98–119

Ricca V, Mannucci E, Moretti S, Di Bernardo M, Zucchi T, Cabras P et al (2000) Screening for binge eating disorder in obese outpatients. Compr Psychiatry 41:111–115

Samuels F, Zimmerli EJ, Devlin MJ et al (2009) The development of hunger and fullness during a laboratory meal in patients with binge eating disorder. Int J Eat Disord 42(2):125–129

Smyth J, Wonferlich S, Crosby R et al (2001) The use of ecological momentary assessment approaches in eating disorder research. Int J Eat Disord 30(1):83–95

Stice E, Telch CF, Rizvi SL (2000) Development and validation of the eating disorder diagnostic scale: a brief self-report measure of anorexia, bulimia, and binge-eating disorder. Psychol Assess 12(1):123–131

Stice E, Fisher M, Martinez E (2004) Eating disorder diagnostic scale: additional evidence of reliability and validity. Psychol Assess 16(1):60–71

Sysko R, Alavi S (2018) Eating disorder. In: Hunsley J, Mash EJ (eds) A guide to assessments that work. Oxford University Press, New York, pp 541–562

Sysko R, Devlin MJ, Walsh BT et al (2007a) Satiety and test meal intake among women with binge eating disorder. Int J Eat Disord 40(6):554–561

Sysko R, Walsh BT, Wilson GT (2007b) Expectancies, dietary restraint, and test meal intake among undergraduate women. Appetite 49:30–37

Sysko R, Devlin MJ, Schebendach J et al (2013) Hormonal responses and test meal intake among obese teenagers before and after laparoscopic adjustable gastric banding. Am J Clin Nutr 98(5):1151–1161

Sysko R, Glasofer DR, Hildebrandt T et al (2015) The eating disorder assessment for DSM-5 (EDA-5): development and validation of a structured interview for feeding and eating disorders. Int J Eat Disord 48(5):452–463

Sysko R, Ojserkis R, Hildebrandt T et al (2017) Impulsivity: a common substrate for eating and alcohol use disorders? Appetite 112:1–8

Sysko R, Steinglass J, Schebendach J, Mayer LES, Walsh BT (2018) Rigor and reproducibility via laboratory studies of eating behavior: A focused update and conceptual review. Int J Eat Disord 51(7):608–616. https://doi.org/10.1002/eat.22900

Tanofsky-Kraff M, Shomaker LB, Olsen C et al (2011) A prospective study of pediatric loss of control eating and psychological outcomes. J Abnorm Psychol 120(1):108–118

Thelen MH, Farmer J, Wonderlich S, Smith M (1991) A revision of the Bulimia test: the BULIT-R. Psychol Assess 3:119–124

Thomas JJ, Roberto CA, Berg KC (2016) Assessment measures, then and now: a look back at seminal measures and a look forward to the brave new world. In: Walsh BT, Attia E, Glasofer DR, Sysko R (eds) Handbook of assessment and treatment of eating disorders. American Psychiatric Association, Arlington, pp 137–156

Van der Ster WG, Noring C, Holmgren S (1994) Binge eating versus nonpurged eating in bulimics: is there a carbohydrate craving after all? Acta Psychiatr Scand 89(6):376–381

Walsh BT, Boudreau G (2003) Laboratory studies of binge eating disorder. Int J Eat Disord 34:S30–S38

Walsh BT, Kissileff HR, Cassidy SM, Dantzic S (1989) Eating behavior of women with bulimia. Arch Gen Psychiatry 46(1):54–58

Walsh BT, Zimmerli E, Devlin MJ et al (2003) A disturbance of gastric function in bulimia nervosa. Biol Psychol 54(9):929–933

Wilfley DE, Schwartz MB, Spurrell EB, Fairburn CG (1997) Assessing the specific psychopathology of binge eating disorder patients: interview or self-report? Behav Res Ther 35(12):1151–1159

Wilson GT, Vitousek KM (1999) Self-monitoring in the assessment of eating disorders. Psychol Assess 11:480–489

Wilson GT, Nonas CA, Rosenblum GD (1993) Assessment of binge eating in obese patients. Int J Eat Disord 13(1):25–33

Yanovski SZ, Leet M, Yanovski JA et al (1992) Food selection and intake of obese women with binge-eating disorder. Am J Clin Nutr 56(6):975–980

Yanovski SZ, Nelson JE, Dubbert BK, Spitzer RL (1993) Association of binge eating disorder and psychiatric comorbidity in obese subjects. Am J Psychiatry 150(10):1472–1479

Yanovski SZ, Marcus MD, Wadden TA, Walsh BT (2015) The questionnaire on eating and weight patterns-5 (QEWP-5): an updated screening instrument for binge eating disorder. Int J Eat Disord 48(3):259–261

Zanarini MC, Bender D, Sanislow C et al (2000) The collaborative longitudinal personality disorders study: reliability of axis I and II diagnoses. J Personal Disord 14(4):291–299

Developmental and Cultural Aspects of Binge Eating

Juan C. Hernandez and Marisol Perez

Abstract

Research on binge eating has focused on construct definition, and development/refinement of etiological models with more of a person-level focus and with less focus on other levels of organization. In childhood and adolescence, defining the consumption of a large amount of food can be difficult due to lack of autonomy and developmental growth, leading some researchers to use loss of control over eating as a more objective criterion. Among youth, existing research suggests parenting style, peer influences, and attachment styles may predispose individuals to the eventual development of binge eating. While, parental comments and food environment increase risk for binge eating. The second half of this chapter is dedicated to discussing binge eating across marginalized populations. Prevalence rates of binge eating across race/ethnicity, food insecurity, and sexual minorities are provided. Research on minority stress framework that highlights perceived environmental threats (e.g., cultural stereotypes, prejudice), minority stressors (e.g., discrimination, weight bias stigma) as contributors to social isolation and rejection, and negative affect, which in turn, leads to binge eating, is discussed. Further, risk factor research on discrimination and weight bias are discussed. It is hoped that future research will focus more on *under what conditions* for *whom* does binge eating develop.

Keywords

Binge eating · Children · Adolescents · Parent · Friends · Attachment · Food · Race · Ethnicity · Food insecurity · Sexual minorities

Learning Objectives
In this chapter, you will:

1. Review prevalence rates of binge eating across childhood and adolescence, and among marginalized groups.
2. Identify distal and proximal risk factor research in childhood from different levels of organizations within a developmental systems theory.
3. Gain knowledge related to minority-specific stressors that put marginalized communities at greater risk for binge eating.

1 Psychological Models of Binge Eating

The last two decades have yielded a dramatic increase in research on binge eating. Two current, predominant psychological models of binge

J. C. Hernandez · M. Perez (✉)
Arizona State University, Tempe, AZ, USA
e-mail: jchern24@asu.edu; Marisol.Perez@asu.edu

eating are the enhanced cognitive behavioral model of eating disorders and the affect regulation model of binge eating. Consistent with the history of clinical psychology and other mental health conditions such as depression, it is not surprising that the two models lie within cognitive-behavioral and interpersonal frameworks (Hunsley and Lee 2014).

The enhanced cognitive behavioral model (CBT-E) of eating disorders conceptualizes individuals with perfectionism and over evaluation of eating, shape, and weight, as leading to strict dieting and other weight-control behaviors, which in turn, leads to the development of binge eating (Fairburn et al. 2003). Further, mood intolerance bidirectionally influences binge eating. Research has supported various aspects of the model in both adults and children (Allen et al. 2012; Byrne and McClean 2002; Lampard et al. 2013). One study examined the enhanced cognitive-behavioral model among children aged 8–13 years, who were assessed annually over a 2-year period and found the model to fit the data adequately but proposed that the addition of affect regulation to the model could improve model fit, variance, and prospective predictability of binge eating (Allen et al. 2012). This is consistent with the adult treatment literature on binge-eating disorder (BED), which highlights that cognitive behavioral therapy treatment outcomes range from 30 to 50% for complete abstinence from binge eating at the end of treatment, and approximately 68% remission rate at long-term follow-up suggesting model refinement in the enhanced cognitive-behavioral model of binge eating could potentially yield better treatment results (Eddy et al. 2016; Hay 2013). Thus, although the enhanced cognitive behavioral model is a predominant model of binge eating, extensive research needs to continue to identify the key most influential risk factors that proximally (e.g., a variable that immediately contributes to binge eating) and distally (e.g., an underlying vulnerability that places someone at risk in the future for developing binge eating) contribute to the development and maintenance of binge eating.

The affect-regulation model posits that binge eating emerges as a learned coping behavior for dealing with high negative affect (Polivy and Herman 1993; Haedt-Matt and Keel 2011). Binge eating reduces high negative affectivity thereby negatively reinforcing the behavior (Polivy and Herman 1993; Haedt-Matt and Keel 2011). Ecological momentary assessment (EMA) of binge eating supports the spiked increase in negative affect prior to a binge and a subsequent decrease in negative affect post binge (Berg et al. 2013, 2015, 2017), although some evidence has not supported this association (Haedt-Matt and Keel 2011). Research supports the affect-regulation model in adults and youth (Burton and Abbott 2019; Czaja et al. 2009; Russell et al. 2017). A recent study cross-sectionally combined the enhanced cognitive-behavioral model and the affect-regulation model to examine the fit of the combined model on a sample of 766 college students (Burton and Abbott 2019). The model that fit the data best included low self-esteem associated with high negative affect, which when combined with difficulties with emotion regulation resulted in a dual path to binge eating. Binge eating was associated with difficulties with emotion regulation either via dietary restraint or eating beliefs (i.e., lack of control over eating, positive beliefs related to coping via eating, and permissive beliefs) (Burton and Abbott 2019). However, further model refinement is needed. For example, treatments targeting emotion regulation yield binge-eating abstinence rates between 64 and 79% post treatment, with only 55% remission rates at 6-month follow-up, leaving room for improvement (Safer et al. 2010; Telch et al. 2001). Model refinement could inform more effective treatment targeting emotion regulation.

Although the last two decades has seen a spike in research developing and refining etiological and psychological models of binge eating, there has been a lack of inclusion of developmental science. In particular, models have only focused on characteristics of the individual without considering the social, cultural, and environmental contexts. A prominent theory in developmental psychopathology is developmental systems theory (Damon and Lerner 2008). Developmental systems theory suggests that there are multiple levels of organization within human development (consisting of genetic, biological, physiological, person-level, family-level, community-level, societal-level, and cultural variables) that are integrated or fused in some way, such that there are unification and bidirectional links among the different levels of organization that contribute to specific outcomes and human development as a whole (Damon and Lerner 2008). The majority of etiological models of binge eating focus entirely on the person-level with less focus or specification of factors that contribute to binge eating at the other levels of organization. For example, Ledoux and colleagues (2015) found that neighborhood presence of a fast food restaurant (community-level) among ethnic minority women (cultural-level) who engaged in binge eating (person-level) displayed higher body mass index (physiological-level). Developmentally, integrating these levels of organization into an etiological model can greatly improve the validity and reliability of the framework.

The aim of this chapter is to begin to synthesize the most relevant risk factor research for binge eating from childhood and across cultures with an emphasis on different levels of organization from developmental systems theory when relevant. Given the infancy of developmental and cultural considerations in binge eating, this literature review will raise more questions than be able to provide answers. However, it is hoped that it will spur future research that will advance our understanding of and further develop etiological models of binge eating. This chapter is organized into two sections, with the first section focusing on prevalence rates of binge eating across childhood and risk factor research informed from developmental science. The second half is focused on binge eating across marginalized populations, with the first section providing prevalence rates of binge eating across income, race/ethnicity, and sexual minorities, and risk factor research informed from a minority stress framework.

2 Binge Eating Across Childhood

According to the *Diagnostic and Statistical Manual of Mental Disorders*—5th Edition (2013;

DSM-5), binge eating is diagnostically defined as eating an objectively large amount of food with a sense of loss of control. According to the International Classification of Disease—Eleventh Revision (ICD-11), only loss of control is required to meet criteria. However, the assessment of binge eating has unique challenges related to development. For example, researchers have pointed out that in childhood and adolescence it can be difficult to objectively define what constitutes a "large amount of food" due to inter-individual differences across children and families in terms of biological appetite regulation and cultural norms (Tanofsky-Kraff et al. 2008). Further, the quantities children eat are often monitored by caregivers limiting the knowledge of how much a youth would have eaten if unregulated. In addition, awareness during a meal and the ability to accurately recall food consumption is impacted by the age of the youth. Given these developmental constraints, researchers suggest loss of control of eating may be a more reliable criterion in youth (Tanofsky-Kraff et al. 2008). Research examining the rates of binge eating and related constructs, like overeating, in children have consisted of self-report questionnaires or interview formats. However, research indicates self-report measures may produce inflated estimates compared to clinical interviews (Goldschmidt et al. 2007; Tanofsky-Kraff et al. 2003). This may be due, in part, to variability in youth's understanding of the construct of binge eating which suggests that estimates of binge eating in children may have higher measurement error when compared to adult samples (Neumark-Sztainer and Story 1998).

2.1 Prevalence Rates

2.1.1 Overeating

Overeating is defined as consuming an objectively large amount of food, regardless of whether an individual experiences a loss of control. While difficult to assess, overeating early in life may put a child at greater risk for later binge eating and thus is a critical variable to examine developmentally. Among nontreatment-seeking samples, prevalence of overeating is 1.4% in children aged 6–8 years, 11.2% in 9–10 years, and 7.7% in children 11–12 years old using clinical interviews (Tanofsky-Kraff et al. 2009a, b). A separate study with clinical interviews, found 5.8% of children aged 8–13 years report overeating (Allen et al. 2008). Among self-report questionnaires, prevalence rates of overeating range from 5.9 to 15% among children aged 8–12 years (Ackard et al. 2003; Matton et al. 2013). The majority of youth who report overeating tend to remit at 1–5 year follow-up assessment (Allen et al. 2008; Goldschmidt et al. 2016).

More research exists on the prevalence rates among adolescents than children. Large epidemiological studies of adolescents estimate rates of overeating from 0.5 to 3.0% among 13–19 years of age (Sonneville et al. 2013). Using clinical interviews, 17% of adolescent high schoolers screened for eating or weight/shape concerns reported overeating (Goldschmidt et al. 2008). Among self-report questionnaires, rates of overeating were found to be 4.2–7.8% among 13–14 year olds, 3.9–5.5% among 15–16 years old, and 5.6–7.9% among 16–17 years old (Ackard et al. 2003). Similarly, Skinner et al. (2012) found rates of overeating from 1.5 to 5.1%. Another study found rates of overeating at 3.2% among adolescents in high school with the majority reporting no overeating 5 years later, but a new subgroup with no previous history of overeating emerging with overeating at a rate of 5.5% in young adulthood (Goldschmidt et al. 2016).

Overall, research demonstrates the majority of youth remit in overeating during follow-up assessments across both childhood and adolescence.

2.1.2 Binge Eating

The definition of binge eating according to the DSM-5 includes the consumption of large amounts of food and a loss of control. Using clinical interviews among nontreatment seeking samples, subjective (i.e., loss of control over eating without consuming a large amount of food in a discrete period of time), and objective binge eating (i.e., loss of control over eating and consuming a large amount of food in a discrete

period of time) is observed in 1.4% of children 6–8 years of age, 3.5% for subjective and objective binge eating in 9–10 years old, and 1.4% for subjective binge eating and 2.1% for objective binge eating in 11–12 years old (Tanofsky-Kraff et al. 2009a, b). A separate study with clinical interviews found 5.0% of children engage in subjective binge eating and 4.2% in objective binge eating between the ages of 8–13 years (Allen et al. 2008). Estimates tend to be higher when using self-report questionnaires with rates of binge eating ranging from 2.3% to 15% among 8–18 years old (Ackard et al. 2003; Chen et al. 2009; Haines et al. 2006, 2010; Matton et al. 2013; Spanos et al. 2010). When parents are administered self-report questionnaires, parents report binge eating in 2.0% of children aged 5–6 years (Lamerz et al. 2005), and 8.4% among children aged 4–7 years (Equit et al. 2013).

Among nontreatment seeking adolescents, large epidemiological studies report binge eating rates range from 1.0 to 5.0% (Allen et al. 2013; Sonneville et al. 2013; Swanson et al. 2011). Studies with clinical interviews report rates of 4.0% for objective binge eating and 2.0% of subjective binge eating with peak risk of age of onset for binge eating occurring at age 16 years (Stice et al. 1998). Among self-report questionnaires, binge eating rates are observed at 2.4–12.0% for adolescent youth (Carter et al. 2001; Johnson et al. 1999; Neumark-Sztainer et al. 2007; Skinner et al. 2012).

Among children and adolescents who are overweight or obese, prevalence rates for binge eating tend to be higher. In a recent meta-analysis that reviewed 36 different studies with a cumulative sample of 9818 children and adolescents, binge eating was prevalent in 25.8% of the youth irrespective of age, gender, BMI, assessment method, and time frame (He et al. 2017). Prevalence rates were higher for loss of control over eating (30%) than for those with binge eating (22%). In addition, clinical interviews yielded lower prevalence rates than other assessment methods and treatment studies had higher rates than other types of research (He et al. 2017).

Overall, research on binge eating suggests prevalence rates increase with age (Allen et al.

2013; Sonneville et al. 2013) and can remain stable through adulthood particularly among adolescent with greater psychosocial problems (Goldschmidt et al. 2014, 2016).

2.1.3 Loss of Control of Eating

Among youth, loss of control over eating, regardless of eating episode size, has garnered attention in the past two decades due to issues assessing the consumption of large amounts of food in children. Prevalence rates ranging from 2.9% in 6–8 years old (Tanofsky-Kraff et al. 2009a, b), 7.1% in 9–10 years old (Tanofsky-Kraff et al. 2009a, b), 3.6% in 11–12 years old (Tanofsky-Kraff et al. 2009a, b), and 23.9–28.2% among adolescents (Shomaker et al. 2010; Tanofsky-Kraff et al. 2009a, b) have been reported in youth. Using clinical interviews, another study found youth reported 28.4% with at least one episode of loss of control of eating in the past month among youth 11–15 years old (Elliott et al. 2010). Similarly, a separate study found 23.6% of children aged 6–13 years old reported loss of control eating at least once in their lifetime (Tanofsky-Kraff et al. 2011). Researchers have observed that loss of control over eating can emerge before binge eating and is predictive of the development of binge-eating disorder and other psychiatric symptoms 4 years later (Sonneville et al. 2013; Tanofsky-Kraff et al. 2011). Further, loss of control of eating, regardless of amount of food consumed, is associated with greater psychopathology, disordered eating, and emotional eating among children and adolescents (Shomaker et al. 2010).

2.2 Developmental Risk Factors for Binge Eating

The following section will cover three distal risk factors (parenting style, peer influences, and attachment style) that may predispose individuals to the eventual development of binge eating. In addition, two proximal risk factors (parental comments and food environment) that can immediately increase risk for binge eating is reviewed. It is important to note, that in general, there is a paucity of research on developmental risk factors.

2.2.1 Parental and Peer Influences

Parents are the primary and consistent socialization driving force of children. As such, there is extensive research examining the multitude of ways parents can influence a child's eating. Parenting style research tends to examine a parent's general approach to parenting. Authoritarian parents are characterized by having high demands (e.g., mistakes made by the child are punished harshly) and low responsiveness (e.g., provide little feedback and nurturance to the child). Evidence suggests that parenting style is a distal risk factor for binge eating where adolescent girls with authoritarian mothers have an increased risk to report binge eating 5 years later when compared to those with mothers with authoritative, permissive, or neglectful mothers (Zubatsky et al. 2015). Parent feeding strategy has also been associated with child eating behavior where, parental restriction of palatable foods has been associated with increased food consumption and eating in the absence of hunger (Faith et al. 2004). Parents' relationships with youth also play a role in binge eating. A large epidemiological study found that higher mother–child and father–child connectedness among female (but not male) adolescents was associated with lower binge eating concerns (Hazzard 2019). Parental comments are a proximal risk factor for binge eating. One study found that the importance of weight to fathers significantly predicted an increased risk to binge eat weekly for girls, while negative comments about weight by fathers significantly predicted an increased risk to binge eat weekly for boys (Field et al. 2008). Evidence demonstrates that mothers and fathers who engage in conversations related to weight and size with adolescents have children who are more likely to diet, use unhealthy weight-control behaviors and binge eat (Berge et al. 2013). In contrast, when parents engage in conversations about healthful eating, their adolescents are less likely to diet and use unhealthy weight-control behaviors (Berge et al. 2013). Overall, parents can influence a youth's risk for binge eating both immediately and distally via parenting style, feeding style, conversations around food and weight, and interpersonal connections.

Friends are an important source of influence during the school-age years. There is evidence that in children as young as 2 years old, friends can influence eating behaviors and attitudes (Salvy et al. 2008, 2009). Given the prevalence of binge eating, the majority of research on friend influences is among adolescent youth. In a study of friendship cliques among adolescent girls, the extent of body talk with friends, body comparisons, and friend encouragement to lose weight were all predictive of binge eating (Paxton et al. 1999). Another study among 173 friendship cliques among adolescent girls, found similar scores among clique members on dieting, extreme weight-loss behaviors, and binge eating (Hutchinson and Rapee 2007). In a study of adolescent boys and girls, conversations about weight loss with peers uniquely predicted binge eating in girls whereas peer encouragement to lose weight (e.g., "you should lose weight") uniquely predicted binge eating in boys (Vincent and McCabe 2000). In a large epidemiological study of adolescents, negative comments about weight by boys, and the importance of weight to peers were associated with an increased risk of binge eating weekly for girls but not boys (Field et al. 2008). Overall, the evidence highlights the significance of the peer environment in creating a vulnerability in youth that increases risk for the development of binge eating.

For both boys and girls, parental- and friend-related variables influence binge eating. However, across studies that are either epidemiological, community, or school-setting, differential risk factors emerge across the genders.

2.2.2 Attachment Style

Attachment theory has also emerged as a potential developmental explanation for the onset of binge eating and other eating pathology. Attachment theory argues that a secure attachment (e.g., self-perceptions of high value and competency due to consistent availability from past caregivers) relationship is necessary to cope with previously mentioned developmental tasks (e.g., physical changes), and can buffer against psychological risks (e.g., media exposure; Gander et al. 2015). Attachment theorists propose that not having a

secure attachment style will lead to developmental problems achieving independence, a potential precursor to the onset of eating pathology (Rhodes and Kroger 1992). Cross-sectional research on adults finds lower binge eating symptoms among those with higher levels of secure attachment (Ugo et al. 2012). Similar to adults, adolescents with an anxious attachment style display greater binge eating symptoms (Boone 2013). Among women with eating disorders, anxious attachment style relates to both depression and eating disorder symptoms via emotional reactivity (Shakory et al. 2015; Tasca et al. 2009). In a treatment study of individuals with BED, reduction in attachment insecurity predicted improved interpersonal relationships and depressive symptoms at an 1-year follow-up post treatment (Maxwell et al. 2014). Although more research is needed, attachment insecurity seems to be an individual characteristic that might predispose individuals to both interpersonal and emotional regulation difficulties leading to binge eating.

2.2.3 Food Environment

Food advertising refers to any form of marketing and communication that is designed to increase a child's familiarity, knowledge, exposure, appeal, or consumption to a specific brand (Cairns et al. 2013). The majority of food advertising aimed at child audiences consists of highly palatable, energy-dense, and/or sugar-dense foods (Cairns et al. 2013). Experimental studies demonstrate that exposure to food advertising increases food intake and snack food choices among children (Folkvord et al. 2013, 2014, 2015). Research suggests that children who are impulsive may be more susceptible to food advertising (Folkvord et al. 2014). In addition, children with an attentional bias toward food cues also demonstrate increased consumption of energy-dense foods after food cue exposure (Folkvord et al. 2015). Cue reactivity theory is used to explain how food cues in marketing increase eating behavior (Jansen 1998). Based on classical conditioning, experimental studies with adults have found that food cues trigger psychological and physiological responses. For example, food cues in advertisements can trigger salivation and increased thoughts about food (Boswell and Kober 2016; Meyer et al. 2015), which subsequently induce eating behavior irrespective of hunger. It is hypothesized that the constant bombardment of food cues contributes to the dysregulation of the body's hunger and satiety signals. Research on treatments designed to reduce food cue reactivity in children have found significant posttreatment reductions in binge eating episodes, loss of control eating, and less reactivity to food cues (Boutelle et al. 2017). Collectively, the research suggests the food environment can be a proximal risk factor for overeating and binge eating.

The majority of risk research on binge eating in youth examines only one or two risk factors at a time. More complex and dynamic modeling is needed to better represent the mutlivariable interactional relationships related to binge eating. For example, very little research has utilized actor–partner interdependence models when examining parent or family influences which more accurately reflect how parents and child independently can influence their own and the other person's eating behaviors, as well as the interaction between the two individuals. In sum, research is needed that better reflects the complex worlds youth live in. More complex and dynamic modeling can begin to advance knowledge in the understanding of under which conditions, who is at risk for or resilient to binge eating.

3 Binge Eating Among the Marginalized: Cultural Considerations

Integrating what we know about the developmental trajectories of binge eating into a cultural framework is necessary for developing appropriate etiological models. While there is still work to be done, the field has made strides toward building an understanding of the unique risk marginalized communities face in the development of binge eating. Minority Stress Theory (see Fig. 1) argues that health disparities among minority communities can be explained by increased exposure to deleterious social factors (e.g., harassment, violence, isolation, and rejection) targeting marginalized groups (Meyer 2003; LeBlanc et al. 2015). Meyer (2003) showed that minority stressors exist on a continuum of proximity to the self, such that distal stressors are objective and found in a minority group member's environment (e.g., stereotypes, prejudice) which eventually lead to internal proximal stressors related to perceived environmental threat (e.g., internalized discrimination, negative attitude toward own minority group). Research has found preliminary evidence that stress from the environment related to one's ethnic identity is distinct from general life stress across Hispanics, African, and Asian American individuals, and predictive of eating disorder symptoms among African American women (Kroon van Diest et al. 2014). To date, the utility of Minority Stress Theory as an explanation for binge eating disparities has largely only been tested among sexual minority groups. Given the shared exposure to deleterious social factors experienced by other minority identities (e.g., racial, ethnic, gender, and disability; Hayes et al. 2011; Mays and Cochran 2001), it is theorized that a similar stress model generalizes to other minority groups.

3.1 Prevalence Rates

3.1.1 Race and Ethnicity

Across adolescent and adult samples, racial and ethnic minority status is considered a risk factor for binge eating (Lee-Winn et al. 2016; Marques et al. 2011). For adolescents, research tends to find higher prevalence rates of binge eating for racial and ethnic minorities. Generally, ethnic minority adolescents display higher prevalence rates of overeating and binge eating when compared to non-Hispanic White youth even after controlling for BMI (Ackard et al. 2003). In a large national epidemiological study, Hispanic (2.4%) and Black (1.5%) adolescents had higher lifetime prevalence rates of BED than non-Hispanic White adolescents (1.4%) (Swanson et al. 2011). Similarly, Hispanic (3%) and Black (3.5%) adolescents had higher lifetime

DISTAL RISK FACTORS   PROXIMAL RISK FACTORS

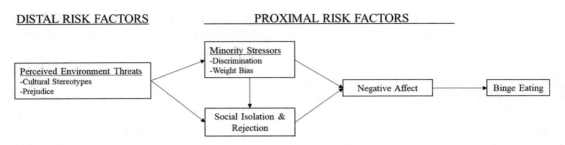

Fig. 1 Conceptual model of minority stress theory adapted from Mason and Lewis (2016) model

subthreshold BED than non-Hispanic White (2.0%) adolescents (Swanson et al. 2011). According to a nationally representative cross-sectional study of adolescents aged 13–18 years, racial/ethnic minorities (Hispanic, non-Hispanic Black) showed more binge eating behavior than non-Hispanic White participants. Additionally, Hispanics reported more fear of weight gain from binge eating, more loss of control, and distress than non-Hispanic Black and non-Hispanic White participants (Lee-Winn et al. 2016). Although, Johnson et al. (2002) suggest that differences may be more prevalent when considering developmental trends rather than general prevalence. For instance, binge-eating rates peak for non-Hispanic Black samples during early adolescence, but steadily increase across adolescence among non-Hispanic White samples, and Hispanic females report the highest rates of binge eating across time points (Croll et al. 2002).

Racial and ethnic differences can also be seen in adult samples. A large epidemiological study of adults found that the prevalence rates of past year binge eating disorder did not differ between Hispanics, African American, Asian, and non-Hispanic White individuals, but there were racial/ethnic group differences on subthreshold BED (Marques et al. 2011). Specifically, both the Hispanic (2.72%) and African American (2.13%) groups reported significantly higher prevalence rates than the non-Hispanic White group (1.04%). In a study examining racial/ethnic differences in adults participating in 11 randomized controlled trials for BED, there were no differences between Hispanics, Blacks, and Whites on frequency of binge eating episodes at baseline (Franko et al. 2012). In another study

of treatment-seeking adults for BED, Hispanic individuals reported the age of onset for dieting and binge eating to occur around the age of 25 years, whereas Blacks and Whites reported the age of onset of dieting first (25 and 20 years, respectively) with binge eating emerging 2–3 years later (age of onset 27 and 22 years, respectively) (Lydecker and Grilo 2016). In a 7-year longitudinal study, cognitions associated with binge eating (such as fear of losing control or embarrassment over amount of food eaten) prospectively predicted depressive symptoms among Asians/Pacific Islanders and non-Hispanic White, but not Hispanic and Black individuals (Hazzard et al. 2019).

3.1.2 Food Insecurity

It was commonly thought that underweight would characterize those with food insecurity (i.e., lack of consistent access to enough food for a healthy lifestyle; United States Department of Agriculture). Research now demonstrates that food insecurity and obesity co-occur (Dinour et al. 2007). In addition, preliminary research is starting to emerge that binge eating may also co-occur among the food insecure. Qualitative research conducted with parents of obese children from low-income households that included food insecure and food secure groups found that parents spontaneously described eating episodes among their children that resembled binge eating within the food insecure groups but not the food secure groups (Tester et al. 2016). In addition, hiding food and nighttime eating emerged as themes from the food insecure groups but not the food secure groups (Tester et al. 2016). Research shows increased rates of objective binge eating, subjective binge eating, overeating, nighttime eating among food insecure adults when compared to food secure adults (Becker et al. 2017). In addition, research has found that as food insecurity increases in severity, rates of clinically significant eating disorder symptoms increase as well, with the most food insecure demonstrating the highest rates of binge eating, overeating, and night eating with distress (Becker et al. 2019). Youth of low socioeconomic status have higher rates of binge eating when compared to middle and high-status groups (Ackard et al. 2003). The research on food insecurity and binge eating is just beginning to emerge. Extensive research is needed to identify how the environmental context of food insecurity might contribute to the development of binge eating, potentially how the emergence of binge eating might contribute to further food insecurity, and how the variation of availability and scarcity of food might alter the expression of eating disorder symptoms.

3.1.3 Sexual Minorities

Sexual minorities will here signify individuals without a heterosexual orientation or who engage in sexual behavior with the same or both sexes. Sexual minorities display higher rates of eating pathology, particularly sexual minorities who are transgender or an ethnic minority (Calzo et al. 2017). Both minority stress and sociocultural models have been offered as potential explanations for these disparities. Minority stress models of binge eating among sexual minorities characterize binge eating as a response to discrimination and internalized negative attitudes or stigma toward sexual minorities (Mason and Lewis 2016). Sociocultural models of eating pathology argue gender and sexual minority appearance ideals and general importance of physical attractiveness among these groups lead to disparities (McClain and Peebles 2016; Tylka and Andorka 2012). In a sample of lesbian and bisexual women, 13.4% engaged in moderate binge eating, and 4.9% engaged in severe binge eating, with those classified as overweight or obese being the most likely to report binge eating (Mason and Lewis 2016). Similarly, recent studies found that bisexual women and women who have sex with women are more likely to report

binge eating than heterosexual women (Laska et al. 2015; Meneguzzo et al. 2018). A sample of undergraduate men and women showed high rates of subjective and objective binge eating among women who have sex with women and high rates of subclinical objective binge eating among men who have sex with men. Additionally, men who have sex with both men and women reported high rates of clinical levels of binge eating (Von Schell et al. 2018). A study conducted among a large cohort of U.S., youth found that binge eating disparities were reported at the youngest ages (age 12 years) and maintained throughout adolescence (Austin et al. 2009). Among females, sexual minorities were more likely to engage in binge eating while their straight counterparts were more likely to engage in purging behavior. Among males, sexual minorities were more likely to engage in both binge eating and purging (Austin et al. 2009).

Overall, sexual minority status is associated with increased risk for binge eating. However, the literature is predominantly descriptive in nature. Research is needed that examines differences in clinical presentation and the generalizability of current etiological models.

3.2 Cultural Risk Factors for Binge Eating

3.2.1 Discrimination

Wallace (2005) argues that the United States maintains a social conditioning process generationally transmitting discrimination, stigmatization, stereotyping, and violence toward diverse and different others. This conditioning process has disproportionate negative impacts on health behaviors among minority groups. Experiencing discrimination and prejudice can increase an individual's stress level (Allison 1998). Ethnic discrimination has been found to potentially play a key role in the development of eating problems through increases of negative body image and low self-esteem (Smolak and Striegel-Moore 2001). Research shows that perceived discrimination and acculturative stress are key risk factors for disordered eating behavior (Kwan et al. 2018; Cheng 2014). Kwan, Gordon and

Minnich (2018) argue that among ethnic minorities, perceived discrimination may lower self-esteem and create a negative mood state that induces binge eating as a maladaptive coping response to escape the negative effect of self-awareness (Heatherton and Baumiester 1991). Among African American women, discriminatory stress is related to the severity of binge eating symptoms (Harrington et al. 2006). In a large epidemiological study of African Americans, perceived discrimination is associated with increasing odds of having binge-eating disorder across both genders when controlling for age, education, employment, and marital status; however, the association between discrimination and binge-eating disorder was greater for women than men (Assari 2018).

3.2.2   Weight Bias

The majority of the current literature examining weight bias and its effect on binge eating has been done with samples of women. Weight bias internalization develops as an individual directs cultural and social stigma regarding weight to themselves (Durso and Latner 2008). Weight bias internalization, in turn, may lead to negative affect and increase the likelihood of binge eating to cope with the negative arousal (Fairburn et al. 2003). While weight stigma is more common among those experiencing overweight or obesity, research demonstrates that all weight categories can experience weight stigma (Puhl et al. 2013). Research shows that weight discrimination is associated with overeating (Sutin et al. 2016). Among undergraduate and community samples, those with higher internalized weight bias report more binge eating (Mehak et al. 2018; Schvey and White 2015). O'Brien et al. (2016) found that after accounting for age, gender, and weight status, weight bias internalization mediated the relationship between weight stigma and binge eating. Wang et al. (2017) found an additional potential risk factor for weight bias internalization among treatment-seeking patients with BED, such that those with greater rumination might spend more time preoccupied with weight-based discrimination and internalize related negative attitudes. Among adolescents, weight bias internalization was significantly higher among those who

reported binge eating and eating to cope with stress (Puhl and Himmelstein 2018). More generally, reviews on weight stigma in daily life show that higher frequency of stigma-related experiences are associated with unhealthy eating behaviors (Vartanian and Porter 2016).

Research conducted with a sample of sexual minority young adults found that sexual minority men and women who experienced weight-based discrimination had a greater risk of binge eating (Gordon et al. 2018). According to a national survey of LGBTQ teens, weight-based victimization from family members was also associated with binge eating, with over half of the participants reported experiencing weight-based victimization from family members and peers (Himmelstein et al. 2019). Gathering a better understanding of the trajectory of internalizing weight-based messages in the context of sexual orientation stigma is key to improving binge-eating behavior among these groups.

The literature on discrimination and weight-bias supports the Minority Stress Theory, which purports that proximal stressors such as discrimination, can contribute to health disparities. Both discrimination and weight bias are associated with binge eating, however, the research is largely cross sectional. Experimental studies are needed that can examine if minority-specific stressors contribute to increases in negative affectivity, which in turn, triggers binge eating.

4   Overall Summary and Future Research

Echoed throughout this entire chapter is the critical need for more research from a developmental and cultural perspective. Research has just begun to scratch the surface in understanding the development of binge eating across contexts. The last two decades of research on binge eating have focused on construct definition, and development and refinement of etiological models. In addition, research focused on examining prevalence rates across demographic groups. Overeating and binge eating have been observed in children as young as 6 years old, and throughout childhood. Loss-of-control eating can precede binge eating and is associated with greater psychopathology. Risk factor research has predominantly focused on examining one or two risk factors at a time, neglecting the complex, interactional dynamics across multiple levels including cultural, environmental, social, biological, physiological, political, and familial. Nonetheless, parents, friends, attachment styles, and food environments have all been associated with binge eating.

Racial, ethnic, and sexual minority status, and food insecurity are all risk factors for binge eating. Yet, this cultural work is in its infancy. Detailed examination of the generalizability of etiological models of binge eating across marginalized groups is needed. Further, incorporating cultural-specific variables and stressors may enhance prospective predictability of these models.

It is hoped that future research on binge eating will focus more on *under what conditions* for *whom* does binge eating develops. This more nuanced and interactional focus can assist in the tailoring of current treatments for binge eating and BED, as well lead to the development of targeted prevention efforts. This is consistent with the recent recommendations by the American Psychological Association (2017):

> We need to understand how the relationships among stressors operate in different populations. Developing new models of the relationships among stressors with direct and indirect effects on susceptibility and resilience over time will help us to develop the best models for prevention and intervention. (p. 45)

They further suggest developing longitudinal research examining the bidirectional relationships among political, social, and physical environments shape individual-level outcomes consistent with the developmental systems theories.

References

Ackard DM, Neumark-Sztainer D, Story M et al (2003) Overeating among adolescents: prevalence and associations with weight-related characteristics and psychological health. Pediatrics 111:67–74

Allen KL, Byrne SM, La Puma M et al (2008) The onset and course of binge eating in 8- to 13-year-old healthy weight, overweight and obese children. Eat Behav 9 (4):438–446

Allen KL, Byrne SM, McLean NJ (2012) The dual-pathway and cognitive-behavioural models of binge eating: prospective evaluation and comparison. Eur Child Adolesc Psychiatry 21(1):51–62

Allen KL, Byrne SM, Oddy WH et al (2013) Early onset binge eating and purging eating disorders: course and outcome in a population-based study of adolescents. J Abnorm Child Psychol 41 (7):1083–1096

Allison KW (1998) Stress and oppressed social category membership. In: Swim JK, Stangor C (eds) Prejudice: the target's perspective. Elsevier, Amsterdam, pp 145–170

American Psychiatric Association. Diagnostic and statistical manual of mental disorders (DSM-5®). American Psychiatric Pub; 22 May 2013

American Psychological Association, APA working group on stress and health disparities (2017) Stress and health disparities: contexts, mechanisms, and interventions among racial/ethnic minority and low-socioeconomic status populations. http://www.apa.org/pi/health-disparities/resources/stress-report.aspx

Assari S (2018) Perceived discrimination and binge eating disorder; gender differences in African Americans. J Clin Med 7(5):89

Austin SB, Ziyadeh NJ, Corliss J et al (2009) Sexual orientation disparities in purging and binge eating from early to late adolescence. J Adolesc Health 45 (3):238–245

Becker CB, Middlemass K, Taylor B et al (2017) Food insecurity and eating disorder pathology. Int J Eat Disord 50(9):1031–1040

Becker CB, Middlemass KM, Gomez F et al (2019) Eating disorder pathology among individuals living with food insecurity: a replication study. Clin Psychol Sci 7 (5):1144–1158. https://doi.org/10.1177/2167702619851811

Berg KC, Crosby RD, Cao L et al (2013) Facets of negative affect prior to and following binge-only, purge-only, and binge/purge events in women with bulimia nervosa. J Abnorm Psychol 122(1):111–118

Berg KC, Crosby RD, Cao L et al (2015) Negative affect prior to and following overeating-only, loss of control eating-only, and binge eating episodes in obese adults. Int J Eat Disord 48(6):641–653

Berg KC, Cao L, Crosby RD et al (2017) Negative affect and binge eating: reconciling differences between two analytic approaches in ecological momentary assessment research. Int J Eat Disord 50(10):1222–1230

Berge JM, MacLehose R, Loth KA et al (2013) Parent conversations about healthful eating and weight: associations with adolescent disordered eating behaviors. JAMA Pediatr 167(8):746–753

Boone L (2013) Are attachment styles differentially related to interpersonal perfectionism and binge eating symptoms? Pers Individ Differ 54(8):931–935

Boswell RG, Kober H (2016) Food cue reactivity and craving predict eating and weight gain: a meta-analytic review. Obes Rev 17(2):159–177

Boutelle KN, Knatz S, Carlson J et al (2017) An open trial targeting food cue reactivity and satiety sensitivity in overweight and obese binge eaters. Cogn Behav Pract 24(3):363–373

Burton AL, Abbott MJ (2019) Processes and pathways to binge eating: development of an integrated cognitive and behavioural model of binge eating. J Eat Disord 7:18. https://doi.org/10.1186/s40337-019-0248-0

Byrne SM, McLean NJ (2002) The cognitive-behavioral model of bulimia nervosa: a direct evaluation. Int J Eat Disord 31:17–31

Cairns G, Angus K, Hastings G et al (2013) Systematic reviews of the evidence on the nature, extent and effects of food marketing to children: a retrospective summary. Appetite 62:209–215

Calzo JP, Blashill AJ, Brown TA et al (2017) Eating disorders and disordered weight and shape control behaviors in sexual minority populations. Curr Psychiatry Rep 19(8):49

Carter JC, Stewart DA, Fairburn CG (2001) Eating disorder examination questionnaire: norms for young adolescent girls. Behav Res Ther 39(5):625–632

Chen EY, McCloskey MS, Keenan KE (2009) Subtyping dietary restraint and negative affect in a longitudinal community sample of girls. Int J Eat Disord 42:275–283

Cheng H-L (2014) Disordered eating among Asian/Asian American women: racial and cultural factors as correlates. Couns Psychol 42(6):821–851

Croll J, Neumark-Sztainer D, Story M et al (2002) Prevalence and risk and protective factors related to disordered eating behaviors among adolescents: relationship to gender and ethnicity. J Adolesc Health 31(2):166–175

Czaja J, Reif W, Hilbert A (2009) Emotion regulation and binge eating in children. Int J Eat Disord 42(4):356–362

Damon W, Lerner RM (2008) The scientific study of child and adolescent development: important issues in the field today. In: Damon W, Lerner RM, Kuhn D et al (eds) Child and adolescent development: an advanced course. Wiley, New Jersey, pp 3–13

Dinour LM, Bergen D, Yeh M-C (2007) The food insecurity-obesity paradox: a review of the literature and the role food stamps may play. J Acad Nutr Diet 107(11):1952–1961

Durso LE, Latner JD (2008) Understanding self-directed stigma: development of the weight bias internalization scale. Obesity 16(S2):S80–S86

Eddy KT, Tabri N, Thomas JJ et al (2016) Recovery from anorexia nervosa and bulimia nervosa at 22 year follow-up. J Clin Psychiatry 78(02):184–189

Elliott C, Tanofsky-Kraff M, Shomaker LB et al (2010) An examination of the interpersonal model of loss of control eating in children and adolescents. Behav Res Ther 5:424–428

Equit M, Pälmke M, Becker N et al (2013) Eating problems in young children – a population based study. Acta Paediatr 102:149–155

Fairburn CG, Cooper Z, Shafran R (2003) Cognitive behavior therapy for eating disorders: a "transdiagnostic" theory and treatment. Behav Res Ther 41(5):509–528

Faith MS, Scanlon KS, Birch LL et al (2004) Parent-child feeding strategies and their relationships to child eating and weight status. Obes Res 12(11):1711–1722

Field AE, Javaras KM, Aneja P et al (2008) Family, peer, and media predictors of becoming eating disordered. Arch Pediatr Adolesc Med 162(6):574–579

Folkvord F, Anschütz DJ, Buijzen M et al (2013) The effect of playing advergames that promote energy-dense snacks or fruit on actual food intake among children. Am J Clin Nutr 97:239–245

Folkvord F, Anschütz DJ, Nederkoorn C et al (2014) Impulsivity, "advergames," and food intake. Pediatrics 133(6):1007–1012

Folkvord F, Anschütz DJ, Wiers RW et al (2015) The role of attentional bias in the effect of food advertising on actual food intake among children. Appetite 84:251–258

Franko DL, Thompson-Brenner H, Thompson DR et al (2012) Racial/ethnic differences in adults in randomized clinical trials of binge eating disorder. J Consult Clin Psychol 80(2):186–195

Gander M, Sevecke K, Buchheim A (2015) Eating disorders in adolescence: attachment issues from a developmental perspective. Front Psychol 6:1136

Goldschmidt AB, Doyle AC, Wilfley DE (2007) Assessment of binge eating in overweight youth using a questionnaire version of the child eating disorder examination with instructions. Int J Eat Disord 40:460–467

Goldschmidt AB, Jones M, Manwaring JL et al (2008) The clinical significance of loss of control over eating in overweight adolescents. Int J Eat Disord 41:153–158

Goldschmidt AB, Wall MM, Loth KA et al (2014) The course of binge eating from adolescence to young adulthood. Health Psychol 33(5):457–460

Goldschmidt AB, Wall MM, Zhang J et al (2016) Overeating and binge eating in emerging adulthood: 10-year stability and risk factors. Dev Psychol 52(3):475–483

Gordon AR, Kenney EL, Haines J et al (2018) Weight-based discrimination and disordered eating behaviors among US sexual minority adolescents and young adults. J Adolesc Health 62(2):S1

Haedt-Matt AA, Keel PK (2011) Revisiting the affect regulation model of binge eating: a meta-analysis of studies using ecological momentary assessment. Psychol Bull 137(4):660–681

Haines J, Neumark-Sztainer D, Eisenberg ME et al (2006) Weight teasing and disordered eating behaviors in adolescents: longitudinal findings from project EAT (eating among teens). Pediatrics 117(2):209–215

Haines J, Gillman MW, Rifas-Shiman S et al (2010) Family dinner and disordered eating behaviors in a large cohort of adolescents. Eat Disord 18(1):10–24

Harrington EF, Crowther JH, Payne Henrickson HC et al (2006) The relationships among trauma, stress, ethnicity, and binge eating. Cultur Divers Ethnic Minor Psychol 12(2):212–229

Hay P (2013) A systematic review of evidence for psychological treatments in eating disorders: 2005–2012. Int J Eat Disord 46(5):462–469

Hayes JA, Chun-Kennedy C, Edens A et al (2011) Do double minority students face double jeopardy? Testing minority stress theory. J Coll Couns 14(2):117–126

Hazzard VM (2019) Family risk and protective factors for binge eating-related concerns in a nationally representative sample of young adults in the United States. https://deepblue.lib.umich.edu/bitstream/handle/2027.42/149973/hazvivie_1.pdf?sequence=1&isAllowed=y

Hazzard VM, Hahn SL, Bauer KW et al (2019) Binge eating-related concerns and depressive symptoms in young adulthood: seven-year longitudinal associations and differences by race/ethnicity. Eat Behav 32:90–94

He J, Cai Z, Fan X (2017) Prevalence of binge and loss of control eating among children and adolescents with overweight and obesity: an exploratory meta-analysis. Int J Eat Disord 50(2):91–103

Heatherton TF, Baumeister RF (1991) Binge eating as escape from self-awareness. Psychol Bull 110(1):86–108

Himmelstein MS, Puhl RM, Quinn DM (2019) Overlooks and understudied: health consequences of weight stigma in men. Obesity 27(10):1598–1605. https://doi.org/10.1002/oby.22599

Hunsley J, Lee CM (2014) Introduction to clinical psychology: an evidence-based approach, 3rd edn. Wiley, New York, pp 399–411

Hutchinson DM, Rapee RM (2007) Do friends share similar body image and eating problems? The role of social networks and peer influences in early adolescence. Behav Res Ther 45(7):1557–1577

Jansen A (1998) A learning model of binge eating: cue reactivity and cue exposure. Behav Res Ther 36:257–272

Johnson WG, Grieve FG, Adams CD et al (1999) Measuring binge eating in adolescents: adolescent and parent versions of the questionnaire of eating and weight patterns. Int J Eat Disord 26(3):301–314

Johnson WG, Rohan KJ, Kirk AA (2002) Prevalence and correlates of binge eating in white and African American adolescents. Eat Behav 3:179–189

Kroon van Diest AM, Tartakovsky M, Stachon C et al (2014) The relationship between acculturative stress and eating disorder symptoms: is it unique from general life stress? J Behav Med 37(3):445–457

Kwan MY, Gordon KH, Minnich AM (2018) An examination of the relationships between acculturative stress, perceived discrimination, and eating disorder symptoms among ethnic minority college students. Eat Behav 28:25–31

Lamerz A, Kuepper-Nybelen J, Bruning N et al (2005) Prevalence of obesity, binge eating, and night eating in a cross-sectional field survey of 6-year old children and their parents in a German urban population. J Child Psychol Psychiatry 46(4):385–393

Lampard AM, Tasca GA, Balfour L et al (2013) An evaluation of the transdiagnostic cognitive-behavioural model of eating disorders. Eur Eat Disord Rev 21(2):99–107

Laska MN, VanKim NA, Erickson DJ et al (2015) Disparities in weight and weight behaviors by sexual orientation in college students. Am J Public Health 105(1):111–121

LeBlanc AJ, Frost DM, Wight RG (2015) Minority stress and stress proliferation among same-sex and other marginalized couples. J Marriage Fam 77(1):40–59

Ledoux T, Adamus-Leach H, O'Connor DP, Mama S, Lee RE (2015) The association of binge eating and neighbourhood fast-food restaurant availability on diet and weight status. Public Health Nutr 18(2):352–360

Lee-Winn AE, Townsend L, Reinblatt SP et al (2016) Associations of neuroticism and impulsivity with binge eating in a nationally representative sample of adolescents in the United States. Pers Individ Differ 90:66–72

Lydecker JA, Grilo CM (2016) Different yet similar: examining race and ethnicity in treatment-seeking adults with binge eating disorder. J Consult Clin Psychol 84(1):88–94

Marques L, Alegria M, Becker AE et al (2011) Comparative prevalence, correlates of impairment, and service utilization for eating disorders across US ethnic groups: implications for reducing ethnic disparities in health care access for eating disorders. Int J Eat Disord 44(5):412–420

Mason TB, Lewis RJ (2016) Minority stress, body shame, and binge eating among lesbian women: social anxiety as a linking mechanism. Psychol Women Q 40(3):428–440

Matton A, Goossens L, Braet C et al (2013) Continuity in primary school children's eating problems and the influence of parental feeding strategies. J Youth Adolesc 42(1):52–66

Maxwell H, Tasca GA, Ritchie K et al (2014) Change in attachment insecurity is related to improved outcomes 1-year post group therapy in women with binge eating disorder. Psychotherapy 51(1):57–65

Mays VM, Cochran SD (2001) Mental health correlates of perceived discrimination among lesbian, gay, and bisexual adults in the United States. Am J Public Health 91(11):1869–1876

McClain Z, Peebles R (2016) Body image and eating disorders among lesbian, gay, bisexual, and transgender youth. Pediatr Clin N Am 63(6):1079–1090

Mehak A, Friedman A, Cassin SE (2018) Self-objectification, weight bias internalization, and binge eating in young women: testing a mediational model. Body Image 24:111–115

Meneguzzo P, Collantoni E, Gallicchio D et al (2018) Eating disorders symptoms in sexual minority women: a systematic review. Eur Eat Disord Rev 26(4):275–292

Meyer IH (2003) Prejudice, social stress, and mental health in lesbian, gay, and bisexual populations: conceptual issues and research evidence. Psychol Bull 129 (5):674–697

Meyer MD, Risbrough VB, Liang J et al (2015) Pavlovian conditioning to hedonic food cues in overweight and lean individuals. Appetite 87:56–61

Neumark-Sztainer D, Story M (1998) Dieting and binge eating among adolescents: what do they really mean? J Am Diet Assoc 98(4):446–450

Neumark-Sztainer DR, Wall MM, Haines JI et al (2007) Shared risk and protective factors for overweight and disordered eating in adolescents. Am J Prev Med 33 (5):359–369

O'Brien KS, Latner JD, Puhl RM et al (2016) The relationship between weight stigma and eating behavior is explained by weight bias internalization and psychological distress. Appetite 102:70–76

Paxton SJ, Schutz HK, Wertheim EH et al (1999) Friendship clique and peer influences on body image concerns, dietary restraint, extreme weight-loss behaviors, and binge eating in adolescent girls. J Abnorm Psychol 108(2):255–266

Polivy J, Herman CP (1993) Etiology of binge eating: psychological mechanisms. In: Polivy J, Herman CP (eds) Binge eating: nature, assessment, and treatment. Guilford Press, New York, pp 173–205

Puhl RM, Himmelstein MS (2018) Weight bias internalization among adolescents seeking weight loss: implications for eating behaviors and parental communication. Front Psychol 9:2271

Puhl R, Peterson JL, Luedicke J (2013) Fighting obesity or obese persons? Public perceptions of obesity-related health messages. Int J Obes 37(6):774–782

Rhodes B, Kroger J (1992) Parental bonding and separation-individuation difficulties among late adolescent eating disordered women. Child Psychiatry Hum Dev 22(4):249–263

Russell SL, Haynos AF, Crow SJ et al (2017) An experimental analysis of the affect regulation model of binge eating. Appetite 110:44–50

Safer DL, Robinson AH, Jo B (2010) Outcome from a randomized controlled trial of group therapy for binge eating disorder: comparing dialectical behavior therapy adapted for binge eating to an active comparison group therapy. Behav Ther 41(1):106–120

Salvy SJ, Vartanian LR, Coelho JS et al (2008) The role of familiarity on modelling of eating and food consumption in children. Appetite 50:514–518

Salvy SJ, Howard M, Read M et al (2009) The presence of friends increases food intake in youth. Am J Clin Nutr 90:282–287

Schvey NA, White MA (2015) The internalization of weight bias is associated with severe eating pathology among lean individuals. Eat Behav 17:1–5

Shakory S, Van Exan J, Mills JS et al (2015) Binge eating in bariatric surgery candidates: the role of insecure attachment and emotion regulation. Appetite 91(1):69–75

Shomaker LB, Tanofsky-Kraff M, Elliott C et al (2010) Salience of loss of control for pediatric binge episodes: does size really matter? Int J Eat Disord 43 (8):707–716

Skinner HH, Haines J, Austin SB et al (2012) A prospective study of overeating, binge eating, and depressive symptoms among adolescent and young adult women. J Adolesc Health 50(5):478–483

Smolak L, Striegel-Moore RH (2001) Challenging the myth of the golden girl: ethnicity and eating disorders. In: Striegel-Moore RH, Smolak L (eds) Eating disorders: innovative directions in research and practice. American Psychological Association, Washington, DC, pp 111–132

Sonneville KR, Horton NJ, Micali N et al (2013) Longitudinal associations between binge eating and overeating and adverse outcomes among adolescents and young adults: does loss of control matter? JAMA Pediatr 167 (2):149–155

Spanos A, Klump KL, Burt SA et al (2010) A longitudinal investigation of the relationship between disordered eating attitudes and behaviors and parent-child conflict: a monozygotic twin differences design. J Abnorm Psychol 119(2):293–299

Stice E, Killen JD, Hayward C et al (1998) Age of onset for binge eating and purging during late adolescence: a 4-year survival analysis. J Abnorm Psychol 107 (4):671–675

Sutin A, Robinson E, Daly M et al (2016) Weight discrimination and unhealthy eating-related behaviors. Appetite 102:83–89

Swanson SA, Crow SJ, Le Grange D et al (2011) Prevalence and correlates of eating disorders in adolescents: results from the national comorbidity survey replication adolescent supplement. Arch Gen Psychiatry 68 (7):714–723

Tanofsky-Kraff M, Morgan CM, Yanovski SZ et al (2003) Comparison of assessments of children's eating-disordered behaviors by interview and questionnaire. Int J Eat Disord 33(2):213–244

Tanofsky-Kraff M, Marcus MD, Yanovski SZ et al (2008) Loss of control eating disorder in children age 12 years and younger: proposed research criteria. Eat Behav 9 (3):360–365

Tanofsky-Kraff M, McDuffie JR, Yanovski SZ et al (2009a) Laboratory assessment of the food intake of children and adolescents with loss of control eating. Am J Clin Nutr 89(3):738–745

Tanofsky-Kraff M, Yanovski SZ, Schvey NA et al (2009b) A prospective study of loss of control eating for body weight gain in children at high risk for adult obesity. Int J Eat Disord 42:26–30

Tanofsky-Kraff M, Shoemaker LB, Olsen C et al (2011) A prospective study of pediatric loss of control eating and psychological outcomes. J Abnorm Psychol 120 (1):108–118

Tasca GA, Szadkowski L, Illing V et al (2009) Adult attachment, depression, and eating disorder symptoms:

the mediating role of affect regulation strategies. Pers Individ Differ 47:662–667

Telch CF, Agras WS, Linehan MM (2001) Dialectical behavior therapy for binge eating disorder. J Consult Clin Psychol 69(6):1061–1065

Tester JM, Lang TC, Laraia BA (2016) Disordered eating behaviours and food insecurity: a qualitative study about children with obesity in low-income households. Obes Res Clin Pract 10(5):544–552

Tylka TL, Andorka MJ (2012) Support for an expanded tripartite influence model with gay men. Body Image 9 (1):57–67

Ugo P, Cacioppo C, Schimmenti A (2012) The moderating role of father's care on the onset of binge eating symptoms among female late adolescents with insecure attachment. Child Psychiatry Hum Dev 43 (2):282–292

Vartanian LR, Porter AM (2016) Weight stigma and eating behavior: a review of the literature. Appetite 102:3–14

Vincent MA, McCabe MP (2000) Gender differences among adolescents in family, and peer influences on body dissatisfaction, weight loss, and binge eating behaviors. J Youth Adolesc 29(2):205–221

Von Schell A, Ohrt TK, Bruening AB et al (2018) Rates of disordered eating behaviors across sexual minority undergraduate men and women. Psychol Sex Orientat Gend Divers 5(3):352–359

Wallace JB (2005) American perceptions of Africa based on media representations: Africa and the world. African Renaissance 2(3):93–96

Wang SB, Lydecker JA, Grilo CM (2017) Rumination in patients with binge-eating disorder and obesity: associations with eating-disorder psychopathology and weight-bias internalization. Eur Eat Disord Rev 25(2):98–103

Zubatsky M, Berge J, Neumark-Sztainer D (2015) Longitudinal associations between parenting style and adolescent disordered eating behaviors. Eat Weight Disord 20(2):187–194

Psychosocial Correlates of Binge Eating

Amy Heard Egbert, Kathryn Smith, and Andrea B. Goldschmidt

Abstract

Binge eating, defined as the consumption of an objectively large amount of food accompanied by a sense of loss of control (LOC) while eating, is a transdiagnostic feature of eating disorders and often presents outside the context of full syndrome eating disorder diagnoses. Individuals with binge eating often report lower levels of psychosocial functioning than those who do not endorse binge eating. This chapter provides a comprehensive review of the existing literature on the relation between binge eating and psychosocial functioning in youth and adults, including cross-sectional, longitudinal, and momentary studies. Overall, the evidence suggests that individuals who endorse binge eating also experience higher levels of psychiatric comorbidity, lower quality of life, and increased internalizing and externalizing symptoms than the general population. Additionally, in some cases, the relation between binge eating and psychosocial impairment may be bidirectional in nature. This evidence underscores the importance of addressing psychosocial impairment in the context of binge eating treatment, as well as the need to assess binge eating symptoms among individuals with heightened psychosocial impairment in clinical settings.

Keywords

Binge eating · Psychosocial factors · Youth · Adults · Loss of control eating · Overeating

A. H. Egbert
Department of Psychology, Loyola University of Chicago, Chicago, IL, USA

K. Smith
Center for Bio-behavioral Research, Sanford Research, Fargo, ND, USA

Department of Psychiatry and Behavioral Science, University of North Dakota School of Medicine and Health Sciences, Fargo, ND, USA

A. B. Goldschmidt (✉)
Department of Psychiatry and Human Behavior, The Warren Alpert Medical School of Brown University, Providence, RI, USA

Weight Control and Diabetes Research Center/The Miriam Hospital, Providence, RI, USA
e-mail: Andrea.Goldschmidt@Lifespan.org

Learning Objectives

In this chapter, you will:

- Learn about the association between binge-type eating disorders and psychosocial impairment.
- Understand the relationship between symptoms of binge eating and psychosocial functioning, outside of a full-scale clinical disorder.
- Learn about specific types of psychosocial impairment that may have

(continued)

G. K.W. Frank, L. A. Berner (eds.), *Binge Eating*, https://doi.org/10.1007/978-3-030-43562-2_4

bidirectional relationships with both binge-type eating disorders and subclinical symptoms of binge eating.

1   Introduction

Binge eating, defined as the consumption of an objectively large amount of food accompanied by a sense of loss of control (LOC) while eating (American Psychiatric Association 2013), is a transdiagnostic feature of eating disorders that is also commonly reported in individuals with overweight or obesity (particularly those seeking weight loss treatment) (Striegel-Moore et al. 1998), as well as community-based individuals with or without an eating or weight disorder (Mustelin et al. 2017). Although binge eating by definition involves two distinct but related components—overeating and LOC—researchers and clinicians have long acknowledged that many individuals with and without eating disorders engage in episodes involving LOC that are *not* objectively large (Beglin and Fairburn 1992). Thus, studies of binge eating often distinguish between LOC episodes that involve objectively and subjectively large amounts of food and may also investigate objectively large eating episodes that do *not* involve LOC (i.e., objective overeating).

In addition to cross-sectional and prospective associations with eating disorders and obesity, binge eating is related to multiple forms of psychosocial impairment, including elevated eating-related and general psychopathology, interpersonal distress, and personality-related dysfunction (Goldschmidt 2017). The purpose of this chapter is to provide a comprehensive review of psychosocial correlates of binge eating in humans across the age spectrum, including both cross-sectional and prospective findings. Although most of the extant literature has utilized interview- and/or questionnaire-based assessment methods, we also include studies that have captured momentary psychosocial correlates of binge eating via laboratory-based paradigms or ecological momentary assessment (EMA), the latter of

which involves assessing features of eating episodes and related factors (e.g., mood, shape and weight concerns) in near real time in the natural environment. Given that binge eating frequently presents with psychosocial comorbidities, even in individuals who do not meet full criteria for an eating disorder, in the following sections, we will first review the correlate of interest with respect to broad eating disorder diagnostic categories, and then provide more detail with respect to specific components of binge eating (i.e., overeating and/or LOC) in subgroups with and without eating disorders (e.g., community, samples individuals seeking treatment for obesity) from cross-sectional, longitudinal, and momentary studies. This chapter will include references to seminal studies that evaluate psychosocial factors related to binge eating as well as review papers that provide a more exhaustive list of references. Although many of the studies included in this chapter are not definitive as to causation, we include a proposed theoretical model of the basic associations between binge eating and psychosocial functioning domains (see Fig. 1).

2   Psychiatric Comorbidity

2.1   Eating-Related Psychopathology

2.1.1   Cross-Sectional Associations
By nature, eating-related psychopathology (e.g., eating, shape, and/or weight concerns, dietary restraint) is prevalent in individuals with eating disorders and is often part of their diagnostic criteria. Overvaluation of shape and weight and other manifestations of shape- and weight-related concerns (e.g., fear of weight gain) are embedded in the diagnostic criteria for anorexia nervosa (AN), while only the former construct is embedded in the criteria for bulimia nervosa (BN). Although overvaluation of shape and weight is not part of the diagnostic criteria for binge-eating disorder (BED), there is evidence that many individuals with BED also present with significant overvaluation (Hrabosky et al. 2007). Additionally, although restraint theories of binge

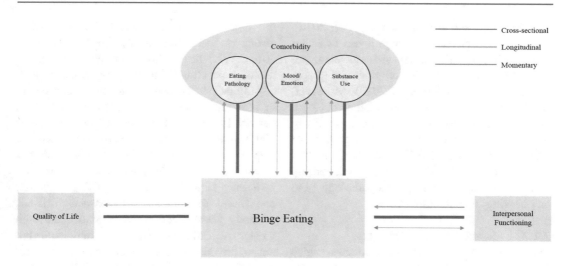

Fig. 1 Visual representation of the psychosocial correlates of binge eating. Note: Environmental and genetic factors not accounted for. There are likely to be bidirectional relationships among the correlates themselves, but these are beyond the scope of the current chapter and thus are not depicted. Thicker lines represent stronger evidence for associations between the variables

eating suggest that the desire for thinness leads to excessive dietary restraint, which then leads to binge eating to compensate for unsustainable caloric deprivation (Davis et al. 1988; Howard and Porzelius 1999), the dietary restraint often seen in AN-binge/purge subtype (AN-BP) and BN is less consistently associated with BED, while other forms of eating-related psychopathology, such as weight concerns and general body dissatisfaction, present similarly across diagnostic categories (Wilfley et al. 2000).

Research suggests that in adults, LOC, regardless of episode size, is a robust correlate of eating-related psychopathology, such as shape and weight concerns, in both community-based and clinical samples (Striegel-Moore et al. 1998; Elder et al. 2008). There is evidence that eating-related psychopathology is indistinguishable in men reporting objectively large binge episodes as compared with those reporting subjectively large binge episodes, and individuals with LOC report higher levels of dietary restraint and excessive exercise than those without LOC, regardless of episode size, even after adjusting for BMI (Kelly et al. 2018). Similar to adults, cross-sectional studies of youth indicate that those with LOC report higher levels of weight and

shape concerns than those without (Goldschmidt et al. 2015), whether they endorse objective binge episodes or subjective binge episodes (Goldschmidt et al. 2008; Shomaker et al. 2010). However, in contrast to research with adults, some studies have found that LOC and overeating are associated with similar levels of dieting and body dissatisfaction in youth, adjusting for BMI (Goldschmidt et al. 2015). Taken together, this research demonstrates that all symptoms of binge eating, including LOC and overeating independent of one another, are associated with higher levels of eating-related psychopathology.

2.1.2 Longitudinal Associations

In addition to these cross-sectional findings, there also are prospective data supporting an association between binge eating and eating disorder attitudes and cognitions. In adolescent girls, overvaluation of shape and weight, elevated dieting behavior, and higher levels of body dissatisfaction all predicted binge eating 2 years later (Stice et al. 2002). Additionally, in both boys and girls, body dissatisfaction in early/middle adolescence predicted both objective binge episodes and overeating in young adulthood, accounting for demographic factors (Goldschmidt et al. 2016). There is also evidence that high levels of eating concerns in early adolescence are associated with the onset of binge eating in both boys and girls in late adolescence (Allen et al. 2016). Moreover, LOC eating in children and adolescents both predicts and is predicted by larger increases in shape and weight concerns over 2–5 years of follow-up (Tanofsky-Kraff et al. 2011; Hilbert et al. 2013), suggesting that the relationship between eating-related psychopathology and binge eating may be bidirectional.

2.1.3   Momentary Associations

In contrast to analogue studies of rodents, which suggest that dietary restriction can promote binge-like eating behavior (Rospond et al. 2015), findings from studies of humans who binge eat have found either no effect (Stice et al. 2004) or an inverse effect (Zambrowicz et al. 2019) of self-reported dietary restraint on energy intake during a laboratory-based test meal. Experimentally induced energy restriction also does not appear to be associated with increased energy intake and/or perceptions of binge eating among adults with AN-BP, BN, BED, or subthreshold binge eating (Hetherington et al. 2000; Chua et al. 2004). To date, only one study has investigated the effects of dietary restraint-like constructs in children with binge eating, reporting no differences in laboratory eating behavior under fasted and non-fasted conditions (Mirch et al. 2006).

Findings from EMA studies of eating-related psychopathology and binge eating have been somewhat mixed. For example, several studies found that self-reported dietary restraint (le Grange et al. 2001) and dietary restriction (Zunker et al. 2011) were elevated prior to binge episodes in adults with BED and BN, respectively, while other studies have found either no momentary association between dietary restraint and binge eating in adults with binge-type eating disorders (Engelberg et al. 2005), or an inverse association (Legenbauer et al. 2018; Pearson et al. 2018). In women with AN, EMA studies reveal that momentary associations between dietary restriction and binge eating appear to depend on the type of restrictive eating behavior (e.g., fasting vs. skipping meals; De Young et al. 2014). With respect to other facets of eating-related psychopathology, appearance-related cognitions have been associated with binge eating in individuals with AN (Mason et al. 2018b) but not BN, BED (Hilbert and Tuschen-Caffier 2007), or a diagnostically heterogeneous sample (Mason et al. 2019). Of the few studies investigating momentary eating-related psychopathology in relation to binge eating in youth, one found that higher between-person levels of body dissatisfaction were associated with greater overeating severity at the univariate (but not multivariate) level in children with overweight/obesity (Goldschmidt et al. 2018b), while another found cognitions related to food, eating, and body image were both antecedents and consequences of LOC episodes in children (Hilbert et al. 2009).

2.2   Quality of Life

2.2.1   Cross-Sectional Associations

Overall, quality of life is often reduced in individuals with binge eating. In adults, BED, BN, and AN-BP all are characterized by lower health-related quality of life, especially in the domain of mental health, as compared with healthy controls (see Ágh et al. 2016 for a review). There also is evidence that the severity of eating-related symptomatology is associated with greater impairment in health-related quality of life across diagnoses (Padierna et al. 2000). A meta-analysis by Winkler et al. (2014) found that health-related quality of life was significantly lower among individuals across eating disorder diagnoses, including those characterized by binge eating, as compared to the general population. Indeed, most individuals with binge-type eating disorders report some degree of health-related quality-of-life impairment (Hudson et al. 2007), and there is evidence that this impairment may be greater than that reported by individuals with AN-R. For example, individuals with BN and AN-BP appear to have lower mental health-related quality of life than those with AN-R, while individuals with BED and AN-BP appear to have lower social/relationship quality of life than those with AN-R (Mond et al. 2005).

With regard to specific symptoms of binge eating, in adults, individuals with LOC experience lower quality of life than healthy controls (Jenkins et al. 2012). Engaging in both subjective binge eating (SBE) and objective binge eating (OBE) is associated with higher weight-related medical comorbidities, such as type-2 diabetes (Kelly et al. 2018), greater functional impairment socially and at work (Pawaskar et al. 2017), and higher levels of job strain (Gralle et al. 2017). Additionally, SBE has been associated with poorer overall quality of life independent of OBE (Latner et al. 2008), emphasizing the importance of considering LOC, regardless of episode size, in understanding associations between binge eating and quality of life. Importantly, there is evidence that this impairment exists even after accounting for BMI, indicating that binge eating in and of itself is associated with lower quality of life above and beyond the influence of comorbid excess weight conditions (Mustelin et al. 2017). In youth with either a full-syndrome or subthreshold eating disorder, both the presence of OBE and/or SBE, as well as higher BMI, have been associated with lower levels of health-related quality of life (Jenkins et al. 2014). However, research also shows that in both overweight and normal-weight youth, individuals reporting OBE have lower quality of life as compared with those who do not endorse OBE (Wilfley et al. 2011; Pasold et al. 2014). Consistent with adult research, youth who experience both OBE and obesity report lower quality of life than those with obesity alone (Ranzenhofer et al. 2012).

2.2.2 Longitudinal Associations

Most prospective studies of quality of life in relation to binge eating have been conducted in the context of intervention trials, making it difficult to tease apart effects of binge eating status from those of treatment itself on quality of life outcomes (e.g., Pohjolainen et al. 2016). In a national survey of nearly 10,000 adult women, disordered eating at baseline, including OBE, was associated with lower health-related quality of life over the course of 9 years as compared with both

the overall study sample and with the general population (Wade et al. 2012). Additionally, there is support for a bidirectional relation between disordered eating, including OBE, and health-related quality of life over time, such that poorer quality of life may also contribute to the development and exacerbation of symptoms such as OBE (Mitchison et al. 2015). Similar to the adult literature, few studies of pediatric samples have assessed longitudinal associations between binge eating and quality of life. Allen et al. (2013) recently investigated a large naturalistic sample of community-based adolescents and found that BED predicted poor mental quality of life over 6 years of follow-up. Furthermore, Mitchell and Steele (2017) found support for a bidirectional relation between disordered eating and health-related quality of life in youth aged 7–10 years over the course of one academic school year but did not specifically assess effects of binge eating. Clearly, more research is needed to investigate the specific impact of binge eating symptoms on quality of life in both youth and adults.

2.2.3 Momentary Associations

Momentary associations between quality of life and binge eating are difficult to characterize since the quality of life is typically conceptualized as a more stable construct that does not change from moment to moment. However, Mason et al. (2018b) recently demonstrated that trait-level work-related quality of life was negatively associated with EMA-assessed binge eating among women with AN. On the contrary, earlier work (Heron et al. 2015) indicated the opposite among college students with disordered eating symptoms: that lower body image-related quality of life was associated with less reporting of overeating and LOC (both at a trend level). Further work is needed to clarify these discrepant results.

2.3 Mood-Related Symptoms and Self-Evaluation

2.3.1 Cross-Sectional Associations

Rates of depression and anxiety are elevated in individuals with binge-type eating disorders.

Adults with BED evidence higher levels of depression, anxiety, and suicidal ideation as compared with individuals without BED (Grucza et al. 2007), and these mood disturbances appear to persist across cultures (Pinheiro et al. 2016). In both BN and AN-BP, depression is one of the most commonly diagnosed comorbid psychiatric disorders (O'Brien and Vincent 2003), and research suggests that individuals with AN-BP may endorse even more depressive symptoms than those with BN (Herzog et al. 1992b). Overall, the lifetime rates of major depressive disorder in individuals with binge-type eating disorders range from 32 to 50% (Hudson et al. 2007).

In addition to depression, estimates suggest that approximately two-thirds of individuals with BN or AN-BP may have had a lifetime diagnosis of an anxiety disorder and even those without a full-syndrome diagnosis exhibit more symptoms of anxiety than healthy controls (Kaye et al. 2004). Such research also suggests that symptoms of anxiety often precede the eating disorder diagnosis and may, in fact, be a risk factor for the development of BN or AN (including AN-BP) (Kaye et al. 2004). Additionally, it is estimated that close to two-thirds of individuals with BED have had a lifetime diagnosis of an anxiety disorder (Hudson et al. 2007). However, it is important to note that many seminal studies investigating rates of anxiety and eating disorders were published prior to the release of DSM-5 when OCD was reclassified as no longer being an anxiety disorder. Therefore, such studies often include OCD when investigating the relation between anxiety disorders and binge eating. Estimates suggest that the lifetime prevalence of OCD in individuals with BN is up to 32% and when subthreshold symptoms of OCD are considered, estimates rise to near 50% (Rubenstein et al. 1993). However, OCD symptoms may lessen when the eating disorder is treated, especially in the case of AN (Pollice et al. 1997). Because of this, more research is needed to examine the relation between anxiety and BED, BN, and AN-BP while considering rates of OCD separately.

Subsyndromal symptoms of binge eating also have been cross-sectionally associated with internalizing symptoms. In adults, LOC is associated with greater symptoms of depression and other mood disturbances (Colles et al. 2008). Additionally, evidence from samples with obesity suggests that OBE is associated with greater mood-related disturbances than either overeating or obesity alone (Striegel-Moore et al. 1998). Furthermore, OBE has been associated with specific correlates of depression, including rumination and suicidality (Forrest et al. 2017; Wang et al. 2017). Symptoms of binge eating in adults have also been associated with anxiety, even after accounting for depressive symptoms (Rosenbaum and White 2015). Similar to the adult literature, research shows that youth with OBE have higher levels of depression and lower levels of self-esteem than those without OBE (Ackard et al. 2003). Additionally, some research suggests that youth who endorse overeating, with or without LOC, experience higher levels of depression and lower self-esteem than those who do not (Goldschmidt et al. 2015). In addition to depressive symptoms, both LOC and OBE in youth have also been associated with higher levels of anxiety as compared with healthy controls (Ackard et al. 2003; Shomaker et al. 2010).

2.3.2 Longitudinal Associations

In addition to cross-sectional findings, there is also evidence that internalizing symptoms may be associated with binge eating over time. Although the rates of depression are high in individuals with binge-type eating disorders, longitudinal evidence is mixed as to whether depression is a cause or consequence of BN and AN-BP (or both), as some studies indicate that the dietary restriction often present in these disorders may lead to depressive symptoms, while others indicate that the relationship between depression and eating disorders may be bidirectional (O'Brien and Vincent 2003).

Longitudinal research focused on binge eating and depressive symptoms outside of a full syndrome eating disorder is limited in adults. Nevertheless, research focused on emerging adulthood has indicated that depressive symptoms during adolescence may be predictive of fear of LOC in young adulthood, adjusting for BMI (Liechty and Lee 2013). Additionally, retrospective research has found that suicidal ideation may precede OBE in adults (Forrest et al. 2017). In contrast, a longitudinal study of ~700 adult women found that baseline symptom of depression and anxiety did not predict binge eating symptoms 5 years later (Vogeltanz-Holm et al. 2000). However, given the dearth of longitudinal research on adults, it is difficult to interpret these mixed results. More research is needed to clarify the role of internalizing symptoms and binge eating over time in order to understand whether depressive symptoms may be a risk factor that can be targeted in the prevention of binge eating, or if they should be considered secondary symptoms that will resolve when binge eating is treated.

In youth, research suggests that the relation between symptoms of binge eating and internalizing symptoms is also complex. For example, studies have found that although OBE may predict depressive symptoms in youth months later, SBE does not (Stice et al. 2002; Sinclair-McBride and Cole 2017). However, there is evidence that, when investigating the inverse direction of association, depressive symptoms predict both OBE and SBE (Sinclair-McBride and Cole 2017). In contrast, other studies have found that, compared to children who only reported LOC at baseline or never reported LOC, children with persistent LOC had higher levels of depressive symptoms almost 5 years later, regardless of whether they experienced SBE or OBE (Tanofsky-Kraff et al. 2011). In addition to depressive symptoms, LOC in youth appears to predict symptoms of anxiety 4 years later (Tanofsky-Kraff et al. 2011). There is also evidence that low self-esteem may also play a role in the development of binge eating. Goldschmidt et al. (2012) found that a combination of dieting, low self-esteem, and depressive symptoms in adolescence were the most salient factors in predicting binge eating in early adulthood, and that low self-esteem and depressive symptoms were significant predictors above and beyond the impact of dieting (Goldschmidt et al. 2012). Although the relation between LOC and internalizing symptoms is well-established, research is mixed as to whether individuals with

overeating (OE) without LOC experience higher levels of internalizing symptoms than healthy controls (Goldschmidt 2017). More prospective research is needed to examine this association.

One underlying dimension that may explain the relation between mood disturbances and binge eating is emotion regulation (Goldschmidt et al. 2017). Difficulties with emotion regulation are a transdiagnostic correlate of eating disorders in both youth and adults. Specifically, both full-syndrome and subthreshold binge eating in adults has been associated with difficulty with emotion regulation, including higher levels of negative urgency (i.e., the tendency to act impulsively when distressed), and suppression of negative thoughts (Dingemans et al. 2017). In youth, binge eating has also been associated with maladaptive emotion regulation strategies, including escape-avoidance coping (Lee-Winn et al. 2016). Indeed, affect regulation models assert that binge eating functions to reduce or regulate negative affect; however, empirical evidence supporting these models is mixed (see Dingemans et al. 2017).

2.3.3 Momentary Associations

A number of laboratory studies in clinical and nonclinical samples have examined the effects of mood inductions on eating behaviors, including binge eating (Cardi et al. 2015; Evers et al. 2018). The methodology of these laboratory studies has varied widely, relatively few have used clinical samples, and most research has focused primarily on caloric intake as a marker of binge eating. A recent review identified only eight laboratory studies in BN and BED (Evers et al. 2018), reporting an overall nonsignificant effect of negative mood inductions compared to neutral mood inductions on food intake. However, there was substantial heterogeneity in results; some studies found increased food intake after negative mood inductions in binge eating samples (Levine and Marcus 1997), while others have found food intake to decrease or remain unchanged compared to neutral conditions (Laessle and Schulz 2009). While positive mood inductions have generally been shown to increase food intake across clinical and community samples, little research exists among individuals with binge eating (Cardi et al. 2015; Evers et al. 2018).

While there is clearly a need for future research in this area, laboratory studies examining emotion-induced eating demonstrate limited and inconclusive evidence for the causal role of affect in objectively measured food intake among individuals with binge eating, though there is some evidence for the causal role of negative affect in other meaningful contextual constructs (i.e., LOC, eating rapidly) thought to be central to binge eating psychopathology. However, laboratory designs preclude examination of the dynamic relationships between affect and binge eating symptoms in daily life. To this end, ambulatory assessment methods have been useful in elucidating micro-temporal processes underlying a range of eating disorder symptoms in the natural environment. EMA has been most commonly applied to study momentary affective antecedents and consequences of eating disorder behaviors, particularly binge eating (Haedt-Matt and Keel 2011). Across eating disorder diagnoses, EMA studies in adult samples have consistently found that negative affect increases and positive affect decreases in the hours prior to binge episodes, after which negative affect decreases and positive affect increases (e.g., Engel et al. 2013). This pattern illustrates a pernicious cycle of negative and positive emotional reinforcement from binge eating, which ultimately serves to maintain the behavior over time. However, despite the large body of evidence in adults linking momentary affect to binge eating, thus far EMA studies of binge eating symptoms in youth remain limited, and contrary to the adult literature, negative affect has not emerged as a consistent predictor of LOC (Hilbert et al. 2009; Ranzenhofer et al. 2014; Goldschmidt et al. 2018b).

In addition to examining affect intensity (i.e., high versus low levels) and valence (i.e., positive versus negative) surrounding binge eating, EMA data in adult samples have been harnessed to provide more fine-grained analyses of emotion-related constructs relevant to binge eating. In particular, these studies have indicated that EMA-measured affect instability (i.e., the degree of momentary shifts in the valence and intensity of emotion, which is thought to be broadly related to emotion dysregulation) is associated with binge eating (Berner et al. 2017), as well as LOC in nonclinical samples (Stevenson et al. 2018). More recent EMA research has also addressed relationships between specific momentary emotion regulation strategies and binge eating, demonstrating that maladaptive strategies (i.e., cognitive rumination) were predictive of binge episodes in BED (Svaldi et al. 2019). Momentary emotion regulation difficulties were also found to increase after, but not before, LOC episodes in a nonclinical sample (Stevenson et al. 2019), suggesting binge eating symptoms may further compound momentary emotion regulation problems, perhaps due to increased rumination about prior eating behaviors. Emerging work has shown that the strength of relationships between momentary negative affect and binge eating may differ based on state-level processes (e.g., momentary expectations about eating, dietary restraint, and cognitive dissociation (Pearson et al. 2018; Mason et al. 2018a)); as well as trait-level characteristics, including eating disorder diagnosis (De Young et al. 2013), co-occurring posttraumatic stress disorder (Karr et al. 2013), and neurobiological and neurocognitive functioning (Goldschmidt et al. 2019; Smith et al. 2019). Taken together, this body of work illustrates that affective processes surrounding binge eating are dynamic, multifaceted, and importantly, not necessarily uniform across all individuals or age groups.

2.4 Substance Use

2.4.1 Cross-Sectional Associations

After mood and anxiety disorders, substance use disorders (SUDs) are among the most common comorbid psychiatric diagnoses in individuals with binge eating. Estimates suggest that lifetime rates of SUDs are close to 25% in adults with BED or AN, including AN-BP, and close to 37% in individuals with BN (Hudson et al. 2007; Bahji et al. 2019). Additionally, there is evidence that individuals with BN and AN-BP have higher rates of SUDs than those with restricting AN (O'Brien and Vincent 2003), suggesting that the comorbidity between eating disorders and SUDs may be specifically related to the presence of binge eating. This is consistent with the idea that both binge eating and substance use are impulsive behaviors driven by difficulties regulating emotions and reward seeking (Dawe and Loxton 2004). Both adults and youth who present for treatment for SUDs have been found to have increased rates of BN than expected (i.e., 28–41%), especially females (Grilo et al. 1997). In youth, research has also found that both AN-BP and BN are cross-sectionally associated with substance use (Wiederman and Pryor 1996).

In adults, binge eating symptoms appear to be common in SUDs treatment-seeking samples. For example, estimates suggest that close to 30% of individuals in treatment for SUDs may also endorse OBE (Cohen et al. 2010). Additionally, a recent study of adults undergoing methadone maintenance treatment for addiction to opioids found that rates of LOC were more than triple that of community samples (Goldschmidt et al. 2018a). In youth, research has found that individuals with OBE tend to report more substance use than those without OBE (Ross and Ivis 1999).

2.4.2 Longitudinal Associations

In addition to cross-sectional studies, there is evidence that binge eating may be prospectively associated with increased substance use. For example, longitudinal studies have found that both BN and BED in adolescence are associated with drug use in early adulthood (Micali et al. 2015). Conversely, research also suggests that individuals with SUDs may have higher rates of BN (Grilo et al. 1998). Increased substance use has also been associated with binge eating outside of a full syndrome eating disorder diagnosis. In adults, research suggests that illicit drug use and greater occurrence of alcoholic intoxication at baseline may be associated with binge eating in young women 5 years later (Vogeltanz-Holm et al. 2000). However, most of the research examining associations between subclinical levels of substance use and binge eating have focused on youth. Indeed, one prospective study found that both OE and SBE in early adolescence were associated with an increased likelihood of using illicit drugs in young adulthood (Sonneville et al. 2013). A significant limitation to our understanding of prospective associations between SUD and binge eating is that most studies have assessed lifetime rates of SUDs, and many do not account for comorbid diagnoses that may be related to both binge eating and SUDs, such as depression (O'Brien and Vincent 2003). Thus, more research is needed to better understand the mechanisms underlying the relation between SUDs and binge eating (Wolfe and Maisto 2000).

2.4.3 Momentary Associations

To date, there have been limited laboratory-based studies that have examined the acute effects of alcohol or drugs on binge eating behaviors. Bruce et al. (2011) found that acute alcohol intoxication reduced urges to binge in women with BN-spectrum disorders, but actual energy intake was not measured. On the contrary, food deprivation did not affect self-administration of alcohol in women with BN (Bulik and Brinded 1993). Consistent with these findings, EMA data reveal no differences in the likelihood of binge eating on days that were and were not characterized by alcohol intoxication among women with BN (Pisetsky et al. 2016). Similarly, among non-eating disordered adults with overweight/ obesity, there were no significant differences among OBE, SBE, OE, or non-pathological eating episodes in terms of the number of alcoholic beverages consumed prior to eating, although there was a trend for OE episodes to be associated with greater alcohol consumption relative to other types of eating episodes (Goldschmidt et al. 2014a). Thus, contrary to literature suggesting a disinhibiting effect of alcohol on food consumption (Yeomans 2004), there appears to be limited data supporting a momentary association between binge eating and alcohol ingestion. No research to date has investigated the acute effects of other substances on binge eating behavior.

3 Interpersonal Functioning

3.1 Cross-Sectional Associations

In addition to associations with intrapersonal factors, there is evidence that binge eating is associated with a range of interpersonal difficulties. Interpersonal theories of binge eating posit that interpersonal problems often lead to negative affect, which in turn leads to binge eating (Wilfley et al. 2002). This theory has been supported in both adult and youth samples (Elliott et al. 2010; Ansell et al. 2012). Additionally, research has focused on identifying which interpersonal difficulties appear to be strongly associated with binge eating. To this end, it appears that weight stigma and teasing often arise as a common source of interpersonal distress, given their strong association with binge eating and overweight/obesity. Cross-sectional research suggests that rates of bullying and victimization are higher among individuals with both BN and BED than they are among healthy controls; however, almost no studies have examined this relation prospectively making it difficult to infer causation (Lie et al. 2019). There is also evidence that bullying and teasing specifically related to appearance is more strongly correlated with BN and BED than is general bullying, and that individuals with binge-type eating disorders may experience more bullying than those with restrictive eating disorders (Lie et al. 2019).

Aside from bullying, binge-type eating disorders are also associated with other interpersonal difficulties that are likely related to certain maladaptive personality traits. There is evidence that the highest prevalence of personality disorders in individuals with eating disorders is found in women with AN-BP, and that borderline personality disorder is the most commonly diagnosed (Herzog et al. 1992a). Other studies have found that histrionic personality disorder is most commonly associated with a diagnosis of BN (Wonderlich et al. 1990), lending support to an association between binge eating and Cluster B personality disorders (narcissistic, histrionic, borderline, and antisocial). However, a recent meta-analysis by Farstad et al. (2016) found that in both AN-BP and BN, Cluster C disorders (avoidant, dependent, obsessive compulsive) were the most commonly diagnosed. Although fewer studies have examined personality disorders in individuals with BED, a study by Becker et al. (2010) found that BED was also most strongly associated with Cluster C personality disorders.

In terms of specific personality traits, research shows that individuals with eating disorders tend to be more interpersonally submissive (Troop et al. 2003; Hartmann et al. 2010). For example, BED has been associated with less dominance and less interpersonal warmth (Luo et al. 2018), BN has been associated with more social avoidance and less satisfaction with social support systems (Grisset and Norvell 1992), and AN-BP has also been associated with interpersonal sensitivity (Solmi et al. 2019). Overall, there is evidence that while AN-BP is associated with avoidance of expressing feelings to others, BN is specifically associated with less social skills competence, greater fear of negative evaluation by others, and less social support seeking (Arcelus et al. 2013).

In both adults and youth, subthreshold binge eating symptoms have also been cross-sectionally associated with interpersonal difficulties. In adults, poorer overall interpersonal functioning has been associated with both LOC and OBE (Ansell et al. 2012). Additionally, bulimic eating patterns, including OBE, have been associated with social reassurance seeking, perceptions of rejection, and interpersonal stress (Kwan et al. 2017). In youth, factors such as parental emotional unresponsiveness and being teased by family members about weight have been associated with both LOC and OBE (Field et al. 2008; Hartmann et al. 2012), especially among youth with higher socioeconomic status (West et al. 2019). Some research also shows that youth with OBE (meeting criteria for either threshold or subthreshold BED) report lower levels of emotional involvement from their families as compared with healthy controls (see Saltzman and Liechty 2016 for a review). Additionally, LOC in youth has also been associated with general interpersonal difficulties (Elliott et al. 2010).

3.2   Longitudinal Associations

Given the limited longitudinal data on interpersonal functioning in individuals with binge eating (Arcelus et al. 2013), it is difficult to understand the nature of the relation, particularly with respect to directionality. The few studies that have been conducted suggest that weight-based teasing (Haines et al. 2006; Neumark-Sztainer et al. 2007) and broad interpersonal problems (Stice et al. 2002) are associated with onset of binge eating in adolescence and young adulthood in community-based and clinical samples. Yet theoretically, a bidirectional association is possible, with interpersonal difficulties predicting the development of binge eating (consistent with interpersonal theory), and binge eating leading to further social withdrawal and difficulties managing interpersonal issues related due to shame and stigma. Further research is clearly needed in this area to inform prevention and intervention efforts.

3.3   Momentary Associations

Findings regarding the impact of interpersonal stress on momentary eating-related processes have been mixed. In adults, induction of social stress predicts urges to binge among college-aged women with subclinical eating disorder symptoms (Cattanach et al. 1988) and individuals with BN and BED (Hilbert et al. 2011; Rosenberg et al. 2013), but is not associated with objectively measured energy intake among women with BN (Levine and Marcus 1997) or BED (Laessle and Schulz 2009). Additionally, evidence suggests that recent social stress (either naturalistic or induced) does not directly impact eating behavior among youth with LOC eating (Jarcho et al. 2015), but rather is mediated by the experience of negative affect (Shank et al. 2017). Indeed, naturalistic family mealtime interactions were found to be more negative in youth with LOC

eating (Czaja et al. 2011); moreover, critical appearance- and eating-related comments by parents in the laboratory predicted energy intake among children, but this was not specific to those with LOC eating (Hilbert et al. 2010).

EMA research in adults with BN revealed that interpersonal stressors predicted an increased likelihood of binge eating, both directly and through increases in negative affect (Goldschmidt et al. 2014b), but were not associated with binge eating among diagnostically heterogeneous women with binge eating (Mason et al. 2019). In adolescents with LOC eating, interpersonal problems predicted LOC eating at the momentary level (Ranzenhofer et al. 2014). However, this finding was only partially replicated in a heterogeneous sample of children and adolescents with overweight/obesity, in that interpersonal problems were concurrently associated with both LOC and overeating severity (separately) at the univariate level, but not after taking into account other contextual factors, such as food hedonics and craving (Goldschmidt et al. 2018b).

4   Conclusion

In summary, binge eating appears to be both concurrently and prospectively linked to suboptimal psychosocial functioning in a wide range of domains in both youth and adults. In addition to broader associations, some psychosocial factors (particularly those related to affect and emotion regulation) appear to be important precipitants and/or consequences of binge eating at the momentary level. In terms of future directions, while the cross-sectional findings clearly support a link between psychosocial functioning and binge eating, additional longitudinal research is needed to investigate directionality of associations. This is particularly the case in areas where momentary findings differ from those from cross-sectional and longitudinal studies, such as in the domain of substance abuse. Taken together, this large body of work suggests that interventions for binge eating should address not only the problematic eating behavior itself, but also the psychosocial context in which such behavior occurs. Additionally, given some evidence of a bidirectional relation between binge eating and psychosocial functioning, especially in the domain of psychiatric comorbidity, it may also be important to assess for the emergence of binge eating in individuals with difficulties with emotion regulation and other internalizing symptoms. While several effective interventions for binge eating exist, in this current era of personalized medicine and treatment optimization, understanding for whom and under what conditions a given intervention will be effective would benefit from incorporation of relevant psychosocial factors at both the general and proximal levels (Fig. 1).

References

Ackard DM, Neumark-Sztainer D, Story M, Perry C (2003) Overeating among adolescents: prevalence and associations with weight-related characteristics and psychological health. Pediatrics 111:67–74. https://doi.org/10.1542/peds.111.1.67

Ágh T, Kovács G, Supina D et al (2016) A systematic review of the health-related quality of life and economic burdens of anorexia nervosa, bulimia nervosa, and binge eating disorder. Eat Weight Disord EWD 21:353–364. https://doi.org/10.1007/s40519-016-0264-x

Allen KL, Byrne SM, Oddy WH, Crosby RD (2013) DSM–IV–TR and DSM-5 eating disorders in adolescents: prevalence, stability, and psychosocial correlates in a population-based sample of male and female adolescents. J Abnorm Psychol 122:720–732. https://doi.org/10.1037/a0034004

Allen KL, Byrne SM, Crosby RD, Stice E (2016) Testing for interactive and non-linear effects of risk factors for binge eating and purging eating disorders. Behav Res Ther 87:40–47. https://doi.org/10.1016/j.brat.2016.08.019

American Psychiatric Association (2013) Diagnostic and statistical manual of mental disorders (DSM-5®). American Psychiatric Publishing

Ansell EB, Grilo CM, White MA (2012) Examining the interpersonal model of binge eating and loss of control over eating in women. Int J Eat Disord 45:43–50. https://doi.org/10.1002/eat.20897

Arcelus J, Haslam M, Farrow C, Meyer C (2013) The role of interpersonal functioning in the maintenance of eating psychopathology: a systematic review and testable model. Clin Psychol Rev 33:156–167. https://doi.org/10.1016/j.cpr.2012.10.009

Bahji A, Mazhar MN, Hudson CC et al (2019) Prevalence of substance use disorder comorbidity among individuals with eating disorders: a systematic review and meta-analysis. Psychiatry Res 273:58–66. https://doi.org/10.1016/j.psychres.2019.01.007

Becker DF, Masheb RM, White MA, Grilo CM (2010) Psychiatric, behavioral, and attitudinal correlates of avoidant and obsessive-compulsive personality pathology in patients with binge-eating disorder. Compr Psychiatry 51:531–537. https://doi.org/10.1016/j.comppsych.2009.11.005

Beglin SJ, Fairburn CG (1992) What is meant by the term "binge"? Am J Psychiatry 149:123–124. https://doi.org/10.1176/ajp.149.1.123

Berner LA, Crosby RD, Cao L et al (2017) Temporal associations between affective instability and dysregulated eating behavior in bulimia nervosa. J Psychiatr Res 92:183–190. https://doi.org/10.1016/j.jpsychires.2017.04.009

Bruce KR, Steiger H, Israel M et al (2011) Effects of acute alcohol intoxication on eating-related urges among women with bulimia nervosa. Int J Eat Disord 44:333–339. https://doi.org/10.1002/eat.20834

Bulik CM, Brinded EC (1993) The effect of food deprivation on alcohol consumption in bulimic and control women. Addiction 88:1545–1551

Cardi V, Leppanen J, Treasure J (2015) The effects of negative and positive mood induction on eating behaviour: a meta-analysis of laboratory studies in the healthy population and eating and weight disorders. Neurosci Biobehav Rev 57:299–309. https://doi.org/10.1016/j.neubiorev.2015.08.011

Cattanach L, Malley R, Rodin J (1988) Psychologic and physiologic reactivity to stressors in eating disordered individuals. Psychosom Med 50:591–599

Chua JL, Touyz S, Hill AJ (2004) Negative mood-induced overeating in obese binge eaters: an experimental study. Int J Obes 28:606. https://doi.org/10.1038/sj.ijo.0802595

Cohen LR, Greenfield SF, Gordon S et al (2010) Survey of eating disorder symptoms among women in treatment for substance abuse. Am J Addict 19:245–251. https://doi.org/10.1111/j.1521-0391.2010.00038.x

Colles SL, Dixon JB, O'Brien PE (2008) Loss of control is central to psychological disturbance associated with binge eating disorder. Obesity 16:608–614. https://doi.org/10.1038/oby.2007.99

Czaja J, Hartmann AS, Rief W, Hilbert A (2011) Mealtime family interactions in home environments of children with loss of control eating. Appetite 56:587–593. https://doi.org/10.1016/j.appet.2011.01.030

Davis R, Freeman RJ, Garner DM (1988) A naturalistic investigation of eating behavior in bulimia nervosa. J Consult Clin Psychol 56:273–279. https://doi.org/10.1037/0022-006X.56.2.273

Dawe S, Loxton NJ (2004) The role of impulsivity in the development of substance use and eating disorders.

Neurosci Biobehav Rev 28:343–351. https://doi.org/10.1016/j.neubiorev.2004.03.007

De Young KP, Lavender JM, Wonderlich SA et al (2013) Moderators of post-binge eating negative emotion in eating disorders. J Psychiatr Res 47:323–328. https://doi.org/10.1016/j.jpsychires.2012.11.012

De Young KP, Lavender JM, Crosby RD et al (2014) Bidirectional associations between binge eating and restriction in anorexia nervosa. An ecological momentary assessment study. Appetite 83:69–74. https://doi.org/10.1016/j.appet.2014.08.014

Dingemans A, Danner U, Parks M (2017) Emotion regulation in binge eating disorder: a review. Nutrients 9:1274. https://doi.org/10.3390/nu9111274

Elder KA, Paris M, Añez LM, Grilo CM (2008) Loss of control over eating is associated with eating disorder psychopathology in a community sample of Latinas. Eat Behav 9:501–503. https://doi.org/10.1016/j.eatbeh.2008.04.003

Elliott CA, Tanofsky-Kraff M, Shomaker LB et al (2010) An examination of the interpersonal model of loss of control eating in children and adolescents. Behav Res Ther 48:424–428. https://doi.org/10.1016/j.brat.2009.12.012

Engel SG, Wonderlich SA, Crosby RD et al (2013) The role of affect in the maintenance of anorexia nervosa: evidence from a naturalistic assessment of momentary behaviors and emotion. J Abnorm Psychol 122:709–719. https://doi.org/10.1037/a0034010

Engelberg MJ, Gauvin L, Steiger H (2005) A naturalistic evaluation of the relation between dietary restraint, the urge to binge, and actual binge eating: a clarification. Int J Eat Disord 38:355–360. https://doi.org/10.1002/eat.20186

Evers C, Dingemans A, Junghans AF, Boevé A (2018) Feeling bad or feeling good, does emotion affect your consumption of food? A meta-analysis of the experimental evidence. Neurosci Biobehav Rev 92:195–208. https://doi.org/10.1016/j.neubiorev.2018.05.028

Farstad SM, McGeown LM, von Ranson KM (2016) Eating disorders and personality, 2004–2016: a systematic review and meta-analysis. Clin Psychol Rev 46:91–105. https://doi.org/10.1016/j.cpr.2016.04.005

Field AE, Javaras KM, Aneja P et al (2008) Family, peer, and media predictors of becoming eating disordered. Arch Pediatr Adolesc Med 162:574–579. https://doi.org/10.1001/archpedi.162.6.574

Forrest LN, Zuromski KL, Dodd DR, Smith AR (2017) Suicidality in adolescents and adults with binge-eating disorder: results from the national comorbidity survey replication and adolescent supplement. Int J Eat Disord 50:40–49. https://doi.org/10.1002/eat.22582

Goldschmidt AB (2017) Are loss of control while eating and overeating valid constructs? A critical review of the literature. Obes Rev 18:412–449. https://doi.org/10.1111/obr.12491

Goldschmidt AB, Jones M, Manwaring JL et al (2008) The clinical significance of loss of control over eating in overweight adolescents. Int J Eat Disord 41:153–158. https://doi.org/10.1002/eat.20481

Goldschmidt AB, Wall M, Loth KA et al (2012) Which dieters are at risk for the onset of binge eating? A prospective study of adolescents and young adults. J Adolesc Health 51:86–92. https://doi.org/10.1016/j.jadohealth.2011.11.001

Goldschmidt AB, Crosby RD, Cao L et al (2014a) Ecological momentary assessment of eating episodes in obese adults. Psychosom Med 76:747–752. https://doi.org/10.1097/PSY.0000000000000108

Goldschmidt AB, Wonderlich SA, Crosby RD et al (2014b) Ecological momentary assessment of stressful events and negative affect in bulimia nervosa. J Consult Clin Psychol 82:30–39. https://doi.org/10.1037/a0034974

Goldschmidt AB, Loth KA, MacLehose RF et al (2015) Overeating with and without loss of control: associations with weight status, weight-related characteristics, and psychosocial health. Int J Eat Disord 48:1150–1157. https://doi.org/10.1002/eat.22465

Goldschmidt AB, Wall MM, Zhang J et al (2016) Overeating and binge eating in emerging adulthood: 10-year stability and risk factors. Dev Psychol 52:475–483. https://doi.org/10.1037/dev0000086

Goldschmidt AB, Lavender JM, Hipwell AE et al (2017) Emotion regulation and loss of control eating in community-based adolescents. J Abnorm Child Psychol 45:183–191. https://doi.org/10.1007/s10802-016-0152-x

Goldschmidt AB, Cotton BP, Mackey S et al (2018a) Prevalence and correlates of loss of control eating among adults presenting for methadone maintenance treatment. Int J Behav Med 25:693–697. https://doi.org/10.1007/s12529-018-9750-z

Goldschmidt AB, Smith KE, Crosby RD et al (2018b) Ecological momentary assessment of maladaptive eating in children and adolescents with overweight or obesity. Int J Eat Disord 51:549–557. https://doi.org/10.1002/eat.22864

Goldschmidt AB, Smith KE, Lavender JM et al (2019) Trait-level facets of impulsivity and momentary, naturalistic eating behavior in children and adolescents with overweight/obesity. J Psychiatr Res 110:24–30. https://doi.org/10.1016/j.jpsychires.2018.12.018

Gralle APBP, Moreno AB, Juvanhol LL et al (2017) Job strain and binge eating among Brazilian workers participating in the ELSA-Brasil study: does BMI matter? J Occup Health 59(3):247–255. https://doi.org/10.1539/joh.16-0157-OA

Grilo CM, Martino S, Walker ML et al (1997) Psychiatric comorbidity differences in male and female adult psychiatric inpatients with substance use disorders. Compr Psychiatry 38:155–159. https://doi.org/10.1016/S0010-440X(97)90068-7

Grilo CM, Becker DF, Fehon DC et al (1998) Psychiatric morbidity differences in male and female adolescent inpatients with alcohol use disorders. J Youth Adolesc 27:29–41. https://doi.org/10.1023/A:1022824730935

Grisset NI, Norvell NK (1992) Perceived social support, social skills, and quality of relationships in bulimic women. J Consult Clin Psychol 60:293–299. https://doi.org/10.1037/0022-006X.60.2.293

Grucza RA, Przybeck TR, Cloninger CR (2007) Prevalence and correlates of binge eating disorder in a community sample. Compr Psychiatry 48:124–131. https://doi.org/10.1016/j.comppsych.2006.08.002

Haedt-Matt AA, Keel PK (2011) Revisiting the affect regulation model of binge eating: a meta-analysis of studies using ecological momentary assessment. Psychol Bull 137:660–681. https://doi.org/10.1037/a0023660

Haines J, Neumark-Sztainer D, Eisenberg ME, Hannan PJ (2006) Weight teasing and disordered eating behaviors in adolescents: longitudinal findings from project EAT (eating among teens). Pediatrics 117:e209–e215. https://doi.org/10.1542/peds.2005-1242

Hartmann A, Zeeck A, Barrett MS (2010) Interpersonal problems in eating disorders. Int J Eat Disord 43:619–627. https://doi.org/10.1002/eat.20747

Hartmann AS, Czaja J, Rief W, Hilbert A (2012) Psychosocial risk factors of loss of control eating in primary school children: a retrospective case-control study. Int J Eat Disord 45:751–758. https://doi.org/10.1002/eat.22018

Heron KE, Mason TB, Sutton TG, Myers TA (2015) Evaluating the real-world predictive validity of the body image quality of life inventory using ecological momentary assessment. Body Image 15:105–108. https://doi.org/10.1016/j.bodyim.2015.07.004

Herzog DB, Keller MB, Lavori PW et al (1992a) The prevalence of personality disorders in 210 women with eating disorders. J Clin Psychiatry 53:147–152

Herzog DB, Keller MB, Sacks NR et al (1992b) Psychiatric comorbidity in treatment-seeking anorexics and bulimics. J Am Acad Child Adolesc Psychiatry 31:810–818. https://doi.org/10.1097/00004583-199209000-00006

Hetherington MM, Stoner SA, Andersen AE, Rolls BJ (2000) Effects of acute food deprivation on eating behavior in eating disorders. Int J Eat Disord 28:272–283

Hilbert A, Tuschen-Caffier B (2007) Maintenance of binge eating through negative mood: a naturalistic comparison of binge eating disorder and bulimia nervosa. Int J Eat Disord 40:521–530. https://doi.org/10.1002/eat.20401

Hilbert A, Rief W, Tuschen-Caffier B et al (2009) Loss of control eating and psychological maintenance in children: an ecological momentary assessment study. Behav Res Ther 47:26–33. https://doi.org/10.1016/j.brat.2008.10.003

Hilbert A, Tuschen-Caffier B, Czaja J (2010) Eating behavior and familial interactions of children with loss of control eating: a laboratory test meal study.

Am J Clin Nutr 91:510–518. https://doi.org/10.3945/ajcn.2009.28843

Hilbert A, Vögele C, Tuschen-Caffier B, Hartmann AS (2011) Psychophysiological responses to idiosyncratic stress in bulimia nervosa and binge eating disorder. Physiol Behav 104:770–777. https://doi.org/10.1016/j.physbeh.2011.07.013

Hilbert A, Hartmann AS, Czaja J, Schoebi D (2013) Natural course of preadolescent loss of control eating. J Abnorm Psychol 122:684–693. https://doi.org/10.1037/a0033330

Howard CE, Porzelius LK (1999) The role of dieting in binge eating disorder: etiology and treatment implications. Clin Psychol Rev 19:25–44. https://doi.org/10.1016/S0272-7358(98)00009-9

Hrabosky JI, Masheb RM, White MA, Grilo CM (2007) Overvaluation of shape and weight in binge eating disorder. J Consult Clin Psychol 75:175–180. https://doi.org/10.1037/0022-006X.75.1.175

Hudson JI, Hiripi E, Pope HG, Kessler RC (2007) The prevalence and correlates of eating disorders in the national comorbidity survey replication. Biol Psychiatry 61:348–358. https://doi.org/10.1016/j.biopsych.2006.03.040

Jarcho JM, Tanofsky-Kraff M, Nelson EE et al (2015) Neural activation during anticipated peer evaluation and laboratory meal intake in overweight girls with and without loss of control eating. NeuroImage 108:343–353. https://doi.org/10.1016/j.neuroimage.2014.12.054

Jenkins PE, Conley CS, Rienecke Hoste R et al (2012) Perception of control during episodes of eating: relationships with quality of life and eating psychopathology. Int J Eat Disord 45:115–119. https://doi.org/10.1002/eat.20913

Jenkins PE, Hoste RR, Doyle AC et al (2014) Health-related quality of life among adolescents with eating disorders. J Psychosom Res 76:1–5. https://doi.org/10.1016/j.jpsychores.2013.11.006

Karr TM, Crosby RD, Cao L et al (2013) Posttraumatic stress disorder as a moderator of the association between negative affect and bulimic symptoms: an ecological momentary assessment study. Compr Psychiatry 54:61–69. https://doi.org/10.1016/j.comppsych.2012.05.011

Kaye WH, Bulik CM, Thornton L et al (2004) Comorbidity of anxiety disorders with anorexia and bulimia nervosa. Am J Psychiatry 161:2215–2221. https://doi.org/10.1176/appi.ajp.161.12.2215

Kelly NR, Cotter E, Guidinger C (2018) Men who engage in both subjective and objective binge eating have the highest psychological and medical comorbidities. Eat Behav 30:115–119. https://doi.org/10.1016/j.eatbeh.2018.07.003

Kwan MY, Minnich AM, Douglas V et al (2017) Bulimic symptoms and interpersonal functioning among college students. Psychiatry Res 257:406–411. https://doi.org/10.1016/j.psychres.2017.08.021

Laessle RG, Schulz S (2009) Stress-induced laboratory eating behavior in obese women with binge eating disorder. Int J Eat Disord 42:505–510. https://doi.org/10.1002/eat.20648

Latner JD, Vallance JK, Buckett G (2008) Health-related quality of life in women with eating disorders: association with subjective and objective binge eating. J Clin Psychol Med Settings 15:148. https://doi.org/10.1007/s10880-008-9111-1

le Grange D, Gorin A, Catley D, Stone AA (2001) Does momentary assessment detect binge eating in overweight women that is denied at interview? Eur Eat Disord Rev 9:309–324. https://doi.org/10.1002/erv.409

Lee-Winn AE, Townsend L, Reinblatt SP, Mendelson T (2016) Associations of neuroticism-impulsivity and coping with binge eating in a nationally representative sample of adolescents in the United States. Eat Behav 22:133–140. https://doi.org/10.1016/j.eatbeh.2016.06.009

Legenbauer T, Radix AK, Augustat N, Schütt-Strömel S (2018) Power of cognition: how dysfunctional cognitions and schemas influence eating behavior in daily life among individuals with eating disorders. Front Psychol 9:2138. https://doi.org/10.3389/fpsyg.2018.02138

Levine MD, Marcus MD (1997) Eating behavior following stress in women with and without bulimic symptoms. Ann Behav Med 19:132–138. https://doi.org/10.1007/BF02883330

Lie SØ, Rø Ø, Bang L (2019) Is bullying and teasing associated with eating disorders? A systematic review and meta-analysis. Int J Eat Disord 52:497–514. https://doi.org/10.1002/eat.23035

Liechty JM, Lee M (2013) Longitudinal predictors of dieting and disordered eating among young adults in the U.S. Int J Eat Disord 46:790–800. https://doi.org/10.1002/eat.22174

Luo X, Nuttall AK, Locke KD, Hopwood CJ (2018) Dynamic longitudinal relations between binge eating symptoms and severity and style of interpersonal problems. J Abnorm Psychol 127:30–42. https://doi.org/10.1037/abn0000321

Mason TB, Lavender JM, Wonderlich SA et al (2018a) Examining a momentary mediation model of appearance-related stress, anxiety, and eating disorder behaviors in adult anorexia nervosa. Eat Weight Disord 23:637–644. https://doi.org/10.1007/s40519-017-0404-y

Mason TB, Wonderlich SA, Crosby RD et al (2018b) Associations among eating disorder behaviors and eating disorder quality of life in adult women with anorexia nervosa. Psychiatry Res 267:108–111. https://doi.org/10.1016/j.psychres.2018.05.077

Mason TB, Smith KE, Crosby RD et al (2019) Examination of momentary maintenance factors and eating disorder behaviors and cognitions using ecological momentary assessment. Eat Disord:1–14. https://doi.org/10.1080/10640266.2019.1613847

Micali N, Solmi F, Horton NJ et al (2015) Adolescent eating disorders predict psychiatric, high-risk behaviors and weight outcomes in young adulthood. J Am Acad Child Adolesc Psychiatry 54:652–659. https://doi.org/10.1016/j.jaac.2015.05.009

Mirch MC, McDuffie JR, Yanovski SZ et al (2006) Effects of binge eating on satiation, satiety, and energy intake of overweight children. Am J Clin Nutr 84:732–738. https://doi.org/10.1093/ajcn/84.4.732

Mitchell TB, Steele RG (2017) Bidirectional associations between disordered eating and health-related quality of life in elementary school-age youth. J Pediatr Psychol 42:315–324. https://doi.org/10.1093/jpepsy/jsw082

Mitchison D, Morin A, Mond J et al (2015) The bidirectional relationship between quality of life and eating disorder symptoms: a 9-year community-based study of Australian women. PLoS One 10:e0120591. https://doi.org/10.1371/journal.pone.0120591

Mond JM, Owen C, Hay PJ et al (2005) Assessing quality of life in eating disorder patients. Qual Life Res 14:171–178. https://doi.org/10.1007/s11136-004-2657-y

Mustelin L, Bulik CM, Kaprio J, Keski-Rahkonen A (2017) Prevalence and correlates of binge eating disorder related features in the community. Appetite 109:165–171. https://doi.org/10.1016/j.appet.2016.11.032

Neumark-Sztainer DR, Wall MM, Haines JI et al (2007) Shared risk and protective factors for overweight and disordered eating in adolescents. Am J Prev Med 33:359–369. https://doi.org/10.1016/j.amepre.2007.07.031

O'Brien KM, Vincent NK (2003) Psychiatric comorbidity in anorexia and bulimia nervosa: nature, prevalence, and causal relationships. Clin Psychol Rev 23:57–74. https://doi.org/10.1016/S0272-7358(02)00201-5

Padierna A, Quintana JM, Arostegui I et al (2000) The health-related quality of life in eating disorders. Qual Life Res 9:667–674. https://doi.org/10.1023/A:1008973106611

Pasold TL, McCracken A, Ward-Begnoche WL (2014) Binge eating in obese adolescents: emotional and behavioral characteristics and impact on health-related quality of life. Clin Child Psychol Psychiatry 19:299–312. https://doi.org/10.1177/1359104513488605

Pawaskar M, Witt EA, Supina D et al (2017) Impact of binge eating disorder on functional impairment and work productivity in an adult community sample in the United States. Int J Clin Pract 71:e12970. https://doi.org/10.1111/ijcp.12970

Pearson CM, Mason TB, Cao L et al (2018) A test of a state-based, self-control theory of binge eating in adults with obesity. Eat Disord 26:26–38. https://doi.org/10.1080/10640266.2018.1418358

Pinheiro AP, Nunes MA, Barbieri NB et al (2016) Association of binge eating behavior and psychiatric comorbidity in ELSA-Brasil study: results from baseline data. Eat Behav 23:145–149. https://doi.org/10.1016/j.eatbeh.2016.08.011

Pisetsky EM, Crosby RD, Cao L et al (2016) An examination of affect prior to and following episodes of getting drunk in women with bulimia nervosa. Psychiatry Res 240:202–208. https://doi.org/10.1016/j.psychres.2016.04.044

Pohjolainen V, Koponen S, Räsänen P et al (2016) Long-term health-related quality of life in eating disorders. Qual Life Res 25:2341–2346. https://doi.org/10.1007/s11136-016-1250-5

Pollice C, Kaye WH, Greeno CG, Weltzin TE (1997) Relationship of depression, anxiety, and obsessionality to state of illness in anorexia nervosa. Int J Eat Disord 21:367–376. https://doi.org/10.1002/(SICI)1098-108X(1997)21:4<367::AID-EAT10>3.0.CO;2-W

Ranzenhofer LM, Columbo KM, Tanofsky-Kraff M et al (2012) Binge eating and weight-related quality of life in obese adolescents. Nutrients 4:167–180. https://doi.org/10.3390/nu4030167

Ranzenhofer LM, Engel SG, Crosby RD et al (2014) Using ecological momentary assessment to examine interpersonal and affective predictors of loss of control eating in adolescent girls. Int J Eat Disord 47:748–757. https://doi.org/10.1002/eat.22333

Rosenbaum DL, White KS (2015) The relation of anxiety, depression, and stress to binge eating behavior. J Health Psychol 20:887–898. https://doi.org/10.1177/1359105315580212

Rosenberg N, Bloch M, Ben Avi I et al (2013) Cortisol response and desire to binge following psychological stress: comparison between obese subjects with and without binge eating disorder. Psychiatry Res 208:156–161. https://doi.org/10.1016/j.psychres.2012.09.050

Rospond B, Szpigiel J, Sadakierska-Chudy A, Filip M (2015) Binge eating in pre-clinical models. Pharmacol Rep 67:504–512. https://doi.org/10.1016/j.pharep.2014.11.012

Ross HE, Ivis F (1999) Binge eating and substance use among male and female adolescents. Int J Eat Disord 26:245–260. https://doi.org/10.1002/(SICI)1098-108X(199911)26:3<245::AID-EAT2>3.0.CO;2-R

Rubenstein CS, Pigott TA, Altemus M et al (1993) High rates of comorbid OCD in patients with bulimia nervosa. Eat Disord 1:147–155. https://doi.org/10.1080/10640269308248282

Saltzman JA, Liechty JM (2016) Family correlates of childhood binge eating: a systematic review. Eat Behav 22:62–71. https://doi.org/10.1016/j.eatbeh.2016.03.027

Shank LM, Crosby RD, Grammer AC et al (2017) Examination of the interpersonal model of loss of control eating in the laboratory. Compr Psychiatry 76:36–44. https://doi.org/10.1016/j.comppsych.2017.03.015

Shomaker LB, Tanofsky-Kraff M, Elliott C et al (2010) Salience of loss of control for pediatric binge episodes: does size really matter? Int J Eat Disord 43:707–716. https://doi.org/10.1002/eat.20767

Sinclair-McBride K, Cole DA (2017) Prospective relations between overeating, loss of control eating, binge eating, and depressive symptoms in a school-based sample of adolescents. J Abnorm Child Psychol 45:693–703. https://doi.org/10.1007/s10802-016-0186-0

Smith KE, Mason TB, Crosby RD et al (2019) A multimodal, naturalistic investigation of relationships between behavioral impulsivity, affect, and binge eating. Appetite 136:50–57. https://doi.org/10.1016/j.appet.2019.01.014

Solmi M, Collantoni E, Meneguzzo P et al (2019) Network analysis of specific psychopathology and psychiatric symptoms in patients with anorexia nervosa. Eur Eat Disord Rev 27:24–33. https://doi.org/10.1002/erv.2633

Sonneville KR, Horton NJ, Micali N et al (2013) Longitudinal associations between binge eating and overeating and adverse outcomes among adolescents and young adults: does loss of control matter? JAMA Pediatr 167:149–155. https://doi.org/10.1001/2013.jamapediatrics.12

Stevenson BL, Dvorak RD, Wonderlich SA et al (2018) Emotions before and after loss of control eating. Eat Disord 26:505–522. https://doi.org/10.1080/10640266.2018.1453634

Stevenson BL, Wilborn D, Kramer MP, Dvorak RD (2019) Real-time changes in emotion regulation and loss of control eating. J Health Psychol:1359105318823242. https://doi.org/10.1177/1359105318823242

Stice E, Presnell K, Spangler D (2002) Risk factors for binge eating onset in adolescent girls: a 2-year prospective investigation. Health Psychol 21:131–138. https://doi.org/10.1037/0278-6133.21.2.131

Stice E, Fisher M, Lowe MR (2004) Are dietary restraint scales valid measures of acute dietary restriction? Unobtrusive observational data suggest not. Psychol Assess 16:51–59. https://doi.org/10.1037/1040-3590.16.1.51

Striegel-Moore RH, Wilson GT, Wilfley DE et al (1998) Binge eating in an obese community sample. Int J Eat Disord 23:27–37. https://doi.org/10.1002/(SICI)1098-108X(199801)23:1<27::AID-EAT4>3.0.CO;2-3

Svaldi J, Werle D, Naumann E et al (2019) Prospective associations of negative mood and emotion regulation in the occurrence of binge eating in binge eating disorder. J Psychiatr Res 115:61–68. https://doi.org/10.1016/j.jpsychires.2019.05.005

Tanofsky-Kraff M, Shomaker LB, Olsen C et al (2011) A prospective study of pediatric loss of control eating and psychological outcomes. J Abnorm Psychol 120:108. https://doi.org/10.1037/a0021406

Troop NA, Allan S, Treasure JL, Katzman M (2003) Social comparison and submissive behaviour in eating disorder patients. Psychol Psychother Theory Res Pract 76:237–249. https://doi.org/10.1348/147608303322362479

Vogeltanz-Holm ND, Wonderlich SA, Lewis BA et al (2000) Longitudinal predictors of binge eating, intense dieting, and weight concerns in a national sample of women. Behav Ther 31:221–235. https://doi.org/10.1016/S0005-7894(00)80013-1

Wade TD, Wilksch SM, Lee C (2012) A longitudinal investigation of the impact of disordered eating on young women's quality of life. Health Psychol 31:352–359. https://doi.org/10.1037/a0025956

Wang SB, Lydecker JA, Grilo CM (2017) Rumination in patients with binge-eating disorder and obesity: associations with eating-disorder psychopathology and weight-bias internalization. Eur Eat Disord Rev 25:98–103. https://doi.org/10.1002/erv.2499

West CE, Goldschmidt AB, Mason SM, Neumark-Sztainer D (2019) Differences in risk factors for binge eating by socioeconomic status in a community-based sample of adolescents: findings from project EAT. Int J Eat Disord 52:659–668. https://doi.org/10.1002/eat.23079

Wiederman MW, Pryor T (1996) Substance use and impulsive behaviors among adolescents with eating disorders. Addict Behav 21:269–272. https://doi.org/10.1016/0306-4603(95)00062-3

Wilfley DE, Schwartz MB, Spurrell EB, Fairburn CG (2000) Using the eating disorder examination to identify the specific psychopathology of binge eating disorder. Int J Eat Disord 27:259–269. https://doi.org/10.1002/(SICI)1098-108X(200004)27:3<259::AID-EAT2>3.0.CO;2-G

Wilfley DE, Welch RR, Stein RI et al (2002) A randomized comparison of group cognitive-behavioral therapy and group interpersonal psychotherapy for the treatment of overweight individuals with binge-eating disorder. Arch Gen Psychiatry 59:713–721. https://doi.org/10.1001/archpsyc.59.8.713

Wilfley D, Berkowitz R, Goebel-Fabbri A et al (2011) Binge eating, mood, and quality of life in youth with type 2 diabetes: baseline data from the today study. Diabetes Care 34:858–860. https://doi.org/10.2337/dc10-1704

Winkler LA-D, Christiansen E, Lichtenstein MB et al (2014) Quality of life in eating disorders: a meta-analysis. Psychiatry Res 219:1–9. https://doi.org/10.1016/j.psychres.2014.05.002

Wolfe WL, Maisto SA (2000) The relationship between eating disorders and substance use: moving beyond co-prevalence research. Clin Psychol Rev 20:617–631. https://doi.org/10.1016/S0272-7358(99)00009-4

Wonderlich SA, Swift WJ, Slotnick HB, Goodman S (1990) DSM-III-R personality disorders in eating-disorder subtypes. Int J Eat Disord 9:607–616. https://doi.org/10.1002/1098-108X(199011)9:6<607::AID-EAT2260090603>3.0.CO;2-0

Yeomans MR (2004) Effects of alcohol on food and energy intake in human subjects: evidence for passive and active over-consumption of energy. Br J Nutr 92(Suppl 1):S31–S34. https://doi.org/10.1079/bjn20041139

Zambrowicz R, Schebendach J, Sysko R et al (2019) Relationship between three factor eating questionnaire-restraint subscale and food intake. Int J Eat Disord 52:255–260. https://doi.org/10.1002/eat.23014

Zunker C, Peterson CB, Crosby RD et al (2011) Ecological momentary assessment of bulimia nervosa: does dietary restriction predict binge eating? Behav Res Ther 49:714–717. https://doi.org/10.1016/j.brat.2011.06.006

Weight Dysregulation, Positive Energy Balance, and Binge Eating in Eating Disorders

Michael R. Lowe, Leora L. Haller, Simar Singh, and Joanna Y. Chen

Abstract

Determinants of binge eating are usually understood in terms of cognitive, behavioral, and affective disturbances. However, many of those with eating disorders (EDs) experienced elevated premorbid weights, lost substantial weight as their disorder developed and/or are currently at low weight. We review evidence that these three aspects of weight regulation are related to binge eating, to loss of control eating in the absence of objective binge eating and/or to positive energy balance and weight gain over time (an outcome that is greatly feared in those with EDs). Thus, there appear to be powerful reciprocal influences between psychological and biological aspects of binge eating but until recently only the psychological features have been incorporated into models of ED psychopathology and treatment. Further, several studies have shown that weight suppression is also related to eating symptoms in nonclinical groups, which indicates that elevated weight suppression can accelerate weight regain and have other negative impacts in the absence of an ED. In sum, the impact of weight history on the perpetuation of EDs indicates that treatment researchers should consider this impact as they try to improve treatment outcomes.

Keywords

Weight · Obesity · Weight loss · Weight suppression · Binge eating · Loss of control eating · Weight gain

Learning Objectives

After reading this chapter, you will:

- Understand the differences between psychological and weight-related sources of eating disorder psychopathology.
- Be able to define the significance of highest past weight, weight suppression, and current weight in influencing energy intake, loss of control eating, positive energy balance, and future weight gain.
- Understand the potential consequences of significant weight loss in non-eating disordered individuals.
- Be able to explain how weight dysregulation variables represent a potential source of innovative changes to eating disorder treatments to improve treatment outcomes.

M. R. Lowe (✉) · L. L. Haller · S. Singh · J. Y. Chen
Drexel University, Philadelphia, PA, USA
e-mail: ml42@drexel.edu; llh62@drexel.edu; ss4776@drexel.edu; jyc52@drexel.edu

© Springer Nature Switzerland AG 2020
G. K.W. Frank, L. A. Berner (eds.), *Binge Eating*, https://doi.org/10.1007/978-3-030-43562-2_5

1   Introduction

Models of eating disorder (ED) psychopathology and psychotherapy are based on social, behavioral, cognitive, and affective influences. Those with EDs also usually have a problematic weight history but current models view this history as relevant only because psychological influences contribute to intense concerns about weight and to changes in weight. We have been developing a Weight Regulation Model of EDs that reverses this equation—that is, which proposes that weight and weight history are causal variables that have a clinically significant impact on ED psychopathology and ED perpetuation. In this chapter, we describe how several facets of weight and weight history impact binge eating specifically, and increased energy intake more generally. We will use the following notation to refer to individuals with eating disorders, anorexia nervosa, or bulimia nervosa: IWED, IWAN, and IWBN, respectively. Because little research has examined weight dysregulation constructs in binge-eating disorder (BED), BED will not be reviewed in this chapter.

This chapter includes a consideration of both objective binge eating and two related concepts: loss of control (LOC) eating and energy intake beyond energy needs. Research has consistently found that LOC eating is associated with severity of ED psychopathology independent of the amount eaten in LOC episodes (Fitzsimmons-Craft et al. 2014; Pratt et al. 1998). Further, bouts of LOC may involve increased energy intake even if the intake does not rise to the level of an objective binge episode. For instance, this could occur if part of the reason an individual experiences LOC is that they are eating a highly energy-dense "forbidden" food. Thus, it is important to consider if weight regulation variables are related not only to objective binge episodes but to subjectively large, or "subjective binge episodes" as well. As for energy intake, when increased intake contributes to weight gain, it becomes relevant to EDs because weight gain is anathema to IWEDs. Therefore, although this book is focused on binge eating, we also make reference to how

weight regulation variables may contribute to LOC per se and positive energy balance, thereby contributing to psychological distress and weight gain independently of objective binge episodes.

Although most of the research on weight history's relation to EDs has focused on weight suppression (WS), some authors have begun to examine additional aspects of weight and weight history and their relationship to disordered eating (e.g., Hessler et al. 2019; Lowe et al. 2018). Therefore, beyond WS itself, the two components of WS—past highest weight and current weight— will be considered here first. We have found in multiple samples of IWEDs that the highest premorbid BMI is highly correlated with WS (with correlations of 0.50–0.70). This means that some of the effects that have been associated with WS in past research could be driven in part by past highest weight (or BMI), not just by WS. In addition, several studies have found interactions between WS and current weight or BMI on eating disorder-related characteristics. There is reason to believe that being at a relatively low current BMI (e.g., a BMI under 22) may contribute to disordered eating characteristics beyond the impact of WS alone. Finally, we review a number of studies that have examined WS specifically among non-eating disordered groups. This research is of interest because it demonstrates that WS may predict lesser forms of ED psychopathology and that WS is not only associated with pathological outcomes when it occurs in those with disordered eating.

2   The Role of Elevated Premorbid BMIs in Binge Eating

Many studies have now recognized that there is a greater than expected history of overweight or obesity preceding the development of eating disorders. Existing data indicate a prevalence rate of overweight or obesity between 20 and 60% in IWAN and IWBN (e.g., Villarejo et al. 2012; Crisp et al. 1980); given that these disorders develop relatively early in life, these rates are indeed high. This rate is especially high in disorders characterized by binge eating, with 33.2% of patients with BN and 87.8% of patients with BED having lifetime obesity (Villarejo et al. 2012). More recently, Muratore and Lowe (2019) showed that all studies that have retrospectively examined premorbid BMIs in treatment-seeking IWAN and IWBN found elevated premorbid BMIs in these populations, relative to the general population or their non-disordered peers (e.g., Berkowitz et al. 2016; Shaw et al. 2012; Swenne 2016). Although these findings suggest a potential role of elevated BMI in the development of EDs, and binge eating in particular, this pattern is less consistent in prospective studies that examine the emergence of ED symptoms in initially healthy, population-based samples. Only a third of such prospective studies found support for the notion that elevated premorbid BMI increases risk for the development of binge eating (Killen et al. 1994; Stice et al. 2002; Reed et al. 2017) and other eating disorder symptoms (e.g., body dissatisfaction and excessive exercise; Allen et al. 2014; Jendrzyca and Warschburger 2016; Wichstrom 2000; Wiklund et al. 2018).

To reconcile this discrepancy between clinical and nonclinical studies, Muratore and Lowe (2019) proposed the possibility that elevated premorbid BMIs may increase the risk of developing eating disorders of greater clinical severity. They posited that elevated premorbid body weights may increase this risk because (1) it would be associated with more intense body dissatisfaction and more vigorous weight loss efforts and (2) such an individual would have to lose more weight to reach his or her ideal body weight, relative to those with lower premorbid weights. Therefore, individuals with higher premorbid weights are likely to be more weight suppressed postmorbidly, which may increase binge eating in two ways. First, higher historical weight suppression may promote more frequent binge eating and weight gain over time (Lowe et al. 2018). Second, weight gain may prompt new weight control behaviors and short-term weight loss, thereby increasing WS and, ultimately, promoting more binge eating.

Research has identified the reasons that clinical samples with elevated premorbid BMIs are more likely to develop and EDs involving binge eating. The tendency toward a higher-than-average premorbid body weight could indicate a greater premorbid susceptibility to eating beyond energy needs and/or to LOC eating. This notion is supported by prospective studies in which higher BMIs predicted the onset, or an increased likelihood, of bulimic symptoms (Killen et al. 1994; Lowe et al. 2019a; Reed et al. 2017; Stice et al. 2002) and a retrospective study that found higher premorbid BMI in treatment-seeking IWBN than in controls at similar ages (Shaw et al. 2012). Further, evidence indicates that at least one gene strongly related to BMI (FTO) is also related to binge eating propensity (Micali et al. 2015). Numerous studies have found childhood BMIs to be consistently predictive of adult BMIs, and that childhood overweight or obesity increases obesity risks in adulthood (e.g., Freedman et al. 2005; Serdula et al. 1993). Thus, having an elevated premorbid BMI early in life may put an individual at greater risk of developing overweight or obesity as an adult. To combat this chronic disposition toward weight gain and overweight, such at-risk individuals may engage in more extreme weight control behaviors, and the resulting weight-suppressed state would ignite both binge eating and eventual weight gain, thereby perpetuating the disorder.

As in IWBN, high premorbid BMI has also been found in some retrospective clinical studies of IWAN (Muratore and Lowe 2019). Moreover, it prospectively predicted the development of binge eating and purging behaviors in IWAN in one study (Lantz et al. 2017), suggesting that the role of elevated premorbid BMI in the development of BN may also apply to IWAN who eventually develop bulimic symptoms. It may be that elevated premorbid BMI reflects a greater appetitive drive and susceptibility to overeating, and that this occurs before significant weight loss and the initial development of binge eating. Nonetheless, there is also some evidence that IWAN who are identified from population based, prospective studies (and therefore less likely to be in treatment) have *lower* premorbid BMIs than control groups during childhood and adolescence (Duncan et al. 2017; Stice 2016), but these studies did not distinguish IWAN with binge eating and those without. Given the mixed findings, further research on the role of premorbid BMI in the development of binge eating pathology in IWAN would be valuable.

3 Evidence Supporting an Association of Weight Suppression with Binge Eating

In one of the first academic investigations of IWBN, Russell (1979) observed that most of his patients historically weighed substantially more than they did when they presented for treatment, indicating a high level of WS. Elevated WS, secondary to diet-induced weight loss, has been consistently demonstrated in IWBN (Lowe et al. 2018; Keel and Heatherton 2010; Bodell and Keel 2015). WS has been shown to range from a mean of 15–25 lbs in IWBN, with higher levels evident in IWAN and in those receiving higher levels of care (Lowe et al. 2018). However, patients who lose substantial weight over the course of developing an ED do so against a biological predisposition to remain at their elevated, premorbid BMI. It has been demonstrated that the higher an individual's BMI at a given time point, the more likely he/she will exhibit elevated weight gain over time (Guo et al. 2002; Kvaavik et al. 2003). Given this background, individuals susceptible to developing a disorder characterized by binge eating may be subject to *two influences* that help maintain binge eating over time. One is an elevated premorbid BMI (which, if not countered by significant weight loss may continue to increase over time) and the other is the long-term weight suppression that is often maintained after an initial, major weight loss. Once binge eating emerges, however, these forces and the positive energy balance they produce may also be responsible for the fact that many IWBN quickly regain a significant amount of weight once binge eating commences (Shaw et al. 2012). Once the disorder is initiated, higher WS is positively related to the frequency of binge eating (Lowe et al. 2007; Butryn et al. 2006, 2011), even after controlling for relevant covariates (e.g., BMI, TFEQ restraint, all EDE subscales, and purging episodes). Notably, a recent study also found that higher WS was related to increased reinforcing value of food and larger binge-episode size among IWBN (Bodell and Keel 2015). In a study by Butryn et al. (2011), WS not only independently predicted objective binge episodes, but also interacted with current BMI to predict binge episode frequency above and beyond WS or BMI alone: participants with low BMI and high WS had rates of binge eating more than double that of any other group (i.e., low WS, low BMI; low WS, high BMI).

WS is further implicated in the long-term maintenance of bulimic symptoms. Several studies using different samples, measures, covariates, and follow-up periods have shown that those who were more highly weight suppressed took a longer time to recover (Lowe et al. 2011), experienced more rapid weight gain (Herzog et al. 2010), tended to continue binge eating and purging after end of treatment (Butryn et al. 2006), and gained more total weight over time (Lowe et al. 2006b; Stice et al. 2011). In one study examining the change in bulimic symptoms over time, Keel and colleagues (2010) found that higher levels of WS predicted both the onset of a bulimic syndrome and, among a group having a bulimic syndrome at baseline, the maintenance of symptoms at a 10-year follow-up. A third study found similar results at a 20-year follow-up, findings which were mediated by drive for thinness at year 10 (Bodell et al. 2017). Similarly, among IWAN, those with higher premorbid BMIs were more likely to develop binge eating over 18 years of follow-up, and those with higher WS were more likely to engage in binge-purge behaviors—supporting the notion that elevated premorbid BMIs, WS, and perhaps their combination, are potent sources of binge eating across

EDs (Lantz et al. 2017; Berner et al. 2013; Wildes and Marcus 2012).

Taken together, these results support the notion that larger discrepancies between an individual's highest past weight and current weight produce psychobiological pressures that fuel binge eating.

4 Evidence Questioning an Association of Weight Suppression with Binge Eating

Despite compelling evidence demonstrating an association between WS and bulimic symptomatology, WS failed to show a predictive association with binge eating in several studies (Lavender et al. 2015; Carter et al. 2008; Dawkins et al. 2013; Van Son et al. 2013). A possible explanation for these contradictory findings may be the relatively low levels of WS in these studies (means ranging from 5.0 to 18.7 lbs), relative to studies that have demonstrated relationships (means generally ranging from 20 to 35 lbs). It may be that the relation of WS to bulimic symptomatology is non-linear, with WS having a disproportionately stronger impact at higher (e.g., above 20 lbs) than at lower levels of WS (Lowe et al. 2011). It is also clear that WS is not a precondition for BN as there are some IWBN with very low levels of WS (e.g., less than 5 lbs). Another possibility is that the current WS measure is not as sensitive as it might be, a topic we address in the final section of this chapter.

5 Weight Dysregulation in Individuals Without Eating Disorders

Examining the impact of weight history in healthy populations has two potential benefits. First, it could shed light on its impact on IWEDs. Second, the potential psychological downsides of existing in a state of significant negative energy balance (outside of eating

disorders) has not received much attention in the literature (Neumark-Sztainer et al. 2011).

The mean level of WS in college students is much lower than that in adult IWBN (about 20 lbs) or IWAN (about 30 lbs; (Lowe et al. 2006a, 2018; Stice et al. 2011)), but a more recent analysis of the general population with an average age of 36 years (SD = 11) found that 61% of the population were weight suppressed by at least 5% compared to highest past weight, with 58% being at least 10% below their highest past body weight (Lantz 2019). Middle-aged samples may be higher in WS because weight gain increases with age, resulting in both a stronger desire to lose weight and a desire to lose a larger amount of weight. Among a community-based sample of middle-aged women, individuals higher in WS reported more binge eating and weight loss attempts (Goodman et al. 2018). This indicates that WS may signal a commonality among both nonclinical and clinical samples that increases vulnerability to binge eating and future weight gain.

WS in clinical samples is associated with characteristics (e.g., binge eating, food restriction) that may impede treatment and recovery from an ED. To date, studies looking at nonclinical samples indicate that WS can be an indicator of problems for healthy individuals as well. A study in college students found that in females with body concerns, WS predicted future increases in BMI (Stice et al. 2011). Those with higher WS also had lower levels of total energy expenditure but this did not account for their greater weight gain. Thus, this weight gain must have stemmed from a chronic positive energy balance due to increased energy intake, presumably because of the appetitive effects of greater WS. In another study in college students, WS was the only variable of several risk factors tested that predicted the amount of body fat gained at a 2-year follow-up (Lowe et al. 2019b). Lowe et al. (2006a, b) found that WS predicted weight gain during a freshman's first year (where the mean WS level was only 4 lbs), independent of current dieting, indicating that WS is predictive even when the overall level of WS in the sample tested is low. In a study of individuals in a behavioral weight loss program, those higher in WS lost less weight than those lower in WS, and also had poorer treatment efficacy perception (Call et al. 2019). Again, this poorer outcome is apparently due to less effective regulation of eating, even if this did not involve binge eating per se.

The studies reviewed in this section suggest that an overlooked consequence of the excess weight epidemic may be the widespread weight loss that so many individuals with obesity undergo. Although such weight loss has health benefits, it could also have untoward consequences. Such WS may contribute to more LOC eating, positive energy balance, weight regain, and the psychological distress that is generated by these consequences.

6   Developmental Weight Suppression and the Timing of Past Highest Weight: New Horizons

Traditionally, WS is conceived as the difference between highest past weight since reaching adult height and current weight. However, a highest past weight of 160 lbs at 13 years old represents a higher degree of overweight (relative to other 13 year olds) than the same highest weight reached at 17 years old (relative to other 17 year olds). This caveat underscores a notable shortcoming in the current conceptualization and calculation of WS: highest past weight and current weight should be indexed relative to the typical weights of one's peers at the same ages. As such, Lowe et al. (2018) proposed a new construct termed "developmental weight suppression" to address this weakness.

Developmental WS is calculated as the difference between highest premorbid z-BMI (i.e., BMI z-score) and current z-BMI. By utilizing z-BMI, developmental WS takes into account an individual's developmental status with respect to age, height, and gender, and is thus a more clinically and biologically meaningful representation of WS. Preliminary data from our research team

indicate that relevant variables (e.g., binge eating) are more strongly associated with developmental WS than traditional WS (Lowe, unpublished). Furthermore, we have found that the two versions of WS correlate only moderately ($r \sim 0.50$) in a sample of IWBN, indicating that most of their variance is not shared. More information about developmental WS, along with a time-saving algorithm to calculate it, will be available from Singh, Apple, Zhang and Lowe (submitted).

There is a further distinction regarding the measurement of weight suppression in those with BN that is important to take into account. Shaw et al. (2012) found in IWBN that most individuals reported reaching their highest past weight *after* their eating disorder began. This creates a potentially important distinction between highest *premorbid* and highest *postmorbid* weights. If the highest weight reached was attained postmorbidly, it is more likely that this weight was reached, at least in part, because of the eating disorder itself. Therefore we recommend that researchers measure highest past weight (and the age at which it was reached) both before and after an individual develops an eating disorder (and that weight suppression based on both calculations be examined in future studies).

7   Implications

There are two questions raised by the foregoing review. One is whether individuals who are in a highly weight-suppressed state (say, over 20 lbs) will necessarily experience challenges in keeping off a large weight loss. A second question is whether any challenges experienced will dissipate with time.

In regard to those with eating disorders, we think the following considerations are relevant. First, the researchers who documented an increase in metabolic efficiency with weight loss in obese individuals (Rosenbaum and Leibel 2010) believe that this increase is long lasting; whether this is also true among those with eating disorders is unknown. Second, individuals from the National Weight Control Registry, who have lost substantial weight and kept it off for extended periods, report that they need to rigorously control their diet and engage in a high level of physical activity to maintain their weight loss. Third, there is considerable evidence that IWAN and IWBN who remit from their disorder do so at BMIs that, though within the normal weight range, are relatively low in two senses. They are low in relation to what is likely to be a considerably higher past highest z-BMI (based on the evidence reviewed above). Also, they are low in relation to the average BMI of their non-eating disordered peers. For instance, evidence suggests that IWAN (Fichter and Quadflieg 1999) and IWBN (Fairburn et al. 2009) tend to recover at average BMIs in the low 20s, whereas the average BMI for similarly aged, non-eating disordered women in the United States is about 26 (Hales et al. 2017). This suggests that most IWEDs who remit are able to relinquish their symptoms while retaining a significant degree of weight suppression. Safer et al. (2004) found evidence consistent with this viewpoint in a study of IWBN successfully treated by CBT. Although their level of extreme dieting behaviors (as measured by the Dietary Restraint Scale from the EDE) decreased dramatically following successful treatment, their scores on a measure of healthier restrained eating (the Cognitive Restraint Scale from the Three-Factor Eating Inventory) did not change (i.e., they remained elevated both before and after successful treatment). Although these patients relinquished many of their strict dieting behaviors, retaining their high level of cognitive restraint may have been necessary to prevent the weight gain that their ongoing weight suppression would have otherwise produced. Future research could determine if evidence supports this account.

8 Summary

Healthy weight regulation involves achieving and maintaining a BMI in the "normal" range. Many IWEDs deviate from this pattern in one or more ways: they gain more weight than their peers before developing an ED, they lose a great deal of weight in the process of developing an ED, and

they function at low BMI. We review evidence that these factors, alone or in combination, are related to objective binge eating, LOC eating more generally, positive energy balance, and weight gain over time. Novel strategies for helping IWEDs deal with their chronic weight dysregulation could enhance the effectiveness of existing ED treatments.

References

Allen KL, Byrne SM, Oddy WH, Schmidt U, Crosby RD (2014) Risk factors for binge eating and purging eating disorders: differences based on age of onset. Int J Eat Disord 47(7):802–812. https://doi.org/10.1002/eat.22299

Berkowitz SA, Witt AA, Gillberg C, Råstam M, Wentz E, Lowe MR (2016) Childhood body mass index in adolescent-onset anorexia nervosa. Int J Eat Disord 49(11):1002–1009. https://doi.org/10.1002/eat.22584

Berner LA, Shaw JA, Witt AA, Lowe MR (2013) The relation of weight suppression and body mass index to symptomatology and treatment response in anorexia nervosa. J Abnorm Psychol 122(3):694

Bodell LP, Keel PK (2015) Weight suppression in bulimia nervosa: associations with biology and behavior. J Abnorm Psychol 124(4):994

Bodell LP, Brown TA, Keel PK (2017) Weight suppression predicts bulimic symptoms at 20-year follow-up: the mediating role of drive for thinness. J Abnorm Psychol 126(1):32

Butryn ML, Lowe MR, Safer DL, Agras WS (2006) Weight suppression is a robust predictor of outcome in the cognitive-behavioral treatment of bulimia nervosa. J Abnorm Psychol 115(1):62

Butryn ML, Juarascio A, Lowe MR (2011) The relation of weight suppression and BMI to bulimic symptoms. Int J Eat Disord 44(7):612–617

Call CC, Piers AD, Wyckoff EP, Lowe MR, Forman EM, Butryn ML (2019) The relationship of weight suppression to treatment outcomes during behavioral weight loss. J Behav Med 42(2):365–375. https://doi.org/10.1007/s10865-018-9978-8

Carter FA, McIntosh VV, Joyce PR, Bulik CM (2008) Weight suppression predicts weight gain over treatment but not treatment completion or outcome in bulimia nervosa. J Abnorm Psychol 117(4):936

Crisp AH, Hsu LKG, Harding B, Hartshorn J (1980) Clinical features of anorexia nervosa: a study of a consecutive series of 102 female patients. J Psychosom Res 24:179–191. https://doi.org/10.1016/0022-3999(80)90040-9

Dawkins H, Watson HJ, Egan SJ, Kane RT (2013) Weight suppression in bulimia nervosa: relationship with cognitive behavioral therapy outcome. Int J Eat Disord 46(6):586–593

Duncan L, Yilmaz Z, Gaspar H, Walters R, Goldstein J, Anttila V, Bulik-Sullivan B, Ripke S, Eating Disorders Working Group of the Psychiatric Genomics Consortium, Thornton L, Hinney A, Daly M, Sullivan PF, Zeggini E, Breen G, Bulik CM (2017) Significant locus and metabolic genetic correlations revealed in genome-wide association study of anorexia nervosa. Am J Psychiatry 174(9):850–858. https://doi.org/10.1176/appi.ajp.2017.16121402

Fairburn CG, Cooper Z, Doll HA, O'Connor ME, Bohn K, Hawker DM, Wales JA, Palmer RL (2009) Transdiagnostic cognitive-behavioral therapy for patients with eating disorders: a two-site trial with 60-week follow-up. Am J Psychiatry 166(3):311–319

Fichter MM, Quadflieg N (1999) Six-year course and outcome of anorexia nervosa. Int J Eat Disord 23 (4):359–385. https://doi.org/10.1002/(SICI)1098-108X(199912)26:4<359::AID-EAT2>3.0.CO;2-7

Fitzsimmons-Craft EE, Ciao AC, Accurso EC, Pisetsky EM, Peterson CB, Byrne CE, Le Grange D (2014) Subjective and objective binge eating in relation to eating disorder symptomatology, depressive symptoms, and self-esteem among treatment-seeking adolescents with bulimia nervosa. Eur Eat Disord Rev 22(4):230–236. https://doi.org/10.1002/erv.2297

Freedman DS, Khan LK, Serdula MK, Dietz WH, Srinivasan SR, Berenson GS (2005) The relation of childhood BMI to adult adiposity: the Bogalusa heart study. Pediatrics 115(1):22–27. https://doi.org/10.1542/peds.2004-0220

Goodman EL, Baker JH, Peat CM, Yilmaz Z, Bulik CM, Watson HJ (2018) Weight suppression and weight elevation are associated with eating disorder symptomatology in women age 50 and older: results of the gender and body image study. Int J Eat Disord 51 (8):835–841. https://doi.org/10.1002/eat.22869

Guo SS, Wu W, Chumlea WC, Roche AF (2002) Predicting overweight and obesity in adulthood from body mass index values in childhood and adolescence. Am J Clin Nutr 76(3):653–658

Hales CM, Carroll MD, Fryar CD, Ogden CL (2017) Prevalence of obesity among adults and youth: United States, 2015–2016. National Center for Health Statistics, Hyattsville, MD

Herzog DB, Thomas JG, Kass AE, Eddy KT, Franko DL, Lowe MR (2010) Weight suppression predicts weight change over 5 years in bulimia nervosa. Psychiatry Res 177(3):330–334

Hessler JB, Schlegl S, Greetfeld M, Voderholzer U (2019) Dimensions within 24 weight history indices and their association with inpatient treatment outcome in adults with anorexia nervosa: analysis of routine data. J Eat Disord 7:19. https://doi.org/10.1186/s40337-019-0249-z

Jendrzyca A, Warschburger P (2016) Weight stigma and eating behaviours in elementary school children: a prospective population-based study. Appetite 102:51–59. https://doi.org/10.1016/j.appet.2016.02.005

Keel PK, Heatherton TF (2010) Weight suppression predicts maintenance and onset of bulimic syndromes at 10-year follow-up. J Abnorm Psychol 119 (2):268–275

Killen JD, Taylor CB, Hayward C, Wilson DM, Haydel KF, Hammer LD, Simmonds B, Robinson TN, Litt I, Varady A, Kraemer H (1994) Pursuit of thinness and onset of eating disorder symptoms in a community sample of adolescent girls: A three-year prospective analysis. Int J Eat Disord 16(3):227–238. https://doi.org/10.1002/1098-108X(199411)16:3<227::AID-EAT2260160303>3.0.CO;2-L

Kvaavik E, Tell GS, Klepp K-I (2003) Predictors and tracking of body mass index from adolescence into adulthood: follow-up of 18 to 20 years in the Oslo youth study. Arch Pediatr Adolesc Med 157 (12):1212–1218

Lantz EL (2019) Understanding the effects of weight suppression by deconstructing its dimensions Drexel University Philadelphia, PA

Lantz EL, Gillberg C, Råstam M, Wentz E, Lowe MR (2017) Premorbid BMI predicts binge-purge symptomatology among individuals with anorexia nervosa. Int J Eat Disord 50(7):852–855

Lavender JM, Shaw JA, Crosby RD, Feig EH, Mitchell JE, Crow SJ, Hill L, Le Grange D, Powers P, Lowe MR (2015) Associations between weight suppression and dimensions of eating disorder psychopathology in a multisite sample. J Psychiatr Res 69:87–93

Lowe MR, Annunziato RA, Markowitz JT, Didie E, Bellace DL, Riddell L, Maille C, McKinney S, Stice E (2006a) Multiple types of dieting prospectively predict weight gain during the freshman year of college. Appetite 47(1):83–90

Lowe MR, Davis W, Lucks D, Annunziato R, Butryn M (2006b) Weight suppression predicts weight gain during inpatient treatment of bulimia nervosa. Physiol Behav 87(3):487–492

Lowe MR, Thomas JG, Safer DL, Butryn ML (2007) The relationship of weight suppression and dietary restraint to binge eating in bulimia nervosa. Int J Eat Disord 40 (7):640–644

Lowe MR, Berner LA, Swanson SA, Clark VL, Eddy KT, Franko DL, Shaw JA, Ross S, Herzog DB (2011) Weight suppression predicts time to remission from bulimia nervosa. J Consult Clin Psychol 79(6):772–776

Lowe MR, Piers AD, Benson L (2018) Weight suppression in eating disorders: a research and conceptual update. Curr Psychiatry Rep 20(10):80. https://doi.org/10.1007/s11920-018-0955-2

Lowe MR, Marmorstein N, Iacono W, Rosenbaum D, Espel-Huynh H, Muratore AF, Lantz EL, Zhang F (2019a) Body concerns and BMI as predictors of disordered eating and body mass in girls: an 18-year longitudinal investigation. J Abnorm Psychol 128 (1):32–43. https://doi.org/10.1037/abn0000394

Lowe MR, Marti CN, Lesser EL, Stice E (2019b) Weight suppression uniquely predicts body fat gain in first-year female college students. Eat Behav 32:60–64. https://doi.org/10.1016/j.eatbeh.2018.11.005

Micali N, Field AE, Treasure JL, Evans DM (2015) Are obesity risk genes associated with binge eating in adolescence? Obesity (Silver Spring) 23(8):1729–1736. https://doi.org/10.1002/oby.21147

Muratore AF, Lowe MR (2019) Why is premorbid BMI consistently elevated in clinical samples, but not in risk factor samples, of individuals with eating disorders? Int J Eat Disord 52(2):117–120. https://doi.org/10.1002/eat.23029

Neumark-Sztainer D, Wall M, Larson NI, Eisenberg ME, Loth K (2011) Dieting and disordered eating behaviors from adolescence to young adulthood: findings from a 10-year longitudinal study. J Am Diet Assoc 111(7):1004–1011. https://doi.org/10.1016/j.jada.2011.04.012

Pratt EM, Niego SH, Agras WS (1998) Does the size of a binge matter? Int J Eat Disord 24(3):307–312. https://doi.org/10.1002/(SICI)1098-108X(199811)24:3<307::AID-EAT8>3.0.CO;2-Q

Reed ZE, Micali N, Bulik CM, Smith GD, Wade KH (2017) Assessing the causal role of adiposity on disordered eating in childhood, adolescence, and adulthood: a Mendelian randomization analysis. Am J Clin Nutr 106(3):764–772. https://doi.org/10.3945/ajcn.117.154104

Rosenbaum M, Leibel R (2010) Adaptive thermogenesis in humans. Int J Obes (Lond) 34(Suppl 1):S47–S55

Russell G (1979) Bulimia nervosa: an ominous variant of anorexia nervosa. Psychol Med 9(3):429–448

Safer DL, Agras WS, Lowe MR, Bryson S (2004) Comparing two measures of eating restraint in bulimic women treated with cognitive-behavioral therapy. Int J Eat Disord 36(1):83–88

Serdula MK, Ivery D, Coates RJ, Freedman DS, Williamson DF, Byers T (1993) Do obese children become obese adults? A review of the literature. Prev Med 22:167–177. https://doi.org/10.1006/pmed.1993.1014

Shaw JA, Herzog DB, Clark VL, Berner LA, Eddy KT, Franko DL, Lowe MR (2012) Elevated pre-morbid weights in bulimic individuals are usually surpassed post-morbidly: implications for perpetuation of the disorder. Int J Eat Disord 45(4):512–523

Singh S, Apple DE, Zhang F, Niu X, Lowe MR (Under review) A new, developmentally-sensitive measure of weight suppression

Stice E (2016) Interactive and mediational etiologic models of eating disorder onset: evidence from prospective studies. Annu Rev Clin Psychol 12:359–381. https://doi.org/10.1146/annurev-clinpsy-021815-093317

Stice E, Presnell K, Spangler D (2002) Risk factors for binge eating onset in adolescent girls: a 2-year prospective investigation. Health Psychol 21(2):131–138. https://doi.org/10.1037/0278-6133.21.2.131

Stice E, Durant S, Burger KS, Schoeller DA (2011) Weight suppression and risk of future increases in body mass: effects of suppressed resting metabolic rate and energy expenditure. Am J Clin Nutr 94(1):7–11

Swenne I (2016) Influence of premorbid BMI on clinical characteristics at presentation of adolescent girls with eating disorders. BMC Psychiatry 16(1):81

Van Son GE, van der Meer PA, Van Furth EF (2013) Correlates and associations between weight suppression and binge eating symptomatology in a population-based sample. Eat Behav 14(2):102–106

Villarejo C, Fernandez-Aranda F, Jimenez-Murcia S, Penas-Lledo E, Granero R, Penelo E, Tinahones FJ, Sancho C, Vilarrasa N, Montserrat-Gil de Bernabe M, Casanueva FF, Fernandez-Real JM, Fruhbeck G, De la Torre R, Treasure J, Botella C, Menchon JM (2012) Lifetime obesity in patients with eating disorders: increasing prevalence, clinical and personality correlates. Eur Eat Disord Rev 20(3):250–254. https://doi.org/10.1002/erv.2166

Wichstrom L (2000) Psychological and behavioral factors unpredictive of disordered eating: a prospective study of the general adolescent population in Norway. Int J Eat Disord 28(1):33–42. https://doi.org/10.1002/(SICI)1098-108X(200007)28:1<33::AID-EAT5>3.0.CO;2-H

Wiklund CA, Kuja-Halkola R, Thornton LM, Bälter K, Welch E, Bulik CM (2018) Childhood body mass index and development of eating disorder traits across adolescence. Eur Eat Disord Rev 26(5):462–471. https://doi.org/10.1002/erv.2612

Wildes JE, Marcus MD (2012) Weight suppression as a predictor of weight gain and response to intensive behavioral treatment in patients with anorexia nervosa. Behav Res Ther 50(4):266–274

Theoretical Development and Maintenance Models of Binge Eating

M. K. Higgins Neyland, Lisa M. Shank, and Jason M. Lavender

Abstract

Numerous conceptualizations addressing the onset or persistence of binge-eating behaviors have been proposed within the literature. This chapter provides an overview of a wide range of theories addressing the development or maintenance of binge eating. The models vary across a number of features, and this chapter is generally organized based on these features: (1) the focus on binge eating as an independent behavior versus in conjunction with other eating disorder behaviors (e.g., restriction and purging) or within a certain diagnostic class (e.g., bulimia nervosa and binge-eating disorder); (2) the explicit association with a specific psychotherapeutic intervention; and (3) the nature of the variables within the model, including cognitive, affective, interpersonal, sociocultural, and/or biological variables. We also acknowledge the range of the types of variables included within each conceptualization reviewed within this chapter. Notably, various risk or maintenance factors are shared across multiple models (e.g., restraint, affect regulation, and self-oriented cognitions), and the more complex conceptualizations often integrate elements or processes from other models.

Keywords

Restraint · Escape · Avoidance · Emotion · Affect · Interpersonal · Cognitions · Body image · Perfectionism · Neurobiology · Addiction

Learning Objectives

In this chapter, you will:

- Become familiar with major theories of the onset and persistence of binge-eating behaviors.
- Understand the range of biopsychosocial variables identified in binge-eating models, including overlapping and unique elements across theories.
- Recognize clinical prevention and intervention approaches that are aligned with certain binge-eating conceptualizations.

M. K. H. Neyland · L. M. Shank · J. M. Lavender (✉)
Military Cardiovascular Outcomes Research Program (MiCOR), Department of Medicine, Uniformed Services University of the Health Sciences, Bethesda, MD, USA

Metis Foundation, San Antonio, TX, USA
e-mail: mary.neyland.ctr@usuhs.edu; lisa.shank.ctr@usuhs.edu; jason.lavender.ctr@usuhs.edu

1 Introduction

The aim of this chapter is to provide an overview of a number of conceptual models that have been proposed to explain the onset or maintenance of binge-eating behavior. It is worth noting the importance and function of theoretical models of binge eating within the literature. Namely, understanding factors that contribute to the onset of binge eating are crucial in informing the accurate identification of individuals at risk for developing binge eating, as well as providing useful insights into targets for prevention programs aimed at reducing the emergence of the behavior or eating pathology more broadly. Similarly, clarifying factors that contribute to the persistence of binge eating is critical with regard to informing appropriate and modifiable targets for clinical interventions among those who already exhibit the behavior. In addition to focused descriptions of each of these major conceptualizations of binge eating, secondary aims of this chapter are to demonstrate the breadth of these models, the types of variables that overlap to varying degrees across them, and the ways in which they have been applied in certain prevention and clinical interventions.

2 Dietary Restraint Theory

Dietary restraint is defined as the intention to restrict food intake to prevent weight gain and/or promote weight loss (Herman and Mack 1975; Herman and Polivy 1980; Tuschl 1990). Dietary restraint theory proposes that, among individuals who chronically exert cognitive control to regulate their food intake, eating behavior is no longer under primary regulation via physiological cues. In particular, by reducing reliance on such physiological cues to guide food intake and attempting to achieve or maintain weight below the "set point" of one's body, biological mechanisms may activate that increase the likelihood that the individual will engage in disinhibited eating behaviors, such as binge eating (Herman and Mack 1975; Herman and Polivy 1980; Polivy and Herman 1985; Tuschl 1990; Wardle 1987). Dietary restraint theory thus proposes a cycle in which dietary restraint promotes binge-eating behavior, which in turn exacerbates engagement in dietary restraint following loss of control over food intake during binge-eating episodes (Burton and Abbott 2017; Polivy and Herman 1985).

Evidence supporting this theory comes from laboratory-based studies demonstrating that individuals who are higher in dietary restraint engage in eating behavior that does not appear to be driven by physiological cues, but rather by other factors. For example, compared to those with low dietary restraint, individuals across the weight spectrum with high dietary restraint consume more food in response to external food cues (e.g., sight or smell of food) and after consumption of high-calorie preloads (Fedoroff et al. 1997, 2003). Restricting food intake has also been shown to increase food-related preoccupations, placing individuals at increased risk of engaging in overeating and binge eating (Coscina and Dixon 1983; Keys et al. 1950; Polivy and Herman 1985; Wardle 1987). Similarly, studies have shown that self-reported dieting predicts the onset of binge eating or certain eating disorders in community and clinical samples (e.g., Goldschmidt et al. 2012; Liechty and Lee 2013; Patton et al. 1999; Stice and Agras 1998; Stice et al. 2002, 2017). However, criticisms of the theory suggest that it is overly simplified (Burton and Abbott 2017; Schaumberg et al. 2016; Waller 2002). In particular, the majority of individuals who engage in dieting or participate in weight-loss treatments do not develop binge eating, and dieting does not always precede the development of binge-eating disorder. Findings suggest that certain subgroups of those who report dietary restraint and/or dieting, such as individuals who also report negative mood or engage in more severe weight control methods, may be at elevated risk of developing binge eating (Goldschmidt et al. 2012; Neumark-Sztainer et al. 1995; Stice and Agras 1998). However, the core original theory does not directly address biological or psychological vulnerabilities that would potentially account for this discrepancy across individuals (Masheb and Grilo 2000; Mussell et al. 1995; Patton et al. 1999; Wilson 1993).

With regard to clinical applications of this theory, minimizing dietary restraint and normalizing eating in response to physiological cues are key components of cognitive-behavioral interventions for eating disorders, including those defined by recurrent episodes of binge eating (e.g., bulimia nervosa, binge-eating disorder; Fairburn et al. 2003; Fairburn 2008). Notably, others have drawn attention to the idea that altered awareness of physiological hunger cues may complicate these traditional cognitive-behavioral strategies, and have thus suggested alternative clinical approaches to specifically address this concern, such as appetite awareness training and intuitive eating (Craighead and Allen 1995; Tylka and Wilcox 2006). Despite this, however, dietary restraint is a factor that is incorporated in numerous more complex models of binge eating, including several of those described in the remainder of this chapter (e.g., dual pathway model and transdiagnostic model).

3 Escape Theory

The core concept of escape theory suggests that binge-eating behavior functions to lower an individual's self-awareness related to the pressures, threats, long-range concerns, and lasting consequences of their experiences (Heatherton and Baumeister 1991). According to the theory, those who engage in binge eating have high expectations of themselves and are intensely aware of the perceived expectations of others. The demanding nature of these standards makes them difficult to achieve, and as a result, individuals become more self-aware of their perceived shortcomings and overly concerned with others' evaluations and opinions of them. In turn, this process promotes increased negative affect, particularly in the form of anxiety and depression. Binge eating is therefore conceptualized as a behavior that serves to narrow cognitive focus and provide an escape from negative affect associated with aversive self-awareness; over time binge eating is maintained through negative reinforcement processes, with the short-term experience of escaping self-awareness and related aversive affect promoting recurrent behavior, despite more long-term negative consequences.

The literature directly testing the multiple components of escape theory is somewhat modest. As described in detail in a recent review of models of binge eating (see Burton and Abbott 2017), certain studies have provided support for the model in both nonclinical and clinical samples of women (e.g., Blackburn et al. 2006; Engelberg et al. 2007). Further, in systematic reviews and meta-analyses of studies investigating certain relationships outlined within escape theory, evidence has supported negative affect as a triggering event for binge eating (Haedt-Matt and Keel 2011; Leehr et al. 2015); however, the extent to which binge eating is maintained through negative reinforcement processes remains unclear, likely due in part to differing methodologies and statistical approaches used in studies addressing this issue (Berg et al. 2017; Haedt-Matt and Keel 2011; Leehr et al. 2015). As is the case with dietary restraint, concepts and hypothesized associations similar to those found within escape theory have been adopted within several other models, including those focused on personality and affect-related processes (e.g., perfectionism, emotion regulation), as well as more clinically and treatment-oriented theories (e.g., integrative cognitive-affective therapy model, transdiagnostic model).

4 Perfectionism Model

Consistent with findings linking perfectionistic tendencies with eating pathology (see Bardone-Cone et al. 2007), the perfectionism model of binge-eating posits specific causal pathways through which perfectionism promotes binge-eating behavior (Sherry and Hall 2009). In its original formulation, the model was based on the conceptualization of perfectionism as described by Hewitt and Flett (1991), particularly emphasizing the socially-focused facets of perfectionism. Specifically, the model proposes that socially prescribed perfectionism, or the perception that others are demanding perfection of oneself, promotes vulnerability to binge eating by increasing interpersonal discrepancies and decreasing interpersonal esteem, with discrepancies also influencing esteem. Two pathways from socially prescribed perfectionism to binge eating are described: (1) increased interpersonal discrepancies promoting greater negative affect, which in turn increases the risk of binge eating, and (2) decreased interpersonal esteem promoting greater dietary restraint, which in turn increases the risk of binge eating. Empirical tests of the original model provided moderate support, although findings were mixed in some regards, including how negative affective states related to binge eating (binge-eating disorder findings differed for depressive versus anxious affect) and the consistency of a relationship between dietary restraint and binge eating (Sherry and Hall 2009; Sherry et al. 2014). A subsequent reformulation by Mackinnon et al. (2011) focused on the 'concern over mistakes' facet of perfectionism as conceptualized by Frost et al. (1990), which is a broader dimension reflecting the tendency to respond negatively to mistakes, view mistakes as indicative of failure, and anticipate loss of respect from others due to mistakes. Results generally supported this reformulated model, and suggest that perfectionistic themes beyond those that are purely socially-oriented are relevant to the occurrence of binge-eating behavior. Additional evidence from empirical tests of the model has also indicated a prospective relationship between perfectionism and binge eating, but not the reverse (Smith et al. 2017).

Taken together, there is some support for the perfectionism model of binge eating, although there remains a need for additional direct empirical tests, particularly with clinical samples. Of note, other perfectionism-based models of broader eating pathology have been proposed and have received empirical support, including an interactive model of bulimic pathology that focuses on the interactions of perfectionism, body dissatisfaction, and self-esteem in relation to bulimic symptoms (Bardone et al. 2000, 2006), Further, other broader models also incorporate perfectionism (e.g., transdiagnostic model) or related constructs (e.g., self-discrepancy in the integrative cognitive-affective therapy model) as core variables involved in either the onset or maintenance of binge eating or broader eating disorder syndromes.

5   Emotion/Affect Regulation Theory

Several early theories focused specifically on the functional role of binge eating in managing, moderating, and avoiding the experience of aversive emotions and affective states. The core concepts shared across these theories are: (1) aversive emotions, moods, or affective experiences prompt binge eating, (2) binge eating provides for a temporary relief from these aversive experiences, and is thus negatively reinforced, and (3) binge eating represents a maladaptive emotion regulation/coping strategy that occurs due in part to a lack of more adaptive skills. For example, Lacey (1986) conceptualized binge eating as serving to reduce awareness of negative affective states, as well as functioning as a maladaptive method of coping with stressful events. The theory posited a number of predisposing environmental, social, and psychological factors, as well as the concept that the onset of binge eating may be triggered by significant life events (Lacey 1986; Lacey et al. 1986). The maintenance of binge eating is viewed within this model as resulting from negative reinforcement, via the temporary reduction of negative emotional arousal or distraction from boredom. Other early binge-eating theories with an emotion/affect focus generally share these central concepts (Polivy and Herman 1993; Hawkins and Clement 1984), as do more recent conceptualizations based on models of emotion regulation drawn from the broader literature (Lavender et al. 2015; Oldershaw et al. 2015). Of note, emotion regulation conceptualizations represent one primary foundation for the application of dialectical behavior therapy in the treatment of eating disorders (Linehan 1993; Safer et al. 2001, 2009), and evidence suggests potential efficacy of this intervention, delivered in various formats, in reducing binge-eating behavior (Masson et al. 2013; Safer et al. 2001, 2010; Telch et al. 2001), although more research is necessary.

Studies utilizing experimental, naturalistic, and momentary research designs have supported the concept that aversive emotional/affective experiences impact binge eating. In particular, strong evidence supports negative affect as an antecedent to binge-eating behavior (Haedt-Matt and Keel 2011; Lavender et al. 2015; Leehr et al. 2015). However, findings regarding the extent to which negative affective experiences diminish following binge eating, consistent with the concept of negative reinforcement, have been comparatively mixed. Several studies using intensive longitudinal methods in the natural environment (e.g., ecological momentary assessment) have reported post-binge eating reductions in negative affect (e.g., Berg et al. 2015; Engel et al. 2013; Smyth et al. 2007); however, other research has not supported this finding (e.g., Hilbert and Tuschen-Caffier 2007), including a meta-analysis of momentary studies addressing the relationship between affect and binge eating (Haedt-Matt and Keel 2011). Discussions of this issue have addressed the timing of affective experiences (e.g., during, immediately after, or in a longer timeframe following the binge-eating episode) as one important consideration. For instance, in their functional assessment of binge eating, McManus and Waller (1995) suggested that affect may improve during binge eating, but subsequently worsen after. Berg et al. (2017) further addressed the issue of momentary data by noting the importance of considering the pattern of change in affect prior to and following binge eating. Specifically, if negative affect rises prior to binge eating and declines after, as suggested by the affect regulation model, it would be important to ensure that assessments of negative affect occur within a reasonably comparable timeframe pre- and post-binge eating.

In sum, emotion regulation theories have generally received support from many studies, although there are likely many nuances involving environmental factors, individual differences, and other salient momentary variables that may influence the experience of emotions around any episode of binge eating. Given strong empirical support for the model, however, it is unsurprising that it, along with dietary restraint, is one of the most commonly overlapping concepts across various binge-eating theories.

6 Schema Theory

Schemas, a collection of core beliefs reflecting deeply held views of the world, others, and oneself, are conceptualized as typically developing at a young age and forming a lens through which an individual views and interprets his or her experiences in the world (Young 1994). These beliefs may often be reaffirmed by the manner in which an individual processes his or her experiences and interprets interactions with others. Among individuals with eating disorders, it is posited that maladaptive schemas are common, such as those involving long-standing beliefs that one is inadequate and vulnerable to harm, or that abandonment by important others is inevitable (Waller et al. 2000; Pugh 2015). As such, schema theory proposes that when core beliefs arising from maladaptive schemas are activated within a given situation, negative affect is generated, which in turn increases the risk of engaging in binge eating to alleviate the aversive affective experience (Waller et al. 2000). Notably, this general theory serves as the foundation for the application of schema therapy in the treatment of eating disorders, for which there is preliminary support indicating efficacy in the treatment of binge eating and eating pathology more broadly (Waller et al. 2007; McIntosh et al. 2016; Pugh 2015), although additional research is needed.

Numerous studies have provided empirical support for the association of schemas and core beliefs with binge eating and eating pathology more generally. For example, among individuals with eating disorders, certain maladaptive schemas (e.g., emotional deprivation, perceived abandonment, and emotional inhibition) have been found to be associated with binge-purge behaviors (Unoka et al. 2010). Further, individuals with binge-eating disorder or bulimia nervosa have been found to report more pathological core beliefs compared to controls, and findings suggest that core beliefs are correlated with binge eating among those with binge-eating disorder (Burton and Abbott 2017; Waller 2003). Additional research has demonstrated that core beliefs about emotional inhibition (i.e., the propensity to inhibit and restrain emotional expression and communication) relate to the severity of binge eating in a clinical sample of women with bulimic symptoms (Waller et al. 2000). However, it should be noted that not all studies have consistently found support for such relationships (see Dingemans et al. 2006; Pugh 2015).

Taken together, despite moderate to strong evidence suggesting the relevance of schemas and core beliefs to binge-eating behavior, there remains a need for further direct empirical investigations of the momentary cognitive and affective processes suggested within the model. Additionally, this model notably shares elements in common with other theories, including the emotion regulation and escape theories, as well as models associated with specific treatments (e.g., integrative cognitive-affective therapy model, described below).

7   Interpersonal Theory

The interpersonal theory of binge eating derives from the broader interpersonal theory (Sullivan 1953), which proposes a crucial role of relationships between people that can either encourage self-esteem or, conversely, promote anxiety and psychopathology. The application of this theory to binge eating focuses on the concept that interpersonal stressors (e.g., conflict, loss and role transitions) can promote negative affect and low self-esteem, which in turn promote binge eating as a means of coping with these stressors (Wilfley et al. 2000). Over time, repeatedly engaging in binge eating as a means of coping with interpersonal stressors, rather than more adaptively addressing the concerns, may maintain the behavior and even exacerbate symptoms due to potentially worsened interpersonal distress. This theory is the basis for interpersonal psychotherapy adapted for the treatment of eating disorders, particularly binge-eating disorder and bulimia nervosa (Weissman et al. 2000; Wilfley et al. 1998, 2002), which focuses on improving relationships and interpersonal functioning and decreasing associated negative affect. Findings from a variety of studies have supported the efficacy of this interpersonally-focused intervention in the treatment of binge-eating-spectrum disorders (e.g., Brown and Keel 2012; Tanofsky-Kraff et al. 2010; Wilfley et al. 1998, 2002).

Support for interpersonal theory in relation to binge eating has been found in community and clinical samples, including in adults and youth, with findings also supporting a reciprocal relationship in which eating disorder symptoms exacerbate interpersonal problems, and vice versa (Ansell et al. 2012; Ivanova et al. 2015; Ranzenhofer et al. 2014). Further, a review focused on emotion regulation in binge-eating disorder found support for interpersonal problems preceding negative mood, which subsequently promoted binge eating (Dingemans et al. 2017). Notably, along with emotion regulation difficulties, interpersonal difficulties were incorporated as an additional potential maintenance mechanism in the transdiagnostic cognitive-behavioral model of eating pathology (Fairburn et al. 2003).

8   Neurocognitive/
Neurobiological Theories

Consistent with behaviorally and psychologically oriented research on binge eating suggesting the salience of certain temperament and personality constructs, researchers have addressed the role of neurocognitive factors in relation to binge eating (Schag et al. 2013; Wu et al. 2013, 2014, 2016). Among the neurocognitive factors that have been most widely considered are executive functioning (e.g., inhibitory control and set-shifting) and reward processing (Giel et al. 2017; Wu et al. 2014, 2016). Within the executive functioning domain, these conceptualizations generally posit that disturbances in cognitive and/or behavioral control processes may increase the risk of loss of control over eating behavior. For instance, difficulties with inhibiting learned responses might increase risk (e.g., losing control in the presence of commonly consumed palatable foods), as could the inability to effectively shift one's attention or cognitive focus away from food-salient content. Such theories are supported by research demonstrating that individuals with binge-eating disorder tend to show deficits in a variety of executive functioning domains (Giel et al. 2017; Lavagnino et al. 2016). Notably, this theoretical and empirical literature serves as the rationale for recent efforts to develop neurocognitive training interventions for binge eating, such as those focused on improving inhibitory and/or attentional control (see Eichen et al. 2017).

Within the reward processing domain, it has been posited that binge eating is related to a number of disturbances such as reward-related decision-making and sensitivity to rewards (especially food-related; e.g., Wu et al. 2016; Stojek and MacKillop 2017). For instance, individuals who are more prone to experience food intake (particularly of highly palatable foods) as rewarding and pleasurable may be at greater risk of developing binge-eating behaviors. Similarly, individuals who are prone to decision-making that overvalues immediate rewards (e.g., eating high-calorie foods) and discounts later potentially negative outcomes (e.g., excess weight gain and associated health consequences) may be more at risk. Taken together, conceptualizations regarding the role of neurocognitive factors in binge-eating behavior point to a number of potential vulnerabilities, although the extent to which these vulnerabilities are more general or specific to eating- and food-related variables remains a source of continuing debate and investigation (see Berner et al. 2017).

Although a review of neurobiological factors implicated in binge-eating behavior is beyond the scope of this chapter, it is notable that much of the research within this area has focused on identifying alterations in neural circuitry underlying many of the processes included in other models described within this chapter (see Balodis et al. 2015). As a representative example neurobiological framework, a recently proposed neurobiological model of binge-eating disorder reflects an effort to synthesize findings of the neural bases of the disorder (Kessler et al. 2016). Consistent with neurocognitive findings and the conceptualizations briefly outlined above, brain regions and neural circuits underlying executive functioning (e.g., prefrontal cortex) and reward-related processes (e.g., ventral striatum) are highlighted within Kessler et al.'s model, as are other regions and circuits involved in processes described within other models reviewed in this chapter, such as affect/emotion (e.g., amygdala). The model focuses on corticostriatal alterations accompanying a shift from typical goal-directed eating behavior to dysregulated food intake of an initially impulsive and eventually compulsive nature. Specifically, paralleling neurobiological theories for substance use disorders (e.g., Everitt and Robbins 2005), this framework posits a pattern of repeated binge eating that is associated with reward-related food intake shifting from being driven by a ventral striatal-dominant reward-based process to a dorsal striatal impulsive–compulsive process, in conjunction with reduced cortical inhibition and other striatal changes. Associated with this shift is a pattern of enhanced attention to food cues (i.e., attention bias), greater experience of wanting and liking, and decreased reward sensitivity, all of which may further promote the persistence of binge eating behavior. Notably, theory and empirical evidence regarding the neural bases of binge-eating disorder and eating pathology more broadly serves as the major foundation for novel treatment approaches including neuromodulation and neurofeedback (see Dalton et al. 2017).

9 The Dual-Pathway Model

The dual-pathway model was proposed in order to account for influences of sociocultural factors, dieting, and negative affect on the onset of bulimic symptoms, including binge eating (Stice and Agras 1998; Stice et al. 1996). This model proposes that sociocultural pressures to be thin and internalization of this thin body ideal both promote body dissatisfaction, which in turn promotes dieting and negative affect, which then represent dual pathways to the development of bulimic symptoms (binge eating in response to caloric deprivation or as maladaptive affect regulation in response to negative affect). Findings from a substantial body of research have largely supported the dual-pathway model either partially or fully within community and clinical samples, as well as in youth and adults (Holmes et al. 2015; Shepherd and Ricciardelli 1998; Stice 2001; Stice and Agras 1998); however, not all studies have found consistent support for the model, and relative to bulimia nervosa, the support for the model's application to binge-eating disorder is comparatively mixed (Burton and Abbott 2017; Stice et al. 2011, 2017; Van Strien et al. 2005).

Consistent with the greater complexity of the syndrome of bulimia nervosa compared to the individual behavior of binge eating, the dual-pathway model reflects a synthesis of factors derived from models of binge eating and eating pathology generally that address sociocultural factors, dietary restraint, and affect regulation (Stice et al. 1996; Stice 2001). The model also serves as the foundation for dissonance-based eating disorder prevention (i.e., the *Body Project*). Dissonance-based programs incorporate a variety of exercises focused on critiquing the sociocultural thin ideal to increase cognitive dissonance and motivate participants to reduce internalization and pursuit of the thin ideal; consistent with the dual-pathway model, this reduction is proposed to decrease body dissatisfaction, dieting, negative affect, and eating disorder symptoms (see Stice et al. 2019).

At the core of the transdiagnostic model is a dysfunctional scheme for evaluating the self, including overvaluation of eating, weight, and shape, as well as a tendency for perfectionism, each of which is conceptualized as also being impacted by low self-esteem (Fairburn et al. 2003). These cognitive vulnerability factors are viewed as promoting extreme and maladaptive weight control behaviors, which in turn may promote excessive weight loss and/or a cycle of binge eating and compensatory behaviors. Of note, elements that were added to this expanded model, building upon the original cognitive-behavioral framework, include factors from several of the models discussed within this chapter: perfectionism, mood intolerance (e.g., emotion regulation), and interpersonal difficulties. Mood intolerance, in particular, is conceptualized within the model as being most directly linked with binge eating and compensatory behaviors, consistent with affect/ emotion regulation conceptualizations. Further, interpersonal difficulties are viewed as having potentially direct impacts on binge eating, with interpersonal events recognized as a common trigger for binge-eating episodes, although the model also views difficulties in this domain as having broader impacts on other elements within the complex maintenance framework.

10 Transdiagnostic Model

The transdiagnostic model of eating disorders represents an expanded conceptualization based on the original cognitive-behavioral theory of bulimia nervosa (Fairburn et al. 1993, 2003). While not specific to binge eating, this maintenance model includes binge eating as a core transdiagnostic behavior, and arguably represents the most well-known theory of eating disorders within the field. Further, it serves as the model underlying the application of enhanced cognitive behavioral therapy for eating disorders (Fairburn 2008; Fairburn et al. 2003, 2009). Therefore, an abbreviated overview of this theory, with a focus on components of particular relevance to binge eating as proposed within the model, is provided here.

11 Integrative Cognitive-Affective Therapy Models

The core treatment targets within Integrative Cognitive-Affective Therapy (ICAT), a more recently developed psychotherapy targeting bulimia nervosa, are based upon two related but distinct models (Wonderlich et al. 2014, 2015). The first focuses on the onset of bulimia nervosa, and incorporates a number of traits/individual-level vulnerability factors, whereas the second focuses on state-oriented factors posited to contribute to momentary occurrence of bulimic symptoms and ultimately to the maintenance of the disorder. While other models of binge eating and eating pathology more generally do address momentary processes (e.g., binge eating serving to regulate aversive affect, interpersonal experiences as triggers of binge eating), the explicit and detailed articulation of state-oriented processes and mechanisms prompting bulimic symptoms as outlined in the ICAT momentary maintenance model is noteworthy.

The ICAT onset model posits a set of vulnerability factors for bulimia nervosa including temperament, historical negative interpersonal experiences, interpersonal problems, self-evaluative concerns (i.e., high self-discrepancy), and self-regulation difficulties. These factors are viewed as contributing to greater negative emotional intensity and poorer emotion regulation, which in conjunction with thin-ideal internalization and expectancies that dieting and binge eating will reduce negative emotions, promote the development of bulimic symptoms. The more focused ICAT momentary maintenance model addresses factors of relevance to momentary bulimic behaviors. This model highlights common triggering situations (i.e., interpersonal situations, eating-related situations, and momentary experiences of self-discrepancy, self-criticism, or self-neglect) that prompt aversive affective experiences. It is posited that momentary expectations that bulimic behavior will reduce the aversive emotions are activated, which then results in greater risk of engaging in bulimic behaviors. Notably, similar to the transdiagnostic model, the ICAT onset and momentary maintenance models reflect an integration of risk and maintenance factors represented in several of the models outlined within this chapter. Included among these overlapping variables and processes are the desire to reduce or escape aversive affect, affect/emotion dysregulation, interpersonal difficulties, thin-ideal internalization, and self-discrepancy and self-criticism.

12 Food Addiction

Although not specifically a theory of binge eating, the concept of food addiction has received modest but sustained attention within the empirical literatures on eating disorders and obesity (e.g., Gearhardt et al. 2011a; Treasure et al. 2018).

Drawing from the broader addiction literature, the food addiction model focuses on the idea that particular foods, especially those that are highly palatable and highly caloric, may have an addictive quality that promotes forms of dysregulated eating such as binge eating as addictive behaviors (Gearhardt et al. 2011a, b). This application of addiction theory to eating behavior is based in part on theoretical and empirical literature indicating that that certain high-calorie palatable foods act on brain reward pathways similarly to addictive drugs (e.g., Gordon et al. 2018; Treasure et al. 2018; Volkow et al. 2017).

Proponents of the theory point to findings from studies utilizing animal models, such as evidence of withdrawal symptoms after opioid antagonist administration following a period of repeated excessive sugar intake (Colantuoni et al. 2002), as well as evidence that animals display addictive-like behaviors and neural changes in reward systems in response to binge-like intake of high sugar and high-fat foods (Avena et al. 2009; Avena and Bocarsly 2012; Treasure et al. 2018). Further, proponents also note the similarities between binge-eating disorder and substance dependence, such as loss of control over consumption, difficulty reducing amounts consumed, continued consumption despite negative emotional and physical consequences, and similar triggers leading to consumption (e.g., low mood and cravings). A systematic review of studies utilizing the Yale Food Addiction Scale (Gearhardt et al. 2009) also found that high scores on the measure were related to higher body mass index and greater likelihood of an eating disorder diagnosis, especially binge-eating disorder (Penzenstadler et al. 2019).

In contrast, several criticisms have been raised about food addiction, including the limited body of rigorous research among humans (Albayrak et al. 2012), and the concept that one cannot be addicted to food as a substance, but rather to the behavior of eating (Hebebrand et al. 2014), similar to other behavioral addictions (e.g., gambling disorder). Other authors argue against the concept of addictive elements within eating behaviors due to important differences between substance abuse disorders and eating disorders (e.g., course of illness and treatment outcomes; Barbarich-Marsteller et al. 2011; Wilson 2010). Further, it has been noted that conceptualizing binge-eating disorder as an addiction may overlook important aspects of the disorder (e.g., weight/shape overvaluation) that may be important factors in binge eating onset and maintenance (Wilson 2010). As such, food addiction theory remains somewhat contentious, particularly within the eating disorder field, and additional rigorous human research is needed.

13   Conclusion

The goal of this chapter was to provide an overview of major theories that have been proposed to explain the onset and persistence of binge eating. In addition to the numerous overlapping types of variables shared across many of the models (see Table 1), there also are a number of common limitations and considerations across the theories that are important to note with regard to understanding the development and maintenance of binge eating. The first of these considerations focuses on the importance of the overeating versus loss of control dimensions of binge eating, which is particularly relevant when considering age as a factor, given that much of the literature on youth tends to focus on loss of control eating.

Second, although binge eating is a defining behavior within binge-eating disorder and bulimia nervosa, it also occurs among those with anorexia nervosa and in other populations without eating disorders (e.g., individuals with obesity). As such, the behavior is present across the weight spectrum, and there may be both overlapping and distinct factors that play a role in binge eating across individuals with different body weights, across eating disorder groups, and within various other subpopulations. Third, eating disorder theories were historically based on the typical presentation most characteristic of White females. Although there has been increased recognition of and attention to eating disorder symptoms, and particularly binge eating, in other populations such as ethnic minorities and males, further conceptual and empirical work accounting for diverse demographics, identities, backgrounds, and contexts is warranted.

In sum, the present chapter highlights major theories of binge eating, with models focusing either on the individual behavior, the behavior in conjunction with other disordered eating symptoms (e.g., compensatory behaviors), or as a broader eating disorder syndrome. As evident across the overviews of the theories, these models incorporate a variety of variables across different biopsychosocial domains (see Table 1), thus reflecting the complexity of the processes

Table 1 Core variable types included across models/theories

Theory/model	Cognitive	Affective	Interpersonal	Sociocultural	Biological
Dietary restraint	++				++
Escape	++	++	+	+	
Perfectionism	++	++	++	+	
Emotion/affect regulation	+	++	+		
Schema	++	++	++		
Interpersonal	+	++	++	+	
Neurocognitive/neurobiological	++	+			++
Dual pathway	+	++		++	+
Transdiagnostic	++	+	+	+	
ICAT	++	++	++	++	
Food addiction	+	+			++

Note: ++ indicates the type of variable is core to or explicitly considered within the model; + indicates the type of variable is has a more ancillary role or is implied within the model. The lack of a marker does not necessarily indicate a total absence of the type of variable within a specific model, but may reflect a less explicit role involved in the onset and maintenance of binge eating. Finally, it is notable that original, more focused theories have been integrated within recent, more complex theoretical and treatment models, further demonstrating the importance of an integrative perspective in best understanding the etiology and persistence of binge eating in both conceptual and applied clinical endeavors.

Disclosure

The opinions and assertions expressed herein are those of the authors and do not necessarily reflect the official policy or position of the Uniformed Services University or the Department of Defense.

References

Albayrak O, Wolfle SM, Hebebrand J (2012) Does food addiction exist? A phenomenological discussion based on the psychiatric classification of substance-related disorders and addiction. Obes Facts 5:165–179

Ansell EB, Grilo CM, White MA (2012) Examining the interpersonal model of binge eating and loss of control over eating in women. Int J Eat Disord 45:43–50

Avena NM, Bocarsly ME (2012) Dysregulation of brain reward systems in eating disorders: neurochemical information from animal models of binge eating, bulimia nervosa, and anorexia nervosa. Neuropharmacology 63:87–96

Avena NM, Rada P, Hoebel BG (2009) Sugar and fat bingeing have notable differences in addictive-like behavior. J Nutr 139:623–628

Balodis IM, Grilo CM, Potenza MN (2015) Neurobiological features of binge-eating disorder. CNS Spectr 2:557–565

Barbarich-Marsteller NC, Foltin RW, Walsh BT (2011) Does anorexia nervosa resemble an addiction? Curr Drug Abuse Rev 4:197–200

Bardone AM, Vohs KD, Abramson LY et al (2000) The confluence of perfectionism, body dissatisfaction, and low self-esteem predicts bulimic symptoms. Behav Ther 31:265–280

Bardone-Cone AM, Abramson LY, Vohs KD et al (2006) Predicting bulimic symptoms: an interactive model of self-efficacy, perfectionism, and perceived weight status. Behav Res Ther 44:27–42

Bardone-Cone AM, Wonderlich SA, Frost RO et al (2007) Perfectionism and eating disorders: current status and future directions. Clin Psychol Rev 27:384–405

Berg KC, Crosby RD, Cao L et al (2015) Negative affect prior to and following overeating-only, loss of control eating-only, and binge eating episodes in obese adults. Int J Eat Disord 48:641–653

Berg KC, Cao L, Crosby RD et al (2017) Negative affect and binge eating: reconciling differences between two analytic approaches in ecological momentary assessment research. Int J Eat Disord 50:1222–1230

Berner LA, Winter SR, Matheson BE et al (2017) Behind binge eating: a review of food-specific adaptations of neurocognitive and neuroimaging tasks. Physiol Behav 176:59–70

Blackburn S, Johnston L, Blampied N et al (2006) An application of escape theory to binge eating. Eur Eat Disord Rev 14:23–31

Brown TA, Keel PK (2012) Current and emerging directions in the treatment of eating disorders. Subst Abus 6:33–61

Burton AL, Abbott MJ (2017) Conceptualising binge eating: a review of the theoretical and empirical literature. Behav Chang 34:168–198

Colantuoni C, Rada P, McCarthy J et al (2002) Evidence that intermittent, excessive sugar intake causes endogenous opioid dependence. Obes Res 10:478–488

Coscina DV, Dixon LM (1983) Body weight regulation in anorexia nervosa: insights from an animal model. In: Darby P, Garfinkel P, Garner D, Coscina D (eds) Anorexia nervosa: recent developments. Allan R. Liss, New York, pp 207–220

Craighead LW, Allen HN (1995) Appetite awareness training: a cognitive behavioral intervention for binge eating. Cogn Behav Pract 2:249–270

Dalton B, Campbell IC, Schmidt U (2017) Neuromodulation and neurofeedback treatments in eating disorders and obesity. Curr Opin Psychiatry 30:458–473

Dingemans A, Spinhoven P, van Furth EF (2006) Maladaptive core beliefs and eating disorder symptoms. Eat Behav 7:258–265

Dingemans A, Danner U, Parks M (2017) Emotion regulation in binge-eating disorder: a review. Nutrients 9:1274

Eichen DM, Matheson BE, Appleton-Knapp SL et al (2017) Neurocognitive treatments for eating disorders and obesity. Curr Psychiatry Rep 19:62

Engel SG, Wonderlich SA, Crosby RD et al (2013) The role of affect in the maintenance of anorexia nervosa: evidence from a naturalistic assessment of momentary behaviors and emotion. J Abnorm Psychol 122:709–719

Engelberg MJ, Steiger H, Gauvin L et al (2007) Binge antecedents in bulimic syndromes: an examination of dissociation and negative affect. Int J Eat Disord 40:531–536

Everitt BJ, Robbins TW (2005) Neural systems of reinforcement for drug addiction: from actions to habits to compulsion. Nat Neurosci 8:1481–1489

Fairburn CG (2008) Cognitive behavior therapy and eating disorders. Guilford, New York

Fairburn CG, Marcus MD, Wilson GT (1993) Cognitive-behavioral therapy for binge eating and bulimia nervosa: a comprehensive treatment manual. In: Fairburn CG, Wilson GT (eds) Binge eating: nature, assessment and treatment. Guilford Press, New York, pp 361–404

Fairburn CG, Cooper Z, Shafran R (2003) Cognitive behaviour therapy for eating disorders: a "transdiagnostic" theory and treatment. Behav Res Ther 41:509–528

Fairburn C, Cooper Z, Doll H et al (2009) Transdiagnostic cognitive-behavioral therapy for patients with eating disorders. Am J Psychiatry 166:311–319

Fedoroff ID, Polivy J, Herman CP (1997) The effect of pre-exposure to food cues on the eating behavior of restrained and unrestrained eaters. Appetite 28:33–47

Fedoroff ID, Polivy J, Herman CP (2003) The specificity of restrained versus unrestrained eaters' responses to food cues: general desire to eat, or craving for the cued food? Appetite 41:7–13

Frost R, Marten P, Lahart C et al (1990) The dimensions of perfectionism. Cogn Ther Res 14:449–468

Gearhardt AN, Corbin WR, Brownell KD (2009) Preliminary validation of the Yale food addiction scale. Appetite 52:430–436

Gearhardt AN, White MA, Potenza MN (2011a) Binge-eating disorder and food addiction. Curr Drug Abuse Rev 4:201–207

Gearhardt AN, Yokum S, Orr PT et al (2011b) Neural correlates of food addiction. Arch Gen Psychiatry 68:808–816

Giel KE, Teufel M, Junne F et al (2017) Food-related impulsivity in obesity and binge-eating disorder-a systematic update of the evidence. Nutrients 9(11): E1170

Goldschmidt AB, Wall M, Loth K et al (2012) Which dieters are at risk for the onset of binge eating? A prospective study of adolescents and young adults. J Adolesc Health 51:86–92

Gordon E, Ariel-Donges A, Bauman V et al (2018) What is the evidence for "food addiction?" a systematic review. Nutrients 10:477

Haedt-Matt AA, Keel PK (2011) Revisiting the affect regulation model of binge eating: a meta-analysis of studies using ecological momentary assessment. Psychol Bull 137:660–681

Hawkins RCH, Clement PF (1984) Binge eating: measurement problems and a conceptual model. In: Hawkins RCH, Fremouw WJ, Clement PF (eds) The binge purge syndrome: diagnosis, treatment, and research. Springer, New York, pp 229–251

Heatherton TF, Baumeister RF (1991) Binge eating as escape from self-awareness. Psychol Bull 110:86–108

Hebebrand J, Albayrak O, Adan R et al (2014) "Eating addiction", rather than "food addiction", better captures addictive-like eating behavior. Neurosci Biobehav Rev 47:295–306

Herman CP, Mack D (1975) Restrained and unrestrained eating. J Pers 43:647–660

Herman CP, Polivy J (1980) Restrained eating. In: Stunkard JA (ed) Obesity. W.B. Saunders Company, Philadelphia, PA, pp 208–225

Hewitt PL, Flett GL (1991) Perfectionism in the self and social contexts: conceptualization, assessment, and association with psycho-pathology. J Pers Soc Psychol 60:456–470

Hilbert A, Tuschen-Caffier B (2007) Maintenance of binge eating through negative mood: a naturalistic comparison of binge-eating disorder and bulimia nervosa. Int J Eat Disord 40:521–530

Holmes M, Fuller-Tyszkiewicz M, Skouteris H et al (2015) Understanding the link between body image and binge eating: a model comparison approach. Eat Weight Disord 20:81–89

Ivanova IV, Tasca GA, Hammond N (2015) Negative affect mediates the relationship between interpersonal problems and binge-eating disorder symptoms and psychopathology in a clinical sample: a test of the interpersonal model. Eur Eat Disord Rev 23:133–138

Kessler RM, Hutson PH, Herman BK et al (2016) The neurobiological basis of binge-eating disorder. Neurosci Biobehav Rev 63:223–238

Keys A, Brozek J, Henschel A et al (1950) The biology of human starvation, vol 2. University of Minnesota Press, Minneapolis

Lacey J (1986) Pathogenesis. In: Downey LJ, Malkin JC (eds) Current approaches: bulimia nervosa. Duphar, Southhampton, UK, pp 17–26

Lacey JH, Coker S, Birtchnell SA (1986) Bulimia: factors associated with its etiology and maintenance. Int J Eat Disord 5:475–487

Lavagnino L, Arnone D, Cao B et al (2016) Inhibitory control in obesity and binge-eating disorder: a systematic review and meta-analysis of neurocognitive and neuroimaging studies. Neurosci Biobehav Rev 68:714–726

Lavender JM, Wonderlich SA, Engel SG et al (2015) Dimensions of emotion dysregulation in anorexia nervosa and bulimia nervosa: a conceptual review of the empirical literature. Clin Psychol Rev 40:111–122

Leehr EJ, Krohmer K, Schag K et al (2015) Emotion regulation model in binge-eating disorder and obesity – a systematic review. Neurosci Biobehav Rev 49:125–134

Liechty JM, Lee MJ (2013) Longitudinal predictors of dieting and disordered eating among young adults in the U.S. Int J Eat Disord 46:790–800

Linehan MM (1993) Cognitive-behavioral treatment of borderline personality disorder. Guilford Press, New York

Mackinnon SP, Sherry SB, Graham AR et al (2011) Reformulating and testing the perfectionism model of binge eating among undergraduate women: a short-term, three-wave longitudinal study. J Couns Psychol 58:630–646

Masheb RM, Grilo CM (2000) Onset of dieting vs binge eating in outpatients with binge-eating disorder. Int J Obes 24:404–409

Masson PC, von Ranson KM, Wallace LM et al (2013) A randomized wait-list controlled pilot study of dialectical behaviour therapy guided self-help for binge-eating disorder. Behav Res Ther 51:723–728

McIntosh VVW, Jordan J, Carter JD et al (2016) Psychotherapy for transdiagnostic binge eating: a randomized controlled trial of cognitive-behavioural therapy, appetite-focused cognitive-behavioural therapy, and schema therapy. Psychiatry Res 240:412–420

McManus F, Waller G (1995) A functional analysis of binge-eating. Clin Psychol Rev 15:845–863

Mussell MP, Mitchell JE, Weller CL et al (1995) Onset of binge eating, dieting, obesity, and mood disorders among subjects seeking treatment for binge-eating disorder. Int J Eat Disord 17:395–401

Neumark-Sztainer D, Butler R, Palti H (1995) Dieting and binge eating. J Am Diet Assoc 95:586–589

Oldershaw A, Lavender T, Sallis H et al (2015) Emotion generation and regulation in anorexia nervosa: a systematic review and meta-analysis of self-report data. Clin Psychol Rev 39:83–95

Patton GC, Selzer R, Coffey C et al (1999) Onset of adolescent eating disorders: population based cohort study over 3 years. BMJ 318(7186):765–768

Penzenstadler L, Soares C, Karila L et al (2019) Systematic review of food addiction as measured with the Yale food addiction scale: implications for the food addiction construct. Curr Neuropharmacol 17:526–538

Polivy J, Herman CP (1985) Dieting and binging: a causal analysis. Am Psychol 40:193–201

Polivy J, Herman CP (1993) Etiology of binge eating: psychological mechanisms. In: Fairburn CG (ed) Binge eating: nature, assessment and treatment. Guilford Press, New York, pp 173–205

Pugh M (2015) A narrative review of schemas and schema therapy outcomes in the eating disorders. Clin Psychol Rev 39:30–41

Ranzenhofer LM, Engel SG, Crosby RD et al (2014) Using ecological momentary assessment to examine interpersonal and affective predictors of loss of control eating in adolescent girls. Int J Eat Disord 47:748–757

Safer DL, Telch CF, Agras WS (2001) Dialectical behavior therapy for bulimia nervosa. Am J Psychiatry 158:632–634

Safer DL, Telch CF, Chen EY (2009) Dialectical behavior therapy for binge-eating disorder. Guilford Press, New York

Safer DL, Robinson AH, Jo B (2010) Outcome from a randomized controlled trial of group therapy for binge-eating disorder: comparing dialectical behavior therapy adapted for binge eating to an active comparison group therapy. Behav Ther 41:106–120

Schag K, Schönleber J, Teufel M et al (2013) Food-related impulsivity in obesity and binge-eating disorder – a systematic review. Obes Rev 14:477–495

Schaumberg K, Anderson DA, Anderson LM (2016) Dietary restraint: what's the harm? A review of the relationship between dietary restraint, weight trajectory and the development of eating pathology. Clin Obes 6:89–100

Shepherd H, Ricciardelli LA (1998) Test of Stice's dual pathway model: dietary restraint and negative affect as mediators of bulimic behavior. Behav Res Ther 36:345–352

Sherry SB, Hall PA (2009) The perfectionism model of binge eating: tests of an integrative model. J Pers Soc Psychol 96:690–709

Sherry SB, Sabourin BC, Hall PA et al (2014) The perfectionism model of binge eating: testing unique contributions, mediating mechanisms, and cross-cultural similarities using a daily diary methodology. Psychol Addict Behav 28:1230–1239

Smith MM, Sherry SB, Gautreau CM et al (2017) Are perfectionistic concerns an antecedent of or a consequence of binge eating, or both? A short-term four-wave longitudinal study of undergraduate women. Eat Behav 26:23–26

Smyth JM, Wonderlich SA, Heron KE et al (2007) Daily and momentary mood and stress are associated with binge eating and vomiting in bulimia nervosa patients in the natural environment. J Consult Clin Psychol 75:629–638

Stice E (2001) A prospective test of the dual-pathway model of bulimic pathology: mediating effects of dieting and negative affect. J Abnorm Psychol 110:124–135

Stice E, Agras WS (1998) Predicting onset and cessation of bulimic behaviors during adolescence: a longitudinal grouping analysis. Behav Ther 29:257–276

Stice E, Nemeroff C, Shaw H (1996) A test of the dual pathway model of bulimia nervosa: evidence for restrained-eating and affect-regulation mechanisms. J Soc Clin Psychol 15:340–363

Stice E, Presnell K, Spangler D (2002) Risk factors for binge eating onset in adolescent girls: a 2-year prospective investigation. Health Psychol 21:131–138

Stice E, Marti CN, Durant S (2011) Risk factors for onset of eating disorders: evidence of multiple risk pathways from an 8-year prospective study. Behav Res Ther 49:622–627

Stice E, Gau JM, Rohde P et al (2017) Risk factors that predict future onset of each DSM-5 eating disorder: predictive specificity in high-risk adolescent females. J Abnorm Psychol 126:38–51

Stice E, Marti CN, Shaw H, Rohde P (2019) Meta-analytic review of dissonance-based eating disorder prevention programs: intervention, participant, and facilitator features that predict larger effects. Clin Psychol Rev 70:91–107

Stojek MMK, MacKillop J (2017) Relative reinforcing value of food and delayed reward discounting in obesity and disordered eating: a systematic review. Clin Psychol Rev 55:1–11

Sullivan HS (1953) The interpersonal theory of psychiatry. Norton, New York

Tanofsky-Kraff M, Wilfley DE, Young JF (2010) A pilot study of interpersonal psychotherapy for preventing excess weight gain in adolescent girls at-risk for obesity. Int J Eat Disord 43:701–706

Telch CF, Agras WS, Linehan MM (2001) Dialectical behavior therapy for binge-eating disorder. J Consult Clin Psychol 69:1061–1065

Treasure J, Leslie M, Chami R (2018) Are trans diagnostic models of eating disorders fit for purpose? A consideration of the evidence for food addiction. Eur Eat Disord Rev 26:83–91

Tuschl R (1990) From dietary restraint to binge eating: some theoretical considerations. Appetite 14:105–109

Tylka TL, Wilcox JA (2006) Are intuitive eating and eating disorder symptomatology opposite poles of the same construct? J Couns Psychol 53:474–485

Unoka Z, Tölgyes T, Czobor P et al (2010) Eating disorder behavior and early maladaptive schemas in subgroups of eating disorders. J Nerv Ment Dis 198:425–431

Van Strien T, Engels RC, Van Leeuwe J et al (2005) The Stice model of overeating: tests in clinical and non-clinical samples. Appetite 45:205–213

Volkow ND, Wise RA, Baler R (2017) The dopamine motive system: implications for drug and food addiction. Nat Rev Neurosci 18:741–752

Waller G (2002) The psychology of binge eating. In: Fairburn CG, Brownell KD (eds) Eating disorders and obesity: a comprehensive handbook. Guilford Press, New York, pp 98–102

Waller G (2003) Schema-level cognitions in patients with binge-eating disorder: a case control study. Int J Eat Disord 33:458–464

Waller G, Ohanian V, Meyer C et al (2000) Cognitive content among bulimic women: the role of core beliefs. Int J Eat Disord 28:235–241

Waller G, Kennerley H, Ohanian V (2007) Schema-focused cognitive-behavioral therapy for eating disorders. In: Riso LP, de Toit PL, Stein DJ, Young JE (eds) Cognitive schemas and core beliefs in psychological problems: a scientist-practitioner guide. American Psychological Association, Washington, DC, pp 139–175

Wardle J (1987) Compulsive eating and dietary restraint. Br J Clin Psychol 26:41–55

Weissman MM, Markowitz J, Klerman GL (2000) Comprehensive guide to interpersonal psychotherapy. Basic Behavioral Science Books, New York

Wilfley DE, Frank M, Welch R et al (1998) Adapting interpersonal psychotherapy to a group format (IPT-G) for binge-eating disorder: toward a model for adapting empirically supported treatments. Psychother Res 8(4):379–391

Wilfley DE, MacKenzie KR, Welch RR et al (2000) Interpersonal psychotherapy for group. Basic Books, New York

Wilfley DE, Welch RR, Stein RI et al (2002) A randomized comparison of group cognitive-behavioral therapy and group interpersonal psychotherapy for the treatment of overweight individuals with binge-eating disorder. Arch Gen Psychiatry 59:713–721

Wilson GT (1993) Relation of dieting and voluntary weight loss to psychological functioning and binge eating. Ann Intern Med 119:727–730

Wilson GT (2010) Eating disorders, obesity and addiction. Eur Eat Disord Rev 18:341–351

Wonderlich SA, Peterson CB, Crosby RD et al (2014) A randomized controlled comparison of integrative cognitive-affective therapy (ICAT) and enhanced cognitive-behavioral therapy (CBT-E) for bulimia nervosa. Psychol Med 44:543–553

Wonderlich SA, Peterson CB, Smith TL et al (2015) Integrative cognitive-affective therapy for bulimia nervosa: a treatment manual. Guilford Press, New York

Wu M, Giel KE, Skunde M et al (2013) Inhibitory control and decision making under risk in bulimia nervosa and binge-eating disorder. Int J Eat Disord 46:721–728

Wu M, Brockmeyer T, Hartmann M et al (2014) Set-shifting ability across the spectrum of eating disorders and in overweight and obesity: a systematic review and meta-analysis. Psychol Med 44:3365–3385

Wu M, Brockmeyer T, Hartmann M et al (2016) Reward-related decision making in eating and weight disorders: a systematic review and meta-analysis of the evidence from neuropsychological studies. Neurosci Biobehav Rev 61:177–196

Young JE (1994) Cognitive therapy for personality disorders: a schema-focused approach, 2nd edn. Professional Resource Press, Sarasota, FL

Part II

Preclinical Studies

Preclinical Models of Stress and Environmental Influences on Binge Eating

Maria Vittoria Micioni Di Bonaventura, Emanuela Micioni Di Bonaventura, Carlo Polidori, and Carlo Cifani

Abstract

Preclinical models cannot explain all of the complex internal and external factors that influence eating behaviors in humans. Still, they represent an essential tool to investigate the underlying neuro- and psychobiology implicated in disorders that are associated with binge eating. Several environmental conditions induce aberrant feeding behavior on calorie-dense food in animal models of binge eating. Various kinds of stress (acute or chronic), the combination of repeated cycles of food restriction and refeeding plus stress, food deprivation, and limited access to palatable food have been used to elicit binge-like eating episodes to model human behaviors. Animal studies have revealed the involvement of different neurotransmitter pathways, especially dopamine, opioids, CRF, serotonin, orexin, and GABAergic systems in binge-like eating. They may aid in the ultimate goal of identifying novel, safe, and effective therapeutic targets.

Keywords

Binge-eating disorder · Palatable food · Stress · Environment · Food restriction · Food addiction

Learning Objectives

After reading this chapter, you will:

- Understand how stress in animal models can induce binge-like eating behavior.
- Be able to identify the main environmental factors that influence binge-like eating behavior.
- Understand the neurotransmitter systems implicated in binge-like eating behavior.

1 Introduction: Stress as an Inducer of Binge-Eating Behavior

Binge-eating disorder (BED) was formally recognized as a distinct eating disorder in the *Diagnostic and Statistical Manual of Mental Disorders*, 5th Edition (DSM-5) (APA 2013). BED disproportionately affects females compared to males (Swanson et al. 2011). Binge-eating episodes are also a diagnostic criterion in bulimia nervosa (Waters et al. 2001) and can be associated with obesity (Yanovski 1993). For the

M. V. Micioni Di Bonaventura · E. Micioni Di Bonaventura · C. Polidori · C. Cifani (✉)
School of Pharmacy, Pharmacology Unit, University of Camerino, Camerino, Italy
e-mail: mariavittoria.micioni@unicam.it; emanuela.micioni@unicam.it; carlo.polidori@unicam.it; carlo.cifani@unicam.it

© Springer Nature Switzerland AG 2020
G. K.W. Frank, L. A. Berner (eds.), *Binge Eating*, https://doi.org/10.1007/978-3-030-43562-2_7

first time in the DSM-5, it was indicated that "some individuals with eating disorders report eating-related symptoms resembling those typically endorsed by individuals with substance use disorders" and promoted the concept of *food addiction* (APA 2013). The concept of food addiction, used for the first time by Randolph in 1956 (Randolph 1956) suggests that eating disorders share important commonalities with drug abuse, such as reward seeking and excessive consumption despite the negative consequences (Gold et al. 2003) as well as alterations of brain reward pathways (D'Addario et al. 2014; Johnson and Kenny 2010). Even if the debate remains open (Ziauddeen et al. 2012), pharmacological evidence supports the above observations. In fact, binge eating was reduced by a plethora of chemical molecules that affect also drug abuse intake such as naltrexone, baclofen (Buda-Levin et al. 2005; Corwin et al. 2012), topiramate (Cifani et al. 2009), *Rhodiola rosea* and *Hypericum perforatum* extracts (Cifani et al. 2010; Micioni Di Bonaventura et al. 2012c), A_{2A} adenosine Receptor ($A_{2A}AR$) stimulation (Micioni Di Bonaventura et al. 2012a, b), orexin receptor 1 (OXR1) antagonists (Piccoli et al. 2012) and σ_1 receptor antagonist (Cottone et al. 2012; Del Bello et al. 2019). Further, other studies revealed the strong involvement of stress in the initiation of binge episodes with the inhibitory effect of corticotropin-releasing factor $(CRF)_1$ receptor antagonists on food intake (Iemolo et al. 2013; Micioni Di Bonaventura et al. 2017b), suggesting that palatable food ingestion or drug abuse might represent an escape from distress, anxiety, sadness, and fear (Cottone et al. 2009a). Calorie-dense foods, therefore, may become a "comfort food" and perpetuate binge eating, even in the absence of hungriness, mitigating anxiety and stress responses, reducing the hypothalamic–pituitary–adrenal (HPA) axis activity (Dallman et al. 2003; Pecoraro et al. 2004).

Studies on the intensity of the stress in humans have demonstrated that severe acute stress, activating the HPA axis, suppresses appetite (Adam and Epel 2007), while mild-to-moderate stress could increase food intake (Rutters et al. 2009). This eating behavior induced by less intense stress has been classified as emotional eating and can be measured with a specially designed scale (Arnow and Kenardy 1995) and is strongly correlated to binge eating (Stice et al. 2002).

Taking into account the above considerations, we will describe in the following paragraphs, how environmental conditions affect feeding behavior and elicit binge eating in animals. Specifically, we will focus on stress (acute or chronic), food availability (energy deprivation or intermittent access), and enriched environment (EE) conditions (see Table 1).

2 Stress-Induced Binge-Eating Episode in Preclinical Models

To model the association of binge eating with stress described in the human literature, stress induction has been necessary for preclinical models to increase the consumption of highly palatable food to promote a binge-like eating episode. Several kinds of stress (physical or psychological) have been used to stimulate food intake, differing in severity and time of application (acute or chronic). However, in accordance with the "comfort food" hypothesis (Dallman et al. 2003, 2007) and together with the activation of brain reward systems (Avena et al. 2008; Berridge et al. 2010), all models used palatable food, such as sugar solutions (Colantuoni et al. 2001), chocolate (Cifani et al. 2013), cookies (Hagan et al. 2002), or chocolate-high sucrose pellets (Cottone et al. 2012) to develop the aberrant feeding behavior and the motivation for food overconsumption.

2.1 Acute Stress

One of the first acute stress model of binge eating, now uncommon due to the conflicting pharmacological results (Morley et al. 1983), was the tail-pinch model (Rowland and Antelman 1976), in which pinching a rat's tail increased the food intake. Another method was to use foot shock in combination with three cycles of a food (chow) restriction/refeeding (R–R) protocol (Hagan et al.

Table 1 A suitable animal models to study binge-like eating induced by stress (acute or chronic) or the combination of repeated cycles of food restriction and refeeding plus stress or limited access to palatable food

	Species	Sex	References
Acute stress protocol			
Tail-pinch	Rats	F	Rowland and Antelman (1976)
Food restricting/refeeding plus foot shock	Rats	F	Hagan et al. (2002)
Food restricting/refeeding plus frustration stress	Rats	F	Cifani et al. (2009)
Food restricting/refeeding plus forced swimming stress	Mice	F	Consoli et al. (2009)
Chronic stress protocol			
Flashing LED irradiation stress	Mice	M	Wei et al. (2019)
Exposure to chronic variable stress	Mice	M	Pankevich et al. (2010)
	Mice	M	Teegarden and Bale (2008)
Social stress	Rats	M	Tamashiro et al. (2006)
	Rats	M	Melhorn et al. (2010)
	Mice	M	Bartolomucci et al. (2009)
	Mice	M	Patterson et al. (2013)
	Mice	M	Keenan et al. (2018)
	Mice	M	Razzoli et al. (2015)
Maternal separation	Rats	F	Hancock et al. (2005)
	Rats	M, F	Iwasaki et al. (2000)
	Rats	M, F	McIntosh et al. (1999)
	Rats	M	Penke et al. (2001)
	Mice	F	Schroeder et al. (2017)
Limited-access model to palatable food			
Sugar addiction model	Rats	M	Avena et al. (2008)
Limited access to a dietary fat	Rats	M	Corwin et al. (1998)
	Mice	M	Czyzyk et al. (2010)
Limited access to palatable food	Rats	M	Bello et al. (2009)
	Rats	F	Cottone et al. (2008)

2002). In this model, rats lost weight during the restriction phase but fully recovered their body weight during the periods of *ad libitum* food access, similar to human "yo-yo" dieting. After the foot shock session, food intake was measured and it was found that only the combination of restriction plus foot shock significantly increased the palatable food intake (binge episode) compared to the other three groups (rats with no food restriction and no stress, rats in the stress only condition, or rats in the food restriction without stress condition) (Hagan et al. 2002). The work showed that at least three cycles of R–R plus stress were sufficient to elicit overconsumption of palatable food and the body weight loss during restriction could be considered a physiological stress response that promotes the neurobiological changes that set the organism up for binge eating. The interaction between chronic

food restriction and stress was also found in a model from our group (Cifani et al. 2009), in which the feeding schedule is similar to the Hagan model, but the foot shock was replaced by "frustrative non-reward" or "frustration stress" manipulation (Amsel 1958). This procedure consists of 15 min exposure to the odor and sight of a familiar chocolate paste, without access to it, just before offering the palatable food for 2 h. Similarly, in humans, environmental cues like the sight or smell of palatable food have been associated with binge eating (van der Ster et al. 1994; Waters et al. 2001). Moreover, the frustrative non-reward and similarly aversive experiences are known to activate the negative-valence system, such as the neuroendocrine stress system and cortico-limbic neural circuits that produce anxiety, fear, which contribute to binge-eating behavior as well as their animal homologs

(Sanislow et al. 2010). In addition, the frustrative non-reward stress, unlike the electric foot shock, never induces fear and freezing or conflicting reports on food intake (Sterritt 1962), but elicits a robust behavioral activation, such as repeated movements of the forepaws, head, and trunk of the rats to attempt to reach the palatable food (Micioni Di Bonaventura et al. 2017a), and increased plasma corticosterone levels (Cifani et al. 2009). Although the "frustration stress" or foot shock are important triggers for binge-eating episodes, rats need to be exposed to episodic caloric restriction and to calorie-dense food availability to express binge-eating behavior. Presentation of palatable food per se did not induce binge eating, but it becomes a key factor in food-restricted and stressed rats, underlining the importance of the diet schedule and access (see also below in Sect. 3). Furthermore, if only chow was available, overeating did not occur, consistent with eating for reward as opposed to metabolic need (Hagan et al. 2002), confirming that there is no caloric deficit from R–R. In these animal models, it is also important to point out that the rats that develop binge eating are satiated, without energy deficits, in line with the DSM-5 description of binge-eating episodes in BED. Other binge-eating models use modified versions of this R–R protocol (Consoli et al. 2009; Hancock et al. 2005), and all the current studies, in keeping with the high prevalence of eating disorders among young adolescents and young adult women (Swanson et al. 2011), have been studying female rodents.

Previous studies demonstrated that caloric restriction was able to reprogram stress and orexigenic pathways (Micioni Di Bonaventura et al. 2013; Pankevich et al. 2010) and during the deprivation, the extracellular dopamine (DA) levels in the nucleus accumbens (NAc) were drastically reduced (Pothos et al. 1995). The followed refeeding restored this state, producing reward stimulation by releasing DA (Hernandez and Hoebel 1988; Kelley 2004). These results associated the reward value for the eaten food with the activation of brain reward pathways and a reduction in stress response, as mentioned before, reducing central CRF (Dallman et al. 2003). Then, just as in drug abuse, during

dietary withdrawal, this relief from negative states is reversed, driving the overconsumption of energy-dense food (Pecoraro et al. 2004; Teegarden and Bale 2008) and to the "dark" side of food addiction (Parylak et al. 2011). Furthermore, striatal DA D2 receptors were downregulated in obese rats (Johnson and Kenny 2010), in response to overconsumption of palatable foods. The same brain reward deficit is induced by cocaine or heroin, and it is considered a critical trigger in the transition from casual to compulsive intake of drugs or food. The involvement of reward circuits promotes the *food addiction* hypothesis and the study on a potential treatment of binge eating targeting the HPA axis and other systems.

2.2 Chronic Stress

Chronic stress contributes to the development, maintenance of binge-eating episodes in both animals and humans, enhancing the consumption of palatable foods (Dallman et al. 2003; Pecoraro et al. 2004). The effect of chronic stress to increase food is related to the intensity of the stress experienced: milder stress led to a moderate and short-lived increased preference for palatable foods compared to a higher intensity version, in which the preference was more pronounced (Thompson et al. 2015). In humans, chronic stress shifts consumption toward high carbohydrate and high saturated fat foods (Roberts et al. 2014), and animal research has used a variety of experimental stressors to motivate for food, such as chronic variable stress (CVS), social, and early life stress.

Recently mice were stressed with 2 h flashing LED irradiation stress to mimic the effect of chronic stress on standard food and palatable food intake (Wei et al. 2019). After 1 month of this manipulation, the stressed mice increased in measured food addiction score, showed dysregulation of CRF signaling in the reward system, and increased expression of DA D2 and mu-opioid receptors in the NAc.

2.2.1 Chronic Variable Stress
Several CSV models were studied, usually consisting of one stressor per day for around

10 days in an unpredictable pattern, including restraint, darkness during the light cycle, exposure to novel objects overnight, damp bedding overnight, cage changes, or novel noise overnight (McEuen et al. 2009; Pankevich et al. 2010; Teegarden and Bale 2008).

Those chronic stress models induce not only the dysregulation of reward and stress pathways but also hormonal pathways that regulate the central feeding circuitry, including the hypothalamus, such as melanin-concentrating hormone (MCH) and orexin (Pankevich et al. 2010). After CVS, binge eating increased in previously restricted mice, and only these mice had significantly elevated expression of those two hormones. MCH and orexin interact with the mesolimbic DA system, and it is known that they could promote the consumption of palatable food (Borgland et al. 2009; Georgescu et al. 2005). Interestingly, the activation of orexin-containing neurons was found in response to the injection of neuropeptide S (NPS) (Niimi 2006). This neurotransmitter modulates arousal, anxiety, and food intake (Cifani et al. 2011; Reinscheid et al. 2005), and it could regulate motivation for palatable food since its receptors are expressed in hypothalamic and extrahypothalamic regions (Pandit et al. 2014).

2.2.2 Social Stress

Social, environmental stress, compared to other stress, induces the most powerful hormonal and cardiovascular stress response in animal models and mimics the impact of psychosocial stress on humans (Sgoifo et al. 1999). Different variants of social stress models and the association between social stress and obesity were reviewed in the past (Bartolomucci et al. 2001, 2009; Coccurello et al. 2009; Razzoli et al. 2017). The social stress protocols in animal studies established a social hierarchy (dominant vs. subordinate), introducing an aggressive rodent into a cage housing male rodents that had already naturally established a social hierarchy. The aggressive "intruder" disrupts the social hierarchy, causing the residents' social stress (Stark et al. 2001). In the social defeat protocol, there is an aggressive interaction between the "intruder" and a nonaggressive male rodent (subordinate) and they fight. The subordinate, during the recovery from the stress procedure, become hyperphagic and regain the previously lost weight. The subordinate animals showed increased adiposity, plasma leptin, and insulin levels (Tamashiro et al. 2006, 2011) and larger meal size (Melhorn et al. 2010). Social stress induced persistent overstimulation on food intake and elevations of circulating corticosterone by upregulating HPA axis activity. Also, glucose, ghrelin, hypothalamic expression of orexigenic neuropeptides agouti-related protein (AGRP), neuropeptide Y (NPY) mRNA were elevated, and depression and anxiety were increased in subordinate mice (Bartolomucci et al. 2010; Hardaway et al. 2015; Patterson et al. 2013). AGRP and NPY are hypothalamic neuropeptides involved in the control of energy balance and obesity (Barsh and Schwartz 2002; Cifani et al. 2015). Thus, the subordinate animal's meal pattern could be useful to study mechanisms regulating stress-induced changes in BED associated with obesity, as Razzoli et al. (2015) and Keenan et al. (2018) recently proposed. Keenan et al. (2018) used an alternative social defeat model with unpredictable periods of overcrowding with other defeated mice. In both studies, the subordinate mice developed robust obesity, consistent hyperphagia and insulin resistance, compared to dominant mice. A limitation of the above-described models is that they were performed only in males, and the field is just slowly recognizing the need for studies that differentiate sex effects.

2.2.3 Early Life Stress: Role of Maternal Care

Early life pre- and postnatal stress experiences increase the vulnerability to different diseases (Davis and Pfaff 2014; Pucci et al. 2018a), including eating disorders (Rayworth et al. 2004) and especially binge eating onset (Striegel-Moore et al. 2002).

Using the Hagan model, Hancock et al. (2005) demonstrated that low levels of maternal care (measured using the frequencies of maternal licking and grooming reviewed in Champagne et al. (2003)) influenced vulnerability to the later development of binge-eating behavior. Studying the effect of maternal separation is another method that may help to investigate the effects of early life stressors on later disordered eating. Preclinical experiments revealed that early life environments are linked to long-lasting individual differences in offspring sensitivity to stress (McIntosh et al. 1999; Penke et al. 2001), and HPA responsivity (Penke et al. 2001). Levine (2002) suggested that short separations from the mother were a "psychological" stressor for the pups, while repeated long separation could be considered as "physiological" stressor with different impact on the stress response; thus the duration of mother deprivation in the protocol is a key point. The HPA axis in rats and mice is hyporesponsive in the first 2 weeks of life, but maternal separation (2–24 h) could disinhibit this blockade (Levine 2002), inducing body weight increase throughout the life cycle (McIntosh et al. 1999). Maternal separation also promotes other changes, such as a reduction in plasma glucose and leptin, together with an increase in plasma ghrelin and hypothalamic changes in NPY and CRH mRNAs (Schmidt et al. 2006). Moreover, maternal deprivation elicits both hyperphagia following restricted feeding schedules (Iwasaki et al. 2000) and preference for palatable food later in life (McIntosh et al. 1999; Penke et al. 2001), in particular in females (Iwasaki et al. 2000; McIntosh et al. 1999). Even if females reacted successfully to early adversity, they show increased vulnerability to, for instance, developing anxiety later in life (Sandman et al. 2013). Interestingly, recent work showed that late gestation prenatal stress rewires neural circuits in female mice, leading to binge-like behavior phenotypes (Schroeder et al. 2017).

3   Limited-Access Model to Palatable Food to Elicit Binge-Like Intake

Other rodent models showed that chronic intermittent access to sweet fat food elicited binge-eating behavior in food-deprived or not deprived animals, leading to negative emotional states when the rewarding food is no longer available, with signs of depression or withdrawal similar to drug abuse.

In the "sugar addiction model" proposed by Hoebel's seminal studies, rats were maintained on daily 12 h food deprivation, followed by equal time access to a sugar solution (10% sucrose or 25% glucose) and standard chow (Avena et al. 2008). After several days, the rats showed an escalation in their daily intake and binge eating on the solution during the first hour of access. In addition, the rats changed their feeding pattern by taking larger meals of sugar within the limited-access compared to control rats (with sugar ad libitum). This intermittent access to a sugar solution induced brain changes that are similar to the effects caused by some drugs of abuse and the opioid antagonist naloxone was able to induce withdrawal from the sugar solution in binge rats (Avena et al. 2008; Colantuoni et al. 2001). In those studies, rats received the palatable food during a period of food deprivation, which was different compared to the R–R plus stress models (Cifani et al. 2009; Hagan et al. 2002) or the Corwin model (Corwin et al. 1998). Corwin's model used brief intermittent access to vegetable shortening, only on intermittent days, without energy restriction, to drive escalation in fat intake compared to control rats with food access every day. Although no food restrictions were used, the binge rats restrict eating standard chow on non-binge days relative to controls to compensate for the binge pattern that had developed and resulted in fluctuation of their body weight (Corwin and Buda-Levin 2004). Similarly to the Corwin model, Czyzyk et al. (2010) induced binge-like eating behavior in mice on a high-energy diet using a model with weekly 24 h access to a nutritionally complete high-energy diet along with continuous access to standard chow, compared to mice that had continuous access to both diets. Bello and Cottone used a combination of approaches: Bello and colleagues combined intermittent and limited access to sweetened vegetable shortening and food deprivation to stimulate binge-type intakes in rats (Bello et al. 2009); Cottone and Sabino provided intermittent availability of a highly palatable sucrose-rich diet after 23 h of food abstinence, inducing a progressive escalation of operant responding for the sugary food (Cottone et al. 2008, 2009b).

All the models mentioned highlight important commonalities, representing strong predictors of overeating, for example, the availability of highly palatable, energy-dense food or caloric restriction (as experimental protocol or rats self-restricted their intake of standard chow between periods of access to sweet-fat chow). Intermittent food restriction can be observed in humans who also binge eat, and those models could provide neurobiological information that applies to human research.

4 Environmental Enrichment

Preclinical models describe the importance of environmental enrichment (EE) in animal models to improve neural plasticity (Nithianantharajah and Hannan 2006) and ameliorate pathological processes, including drug addiction and especially alcohol abuse (Bahi 2017; Nithianantharajah and Hannan 2006). Research also showed that EE helps to prevent the behavior in animals related to addiction and depression, and improves recovery from various neurological disorders (Crofton et al. 2015). EE decreased anhedonia-type behavior, social withdrawal, behavioral despair, grooming time in a social contact test, and mobility time in a forced swim test on depression-like behavior in rats (Crofton et al. 2015).

Usually, EE in rodent studies includes a plastic shelter and a running wheel always available and every 5 days a new set of three objects to be introduced to the cages. EE showed to affect ingestive behavior, and consequently, it could attenuate binge-eating behavior (Council 2010; Office of Laboratory Animal Welfare 2002). EE could modify sucrose consumption; for example, Sprague Dawley rats, under a 2-bottle choice sucrose protocol, housed in social or EE condition consumed less sucrose than their isolated counterparts (Brenes and Fornaguera 2008). EE

manipulations in Long Evans rats decreased sucrose seeking after sucrose self-administration training (Grimm et al. 2008). Later, the same group published that both chronic and acute EE reduced sucrose cue-reactivity after 1 or 30 days of forced abstinence from self-administration (Grimm et al. 2013). EE also revealed to decrease the resistance to extinction of responding for sucrose or food in Sprague Dawley EE-housed rats compared to isolated rats (Stairs et al. 2006) and the anticipatory responding for sucrose paired stimuli (van der Harst et al. 2003; Wood et al. 2006). These results indicated that the anticipatory response for sucrose reward is stronger in standard housed animals, and these animals were more sensitive to reward than EE animals. Despite those promising findings, only three studies, to our knowledge, specifically exist on the effects of EE on binge-type eating (Preston et al. 2018; Rodriguez-Ortega et al. 2019a, b). Preston et al. (2018), using the Corwin model with EE and non-EE rats, showed that mild enrichment might not be helpful once bingeing has been ongoing for an extended period of time. Instead, Rodriguez-Ortega et al. (2019a) indicated not only that early EE exposure protected mice from binge-like excessive sucrose intake during adulthood, but also that EE reduced high sucrose intake in mice that were first under standard conditions.

Another study by Rodriguez-Ortega et al. (2019b) provided further positive evidence for EE on binge eating. They found that when adult mice, reared in a standard environment and exposed to sucrose drinking in the dark, were switched to EE conditions, these mice strongly reduced sucrose binge-like intake, suggesting the therapeutic role of EE exposure. Moreover, they found that the OXR antagonist SB-334867 decreased binge-like sucrose intake in both high and low drinking phenotypes mice exposed to standard conditions. On the other hand, exposure to EE conditions blunted the inhibitory effect of SB-334867 on sucrose binge consumption on both phenotypes, suggesting that EE exposure might impact the orexin system (Rodriguez-Ortega et al. 2019b).

These findings suggest that EE exposure might be a promising tool to prevent binge eating, and more studies are needed to evaluate the various degrees of enrichment effect for both prevention and treatment of binge-eating behavior.

5 Neurotransmitter Systems Implicated in Binge-Eating Behavior

Using validated animal models, stress, repeated cycles of restriction/refeeding, and limited-access to palatable food induced various neurobiological changes, especially in DA, opioid, CRF, serotonin, orexin, and GABAergic systems.

The sugar binge model (Avena et al. 2008) showed behavioral cross-sensitization to amphetamine (Avena and Hoebel 2003), increased intake of alcohol (Avena et al. 2004) and extracellular DA in the NAc shell (Rada et al. 2005), upregulation of opioid μ and DA D1 receptors in the NAc shell (Colantuoni et al. 2001), opioid-like withdrawal (Colantuoni et al. 2002) and a significant drop in body temperature after glucose removal (Wideman et al. 2005). According to this, other studies indicated altered DA signaling following chronic overconsumption of food (Johnson and Kenny 2010) and binge eating (Wang et al. 2011). Reduced DA D2 receptors expression in the striatum could be a neuroadaptive response to the overconsumption of palatable foods, possibly desensitization comparable to substance use models and leading to eating disorder behaviors that are associated with obesity.

Peripheral administration of the DA D1-like antagonist SCH23390 (Corwin and Wojnicki 2009) showed a generalized suppression of fat or chow intake of both binge and control rats, while the D2-like antagonist raclopride had different effects. Raclopride at high doses generally reduced food intake, like SCH23390, but at lower doses, it did not affect food intake in daily-access studies in rats or stimulated fat intake in intermittent access rats, suggesting differential pre- and postsynaptic D2 signaling under binge and control conditions. Since the functional interaction observed between the DA D2 receptors and $A_{2A}ARs$ (Hillion et al. 2002) and the blockade of $A_{2A}AR$ can mimic the action of DA D2 agonists (Fenu et al. 1997), it would be interesting to develop a deeper understanding of the strong inhibiting effect of $A_{2A}AR$ agonists on palatable food (Micioni Di Bonaventura et al. 2019b).

The opioid system has also been implicated in the transition from overeating to binge eating. In the sugar addiction model, binge eating decreased enkephalin mRNA in the NAc (Spangler et al. 2004), and μ-opioid receptor binding was enhanced in the NAc shell, cingulate, hippocampus, and locus coeruleus, compared with chow controls (Colantuoni et al. 2001). Moreover, the binge-eating rats are sensitive to the opioid antagonist naloxone, which precipitated signs of withdrawal, indicating that chronic sugar intake can alter the brain opioid system (Colantuoni et al. 2002).

In other models, binge-eating episodes were blocked by the nonselective μ/κ opioid antagonist nalmefene (Cottone et al. 2008) and naloxone (Hagan et al. 2002), consistent with human research that showed that the selective μ-opioid receptor antagonist GSK1521498 reduced motivational response in the striatum to viewing food images in binge-eating obese individuals (Cambridge et al. 2013). Recently, the opioid antagonist naltrexone showed to be well tolerated and effective for the treatment of adolescents with binge eating or purging in an eating disorder treatment program (Stancil et al. 2019). One mechanism to explain these results was the sensitization of opioid receptors (Hagan and Moss 1991), maybe due to a repeated endogenous opioid release, caused by prolonged palatable food consumption (Chang et al. 2010; Kelley et al. 2003). Taking into account the role of the endogenous opioid system in mediating food intake regulation and the discovery that the activation of Nociceptin/Orphanin (N/OFQ) FQ peptide (NOP) receptor (opioid receptor-like 1, ORL1), increased food intake in rats (Micioni Di Bonaventura et al. 2019a), many NOP agonists and antagonists were tested on binge-eating behavior in rodents (Calo et al. 2011; Hardaway et al. 2016; Micioni Di Bonaventura et al. 2013; Pucci et al. 2016; Vitale et al. 2017). Recent studies suggested that N/OFQ system, a functional antagonist of CRF (Filaferro et al. 2014; Rodi et al. 2008), could be a therapeutic target for obesity or BED. In fact, LY2940094 (Statnick et al. 2016) and SB 612111 (Hardaway et al. 2016), both N/OFQ antagonists, attenuated the aberrant feeding behavior in rodents and epigenetic modifications affecting the N/OFQ and CRF systems in response to food restriction and stress exposure (Pucci et al. 2016). The working hypothesis is that N/OFQ may stimulate feeding via inhibition of anorexigenic pathways, resulting in disinhibition of orexigenic mechanisms (Statnick et al. 2016) and such altered N/OFQ-CRF mechanisms may contribute to the development of binge eating.

Growing preclinical evidence showed that microinfusion or systemic administration of a CRF_1 receptor antagonist selectively blocked the binge-eating episode with no effect in control rats (Cottone et al. 2009a; Iemolo et al. 2013; Micioni Di Bonaventura et al. 2014, 2017b; Parylak et al. 2012). Under chronic intermittent dietary restriction of a palatable diet, immunohistochemistry revealed elevated CRF positive cells in the central nucleus of the amygdala (CeA) in binge rats but not in control rats (Cottone et al. 2009a; Iemolo et al. 2013). Indeed, R–R plus frustration stress manipulation increased bed nucleus of the stria terminalis (BNST) neuronal activity and significant upregulation of crhr1 mRNA levels in the dorsal portion of the BNST and in the CeA only in binge-eating rats (Micioni Di Bonaventura et al. 2014, 2017b). These findings extend the role of CRF. It could be possible that CRF induced modulation of binge eating is not only linked to peripheral corticosterone release, after HPA activation, but also directly linked to extra-hypothalamic mechanisms. In humans, the CRF_1 receptor antagonist pexacerfont revealed promising anti-craving properties, particularly in the presence of palatable food in restrained eaters (Epstein et al. 2016).

Unfortunately, this clinical study was interrupted because of administrative reasons but not for the quality of the study or safety concerns. In the future, it would be important to continue exploring the mutual relationship among DA, CRF stress system, and oxytocin on emotional processing, which have been found to trigger binge-eating episodes (Giel et al. 2018; Romano et al. 2015).

There are other novel pharmacological approaches that may help to understand the neuro pathways implicated in BED. For instance, NPY-1R antagonists, that as mentioned before, demonstrated to reduce stress-induced eating (Goebel-Stengel et al. 2014). Miller et al. (2002) showed that overconsumption of high palatable foods increased NPY in the hypothalamus, while reducing anxiety-like behavior in the open field test, showing interactions between NPY, stress, and feeding.

Selective serotonin reuptake inhibitors (SSRIs) were also studied in several studies. Fluoxetine non-selectively reduced palatable food in binge-eating and control rats, and this general reduction was also found with sibutramine, a monoamine reuptake inhibitor in the same animal model for binge eating (Cifani et al. 2009). More recently, serotonin releasing agents (fluoxetine and D-fenfluramine) and the selective serotonin (5-hydroxytryptamine, 5-HT) $5-HT_{2C}$ receptor agonist lorcaserin reduced binge eating, enhancing midbrain dopamine neuronal activity (Xu et al. 2017). Specifically, 5-HT stimulates DA neural activity through a $5-HT_{2C}R$, and activation of this midbrain circuit inhibits binge-like eating behavior in mice.

As previously described, the orexin system plays a role in eating disorders characterized by compulsive binge-type episodes (Grafe and Bhatnagar 2018). Orexin interacts with the reward system to selectively regulate the intake of highly palatable food, and thus it might represent a good target for control of compulsive eating (Piccoli et al. 2012; Vickers et al. 2015). The OXR antagonist (dual OX1/OX2 receptor antagonist) SB-649868 selectively blocked the occurrence of binge-eating episodes without affecting normal food intake in rats (Piccoli et al. 2012). Instead, JNJ-10397049, a selective orexin receptor 2 (OXR2) antagonist, failed to affect binge eating of palatable food. These results indicate an important role of the OX1R in binge-eating behavior, suggesting that selective antagonism at the OX1R could represent an important target for BED and other eating disorders with a compulsive and stress component driving binge-eating episodes.

Modulation of the gamma-aminobutyric acid (GABA) neural systems also represents a possible approach. Baclofen, a GABA(B) receptor agonist, attenuated the intake of highly palatable food in several different animal models of binge eating (Berner et al. 2009; Buda-Levin et al. 2005; Wong et al. 2009). GABA(B) receptors are expressed on both GABA and DA neurons in the ventral tegmental area (VTA), where they inhibit both GABAergic and dopaminergic neurons (Cruz et al. 2004). Those studies suggest that modulating the DA system via the GABA (B) receptors could be a potential target to reduce motivation for highly palatable food.

Several groups have been investigating the endocannabinoid system (Capasso et al. 2018; Monteleone et al. 2017; Pucci et al. 2018b, 2019), because of its known interactions with other pathways involved in the hedonic aspect of food, including DA reward circuitry (D'Addario et al. 2014; Lau et al. 2017). For instance, CB1 receptor antagonists reduced binge eating in different rodent models (Dore et al. 2014; Parylak et al. 2012; Scherma et al. 2013). Unfortunately, rimonabant was the only CB1 receptor antagonist to be approved for use in humans, but it was quickly withdrawn for significant side effects, especially severe negative effects on mood (Sam et al. 2011). However, it would be interesting in the future to evaluate and target other compounds, such as fatty acid amide hydrolase (FAAH) or monoacylglycerol lipase (MAGL) inhibitors, which modulate the concentration of endogenous cannabinoid availability (Petrosino and Di Marzo 2010).

Of note is that binge-eating frequency was found to change through the menstrual cycle in humans (Edler et al. 2007; Schoofs et al. 2011), and similarly, cycle variations occur in female rats (Alboni et al. 2017; Yu et al. 2011; Micioni Di Bonaventura et al. 2017a). Those studies implicate sex hormones in binge eating, but research is lacking to understand better how they mechanistically impact binge-eating behavior.

6   Summary

Animal models support the hypothesis that stress has strong effects on food intake. This line of research has identified a multitude of alterations in neurotransmitter and hormone release and receptor function that could contribute to binge eating. The animal models described herein are based on modifications in the pattern of distribution of the meals throughout the day, with periods of food restriction or limited food access in between, and exposure to chronic or acute stress as variable to induce or affect those patterns. Behaviorally, those biological alterations are associated with stress-related emotional states such as elevated anxiety and depression that contribute to the development and course of binge eating in the animal model. This has important implications for understanding the pathophysiology of BED in humans and deserves further study. BED is the most prevalent eating disorder among adults population, with a strong predominance in females and associated with obesity (Hudson et al. 2007). Only one medication, lisdexamfetamine dimesylate, a prodrug of dextroamfetamine, is currently approved in the United States for the treatment of moderate-to-severe BED in adults. However, the underlying mechanisms that help to reduce binge eating are not clear (Ward and Citrome 2018). Thus, preclinical models that can be translated into human research, continue to be important to understand the neural mechanisms underlying binge-eating behaviors and to promote the development of innovative drugs.

References

Adam TC, Epel ES (2007) Stress, eating and the reward system. Physiol Behav 91:449–458. https://doi.org/10.1016/j.physbeh.2007.04.011

Alboni S, Micioni Di Bonaventura MV, Benatti C, Giusepponi ME, Brunello N, Cifani C (2017) Hypothalamic expression of inflammatory mediators in an animal model of binge eating. Behav Brain Res 320:420–430. https://doi.org/10.1016/j.bbr.2016.10.044

Amsel A (1958) The role of frustrative nonreward in noncontinuous reward situations. Psychol Bull 55:102–119

Arnow B, Kenardy J, Agras WS (1995) The emotional eating scale: the development of a measure to assess coping with negative affect by eating. Int J Eat Disord 18:79–90

Association AP (2013) Diagnostic and statistical manual of mental disorders (DSM-5®). American Psychiatric Publishing, Washington, DC

Avena NM, Hoebel BG (2003) A diet promoting sugar dependency causes behavioral cross-sensitization to a low dose of amphetamine. Neuroscience 122:17–20

Avena NM, Carrillo CA, Needham L, Leibowitz SF, Hoebel BG (2004) Sugar-dependent rats show enhanced intake of unsweetened ethanol. Alcohol 34:203–209

Avena NM, Rada P, Hoebel BG (2008) Evidence for sugar addiction: behavioral and neurochemical effects of intermittent, excessive sugar intake. Neurosci Biobehav Rev 32:20–39. https://doi.org/10.1016/j.neubiorev.2007.04.019

Bahi A (2017) Environmental enrichment reduces chronic psychosocial stress-induced anxiety and ethanol-related behaviors in mice. Prog Neuro-Psychopharmacol Biol Psychiatry 77:65–74. https://doi.org/10.1016/j.pnpbp.2017.04.001

Barsh GS, Schwartz MW (2002) Genetic approaches to studying energy balance: perception and integration. Nat Rev Genet 3:589–600. https://doi.org/10.1038/nrg862

Bartolomucci A et al (2001) Social status in mice: behavioral, endocrine and immune changes are context dependent. Physiol Behav 73:401–410. https://doi.org/10.1016/s0031-9384(01)00453-x

Bartolomucci A et al (2010) Increased vulnerability to psychosocial stress in heterozygous serotonin transporter knockout mice. Dis Model Mech 3:459–470. https://doi.org/10.1242/dmm.004614

Bartolomucci A, Cabassi A, Govoni P, Ceresini G, Cero C, Berra D, Dadomo H, Franceschini P, Dell'Omo G, Parmigiani S, Palanza P, Baune B (2009) Metabolic consequences and vulnerability to diet-induced obesity in male mice under chronic social stress. PLoS One 4(1):e4331

Bello NT, Guarda AS, Terrillion CE, Redgrave GW, Coughlin JW, Moran TH (2009) Repeated binge access to a palatable food alters feeding behavior, hormone profile, and hindbrain c-Fos responses to a test meal in adult male rats. Am J Physiol Regul Integr Comp Physiol 297:R622–R631. https://doi.org/10.1152/ajpregu.00087.2009

Berner LA, Bocarsly ME, Hoebel BG, Avena NM (2009) Baclofen suppresses binge eating of pure fat but not a sugar-rich or sweet-fat diet. Behav Pharmacol 20:631–634. https://doi.org/10.1097/FBP.0b013e328331ba47

Berridge KC, Ho CY, Richard JM, DiFeliceantonio AG (2010) The tempted brain eats: pleasure and desire circuits in obesity and eating disorders. Brain Res 1350:43–64. https://doi.org/10.1016/j.brainres.2010.04.003

Borgland SL et al (2009) Orexin A/hypocretin-1 selectively promotes motivation for positive reinforcers. J Neurosci Off J Soc Neurosci 29:11215–11225. https://doi.org/10.1523/JNEUROSCI.6096-08.2009

Brenes JC, Fornaguera J (2008) Effects of environmental enrichment and social isolation on sucrose consumption and preference: associations with depressive-like behavior and ventral striatum dopamine. Neurosci Lett 436:278–282. https://doi.org/10.1016/j.neulet.2008.03.045

Buda-Levin A, Wojnicki FH, Corwin RL (2005) Baclofen reduces fat intake under binge-type conditions. Physiol Behav 86:176–184. https://doi.org/10.1016/j.physbeh.2005.07.020

Cambridge VC et al (2013) Neural and behavioral effects of a novel mu opioid receptor antagonist in binge-eating obese people. Biol Psychiatry 73:887–894. https://doi.org/10.1016/j.biopsych.2012.10.022

Calo G, Rizzi A, Cifani C, Di Bonaventura MVM, Regoli D, Massi M, Salvadori S, Lambert DG, Guerrini R (2011) UFP-112 a potent and long-lasting agonist selective for the Nociceptin/Orphanin FQ receptor. CNS Neurosci Ther 17(3):178–198

Capasso A, Milano W, Cauli O (2018) Changes in the peripheral endocannabinoid system as a risk factor for the development of eating disorders. Endocr Metab Immune Disord Drug Targets 18:325–332. https://doi.org/10.2174/1871530318666180213112406

Champagne FA, Francis DD, Mar A, Meaney MJ (2003) Variations in maternal care in the rat as a mediating influence for the effects of environment on development. Physiol Behav 79:359–371. https://doi.org/10.1016/s0031-9384(03)00149-5

Chang G-Q, Karatayev O, Barson J, Chang S-Y, Leibowitz S (2010) Increased enkephalin in brain of rats prone to overconsuming a fat-rich diet. Physiol Behav 101:360–369

Cifani C, Polidori C, Melotto S, Ciccocioppo R, Massi M (2009) A preclinical model of binge eating elicited by yo-yo dieting and stressful exposure to food: effect of sibutramine, fluoxetine, topiramate, and midazolam. Psychopharmacology 204:113–125. https://doi.org/10.1007/s00213-008-1442-y

Cifani C, Micioni Di BM, Vitale G, Ruggieri V, Ciccocioppo R, Massi M (2010) Effect of salidroside, active principle of *Rhodiola rosea* extract, on binge eating. Physiol Behav 101:555–562. https://doi.org/10.1016/j.physbeh.2010.09.006

Cifani C et al (2011) Effect of neuropeptide S receptor antagonists and partial agonists on palatable food consumption in the rat. Peptides 32:44–50. https://doi.org/10.1016/j.peptides.2010.10.018

Cifani C, Di Bonaventura MVM, Ciccocioppo R, Massi M (2013) Binge eating in female rats induced by yo-yo dieting and stress. In: Animal models of eating disorders. Springer, Dordrecht, pp 27–49

Cifani C et al (2015) Regulation of hypothalamic neuropeptides gene expression in diet induced obesity resistant rats: possible targets for obesity prediction? Front Neurosci 9:187. https://doi.org/10.3389/fnins.2015.00187

Coccurello R, D'Amato FR, Moles A (2009) Chronic social stress, hedonism and vulnerability to obesity: lessons from rodents. Neurosci Biobehav Rev 33:537–550. https://doi.org/10.1016/j.neubiorev.2008.05.018

Colantuoni C et al (2001) Excessive sugar intake alters binding to dopamine and mu-opioid receptors in the brain. Neuroreport 12:3549–3552. https://doi.org/10.1097/00001756-200111160-00035

Colantuoni C, Rada P, McCarthy J, Patten C, Avena NM, Chadeayne A, Hoebel BG (2002) Evidence that intermittent, excessive sugar intake causes endogenous opioid dependence. Obes Res 10:478–488

Consoli D, Contarino A, Tabarin A, Drago F (2009) Binge-like eating in mice. Int J Eat Disord 42:402–408. https://doi.org/10.1002/eat.20637

Corwin RL, Buda-Levin A (2004) Behavioral models of binge-type eating. Physiol Behav 82:123–130

Corwin RL, Wojnicki FH (2009) Baclofen, raclopride, and naltrexone differentially affect intake of fat and sucrose under limited access conditions. Behav Pharmacol 20:537–548. https://doi.org/10.1097/FBP.0b013e3283313168

Corwin RL, Wojnicki FH, Fisher JO, Dimitriou SG, Rice HB, Young MA (1998) Limited access to a dietary fat option affects ingestive behavior but not body composition in male rats. Physiol Behav 65:545–553. https://doi.org/10.1016/s0031-9384(98)00201-7

Corwin RL, Boan J, Peters KF, Ulbrecht JS (2012) Baclofen reduces binge eating in a double-blind, placebo-controlled, crossover study. Behav Pharmacol 23:616–625. https://doi.org/10.1097/FBP.0b013e328357bd62

Cottone P, Sabino V, Steardo L, Zorrilla EP (2008) Opioid-dependent anticipatory negative contrast and binge-like eating in rats with limited access to highly preferred food. Neuropsychopharmacology 33:524–535. https://doi.org/10.1038/sj.npp.1301430

Cottone P et al (2009a) CRF system recruitment mediates dark side of compulsive eating. Proc Natl Acad Sci U S A 106:20016–20020. https://doi.org/10.1073/pnas.0908789106

Cottone P, Sabino V, Steardo L, Zorrilla EP (2009b) Consummatory, anxiety-related and metabolic adaptations in female rats with alternating access to preferred food. Psychoneuroendocrinology 34:38–49. https://doi.org/10.1016/j.psyneuen.2008.08.010

Cottone P et al (2012) Antagonism of sigma-1 receptors blocks compulsive-like eating. Neuropsychopharmacology 37:2593–2604. https://doi.org/10.1038/npp.2012.89

Council NR (2010) Guide for the care and use of laboratory animals. National Academies Press, Washington, DC

Crofton EJ, Zhang Y, Green TA (2015) Inoculation stress hypothesis of environmental enrichment. Neurosci Biobehav Rev 49:19–31. https://doi.org/10.1016/j.neubiorev.2014.11.017

Cruz HG, Ivanova T, Lunn ML, Stoffel M, Slesinger PA, Luscher C (2004) Bi-directional effects of GABA (B) receptor agonists on the mesolimbic dopamine system. Nat Neurosci 7:153–159. https://doi.org/10.1038/nn1181

Czyzyk TA, Sahr AE, Statnick MA (2010) A model of binge-like eating behavior in mice that does not require food deprivation or stress. Obesity 18:1710–1717. https://doi.org/10.1038/oby.2010.46

D'Addario C et al (2014) Endocannabinoid signaling and food addiction. Neurosci Biobehav Rev 47:203–224. https://doi.org/10.1016/j.neubiorev.2014.08.008

Dallman MF et al (2003) Chronic stress and obesity: a new view of "comfort food". Proc Natl Acad Sci U S A 100:11696–11701. https://doi.org/10.1073/pnas.1934666100

Dallman MF, Warne JP, Foster MT, Pecoraro NC (2007) Glucocorticoids and insulin both modulate caloric intake through actions on the brain. J Physiol 583:431–436. https://doi.org/10.1113/jphysiol.2007.136051

Davis EP, Pfaff D (2014) Sexually dimorphic responses to early adversity: implications for affective problems and autism spectrum disorder. Psychoneuroendocrinology 49:11–25. https://doi.org/10.1016/j.psyneuen.2014.06.014

Del Bello F et al (2019) Investigation of the role of chirality in the interaction with sigma receptors and effect on binge eating episode of a potent sigma1 antagonist analogue of spipethiane. ACS Chem Neurosci 10 (8):3391–3397. https://doi.org/10.1021/acschemneuro.9b00261

Dore R, Valenza M, Wang X, Rice KC, Sabino V, Cottone P (2014) The inverse agonist of CB1 receptor SR141716 blocks compulsive eating of palatable food. Addict Biol 19:849–861. https://doi.org/10.1111/adb.12056

Edler C, Lipson SF, Keel PK (2007) Ovarian hormones and binge eating in bulimia nervosa. Psychol Med 37:131–141. https://doi.org/10.1017/S0033291706008956

Epstein DH, Kennedy AP, Furnari M, Heilig M, Shaham Y, Phillips KA, Preston KL (2016) Effect of the CRF1-receptor antagonist pexacerfont on stress-induced eating and food craving. Psychopharmacology 233:3921–3932. https://doi.org/10.1007/s00213-016-4424-5

Fenu S, Pinna A, Ongini E, Morelli M (1997) Adenosine A2A receptor antagonism potentiates L-DOPA-induced turning behaviour and c-fos expression in 6-hydroxydopamine-lesioned rats. Eur J Pharmacol 321:143–147. https://doi.org/10.1016/s0014-2999(96)00944-2

Filaferro M et al (2014) Functional antagonism between nociceptin/orphanin FQ and corticotropin-releasing factor in rat anxiety-related behaviors: involvement of the serotonergic system. Neuropeptides 48:189–197. https://doi.org/10.1016/j.npep.2014.05.001

Georgescu D et al (2005) The hypothalamic neuropeptide melanin-concentrating hormone acts in the nucleus accumbens to modulate feeding behavior and forced-swim performance. J Neurosci Off J Soc Neurosci 25:2933–2940. https://doi.org/10.1523/JNEUROSCI.1714-04.2005

Giel K, Zipfel S, Hallschmid M (2018) Oxytocin and eating disorders: a narrative review on emerging findings and perspectives. Curr Neuropharmacol 16:1111–1121. https://doi.org/10.2174/1570159X15666171128143158

Goebel-Stengel M, Stengel A, Wang L, Tache Y (2014) Orexigenic response to tail pinch: role of brain NPY (1) and corticotropin releasing factor receptors. Am J Physiol Regul Integr Comp Physiol 306:R164–R174. https://doi.org/10.1152/ajpregu.00335.2013

Gold MS, Frost-Pineda K, Jacobs WS (2003) Overeating, binge eating, and eating disorders as addictions. Psychiatr Ann 33:117–122

Grafe LA, Bhatnagar S (2018) Orexins and stress. Front Neuroendocrinol 51:132–145. https://doi.org/10.1016/j.yfrne.2018.06.003

Grimm JW, Osincup D, Wells B, Manaois M, Fyall A, Buse C, Harkness JH (2008) Environmental enrichment attenuates cue-induced reinstatement of sucrose seeking in rats. Behav Pharmacol 19:777–785. https://doi.org/10.1097/FBP.0b013e32831c3b18

Grimm JW, Weber R, Barnes J, Koerber J, Dorsey K, Glueck E (2013) Brief exposure to novel or enriched environments in rats reduces sucrose cue-reactivity and consumption in rats after 1 or 30 days of forced abstinence from self-administration. PLoS One 8:e54164. https://doi.org/10.1371/journal.pone.0054164

Hagan M, Moss D (1991) An animal model of bulimia nervosa: opioid sensitivity to fasting episodes pharmacology. Biochem Behav 39:421–422

Hagan MM, Wauford PK, Chandler PC, Jarrett LA, Rybak RJ, Blackburn K (2002) A new animal model of binge eating: key synergistic role of past caloric restriction and stress. Physiol Behav 77:45–54. https://doi.org/10.1016/s0031-9384(02)00809-0

Hancock SD, Menard JL, Olmstead MC (2005) Variations in maternal care influence vulnerability to stress-induced binge eating in female rats. Physiol Behav 85:430–439. https://doi.org/10.1016/j.physbeh.2005.05.007

Hardaway JA, Crowley NA, Bulik CM, Kash TL (2015) Integrated circuits and molecular components for stress and feeding: implications for eating disorders. Genes Brain Behav 14:85–97. https://doi.org/10.1111/gbb.12185

Hardaway JA et al (2016) Nociceptin receptor antagonist SB 612111 decreases high fat diet binge eating. Behav Brain Res 307:25–34. https://doi.org/10.1016/j.bbr.2016.03.046

Hernandez L, Hoebel BG (1988) Feeding and hypothalamic stimulation increase dopamine turnover in the accumbens. Physiol Behav 44:599–606. https://doi.org/10.1016/0031-9384(88)90324-1

Hillion J et al (2002) Coaggregation, cointernalization, and codesensitization of adenosine A2A receptors and dopamine D2 receptors. J Biol Chem 277:18091–18097. https://doi.org/10.1074/jbc.M107731200

Hudson JI, Hiripi E, Pope HG Jr, Kessler RC (2007) The prevalence and correlates of eating disorders in the national comorbidity survey replication. Biol Psychiatry 61:348–358. https://doi.org/10.1016/j.biopsych.2006.03.040

Iemolo A, Blasio A, St Cyr SA, Jiang F, Rice KC, Sabino V, Cottone P (2013) CRF-CRF1 receptor system in the central and basolateral nuclei of the amygdala differentially mediates excessive eating of palatable food. Neuropsychopharmacology 38:2456–2466. https://doi.org/10.1038/npp.2013.147

Iwasaki S, Inoue K, Kiriike N, Hikiji K (2000) Effect of maternal separation on feeding behavior of rats in later life. Physiol Behav 70:551–556. https://doi.org/10.1016/s0031-9384(00)00305-x

Johnson PM, Kenny PJ (2010) Dopamine D2 receptors in addiction-like reward dysfunction and compulsive eating in obese rats. Nat Neurosci 13:635–641. https://doi.org/10.1038/nn.2519

Keenan RJ, Chan J, Donnelly PS, Barnham KJ, Jacobson LH (2018) The social defeat/overcrowding murine psychosocial stress model results in a pharmacologically reversible body weight gain but not depression - related behaviours. Neurobiol Stress 9:176–187. https://doi.org/10.1016/j.ynstr.2018.09.008

Kelley AE (2004) Ventral striatal control of appetitive motivation: role in ingestive behavior and reward-related learning. Neurosci Biobehav Rev 27:765–776. https://doi.org/10.1016/j.neubiorev.2003.11.015

Kelley A, Will M, Steininger T, Zhang M, Haber S (2003) Restricted daily consumption of a highly palatable food (chocolate Ensure®) alters striatal enkephalin gene expression. Eur J Neurosci 18:2592–2598

Lau BK, Cota D, Cristino L, Borgland SL (2017) Endocannabinoid modulation of homeostatic and non-homeostatic feeding circuits. Neuropharmacology 124:38–51

Levine S (2002) Regulation of the hypothalamic-pituitary-adrenal axis in the neonatal rat: the role of maternal behavior. Neurotox Res 4:557–564. https://doi.org/10. 1080/10298420290030569

McEuen JG, Semsar KA, Lim MA, Bale TL (2009) Influence of sex and corticotropin-releasing factor pathways as determinants in serotonin sensitivity. Endocrinology 150:3709–3716. https://doi.org/10. 1210/en.2008-1721

McIntosh J, Anisman H, Merali Z (1999) Short- and long-periods of neonatal maternal separation differentially affect anxiety and feeding in adult rats: gender-dependent effects. Brain Res Dev Brain Res 113:97–106. https://doi.org/10.1016/s0165-3806(99) 00005-x

Melhorn SJ, Krause EG, Scott KA, Mooney MR, Johnson JD, Woods SC, Sakai RR (2010) Meal patterns and hypothalamic NPY expression during chronic social stress and recovery. Am J Physiol Regul Integr Comp Physiol 299:R813–R822. https://doi.org/10.1152/ ajpregu.00820.2009

Micioni Di Bonaventura MV, Cifani C, Lambertucci C, Volpini R, Cristalli G, Froldi R, Massi M (2012a) Effects of A(2)A adenosine receptor blockade or stimulation on alcohol intake in alcohol-preferring rats. Psychopharmacology 219:945–957. https://doi.org/ 10.1007/s00213-011-2430-1

Micioni Di Bonaventura MV, Cifani C, Lambertucci C, Volpini R, Cristalli G, Massi M (2012b) A2A adenosine receptor agonists reduce both high-palatability and low-palatability food intake in female rats. Behav Pharmacol 23:567–574. https://doi.org/10.1097/FBP. 0b013e3283566a60

Micioni Di Bonaventura MV, Vitale G, Massi M, Cifani C (2012c) Effect of *Hypericum perforatum* extract in an experimental model of binge eating in female rats. J Obes 2012:956137. https://doi.org/10.1155/2012/ 956137

Micioni Di Bonaventura MV, Ubaldi M, Liberati S, Ciccocioppo R, Massi M, Cifani C (2013) Caloric restriction increases the sensitivity to the hyperphagic effect of nociceptin/orphanin FQ limiting its ability to reduce binge eating in female rats. Psychopharmacology 228:53–63. https://doi.org/10.1007/s00213-013-3013-0

Micioni Di Bonaventura MV et al (2014) Role of bed nucleus of the stria terminalis corticotrophin-releasing factor receptors in frustration stress-induced binge-like palatable food consumption in female rats with a history of food restriction. J Neurosci Off J Soc Neurosci 34:11316–11324. https://doi.org/10.1523/ JNEUROSCI.1854-14.2014

Micioni Di Bonaventura MV, Lutz TA, Romano A, Pucci M, Geary N, Asarian L, Cifani C (2017a) Estrogenic suppression of binge-like eating elicited by cyclic food restriction and frustrative-nonreward stress in female rats. Int J Eat Disord 50:624–635. https://doi. org/10.1002/eat.22687

Micioni Di Bonaventura MV, Ubaldi M, Giusepponi ME, Rice KC, Massi M, Ciccocioppo R, Cifani C (2017b) Hypothalamic CRF1 receptor mechanisms are not

sufficient to account for binge-like palatable food consumption in female rats. Int J Eat Disord 50:1194–1204. https://doi.org/10.1002/eat.22767

Micioni Di Bonaventura MV, Micioni Di Bonaventura E, Cifani C, Polidori C (2019a) N/OFQ-NOP system in food intake. Handb Exp Pharmacol 254:279–295. https://doi.org/10.1007/164_2019_212

Micioni Di Bonaventura MV et al (2019b) Regulation of adenosine A2A receptor gene expression in a model of binge eating in the amygdaloid complex of female rats. J Psychopharmacol:269881119845798. https://doi. org/10.1177/0269881119845798

Miller CC, Holmes PV, Edwards GL (2002) Area postrema lesions elevate NPY levels and decrease anxiety-related behavior in rats. Physiol Behav 77:135–140. https:// doi.org/10.1016/s0031-9384(02)00847-8

Monteleone AM, Piscitelli F, Dalle Grave R, El Ghoch M, Di Marzo V, Maj M, Monteleone P (2017) Peripheral endocannabinoid responses to hedonic eating in binge-eating disorder. Nutrients 9:1377. https://doi.org/10. 3390/nu9121377

Morley JE, Levine AS, Rowland NE (1983) Minireview. Stress induced eating. Life Sci 32:2169–2182. https:// doi.org/10.1016/0024-3205(83)90415-0

Niimi M (2006) Centrally administered neuropeptide S activates orexin-containing neurons in the hypothalamus and stimulates feeding in rats. Endocrine 30:75–79. https://doi.org/10.1385/ENDO:30:1:75

Nithianantharajah J, Hannan AJ (2006) Enriched environments, experience-dependent plasticity and disorders of the nervous system. Nat Rev Neurosci 7:697–709. https://doi.org/10.1038/nrn1970

Office of Laboratory Animal Welfare (2002) Institutional animal care and use committee guidebook. Office of Laboratory Animal Welfare, National Institutes of Health

Pandit R, Luijendijk MC, Vanderschuren LJ, la Fleur SE, Adan RA (2014) Limbic substrates of the effects of neuropeptide Y on intake of and motivation for palatable food. Obesity 22:1216–1219. https://doi.org/10. 1002/oby.20718

Pankevich DE, Teegarden SL, Hedin AD, Jensen CL, Bale TL (2010) Caloric restriction experience reprograms stress and orexigenic pathways and promotes binge eating. J Neurosci Off J Soc Neurosci 30:16399–16407. https://doi.org/10.1523/ JNEUROSCI.1955-10.2010

Parylak SL, Koob GF, Zorrilla EP (2011) The dark side of food addiction. Physiol Behav 104:149–156. https:// doi.org/10.1016/j.physbeh.2011.04.063

Parylak SL, Cottone P, Sabino V, Rice KC, Zorrilla EP (2012) Effects of CB1 and CRF1 receptor antagonists on binge-like eating in rats with limited access to a sweet fat diet: lack of withdrawal-like responses. Physiol Behav 107:231–242. https://doi.org/10.1016/ j.physbeh.2012.06.017

Patterson ZR, Khazall R, Mackay H, Anisman H, Abizaid A (2013) Central ghrelin signaling mediates the metabolic response of C57BL/6 male mice to chronic social defeat stress. Endocrinology 154:1080–1091. https://doi.org/10.1210/en.2012-1834

Pecoraro N, Reyes F, Gomez F, Bhargava A, Dallman MF (2004) Chronic stress promotes palatable feeding, which reduces signs of stress: feedforward and feedback effects of chronic stress. Endocrinology 145:3754–3762. https://doi.org/10.1210/en.2004-0305

Penke Z, Felszeghy K, Fernette B, Sage D, Nyakas C, Burlet A (2001) Postnatal maternal deprivation produces long-lasting modifications of the stress response, feeding and stress-related behaviour in the rat. Eur J Neurosci 14:747–755. https://doi.org/10.1046/j.0953-816x.2001.01691.x

Petrosino S, Di Marzo V (2010) FAAH and MAGL inhibitors: therapeutic opportunities from regulating endocannabinoid levels. Curr Opin Investig Drugs 11:51–62

Piccoli L et al (2012) Role of orexin-1 receptor mechanisms on compulsive food consumption in a model of binge eating in female rats. Neuropsychopharmacology 37:1999–2011. https://doi.org/10.1038/npp.2012.48

Pothos EN, Creese I, Hoebel BG (1995) Restricted eating with weight loss selectively decreases extracellular dopamine in the nucleus accumbens and alters dopamine response to amphetamine, morphine, and food intake. J Neurosci Off J Soc Neurosci 15:6640–6650

Preston KE, Corwin RL, Bader JO, Crimmins SL (2018) Relatively enriched housing conditions delay binge onset but do not attenuate binge size. Physiol Behav 184:196–204. https://doi.org/10.1016/j.physbeh.2017.11.018

Pucci M et al (2016) Epigenetic regulation of nociceptin/orphanin FQ and corticotropin-releasing factor system genes in frustration stress-induced binge-like palatable food consumption. Addict Biol 21:1168–1185. https://doi.org/10.1111/adb.12303

Pucci M et al (2018a) Environmental stressors and alcoholism development: focus on molecular targets and their epigenetic regulation. Neurosci Biobehav Rev 106:165–181. https://doi.org/10.1016/j.neubiorev.2018.07.004

Pucci M, Micioni Di Bonaventura MV, Zaplatic E, Bellia F, Maccarrone M, Cifani C, D'Addario C (2018b) Transcriptional regulation of the endocannabinoid system in a rat model of binge-eating behavior reveals a selective modulation of the hypothalamic fatty acid amide hydrolase gene. Int J Eat Disord 52(1):51–60. https://doi.org/10.1002/eat.22989

Pucci M et al (2019) Preclinical and clinical evidence for a distinct regulation of Mu opioid and type 1 cannabinoid receptor genes expression in obesity. Front Genet 10:523. https://doi.org/10.3389/fgene.2019.00523

Rada P, Avena N, Hoebel B (2005) Daily bingeing on sugar repeatedly releases dopamine in the accumbens shell. Neuroscience 134:737–744

Randolph TG (1956) The descriptive features of food addiction; addictive eating and drinking. Q J Stud Alcohol 17:198–224

Rayworth BB, Wise LA, Harlow BL (2004) Childhood abuse and risk of eating disorders in women. Epidemiology 15:271–278

Razzoli M, Sanghez V, Bartolomucci A (2015) Chronic subordination stress induces hyperphagia and disrupts eating behavior in mice modeling binge-eating-like disorder. Front Nutr 1:00030. https://doi.org/10.3389/fnut.2014.00030

Razzoli M, Pearson C, Crow S, Bartolomucci A (2017) Stress, overeating, and obesity: insights from human studies and preclinical models. Neurosci Biobehav Rev 76:154–162. https://doi.org/10.1016/j.neubiorev.2017.01.026

Reinscheid RK, Xu YL, Civelli O (2005) Neuropeptide S: a new player in the modulation of arousal and anxiety. Mol Interv 5:42–46. https://doi.org/10.1124/mi5.1.8

Roberts CJ, Campbell IC, Troop N (2014) Increases in weight during chronic stress are partially associated with a switch in food choice towards increased carbohydrate and saturated fat intake. Eur Eat Disord Rev 22:77–82. https://doi.org/10.1002/erv.2264

Rodi D, Zucchini S, Simonato M, Cifani C, Massi M, Polidori C (2008) Functional antagonism between nociceptin/orphanin FQ (N/OFQ) and corticotropin-releasing factor (CRF) in the rat brain: evidence for involvement of the bed nucleus of the stria terminalis. Psychopharmacology 196:523–531. https://doi.org/10.1007/s00213-007-0985-7

Rodriguez-Ortega E, Alcaraz-Iborra M, de la Fuente L, Cubero I (2019a) Protective and therapeutic benefits of environmental enrichment on binge-like sucrose intake in C57BL/6J mice. Appetite 138:184–189. https://doi.org/10.1016/j.appet.2019.03.033

Rodriguez-Ortega E, Alcaraz-Iborra M, de la Fuente L, de Amo E, Cubero I (2019b) Environmental enrichment during adulthood reduces sucrose binge-like intake in a high drinking in the dark phenotype (HD) in C57BL/6J mice. Front Behav Neurosci 13:27. https://doi.org/10.3389/fnbeh.2019.00027

Romano A, Tempesta B, Micioni Di Bonaventura MV, Gaetani S (2015) From autism to eating disorders and more: the role of oxytocin in neuropsychiatric disorders. Front Neurosci 9:497. https://doi.org/10.3389/fnins.2015.00497

Rowland NE, Antelman SM (1976) Stress-induced hyperphagia and obesity in rats: a possible model for understanding human obesity. Science 191:310–312. https://doi.org/10.1126/science.1246617

Rutters F, Nieuwenhuizen AG, Lemmens SG, Born JM, Westerterp-Plantenga MS (2009) Acute stress-related changes in eating in the absence of hunger. Obesity 17:72–77. https://doi.org/10.1038/oby.2008.493

Sam AH, Salem V, Ghatei MA (2011) Rimonabant: from RIO to ban. J Obes 2011:432607. https://doi.org/10.1155/2011/432607

Sandman CA, Glynn LM, Davis EP (2013) Is there a viability-vulnerability tradeoff? Sex differences in fetal programming. J Psychosom Res 75:327–335. https://doi.org/10.1016/j.jpsychores.2013.07.009

Sanislow CA et al (2010) Developing constructs for psychopathology research: research domain criteria. J Abnorm Psychol 119:631–639. https://doi.org/10.1037/a0020909

Scherma M et al (2013) Pharmacological modulation of the endocannabinoid signalling alters binge-type eating behaviour in female rats. Br J Pharmacol 169:820–833. https://doi.org/10.1111/bph.12014

Schmidt MV et al (2006) Metabolic signals modulate hypothalamic-pituitary-adrenal axis activation during maternal separation of the neonatal mouse. J Neuroendocrinol 18:865–874. https://doi.org/10.1111/j.1365-2826.2006.01482.x

Schoofs N, Chen F, Braunig P, Stamm T, Kruger S (2011) Binge eating disorder and menstrual cycle in unmedicated women with bipolar disorder. J Affect Disord 129:75–78. https://doi.org/10.1016/j.jad.2010.08.016

Schroeder M et al (2017) A methyl-balanced diet prevents CRF-induced prenatal stress-triggered predisposition to binge eating-like phenotype. Cell Metab 25 (1269–1281):e1266. https://doi.org/10.1016/j.cmet.2017.05.001

Sgoifo A, Koolhaas J, De Boer S, Musso E, Stilli D, Buwalda B, Meerlo P (1999) Social stress, autonomic neural activation, and cardiac activity in rats. Neurosci Biobehav Rev 23:915–923

Spangler R, Wittkowski KM, Goddard NL, Avena NM, Hoebel BG, Leibowitz SF (2004) Opiate-like effects of sugar on gene expression in reward areas of the rat brain brain research. Mol Brain Res 124:134–142. https://doi.org/10.1016/j.molbrainres.2004.02.013

Stairs DJ, Klein ED, Bardo MT (2006) Effects of environmental enrichment on extinction and reinstatement of amphetamine self-administration and sucrose-maintained responding. Behav Pharmacol 17:597–604. https://doi.org/10.1097/01.fbp.0000236271.72300.0e

Stancil SL, Adelman W, Dietz A, Abdel-Rahman S (2019) Naltrexone reduces binge eating and purging in adolescents in an eating disorder program. J Child Adolesc Psychopharmacol 29(9):721–724. https://doi.org/10.1089/cap.2019.0056

Stark JL, Avitsur R, Padgett DA, Campbell KA, Beck FM, Sheridan JF (2001) Social stress induces glucocorticoid resistance in macrophages. Am J Physiol Regul Integr Comp Physiol 280:R1799–R1805. https://doi.org/10.1152/ajpregu.2001.280.6.R1799

Statnick MA et al (2016) A novel nociceptin receptor antagonist LY2940094 inhibits excessive feeding behavior in rodents: a possible mechanism for the treatment of binge eating disorder. J Pharmacol Exp Ther 356:493–502. https://doi.org/10.1124/jpet.115.228221

Sterritt GM (1962) Inhibition and facilitation of eating by electric shock. J Comp Physiol Psychol 55:226–229. https://doi.org/10.1037/h0041388

Stice E, Presnell K, Spangler D (2002) Risk factors for binge eating onset in adolescent girls: a 2-year prospective investigation. Health Psychol 21:131–138

Striegel-Moore RH, Dohm FA, Pike KM, Wilfley DE, Fairburn CG (2002) Abuse, bullying, and discrimination as risk factors for binge eating disorder. Am J Psychiatry 159:1902–1907. https://doi.org/10.1176/appi.ajp.159.11.1902

Swanson SA, Crow SJ, Le Grange D, Swendsen J, Merikangas KR (2011) Prevalence and correlates of eating disorders in adolescents: results from the national comorbidity survey replication adolescent supplement. Arch Gen Psychiatry 68:714–723. https://doi.org/10.1001/archgenpsychiatry.2011.22

Tamashiro KL, Hegeman MA, Sakai RR (2006) Chronic social stress in a changing dietary environment. Physiol Behav 89:536–542. https://doi.org/10.1016/j.physbeh.2006.05.026

Tamashiro KL, Sakai RR, Shively CA, Karatsoreos IN, Reagan LP (2011) Chronic stress, metabolism, and metabolic syndrome. Stress 14:468–474. https://doi.org/10.3109/10253890.2011.606341

Teegarden SL, Bale TL (2008) Effects of stress on dietary preference and intake are dependent on access and stress sensitivity. Physiol Behav 93:713–723. https://doi.org/10.1016/j.physbeh.2007.11.030

Thompson AK, Fourman S, Packard AE, Egan AE, Ryan KK, Ulrich-Lai YM (2015) Metabolic consequences of chronic intermittent mild stress exposure. Physiol Behav 150:24–30. https://doi.org/10.1016/j.physbeh.2015.02.038

van der Harst JE, Baars AM, Spruijt BM (2003) Standard housed rats are more sensitive to rewards than enriched housed rats as reflected by their anticipatory behaviour. Behav Brain Res 142:151–156. https://doi.org/10.1016/s0166-4328(02)00403-5

van der Ster WG, Norring C, Holmgren S (1994) Binge eating versus nonpurged eating in bulimics: is there a carbohydrate craving after all? Acta Psychiatr Scand 89:376–381. https://doi.org/10.1111/j.1600-0447.1994.tb01532.x

Vickers SP, Hackett D, Murray F, Hutson PH, Heal DJ (2015) Effects of lisdexamfetamine in a rat model of binge-eating. J Psychopharmacol 29:1290–1307. https://doi.org/10.1177/0269881115615107

Vitale G et al (2017) Effects of [Nphe(1), Arg(14), Lys(15)] N/OFQ-NH2 (UFP-101), a potent NOP receptor antagonist, on molecular, cellular and behavioural alterations associated with chronic mild stress. J Psychopharmacol 31:691–703. https://doi.org/10.1177/0269881117691456

Wang GJ et al (2011) Enhanced striatal dopamine release during food stimulation in binge eating disorder. Obesity 19:1601–1608. https://doi.org/10.1038/oby.2011.27

Ward K, Citrome L (2018) Lisdexamfetamine: chemistry, pharmacodynamics, pharmacokinetics, and clinical efficacy, safety, and tolerability in the treatment of binge eating disorder. Expert Opin Drug Metab Toxicol 14:229–238. https://doi.org/10.1080/17425255.2018. 1420163

Waters A, Hill A, Waller G (2001) Internal and external antecedents of binge eating episodes in a group of women with bulimia nervosa. Int J Eat Disord 29:17–22

Wei NL et al (2019) Chronic stress increases susceptibility to food addiction by increasing the levels of DR2 and MOR in the nucleus accumbens. Neuropsychiatr Dis Treat 15:1211–1229. https://doi.org/10.2147/NDT. S204818

Wideman C, Nadzam G, Murphy H (2005) Implications of an animal model of sugar addiction, withdrawal and relapse for human health. Nutr Neurosci 8:269–276

Wong KJ, Wojnicki FH, Corwin RL (2009) Baclofen, raclopride, and naltrexone differentially affect intake of fat/sucrose mixtures under limited access conditions. Pharmacol Biochem Behav 92:528–536. https://doi.org/10.1016/j.pbb.2009.02.002

Wood DA, Siegel AK, Rebec GV (2006) Environmental enrichment reduces impulsivity during appetitive conditioning. Physiol Behav 88:132–137. https://doi.org/10.1016/j.physbeh.2006.03.024

Xu P et al (2017) Activation of serotonin 2C receptors in dopamine neurons inhibits binge-like eating in mice. Biol Psychiatry 81:737–747. https://doi.org/10.1016/j.biopsych.2016.06.005

Yanovski SZ (1993) Binge eating disorder: current knowledge and future directions. Obes Res 1:306–324

Yu Z, Geary N, Corwin RL (2011) Individual effects of estradiol and progesterone on food intake and body weight in ovariectomized binge rats. Physiol Behav 104(5):687–693

Ziauddeen H, Farooqi IS, Fletcher PC (2012) Obesity and the brain: how convincing is the addiction model? Nat Rev Neurosci 13:279–286. https://doi.org/10.1038/nrn3212

Emerging Translational Treatments to Target the Neural Networks of Binge Eating

Wilder Doucette and Elizabeth B. Smedley

Abstract

It is an exciting time to study the neurobiological basis of binge-eating behavior, as innovative techniques to manipulate the brain are allowing for novel and more targeted treatment approaches to be evaluated. While some of these techniques are currently being tested in patients, others are at the stage of evaluation in animal models of binge eating. Animal models are useful because the interconnected brain networks that underlie binge eating are highly conserved across mammalian species. Discussed here are a brief and simplified take on three major systems of interconnected networks and their respective roles in binge-eating behavior: (1) *metabolism and caloric homeostasis*; (2) *reward and motivation*; and (3) *cognitive control and decision-making*. Also presented is a brief review of the types of innovative techniques to manipulate the brain that are currently available to translational neuroscientists. Finally, examples of their application in animal models and clinical populations are presented, with consideration of their translational potential.

Keywords

Optogenetics · Chemogenetics · Transcranial magnetic stimulation · Deep brain stimulation · Animal models · Reward · Motivation · Homeostasis · Cognition · Decision-making

Learning Objectives

In this chapter, you will:

- Learn about various approaches that can be used in rodents to study aspects of binge eating. Here we discuss models that use combinations of palatable food, dieting, and stress.
- Learn about the brain networks involved in three major systems thought to be integral to binge eating: *metabolism and caloric homeostasis; reward and motivation; and cognitive control and decision-making*.
- Learn about innovative methods of modulating brain networks that are making more poignant and novel hypotheses testable. Discussed methods include: optogenetics; chemogenetics (DREADDs); and brain stimulation (e.g., DBS, TMS, and tDCs).

W. Doucette (✉)
Department of Psychiatry, Geisel School of Medicine at Dartmouth, Hanover, NH, USA
e-mail: Wilder.T.Doucette@hitchcock.org

E. B. Smedley
Department of Psychological and Brain Sciences, Dartmouth College, Hanover, NH, USA

G. K.W. Frank, L. A. Berner (eds.), *Binge Eating*, https://doi.org/10.1007/978-3-030-43562-2_8

1   Introduction

Binge eating is a type of non-homeostatic feeding (i.e., the consumption of palatable food for reasons other than a caloric need). Binge-eating disorder (BED) is defined by the frequency and persistence of this behavior despite feelings of guilt and distress, and in the absence of compensatory behaviors (such as purging in bulimia nervosa [BN] (American Psychiatric Association and American Psychiatric Association. DSM-5 Task Force 2013). These patterns of feeding presumably emerge from a distributed and interconnected network that controls feeding behavior and has conserved elements across mammals—from rodents to humans. The capacity of these systems to drive non-homeostatic feeding behaviors, like binge eating, has become a liability to humans in the developed world. Cheap, easy access to an abundance of heavily advertised, high calorie, particularly palatable foods have contributed to an increased prevalence of a continuum of dysregulated feeding behavior (Bulik et al. 2003; Leigh and Morris 2018). These behaviors range from mild chronic caloric overconsumption, gradually leading to obesity over time, to more pathologic loss-of-control eating, food addiction, and BED.

In the chapter that follows, we briefly review rodent models of binge eating that are used to evaluate novel treatment approaches and discuss these models' strengths and limitations related to the clinical diagnoses that involve binge eating. We then briefly summarize the conserved mammalian brain systems that, in part, underpin binge eating and provide examples of these systems' dysregulation in patients and rodent models. Finally, we review the strengths and limitations of emerging methods of manipulating brain activity with treatment potential for binge eating.

2   Rodent Models of Binge Eating

Rodent models of binge eating tend to utilize foods high in sugar and fat, similar to the foods preferred by people who binge eat. Non-homeostatic, binge-like consumption of these palatable foods can be modeled in rodents in a number of different ways. Most models provide limited access to the palatable food, and various schedules of food deprivation, caloric restriction, and acute or chronic stress, either alone or in combination. These manipulations yield a variety of behavioral outcomes, each with their own merits and limitations. Below are examples of rodent models used to study binge eating—please see Chap. 7 and (Corwin et al. 2011; Corwin and Buda-Levin 2004) for comprehensive reviews of these models.

A model of sugar addiction developed by Avena and Hoebel maintains animals on a daily schedule of 12 h food deprivation followed by 12 h access to a sugar solution and normal rodent chow. In this paradigm rodents quickly show increased daily intake of the sugar solution (Avena et al. 2006). A disadvantage of this model is that it requires food deprivation, which can contribute to hyperphagia due to a caloric deficit. However, this model is advantageous in studying the "addictive" aspects of intermittent, excessive consumption of highly palatable food because animals in this model develop behaviors and brain changes paralleling the effects of addictive drugs. These similarities are demonstrated by opiate-like withdrawal both spontaneously and induced by an opioid receptor antagonist (Colantuoni et al. 2002), along with cross-sensitization to amphetamine (Avena and Hoebel 2003).

In another model that replicates the dieting and stress that often accompanies binge eating in humans, animals undergo cycles of food restriction (not deprivation) and refeeding of palatable food, as well as non-hunger-related stress via foot shocks or restraint stress prior to feeding (Artiga et al. 2007; Corwin et al. 2011). This history of dieting and stress can lead to continued reward-driven binge eating of both palatable and non-palatable food. Other rodent models have been used in which animals have unrestricted access to normal chow and limited access to a palatable sugar and/or fat-rich food (Berner et al. 2008; Corwin 2004). Importantly, these models do not require caloric restriction to trigger bouts of overconsumption.

Overall, rodent models of binge eating can result in reliable hallmarks of eating disorders like BED including weight gain (Berner et al. 2008), compulsive feeding behavior (Heal et al. 2016; Oswald et al. 2011), and increased impulsivity (Anastasio et al. 2019; Vickers et al. 2017). These traits are also commonly observed in related appetitive disorders like substance use disorders. It is important to acknowledge, however, that all of these models are preclinical approximations of various components of the clinical condition. Thus, while some pharmacologic trials using rodent/rat models have translated clinically [e.g., lisdexamfetamine—(McElroy et al. 2016; Vickers et al. 2015)], others have not.

binge eating arises from dysregulation within distributed and interconnected networks that control feeding behavior (Leigh and Morris 2018; Michaelides et al. 2012). Here, we discuss these brain networks in a simplified framework to organize a rich literature describing the impact of modulating the activity of different network nodes on feeding behavior (Leigh and Morris 2018; Rossi and Stuber 2018). These systems, and the network nodes that comprise them, are commonly divided into groups based on their (oversimplified) functional attributes: metabolism and caloric homeostasis; reward and motivation; cognitive control; and decision-making (Fig. 1).

3 The Brain Systems Underlying Binge Eating

Considerable evidence from both preclinical models and human studies suggests that the non-homeostatic, hedonic feeding component of

3.1 Metabolism and Caloric Homeostasis

Within the hindbrain, the nucleus tractus solitarius (NTS), the major visceral sensory nucleus in the brainstem, is a critical hub for the reception and projection of somatosensory information via the

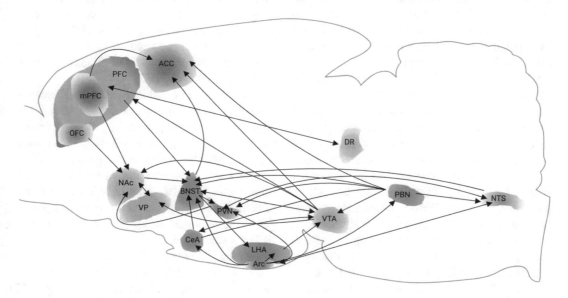

Fig. 1 Brain systems in binge eating displayed in a midsagittal rat brain. Metabolism and caloric homeostasis (teal), Reward and motivation (green), and cognitive control and decision-making (pink). Areas that contain multiple colors are mentioned in connection to multiple circuits. Abbreviations: *OFC* Orbital frontal cortex, *mPFC* Medial prefrontal cortex, *PFC* Prefrontal cortex, *ACC* Anterior cingulate cortex, *NAc* Nucleus accumbens, *VP* Ventral pallidum, *BNST* Bed nucleus of the stria terminalis, *PVN* Paraventricular nucleus of the hypothalamus, *CeA* Central amygdala, *LHA* Lateral hypothalamic area, *Arc* Arcuate nucleus, *VTA* Ventral tegmental area, *DR* Dorsal raphe nucleus, *PBN* Parabrachial nucleus, NTS Nucleus tractus solitarius

vagus, thoracic, and other cranial nerves (Grill and Hayes 2009). The NTS participates in the production of gastrointestinal and feeding responses as well as thermogenic responses mediated by the intermediolateral column of the spinal cord. The NTS also has reciprocal connections with the hypothalamus, predominantly with the paraventricular nucleus of the hypothalamus (PVN), which regulates neuroendocrine responses. Also located in the hypothalamus, adjacent to the third ventricle, is the arcuate nucleus (Arc). Cell types within the Arc (e.g., neurons that express agouti-related peptide [AgRP] or proopiomelanocortin [POMC]) express receptors for, and are strongly influenced by, many circulating molecules associated with homeostatic feeding (e.g., leptin, ghrelin, and insulin) (Cone 2005; Varela and Horvath 2012). Further, the manipulation of either Arc or PVN neuronal activity can robustly influence food intake (Cone et al. 2001; Shor-Posner et al. 1985). Both Arc and PVN neurons project to brain areas that influence a variety of motivated behaviors and ultimately interface with midbrain dopamine (DA) neurons (Baskerville and Douglas 2010). Through these types of interactions with neurons in reward and motivational networks, the hypothalamic neurons gain functions that extend far beyond homeostatic feeding.

Bridging connectivity between ventricular neurons (like the Arc and PVN populations described above) and the networks of the reward and motivational system are neurons within regions that receive axonal innervation from and provide feedback to ventricular cells. These cell groups are located within the bed nucleus of the stria terminalis (BNST), central amygdala (CeA), parabrachial nucleus (PBN), and lateral hypothalamic area (LHA). These regions are implicated in both reward processing and feeding as well as other processes (Alhadeff et al. 2014; Berthoud and Munzberg 2011; Ch'ng et al. 2018; Csaki et al. 2000; Dong and Swanson 2006; Herman et al. 2005). Currently, these regions are poorly understood in terms of the various component cell types and the corresponding connectivity that those cells exhibit. It is clear though, through manipulation of neurons within these regions

(e.g., by ablation, electrical stimulation, or pharmacological perturbation) that these regions affect feeding and reward phenotypes as well as stress and anxiety (Alhadeff et al. 2014; Berthoud and Munzberg 2011; Ch'ng et al. 2018; Csaki et al. 2000; Dong and Swanson 2006; Herman et al. 2005).

3.2   Reward and Motivation

As direct anatomical connections between ventricular neurons and reward-related regions (e.g., ventral tegmental area [VTA] and dorsal raphe nucleus [DR]) are sparse, indirect connections with the VTA likely orchestrate functional interactions. Dopaminergic VTA neurons receive input from PBN, LHA, CeA, and BNST (along with many other areas not addressed here) and send dense projections to the nucleus accumbens (NAc) and prefrontal cortex (PFC) (Beier et al. 2015; Watabe-Uchida et al. 2012). Both of these primary efferent targets of VTA DA fibers exert high-level control over food-seeking and goal-directed action. In addition, peripheral regulators of caloric homeostasis (e.g., ghrelin, leptin, and insulin) also have the capacity to directly modulate the activity of cells within the VTA and DR (Figlewicz et al. 2003; Hommel et al. 2006).

Several neurotransmitter systems involved in feeding behavior and motivation have been implicated in binge-like eating. These include endocannabinoid, opioid, and GABAergic systems that are critical for perceptions of pleasure—a fundamental component to the reward experience (Berridge et al. 2010)—and have been the target of pharmacologic agents that have shown potential in the treatment of appetitive disorders like BED (Goracci et al. 2015). However, the neurotransmitter dopamine (DA) and the DA-rich VTA have been well studied in motivated, goal-directed behavior, and has been the most studied in animal models of binge eating. The direct relationship between DA and food consumption is complex. For example, DA-deficient mice will starve to death without intervention (Zhou and Palmiter 1995), yet amphetamine and other psychostimulant drugs that increase DA lead to reduced appetite and food intake in rodents (Boekhoudt et al. 2017; Davis et al. 2012; Foltin 1989; Janhunen et al. 2013; Leibowitz 1986). Further, blockade of DA receptors can lead to weight gain and obesity as seen with the use of certain antipsychotics (Reynolds and McGowan 2017). Clearly, DA has the ability to drastically alter food intake as well as subsequent weight changes, but the exact mechanisms remain elusive (Runegaard et al. 2019).

The consumption of large quantities of palatable food tends to over-activate the reward and motivation brain systems that dopamine modulates, alter reward salience of food cues, and inhibit cognitive control networks (Volkow et al. 2008). Similarly, animal models of binge eating also demonstrate dysregulation of DA signaling (Avena and Bocarsly 2012; Smith and Robbins 2013). Decreased availability of striatal D2 receptors (D2R) is significantly correlated with binge eating scores in obese patients as well as increased hunger and desire for food in control subjects (Michaelides et al. 2012), which parallel findings in rats (Johnson and Kenny 2010). The strength of data implicating altered striatal DA signaling in the NAc in obesity, binge-eating disorders, and substance use disorders has made it a common target for therapeutic manipulation.

3.3 Cognitive Control and Decision-Making

Our understanding is still limited in how the complex functions of frontal brain structures, such as decision-making, cognitive control, and learning (Miller and Cohen 2001), exert control over food intake and contribute to binge eating behavior. Unlike the previous two brain systems, there is less homology between rodents and humans in these frontal brain structures. However, despite the lower complexity of rodent frontal structures, they possess a similar capacity to regulate appetitive behavior to human structures. Evidence from multiple studies have implicated neural processes occurring in rodent medial prefrontal cortex (Anastasio et al. 2019; Blasio et al. 2014;

Ferragud et al. 2017; Mena et al. 2013) and orbitofrontal cortex (Chawla et al. 2017), in part, through connectivity with metabolic and reward systems as (Balleine and O'Doherty 2010; Heilbronner et al. 2016). While another chapter in this book will focus on the potential relationship between frontal brain regions and binge eating through the lens of executive function/dysfunction, a few examples from the human literature are discussed below.

Functional imaging studies have shown that visual food stimuli have been associated with increased activation in the OFC, anterior cingulate cortex (ACC), and insula across all participants, while BED or obese patients showed increased activation of the medial OFC and striatum compared to control participants (Michaelides et al. 2012; Schienle et al. 2009). Increased suppression of the OFC also tends to parallel cognitive inhibition of food-related craving (Wang et al. 2009), establishing the OFC as a promising target area for selective modulation. Meanwhile, widespread literature demonstrates that BED patients, and patients with substance use disorder, display dysfunctional reward evaluation in the medial prefrontal cortex (mPFC) (Boeka and Lokken 2011). Increased functional activation of the mPFC in response to symptom-provoking stimuli in patients with anorexia nervosa and BN has been demonstrated, mirroring mPFC circuit response to symptom-related stimuli in eating, obsessive-compulsive disorder, as well as in addiction (Uher et al. 2004).

3.4 Summary

A subset of the dysfunction within feeding networks that predispose to binge eating may predate exposure to highly palatable food through the accumulation of genetic and epigenetic risk (Boggiano et al. 2007; Vucetic et al. 2011). It is also known that repeated consumption of high fat/high caloric foods (much like substances of abuse) can lead to cellular and molecular changes that result in altered neural processes within feeding networks (Barry et al. 2009; Luscher and Malenka 2011; Michaelides et al. 2012; Volkow et al. 2012). Thus, a combination of innate heterogeneity (Kirkpatrick et al. 2017; Naef et al. 2011; Vucetic et al. 2011) and its intersection with the plasticity that feeding networks undergo with repeated exposure to palatable food (Johnson and Kenny 2010; Vucetic et al. 2011) and other environmental influences (stress), likely gives rise to variation in the pathophysiology that leads to and sustains binge eating. This variation likely influences the variable response to a given treatment across individuals that are observed in patients, and even some rodent models. Below, we discuss some of the recent technological advances which make it possible to more specifically target the brain networks that contribute to binge eating and then provide examples of their evaluation in rodents and, for some treatment types, in patients.

4 Emerging Treatments for Binge Eating

Since the brain operates across multiple levels of organization, manipulations of the brain can take several forms—from genetic to neural network manipulations (altering the interaction of populations of cells). Because the networks that control feeding behavior greatly overlap with those contributing to other appetitive disorders like addiction, outcomes of emerging interventions in models of other appetitive disorders likely have some relevance for binge eating (Berridge et al. 2010; Johnson and Kenny 2010; Lutter and Nestler 2009). Outlined below is a brief introduction to novel interventions that manipulate the brain networks reviewed above and hold promise for the treatment of binge eating and other forms of dysregulated feeding.

4.1 Optogenetics and Chemogenetics

4.1.1 Optogenetics Methods
Optogenetics is a technique that leverages targeted viral-mediated delivery of opsin-coding DNA into neural tissue (Bernstein and Boyden

2011; Fenno et al. 2011). Once delivered, the opsins are expressed via transcription and translation machinery endogenous to the cells of interest. Opsins, found naturally among a variety of species, are receptors activated by light that drive changes in membrane voltage at the sub-second timescale (Fig. 2). A wide variety of opsins exist that can induce excitatory or inhibitory changes (Fenno et al. 2011; Yizhar et al. 2011). The viral-mediated delivery can also be manipulated to target-specific neuronal types and pathways by leveraging cell type-specific promotors. More complex expression of opsins [pathway specific (Kim et al. 2017; Malvaez et al. 2019) or activity-driven expression (Armstrong et al. 2013; Delbeke et al. 2017; Emiliani et al. 2015; Grosenick et al. 2015; Liu et al. 2012; Paz et al. 2013)] requires multiple viral vectors targeted to different brain regions.

Limitations of this approach include the need for brain surgery to inject the viral vectors and a permanently implanted light source with power and hardware to control the timing of stimulation. The fine temporal and spatial precision of optogenetic stimulation would make it amenable to closed-loop and adaptive stimulation, with modifiable stimulation parameters based on feedback from external data (e.g., location) or internal data (e.g., central or peripheral nervous system activity). This intervention would have the advantage of being constantly available to patients as they go about their lives and encounter triggers for binge eating (similar to current adaptive deep brain stimulation systems).

4.1.2 Optogenetic Example: Targeting the Cognitive Control and Decision-Making Network in Mice
As described earlier, DA can have powerful control over feeding behaviors and yet its mechanism in driving feeding remains elusive. Land and colleagues investigated the activity within D1-type DA receptor-expressing neurons in the mPFC of mice (Land et al. 2014). They determined that food intake leads to increased activity of these D1 mPFC cells and that optogenetic excitation leads to increased feeding while optogenetic

Viral Infusion

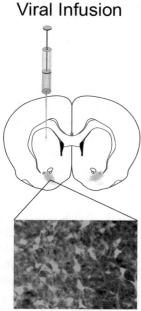

Transient Manipulation

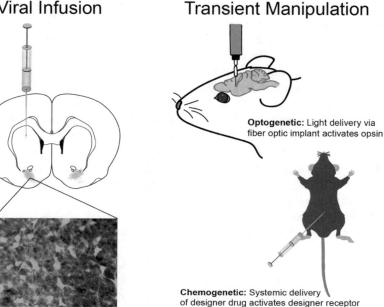

Optogenetic: Light delivery via fiber optic implant activates opsin

Chemogenetic: Systemic delivery of designer drug activates designer receptor

Fig. 2 Optogenetic and chemogenetic (DREADD) methods. (Left) A surgical procedure introduces the desired virus into the brain region(s) of interest. An example of viral expression in the ventral striatum is shown below [adapted from Smedley et al. (2019)]. (Right) Manipulation of the transfected cells in the case of an optogenetic virus, will require light delivery, typically via an implanted fiber optic placed above the region (above). In the case of a chemogenetic (DREADD) virus, delivery of a designer drug via systemic administration (injections in rodents or orally in patients) is required (below)

inhibition resulted in decreased feeding. They further demonstrated that the medial basolateral amygdala (mBLA) was a downstream target of these D1 mPFC cells and stimulation of the D1 mPFC projecting terminals in mBLA (rather than the cell bodies located in mPFC) was sufficient to drive food consumption. Similar pathway- and cell type-specific manipulations that have been performed across the interconnected networks controlling feeding have shown a capacity to alter feeding behavior. To date, no studies have attempted to implement an optogenetic intervention in a translational model of binge eating as most studies have been more mechanistic—attempting to disentangle the complex connectivity and behavioral influence of various cell types within the network nodes involved in the control of feeding behavior (described briefly above). In the future, the capacity to determine how and where specific cell types are dysregulated in a given individual could guide the personalization of an intervention like optogenetics.

4.1.3 Chemogenetics (DREADDs) Methods

Designer receptors exclusively activated by designer drugs (DREADDs or chemogenetics) utilizes a similar viral-mediated delivery of DNA coding for manipulated forms of proteins (receptors) into brain areas and cell types of interest (Fig. 2). These receptors are activated by a designer ligand that, in theory, has no other action in the human body (i.e., no off-target effects) [for reviews see (Roth 2016; Smith et al. 2016; Sternson and Roth 2014)]. The most commonly used proteins are the G-protein-coupled receptors (GCPRs) that include the excitatory Gq-DREADD hM3Dq and the inhibitory Gi-DREADDs hM2Di and hM4Di, all of which are activated by clozapine-N-oxide (CNO). CNO is an inert ligand that only binds to the DREADD receptor (Armbruster et al. 2007). Unlike some other pharmacological manipulations, CNO can be used repeatedly in the same animals without lingering behavioral effects (Mahler and Aston-Jones 2018;

Mahler et al. 2014). However, CNO is a metabolite of the drug clozapine, used therapeutically as an atypical antipsychotic, and may not be as inert as intended (Gomez et al. 2017). Recent studies emphasize the importance of proper dosing and controls or the use of another, albeit less studied, ligand—Compound 21 (Chen et al. 2015; Roth 2016). In the future, a truly inert ligand may be validated that will be able to selectively target DREADDs with no off-target effects. Unlike the subsecond time scale of optogenetics, DREADD activation occurs over a longer time course and depends on the pharmacokinetic properties of ligand concentrations in the brain (Urban and Roth 2015; Zhu and Roth 2014). Most studies have used acute activations that last less than 2 h and investigate behaviors occurring within that timeframe. To date, few studies have investigated the effects of chronic activation of DREADDs in models of chronic neuropsychiatric illness (Hao et al. 2019). Similar to optogenetics, the delivery of the DREADD virus requires an invasive intervention to target brain areas. Unlike optogenetics, following expression, DREADDs are able to be activated with relatively noninvasive approaches. This could theoretically occur with oral medication or with a long-acting injectable form of the activating ligand. Limitations of this approach are that it could not be personalized once expressed, and the timing of activation would have a relatively low temporal resolution (hours) depending on the mode of administration of the activating ligand. However, this degree of temporal control may allow patients experiencing persistent binge eating cravings to instead take a pill to reduce the likelihood of acting on those cravings.

4.1.4 Chemogenetic Example: Targeting the Reward and Motivation Network in Rats

Multiple investigators have explored the impact of altering activity in the ventral pallidum (VP), a structure within the ventral striatum that is a part of the reward and motivation system. DREADD-mediated inhibition of the VP has been shown to reduce palatable food consumption, while manipulations that stimulate the VP have been shown to elevate food consumption (Castro

et al. 2015; Chang et al. 2017; Creed et al. 2016; Richard et al. 2018; Root et al. 2015; Smith et al. 2009; Tooley et al. 2018). Moreover, DREADD-mediated inhibition of the VP can be effective at reducing motivational attraction to food cues (Chang et al. 2015) and motivation to seek other rewards (Chang et al. 2017; Mahler et al. 2014; Prasad and McNally 2016). These examples highlight the potential of DREADD-based manipulations to impact palatable food consumption. Future translationally focused studies would need to test these interventions in rodent models of binge eating and begin to explore how DREADDs could be employed to treat chronic illness like BED or BN.

4.1.5 Considerations for Optogenetic and Chemogenetic Manipulations

It is important to note that both optogenetic and chemogenetic manipulations are distinct from other pharmacological or electrical manipulations of tissue in a number of ways. When compared to a permanent electrolytic lesion, for example, optogenetic or DREADD inactivation is phasic—be it on a short time scale (optogenetics) or a longer time scale (DREADDs). Theoretically, this means that the tissue of interest will maintain its normal functioning when it is not activated. For therapeutic potential, this is likely a useful feature as neural tissue serves many purposes and permanent suppression of activity that would treat binge eating may have many other significant and adverse side effects because of that regions' likely involvement in other complex behaviors. Thus, there remain many unanswered questions about how to successfully implement and personalize these manipulations (i.e., Just prior to a binge onset? Just after a binge? Should activation last for a long time, or is just a short time needed? Are all of these parameters unique to each patient?). Additionally, even the best viral delivery is unlikely to fully express in all of the targeted tissue—that is, DREADD or optogenetic expression is limited to only a percentage of the targeted cells of interest whereas ablations tend to be more complete (Gremel and Costa 2013; Smith et al. 2016). Additionally, with a few exceptions (Busskamp et al. 2010; Stauffer et al. 2016) studies using these technologies have been predominantly limited to rodent work. Use in primates is limited because of technical hurdles, including providing sufficient light to activate a much larger brain area without overheating and the ability to target certain cell populations (Izpisua Belmonte et al. 2015) that have not been as prohibitive in small rodents [for review see Galvan et al. (2018), Wykes et al. (2016)]. Despite these important considerations, optogenetic and DREADD technology allows for outstanding flexibility in targeting regions, connections, and cell subtypes at various timescales. This level of precision creates real opportunities for future therapeutic development.

4.2 Electrical and Magnetic Stimulation

Electrical and magnetic stimulation approaches are currently emerging treatment modalities in which a modulating external stimulus alters the electrical activity of cell bodies and axonal fibers within a targeted brain region. These interventions exist in both invasive and noninvasive forms which then occur in either discrete sessions or are available continuously as wearable/implanted devices (Luigjes et al. 2019; Spagnolo and Goldman 2017).

4.2.1 Noninvasive Methods

One noninvasive approach, transcranial magnetic stimulation (TMS), uses pulsed magnetic fields to stimulate a brain region within a few centimeters of the scalp (thus limiting brain targets to cortical locations). TMS is somewhat effective in treating mood-related disorders, particularly depression (Dunlop et al. 2017), and its potential to treat other neuropsychiatric disorders, including binge eating and addiction is under investigation (McClelland et al. 2013; Rachid 2018). Other noninvasive approaches include the passage of low energy electrical current of varying shapes. Transcranial direct current stimulation (tDCs) uses square pulses of current and transcranial alternating current (tACs) uses sine waves to alter neural activity. These low energy interventions are believed to alter the membrane potential of cells below the threshold for action potential generation whereas TMS is able to cause larger changes in membrane potential. To date, most clinical studies that have used tDCs and tACs have targeted large brain volumes, but advances in electrode designs are allowing for the targeting of more discrete brain regions (Huang and Parra 2019; Woods et al. 2016). All of the noninvasive stimulation approaches have been tried in patients with eating disorders including BN and BED (discussed below).

A major advantage of noninvasive stimulation is that it has a low risk of morbidity, and can occur in an outpatient clinic setting. These interventions have some capacity to be personalized, in terms of targeting and stimulation parameters, based on an individual's unique brain dysfunction. While the technology exists for tDCs and tACs to be wearable devices or administered at home, the power demands of a TMS device make the possibility of a wearable device less likely. This means that the changes in brain activity that are induced with interventions that are not wearable must be durable—persisting from the clinic into everyday life. Maintaining sustained efficacy for noninvasive interventions like TMS has been a major hurdle and limitation of this treatment type.

4.2.2 Noninvasive Stimulation Example: Targeting the Cognitive Control and Decision-Making Network in Humans

The outcomes of randomized, sham-controlled trials of TMS for BN have been mixed. For instance, a 38-patient trial of single session left dorsolateral prefrontal cortex (dlPFC) TMS reported a decreased urge to eat and fewer binge eating episodes at 24 h post-treatment compared to sham stimulation (Van den Eynde et al. 2010). However, other studies utilizing 10–15 days of treatment to the left dlPFC did not find any significant differences in binge eating or purging symptoms between active and sham repetitive (r)TMS (Gay et al. 2016; Walpoth et al. 2008). Another study that used 20 rTMS sessions targeting dorsomedial, not dorsolateral, PFC in a transdiagnostic sample with binge eating and purging (anorexia nervosa, binge eating/purging subtype, and BN) showed that just over half of the participants were classified as treatment responders with a > 50% reduction in binge and purge frequency at follow-up. Interestingly, this study was able to identify differences in cortical–striatal connectivity between patients that had a response to treatment compared to those that did not (Dunlop et al. 2015). This suggests, as has been supported in rodent models (Doucette et al. 2018), that individual differences in neural network activity correlates with variable treatment outcomes (Dunlop et al. 2016). Thus, brain activity measures from relevant networks could be used to identify likely treatment responders and to personalize interventions.

4.2.3   Invasive Methods

A current invasive intervention is deep brain stimulation (DBS), a neurosurgical procedure in which a stimulating electrode is implanted in a specific brain region or regions. The targets are usually nodes of a dysfunctional network and can include deep structures within the brain that are beyond the reach of TMS. Electric current is passed through the electrodes to ameliorate the dysfunctional activity of the targeted network. While DBS has a well validated and established role in the treatment of Parkinson's disease (Mansouri et al. 2018), it has an emerging utility in other neurologic disorders (e.g., epilepsy) (Deeb et al. 2016). There has been more limited use of DBS in patients with treatment-resistant obsessive-compulsive disorder (OCD) (Deeb et al. 2016; Greenberg et al. 2006). The use of DBS to treat other neuropsychiatric disorders (including depression, addiction, obesity, and eating disorders) are under ongoing development. DBS is useful in rodent models because it can be scaled down to target small rodent brain volumes that correspond to the structures targeted by TMS or DBS in humans. While current TMS systems are unable to target discrete structures in the rodent brain due to the large volume of activation, this technology will likely improve to allow for

the targeting of deeper structures in humans and smaller structures in animal models.

4.2.4   Invasive Stimulation Examples: Targeting the Reward and Motivation Network in Rodents and Humans

Multiple groups have investigated DBS targeted to the NAc core or shell, two ventral striatal regions with known differences in anatomical and functional connectivity and different functional roles across an array of reward-related behaviors (Burton et al. 2015; Haber 2016). Work in mice initially supported the potential of DBS targeted to the NAc shell to selectively attenuate binge-like eating (Halpern et al. 2013). Subsequent work in rats demonstrated a similar potential to reduce binge-like eating with DBS targeted to the NAc core (Doucette et al. 2015). However, similar to the variable treatment outcomes that have been observed in patients receiving TMS (Dunlop et al. 2015), substantial individual variability in response to both types of stimulation are well documented. The DBS treatment outcome variation observed with NAc core stimulation (Doucette et al. 2015) was also seen with NAc shell stimulation by multiple groups (Doucette et al. 2018; Wu et al. 2018). Interestingly, work in humans (Dunlop et al. 2015) and rodents (Doucette et al. 2018) suggests that individual differences in brain activity could be used to predict treatment outcomes. These findings highlight a potential avenue to personalize these types of interventions for the treatment of binge eating (Dunlop et al. 2016).

Human studies have utilized DBS to stimulate network nodes within the reward and motivation system, specifically the NAc, as supported by the preclinical data. In two separate case reports, bilateral NAc DBS resulted in a reduction in body mass index in patients with obesity (Harat et al. 2016; Mantione et al. 2010). While these data suggest the potential of reward system-targeted DBS for overeating, a pilot clinical study is the logical next step to examine whether the treatment can effectively reduce binge eating in patients.

4.3 Next Steps

For all of the above intervention types, there remain multiple areas for further development and testing. Across these translational intervention types, there is an opportunity to personalize the intervention for individuals. For some interventions, this could include the selection of an optimal brain target for stimulation or tuning the parameters of stimulation (e.g., frequency). This could even include timing stimulation pulses to optimally interface with ongoing brain activity. In the case of noninvasive interventions that are not wearable/implantable, advances are needed to extend the duration of effect in order to have a meaningful clinical impact. In the case of invasive and wearable interventions, there is a need to identify states (days to hours) or moments (seconds to minutes) that predict high-risk periods of engaging in binge eating behavior. This information could come from external predictors that identify proximity to triggers (people and places) or predictors that identify internal triggers (anxiety, impulsivity, and mood). These predictors could then be used in closed-loop systems to trigger an intervention (DBS, optogenetic) and alter the probability of engaging in a binge episode. Work in mice and rats has demonstrated the capacity of oscillatory brain activity (local field potentials), measured with DBS electrodes, to predict impending binge eating episodes. It was further demonstrated that these predictors could be used to trigger stimulation to decrease binge eating (Dwiel et al. 2019; Wu et al. 2018). Similar interventions for humans have already been developed for other conditions. For example, in epilepsy, closed-loop stimulation systems have been developed to detect neural activity that predicts an imminent seizure and trigger stimulation to prevent seizure onset (Morrell and Group 2011; Wykes et al. 2016). It is likely that such physiologic measures could be identified and used to predict the imminent onset of a binge eating episode. Preclinical work demonstrates the feasibility of such an approach and the accuracy of the prediction could likely be further improved with additional relevant predictors (e.g., GPS-derived location).

Beyond personalized timing of an intervention, it is possible that the intervention itself could be tuned to achieve a therapeutic brain state. A neural activity-based marker of therapeutic efficacy, once validated, could be used to guide the tuning of stimulation parameters (e.g., intensity, frequency, and field shape). This kind of adaptive stimulation has been developed and shown to be advantageous in Parkinson's disease (Little et al. 2016; Rosin et al. 2011). The first step to developing a similar approach in binge eating is to identify a marker of therapeutic brain state(s) (associated with a lower probability of engaging in binge eating). These states could be universal and generalize across individuals, but more than likely will need to be individualized (Dunlop et al. 2016).

5 Conclusion

Preclinical studies that have used the previously described translational interventions to modulate the activity of brain networks in order to regulate binge eating and related appetitive behaviors support their therapeutic potential. However, the considerable variability in treatment outcomes that have been observed across individuals highlights the need to individualize the implementation of the interventions. Clinical studies have also shown both the treatment potential of targeting the networks underlying binge eating with stimulation and treatment response heterogeneity across individuals. In addition, clinical studies have been limited by small sample sizes, questionable placebo/sham-controlled designs, and the use of subjective scales/self-report as the primary outcomes. While the published outcomes of targeted brain stimulation hint at the treatment potential for appetitive disorders like binge eating, ongoing work to improve treatment efficacy through personalization is needed before focal stimulation will significantly contribute to clinical treatment programs.

References

Alhadeff AL, Hayes MR, Grill HJ (2014) Leptin receptor signaling in the lateral parabrachial nucleus contributes to the control of food intake. Am J Physiol Regul Integr Comp Physiol 307(11):R1338–R1344

American Psychiatric Association and American Psychiatric Association. DSM-5 Task Force (2013) Diagnostic and statistical manual of mental disorders: DSM-5, 5th edn. American Psychiatric Association, Washington, DC, p xliv, 947

Anastasio NC, Stutz SJ, Price AE, Davis-Reyes BD, Sholler DJ, Ferguson SM et al (2019) Convergent neural connectivity in motor impulsivity and high-fat food binge-like eating in male Sprague-Dawley rats. Neuropsychopharmacology 44:1752–1761

Armbruster BN, Li X, Pausch MH, Herlitze S, Roth BL (2007) Evolving the lock to fit the key to create a family of G protein-coupled receptors potently activated by an inert ligand. Proc Natl Acad Sci USA 104(12):5163–5168

Armstrong C, Krook-Magnuson E, Oijala M, Soltesz I (2013) Closed-loop optogenetic intervention in mice. Nat Protoc 8(8):1475–1493

Artiga AI, Viana JB, Maldonado CR, Chandler-Laney PC, Oswald KD, Boggiano MM (2007) Body composition and endocrine status of long-term stress-induced binge-eating rats. Physiol Behav 91(4):424–431

Avena NM, Bocarsly ME (2012) Dysregulation of brain reward systems in eating disorders: neurochemical information from animal models of binge eating, bulimia nervosa, and anorexia nervosa. Neuropharmacology 63(1):87–96

Avena NM, Hoebel BG (2003) A diet promoting sugar dependency causes behavioral cross-sensitization to a low dose of amphetamine. Neuroscience 122(1):17–20

Avena NM, Rada P, Hoebel BG (2006) Sugar bingeing in rats. Curr Protoc Neurosci Chapter 9:Unit9 23C

Balleine BW, O'Doherty JP (2010) Human and rodent homologies in action control: corticostriatal determinants of goal-directed and habitual action. Neuropsychopharmacology 35(1):48–69

Barry D, Clarke M, Petry NM (2009) Obesity and its relationship to addictions: is overeating a form of addictive behavior? Am J Addict 18(6):439–451

Baskerville TA, Douglas AJ (2010) Dopamine and oxytocin interactions underlying behaviors: potential contributions to behavioral disorders. CNS Neurosci Ther 16(3):e92–e123

Beier KT, Steinberg EE, DeLoach KE, Xie S, Miyamichi K, Schwarz L et al (2015) Circuit architecture of VTA dopamine neurons revealed by systematic input-output mapping. Cell 162(3):622–634

Berner LA, Avena NM, Hoebel BG (2008) Bingeing, self-restriction, and increased body weight in rats with limited access to a sweet-fat diet. Obesity 16(9):1998–2002

Bernstein JG, Boyden ES (2011) Optogenetic tools for analyzing the neural circuits of behavior. Trends Cogn Sci 15(12):592–600

Berridge KC, Ho CY, Richard JM, DiFeliceantonio AG (2010) The tempted brain eats: pleasure and desire circuits in obesity and eating disorders. Brain Res 1350:43–64

Berthoud HR, Munzberg H (2011) The lateral hypothalamus as integrator of metabolic and environmental needs: from electrical self-stimulation to opto-genetics. Physiol Behav 104(1):29–39

Blasio A, Steardo L, Sabino V, Cottone P (2014) Opioid system in the medial prefrontal cortex mediates binge-like eating. Addict Biol 19(4):652–662

Boeka AG, Lokken KL (2011) Prefrontal systems involvement in binge eating. Eat Weight Disord 16(2):e121–e126

Boekhoudt L, Roelofs TJM, de Jong JW, de Leeuw AE, Luijendijk MCM, Wolterink-Donselaar IG et al (2017) Does activation of midbrain dopamine neurons promote or reduce feeding? Int J Obes 41(7):1131–1140

Boggiano MM, Artiga AI, Pritchett CE, Chandler-Laney PC, Smith ML, Eldridge AJ (2007) High intake of palatable food predicts binge-eating independent of susceptibility to obesity: an animal model of lean vs obese binge-eating and obesity with and without binge-eating. Int J Obes 31(9):1357–1367

Bulik CM, Sullivan PF, Kendler KS (2003) Genetic and environmental contributions to obesity and binge eating. Int J Eat Disord 33(3):293–298

Burton AC, Nakamura K, Roesch MR (2015) From ventral-medial to dorsal-lateral striatum: neural correlates of reward-guided decision-making. Neurobiol Learn Mem 117:51–9

Busskamp V, Duebel J, Balya D, Fradot M, Viney TJ, Siegert S et al (2010) Genetic reactivation of cone photoreceptors restores visual responses in retinitis pigmentosa. Science 329(5990):413–417

Castro DC, Cole SL, Berridge KC (2015) Lateral hypothalamus, nucleus accumbens, and ventral pallidum roles in eating and hunger: interactions between homeostatic and reward circuitry. Front Syst Neurosci 9:90

Ch'ng S, Fu J, Brown RM, McDougall SJ, Lawrence AJ (2018) The intersection of stress and reward: BNST modulation of aversive and appetitive states. Prog Neuropsychopharmacol Biol Psychiatry 87(Pt A):108–125

Chang SE, Todd TP, Bucci DJ, Smith KS (2015) Chemogenetic manipulation of ventral pallidal neurons impairs acquisition of sign-tracking in rats. Eur J Neurosci 42(12):3105–3116

Chang SE, Smedley EB, Stansfield KJ, Stott JJ, Smith KS (2017) Optogenetic inhibition of ventral pallidum neurons impairs context-driven salt seeking. J Neurosci Off J Soc Neurosci 37(23):5670–5680

Chawla A, Cordner ZA, Boersma G, Moran TH (2017) Cognitive impairment and gene expression alterations in a rodent model of binge eating disorder. Physiol Behav 180:78–90

Chen X, Choo H, Huang XP, Yang X, Stone O, Roth BL et al (2015) The first structure-activity relationship studies for designer receptors exclusively activated by designer drugs. ACS Chem Neurosci 6(3):476–484

Colantuoni C, Rada P, McCarthy J, Patten C, Avena NM, Chadeayne A et al (2002) Evidence that intermittent, excessive sugar intake causes endogenous opioid dependence. Obes Res 10(6):478–488

Cone RD (2005) Anatomy and regulation of the central melanocortin system. Nat Neurosci 8(5):571–578

Cone RD, Cowley MA, Butler AA, Fan W, Marks DL, Low MJ (2001) The arcuate nucleus as a conduit for diverse signals relevant to energy homeostasis. Int J Obes Relat Metab Disord 25(Suppl 5):S63–S67

Corwin RL (2004) Binge-type eating induced by limited access in rats does not require energy restriction on the previous day. Appetite 42(2):139–142

Corwin RL, Buda-Levin A (2004) Behavioral models of binge-type eating. Physiol Behav 82(1):123–130

Corwin RL, Avena NM, Boggiano MM (2011) Feeding and reward: perspectives from three rat models of binge eating. Physiol Behav 104(1):87–97

Creed M, Ntamati NR, Chandra R, Lobo MK, Luscher C (2016) Convergence of reinforcing and Anhedonic cocaine effects in the ventral pallidum. Neuron 92 (1):214–226

Csaki A, Kocsis K, Halasz B, Kiss J (2000) Localization of glutamatergic/aspartatergic neurons projecting to the hypothalamic paraventricular nucleus studied by retrograde transport of [3H]D-aspartate autoradiography. Neuroscience 101(3):637–655

Davis C, Levitan RD, Yilmaz Z, Kaplan AS, Carter JC, Kennedy JL (2012) Binge eating disorder and the dopamine D2 receptor: genotypes and sub-phenotypes. Prog Neuro-Psychopharmacol Biol Psychiatry 38(2):328–335

Deeb W, Giordano JJ, Rossi PJ, Mogilner AY, Gunduz A, Judy JW et al (2016) Proceedings of the fourth annual deep brain stimulation think tank: a review of emerging issues and technologies. Front Integr Neurosci 10:38

Delbeke J, Hoffman L, Mols K, Braeken D, Prodanov D (2017) And then there was light: perspectives of Optogenetics for deep brain stimulation and neuromodulation. Front Neurosci 11:663

Dong HW, Swanson LW (2006) Projections from bed nuclei of the stria terminalis, anteromedial area: cerebral hemisphere integration of neuroendocrine, autonomic, and behavioral aspects of energy balance. J Comp Neurol 494(1):142–178

Doucette WT, Khokhar JY, Green AI (2015) Nucleus accumbens deep brain stimulation in a rat model of binge eating. Transl Psychiatry 5:e695

Doucette WT, Dwiel L, Boyce JE, Simon AA, Khokhar JY, Green AI (2018) Machine learning based classification of deep brain stimulation outcomes in a rat model of binge eating using ventral striatal oscillations. Front Psych 9:336

Dunlop K, Woodside B, Lam E, Olmsted M, Colton P, Giacobbe P et al (2015) Increases in frontostriatal connectivity are associated with response to dorsomedial repetitive transcranial magnetic stimulation in refractory binge/purge behaviors. Neuroimage Clin 8:611–618

Dunlop KA, Woodside B, Downar J (2016) Targeting neural Endophenotypes of eating disorders with non-invasive brain stimulation. Front Neurosci 10:30

Dunlop K, Hanlon CA, Downar J (2017) Noninvasive brain stimulation treatments for addiction and major depression. Ann N Y Acad Sci 1394(1):31–54

Dwiel LL, Khokhar JY, Connerney MA, Green AI, Doucette WT (2019) Finding the balance between model complexity and performance: using ventral striatal oscillations to classify feeding behavior in rats. PLoS Comput Biol 15(4):e1006838

Emiliani V, Cohen AE, Deisseroth K, Hausser M (2015) All-optical interrogation of neural circuits. J Neurosci Off J Soc Neurosci 35(41):13917–13926

Fenno L, Yizhar O, Deisseroth K (2011) The development and application of optogenetics. Annu Rev Neurosci 34:389–412

Ferragud A, Howell AD, Moore CF, Ta TL, Hoener MC, Sabino V et al (2017) The trace amine-associated receptor 1 agonist RO5256390 blocks compulsive, binge-like eating in rats. Neuropsychopharmacology 42(7):1458–1470

Figlewicz DP, Evans SB, Murphy J, Hoen M, Baskin DG (2003) Expression of receptors for insulin and leptin in the ventral tegmental area/substantia nigra (VTA/SN) of the rat. Brain Res 964(1):107–115

Foltin RW (1989) Effects of anorectic drugs on the topography of feeding behavior in baboons. J Pharmacol Exp Ther 249(1):101–109

Galvan A, Caiola MJ, Albaugh DL (2018) Advances in optogenetic and chemogenetic methods to study brain circuits in non-human primates. J Neural Transm 125 (3):547–563

Gay A, Jaussent I, Sigaud T, Billard S, Attal J, Seneque M et al (2016) A lack of clinical effect of high-frequency rTMS to dorsolateral prefrontal cortex on bulimic symptoms: a randomised, double-blind trial. Eur Eat Disord Rev 24(6):474–481

Gomez JL, Bonaventura J, Lesniak W, Mathews WB, -Sysa-Shah P, Rodriguez LA et al (2017) Chemogenetics revealed: DREADD occupancy and activation via converted clozapine. Science 357 (6350):503–507

Goracci A, di Volo S, Casamassima F, Bolognesi S, Benbow J, Fagiolini A (2015) Pharmacotherapy of binge-eating disorder: a review. J Addict Med 9 (1):1–19

Greenberg BD, Malone DA, Friehs GM, Rezai AR, Kubu CS, Malloy PF et al (2006) Three-year outcomes in deep brain stimulation for highly resistant obsessive-compulsive disorder. Neuropsychopharmacology 31 (11):2384–2393

Gremel CM, Costa RM (2013) Orbitofrontal and striatal circuits dynamically encode the shift between goal-directed and habitual actions. Nat Commun 4:2264

Grill HJ, Hayes MR (2009) The nucleus tractus solitarius: a portal for visceral afferent signal processing, energy status assessment and integration of their combined effects on food intake. Int J Obes 33(Suppl 1):S11–S15

Grosenick L, Marshel JH, Deisseroth K (2015) Closed-loop and activity-guided optogenetic control. Neuron 86(1):106–139

Haber SN (2016) Corticostriatal circuitry. Dialogues Clin Neurosci 18(1):7–21

Halpern CH, Tekriwal A, Santollo J, Keating JG, Wolf JA, Daniels D et al (2013) Amelioration of binge eating by nucleus accumbens shell deep brain stimulation in mice involves D2 receptor modulation. J Neurosci Off J Soc Neurosci 33(17):7122–7129

Hao S, Yang H, Wang X, He Y, Xu H, Wu X et al (2019) The lateral hypothalamic and BNST GABAergic projections to the anterior ventrolateral periaqueductal gray regulate feeding. Cell Rep 28(3):616–624.e5

Harat M, Rudas M, Zielinski P, Birska J, Sokal P (2016) Nucleus accumbens stimulation in pathological obesity. Neurol Neurochir Pol 50(3):207–210

Heal DJ, Goddard S, Brammer RJ, Hutson PH, Vickers SP (2016) Lisdexamfetamine reduces the compulsive and perseverative behaviour of binge-eating rats in a novel food reward/punished responding conflict model. J Psychopharmacol 30(7):662–675

Heilbronner SR, Rodriguez-Romaguera J, Quirk GJ, Groenewegen HJ, Haber SN (2016) Circuit-based corticostriatal homologies between rat and primate. Biol Psychiatry 80(7):509–521

Herman JP, Ostrander MM, Mueller NK, Figueiredo H (2005) Limbic system mechanisms of stress regulation: hypothalamo-pituitary-adrenocortical axis. Prog Neuro-Psychopharmacol Biol Psychiatry 29(8):1201–1213

Hommel JD, Trinko R, Sears RM, Georgescu D, Liu ZW, Gao XB et al (2006) Leptin receptor signaling in midbrain dopamine neurons regulates feeding. Neuron 51(6):801–810

Huang Y, Parra LC (2019) Can transcranial electric stimulation with multiple electrodes reach deep targets? Brain Stimul 12(1):30–40

Izpisua Belmonte JC, Callaway EM, Caddick SJ, Churchland P, Feng G, Homanics GE et al (2015) Brains, genes, and primates. Neuron 86(3):617–631

Janhunen SK, la Fleur SE, Adan RA (2013) Blocking alpha2A adrenoceptors, but not dopamine receptors, augments bupropion-induced hypophagia in rats. Obesity 21(12):E700–E708

Johnson PM, Kenny PJ (2010) Dopamine D2 receptors in addiction-like reward dysfunction and compulsive eating in obese rats. Nat Neurosci 13(5):635–641

Kim CK, Adhikari A, Deisseroth K (2017) Integration of optogenetics with complementary methodologies in systems neuroscience. Nat Rev Neurosci 18(4):222–235

Kirkpatrick SL, Goldberg LR, Yazdani N, Babbs RK, Wu J, Reed ER et al (2017) Cytoplasmic FMR1-interacting protein 2 is a major genetic factor underlying binge eating. Biol Psychiatry 81(9):757–769

Land BB, Narayanan NS, Liu RJ, Gianessi CA, Brayton CE, Grimaldi DM et al (2014) Medial prefrontal D1 dopamine neurons control food intake. Nat Neurosci 17(2):248–253

Leibowitz SF (1986) Brain monoamines and peptides: role in the control of eating behavior. Fed Proc 45(5):1396–1403

Leigh SJ, Morris MJ (2018) The role of reward circuitry and food addiction in the obesity epidemic: an update. Biol Psychol 131:31–42

Little S, Tripoliti E, Beudel M, Pogosyan A, Cagnan H, Herz D et al (2016) Adaptive deep brain stimulation for Parkinson's disease demonstrates reduced speech side effects compared to conventional stimulation in the acute setting. J Neurol Neurosurg Psychiatry 87(12):1388–1389

Liu X, Ramirez S, Pang PT, Puryear CB, Govindarajan A, Deisseroth K et al (2012) Optogenetic stimulation of a hippocampal engram activates fear memory recall. Nature 484(7394):381–385

Luigjes J, Segrave R, de Joode N, Figee M, Denys D (2019) Efficacy of invasive and non-invasive brain modulation interventions for addiction. Neuropsychol Rev 29(1):116–138

Luscher C, Malenka RC (2011) Drug-evoked synaptic plasticity in addiction: from molecular changes to circuit remodeling. Neuron 69(4):650–663

Lutter M, Nestler EJ (2009) Homeostatic and hedonic signals interact in the regulation of food intake. J Nutr 139(3):629–632

Mahler SV, Aston-Jones G (2018) CNO evil? Considerations for the use of DREADDs in behavioral neuroscience. Neuropsychopharmacology 43(5):934–936

Mahler SV, Vazey EM, Beckley JT, Keistler CR, McGlinchey EM, Kaufling J et al (2014) Designer receptors show role for ventral pallidum input to ventral tegmental area in cocaine seeking. Nat Neurosci 17(4):577–585

Malvaez M, Shieh C, Murphy MD, Greenfield VY, Wassum KM (2019) Distinct cortical-amygdala projections drive reward value encoding and retrieval. Nat Neurosci 22(5):762–769

Mansouri A, Taslimi S, Badhiwala JH, Witiw CD, Nassiri F, Odekerken VJJ et al (2018) Deep brain stimulation for Parkinson's disease: meta-analysis of results of randomized trials at varying lengths of follow-up. J Neurosurg 128(4):1199–1213

Mantione M, van de Brink W, Schuurman PR, Denys D (2010) Smoking cessation and weight loss after chronic deep brain stimulation of the nucleus accumbens: therapeutic and research implications: case report. Neurosurgery 66(1):E218; discussion E218

McClelland J, Bozhilova N, Campbell I, Schmidt U (2013) A systematic review of the effects of neuromodulation on eating and body weight: evidence from human and animal studies. Eur Eat Disord Rev 21(6):436–455

McElroy SL, Hudson J, Ferreira-Cornwell MC, Radewonuk J, Whitaker T, Gasior M (2016)

Lisdexamfetamine dimesylate for adults with moderate to severe binge eating disorder: results of two pivotal phase 3 randomized controlled trials. Neuropsychopharmacology 41(5):1251–1260

Mena JD, Selleck RA, Baldo BA (2013) Mu-opioid stimulation in rat prefrontal cortex engages hypothalamic orexin/hypocretin-containing neurons, and reveals dissociable roles of nucleus accumbens and hypothalamus in cortically driven feeding. J Neurosci Off J Soc Neurosci 33(47):18540–18552

Michaelides M, Thanos PK, Volkow ND, Wang GJ (2012) Dopamine-related frontostriatal abnormalities in obesity and binge-eating disorder: emerging evidence for developmental psychopathology. Int Rev Psychiatry 24(3):211–218

Miller EK, Cohen JD (2001) An integrative theory of prefrontal cortex function. Annu Rev Neurosci 24:167–202

Morrell MJ, Group RNSSiES (2011) Responsive cortical stimulation for the treatment of medically intractable partial epilepsy. Neurology 77(13):1295–1304

Naef L, Moquin L, Dal Bo G, Giros B, Gratton A, Walker CD (2011) Maternal high-fat intake alters presynaptic regulation of dopamine in the nucleus accumbens and increases motivation for fat rewards in the offspring. Neuroscience 176:225–236

Oswald KD, Murdaugh DL, King VL, Boggiano MM (2011) Motivation for palatable food despite consequences in an animal model of binge eating. Int J Eat Disord 44(3):203–211

Paz JT, Davidson TJ, Frechette ES, Delord B, Parada I, Peng K et al (2013) Closed-loop optogenetic control of thalamus as a tool for interrupting seizures after cortical injury. Nat Neurosci 16(1):64–70

Prasad AA, McNally GP (2016) Ventral pallidum output pathways in context-induced reinstatement of alcohol seeking. J Neurosci Off J Soc Neurosci 36 (46):11716–11726

Rachid F (2018) Repetitive transcranial magnetic stimulation in the treatment of eating disorders: a review of safety and efficacy. Psychiatry Res 269:145–156

Reynolds GP, McGowan OO (2017) Mechanisms underlying metabolic disturbances associated with psychosis and antipsychotic drug treatment. J Psychopharmacol 31(11):1430–1436

Richard JM, Stout N, Acs D, Janak PH (2018) Ventral pallidal encoding of reward-seeking behavior depends on the underlying associative structure. Elife 7:e33107

Root DH, Melendez RI, Zaborszky L, Napier TC (2015) The ventral pallidum: subregion-specific functional anatomy and roles in motivated behaviors. Prog Neurobiol 130:29–70

Rosin B, Slovik M, Mitelman R, Rivlin-Etzion M, Haber SN, Israel Z et al (2011) Closed-loop deep brain stimulation is superior in ameliorating parkinsonism. Neuron 72(2):370–384

Rossi MA, Stuber GD (2018) Overlapping brain circuits for homeostatic and hedonic feeding. Cell Metab 27 (1):42–56

Roth BL (2016) DREADDs for neuroscientists. Neuron 89 (4):683–694

Runegaard AH, Fitzpatrick CM, Woldbye DPD, Andreasen JT, Sorensen AT, Gether U (2019) Modulating dopamine signaling and behavior with chemogenetics: concepts, progress, and challenges. Pharmacol Rev 71(2):123–156

Schienle A, Schafer A, Hermann A, Vaitl D (2009) Binge-eating disorder: reward sensitivity and brain activation to images of food. Biol Psychiatry 65(8):654–661

Shor-Posner G, Azar AP, Insinga S, Leibowitz SF (1985) Deficits in the control of food intake after hypothalamic paraventricular nucleus lesions. Physiol Behav 35(6):883–890

Smedley EB, DiLeo A, Smith KS (2019) Circuit directionality for motivation: lateral accumbens-pallidum, but not pallidum-accumbens, connections regulate motivational attraction to reward cues. Neurobiol Learn Mem 162:23–35

Smith DG, Robbins TW (2013) The neurobiological underpinnings of obesity and binge eating: a rationale for adopting the food addiction model. Biol Psychiatry 73(9):804–810

Smith KS, Tindell AJ, Aldridge JW, Berridge KC (2009) Ventral pallidum roles in reward and motivation. Behav Brain Res 196(2):155–167

Smith KS, Bucci DJ, Luikart BW, Mahler SV (2016) DREADDS: use and application in behavioral neuroscience. Behav Neurosci 130(2):137–155

Spagnolo PA, Goldman D (2017) Neuromodulation interventions for addictive disorders: challenges, promise, and roadmap for future research. Brain J Neurol 140(5):1183–1203

Stauffer WR, Lak A, Yang A, Borel M, Paulsen O, Boyden ES et al (2016) Dopamine neuron-specific optogenetic stimulation in rhesus macaques. Cell 166 (6):1564–1571.e6

Sternson SM, Roth BL (2014) Chemogenetic tools to interrogate brain functions. Annu Rev Neurosci 37:387–407

Tooley J, Marconi L, Alipio JB, Matikainen-Ankney B, Georgiou P, Kravitz AV et al (2018) Glutamatergic ventral pallidal neurons modulate activity of the habenula-tegmental circuitry and constrain reward seeking. Biol Psychiatry 83(12):1012–1023

Uher R, Murphy T, Brammer MJ, Dalgleish T, Phillips ML, Ng VW et al (2004) Medial prefrontal cortex activity associated with symptom provocation in eating disorders. Am J Psychiatry 161(7):1238–1246

Urban DJ, Roth BL (2015) DREADDs (designer receptors exclusively activated by designer drugs): chemogenetic tools with therapeutic utility. Annu Rev Pharmacol Toxicol 55:399–417

Van den Eynde F, Claudino AM, Mogg A, Horrell L, Stahl D, Ribeiro W et al (2010) Repetitive transcranial magnetic stimulation reduces cue-induced food craving in bulimic disorders. Biol Psychiatry 67(8):793–795

Varela L, Horvath TL (2012) AgRP neurons: a switch between peripheral carbohydrate and lipid utilization. EMBO J 31(22):4252–4254

Vickers SP, Hackett D, Murray F, Hutson PH, Heal DJ (2015) Effects of lisdexamfetamine in a rat model of binge-eating. J Psychopharmacol 29(12):1290–1307

Vickers SP, Goddard S, Brammer RJ, Hutson PH, Heal DJ (2017) Investigation of impulsivity in binge-eating rats in a delay-discounting task and its prevention by the d-amphetamine prodrug, lisdexamfetamine. J Psychopharmacol 31(6):784–797

Volkow ND, Wang GJ, Fowler JS, Telang F (2008) Overlapping neuronal circuits in addiction and obesity: evidence of systems pathology. Philos Trans R Soc Lond Ser B Biol Sci 363(1507):3191–3200

Volkow ND, Wang GJ, Fowler JS, Tomasi D, Baler R (2012) Food and drug reward: overlapping circuits in human obesity and addiction. Curr Top Behav Neurosci 11:1–24

Vucetic Z, Kimmel J, Reyes TM (2011) Chronic high-fat diet drives postnatal epigenetic regulation of mu-opioid receptor in the brain. Neuropsychopharmacology 36 (6):1199–1206

Walpoth M, Hoertnagl C, Mangweth-Matzek B, Kemmler G, Hinterholzl J, Conca A et al (2008) Repetitive transcranial magnetic stimulation in bulimia nervosa: preliminary results of a single-centre, randomised, double-blind, sham-controlled trial in female outpatients. Psychother Psychosom 77(1):57–60

Wang GJ, Volkow ND, Thanos PK, Fowler JS (2009) Imaging of brain dopamine pathways: implications for understanding obesity. J Addict Med 3(1):8–18

Watabe-Uchida M, Zhu L, Ogawa SK, Vamanrao A, Uchida N (2012) Whole-brain mapping of direct inputs to midbrain dopamine neurons. Neuron 74(5):858–873

Woods AJ, Antal A, Bikson M, Boggio PS, Brunoni AR, Celnik P et al (2016) A technical guide to tDCS, and related non-invasive brain stimulation tools. Clin Neurophysiol 127(2):1031–1048

Wu H, Miller KJ, Blumenfeld Z, Williams NR, Ravikumar VK, Lee KE et al (2018) Closing the loop on impulsivity via nucleus accumbens delta-band activity in mice and man. Proc Natl Acad Sci USA 115(1):192–197

Wykes RC, Kullmann DM, Pavlov I, Magloire V (2016) Optogenetic approaches to treat epilepsy. J Neurosci Methods 260:215–220

Yizhar O, Fenno LE, Davidson TJ, Mogri M, Deisseroth K (2011) Optogenetics in neural systems. Neuron 71 (1):9–34

Zhou QY, Palmiter RD (1995) Dopamine-deficient mice are severely hypoactive, adipsic, and aphagic. Cell 83 (7):1197–1209

Zhu H, Roth BL (2014) DREADD: a chemogenetic GPCR signaling platform. Int J Neuropsychopharmacol 18 (1):1–6

Part III

Human Research

Neuroimaging to Study Brain Reward Processing and Reward-Based Learning in Binge Eating Pathology

Marisa DeGuzman and Guido K.W. Frank

Abstract

This chapter reviews human neuroimaging studies that investigate the neurobiology of reward processing in eating disorders associated with binge eating. Across the relatively small research literature on binge-eating disorder (BED) and bulimia nervosa (BN) using food and nonfood stimuli, neuroimaging studies consistently suggest alterations in brain reward circuit response. Studies tend to identify heightened brain response to visual presentation of reward cues, while reward receipt, including unexpected receipt, is associated with lower brain activation. Those results point toward specific neurotransmitter alterations associated with binge eating pathophysiology. However, there is still extensive heterogeneity across studies due to different study designs and analytic approaches, and research that systematically translates and studies basic science models in humans will have the best chance of identifying neurocircuitry that is specific to this pathology.

Keywords

Reward processing · Reward-based learning · Neuroimaging

Learning Objectives

In this chapter, you will learn:

- To understand the current state of reward-focused neuroimaging research in disorders associated with binge eating.
- To identify the gaps and limitations of the current research.
- To discuss the potential future directions of neuroimaging research on binge eating.

1 Introduction

The motivation to seek and consume food is mediated by the brain's reward circuitry. A form of dysregulated food consumption is binge eating, which has been associated with abnormalities of the reward system in both animal models as well as human research on eating disorders (Avena 2013; Wierenga et al. 2014; Berridge 2009b). Binge eating is a cross-cutting behavior that is characterized by recurrent episodes of eating very large amounts of food in a short period of time with a sense of lack of control (American Psychiatric Association 2013). Binge eating can be

M. DeGuzman (✉)
Department of Psychology, University of the Incarnate Word, San Antonio, TX, USA
e-mail: mdeguzma@uiwtx.edu

G. K.W. Frank
Department of Psychiatry, University of California San Diego, San Diego, CA, USA
e-mail: gfrank@health.ucsd.edu

© Springer Nature Switzerland AG 2020
G. K.W. Frank, L. A. Berner (eds.), *Binge Eating*, https://doi.org/10.1007/978-3-030-43562-2_9

found most characteristically in binge-eating disorder (BED), and it is also a diagnostic criterion for bulimia nervosa (BN) where those episodes are followed by compensatory behaviors to avoid weight gain. Anorexia nervosa can also present with binge eating, but there it only defines a subtype. Notably, an overweight or obese body mass index (BMI) can often be found in BED but it is not a diagnostic prerequisite, while individuals with BN are commonly in the high normal weight range.

Several reviews have surveyed the reward-focused neuroimaging literature on BED, obesity, and other eating disorders characterized by binge eating behavior (Michaelides et al. 2012; Schag et al. 2013; Wu et al. 2016; Boswell and Kober 2016). The research studies reviewed in this chapter (Table 1) will particularly focus on the emerging role of the brain's reward processing circuitry, including the striatum, frontal cortex, and insula. Striatal structures are known to respond to salient stimuli and encode prediction errors during reward learning, contributing to impulsivity in decision-making (Flagel et al. 2011). The medial prefrontal and orbitofrontal cortex are important for reward valuation and sensory-specific satiety, and code when to stop eating a certain food, while other food might still be valued and therefore of interest (Rolls et al. 1981). The insula contains the primary taste cortex and integrates somatosensory and interoceptive processing, emotional, and cognitive regulation, implicating a central role of the insula in appetite control and hedonic food reward processing (Uddin et al. 2017; Craig 2009; Frank 2013; Rolls 2016). The hyper- and hypo-responsivity found across these regions may reflect the imbalance of reward sensitivity and inhibition/impulsivity thought to underlie binge eating pathophysiology (Kessler et al. 2016; Wierenga et al. 2014). A key question that remains to be answered is how neurotransmitter systems drive binge eating and how can they be manipulated to improve outcome of binge eating associated pathology.

2 Functional Magnetic Resonance Imaging (fMRI)

Functional magnetic resonance imaging (fMRI), which measures the blood–oxygen level-dependent (BOLD) signal, is the most common functional neuroimaging modality used to study binge eating neurobiology. This technique allows the comparison of brain activity in response to different stimuli during a brain scan. The following study designs utilized both food-specific stimuli that are directly relevant to binge eating, and nonfood stimuli that explore generalized reward processing.

2.1 Food-Specific Reward Paradigms

2.1.1 Visual Cue Studies

Two studies have used visual and auditory food cues across four groups of women: individuals with subthreshold BED with and without obesity, and individuals without binge eating with and without obesity. The first study demonstrated greater dorsal anterior cingulate cortex response to high-calorie food cues in women with binge eating compared to those without, regardless of weight status (Geliebter et al. 2006). The second study assessed the conservation of regional activation within group rather than comparing regional activation between groups (Geliebter et al. 2016). The five obese binge eaters exhibited strict conservation of right premotor area response to binge food stimuli, which was interpreted to represent motor planning of eating the food stimuli. The design of these studies is strengthened by the inclusion of visual and auditory food cues, different weight status in women with and without binge eating, and a standardized meal 3 h prior to the scan. Problematic though are, as stated above, the small group sizes in those studies. Several other studies have compared visual food reward responses across BED, BN, normal weight, and overweight/obese controls. One group tested the brain response of fasted

Table 1 Functional neuroimaging studies of eating disorders with binge eating that target the reward circuitry

References	Modality and task	Demographic information	Main findings
	Functional magnetic resonance imaging (fMRI)—visual food stimuli		
Geliebter et al. (2006)	Visual and auditory cues for high energy density foods, and nonfood items	10 BE (5 OB, 5 NW), 22.1 ± 2.3 years, 27.4 ± 5.8 kg/m²; 10 non-BE (5 OB, 5 NW), 21.3 ± 0.6 years, 27.7 ± 7.2 kg/m²	Binge eating was associated with greater activation in the ACC during high energy dense versus low-energy dense food cues
Geliebter et al. (2016)	Visual and auditory cues for binge foods, non-binge foods, and non-food items	10 OB: 5 BE, 23.4 ± 2.5 years, 32.3 ± 4.6 kg/m²; 5 non-BE, 21.4 ± 0.5 years, 33.5 ± 6.5 kg/m²; 10 NW: 5 BE, 20.8 ± 1.6 years, 22.4 ± 1.0 kg/m²; 5 non-BE, 21.2 ± 0.8 years, 21.9 ± 1.3 kg/m²	In obese binge eaters, binge food cues activated specifically the premotor area
Lee et al. (2017)	Stroop-match-to-sample with food image interference condition	12 BED, 23.6 ± 2.6 years, 25.6 ± 3.8 kg/m²; 13 BN, 23.7 ± 2.2 years, 21.5 ± 2.2 kg/m²; 14 HC, 23.3 ± 2.2 years, 20.4 ± 2.6 kg/m²	BED had greater activation versus HC to food image in the VS; BN had greater activation versus BED and HC to food images in the premotor cortex and dorsal striatum
Weygandt et al. (2012)	Passive viewing of high caloric food, disgust-inducing or neutral items	17 BED, 26.4 ± 6.4 years, 32.2 ± 4.0 kg/m²; 14 BN, 23.1 ± 3.8 years, 22.1 ± 2.5 kg/m²; 19 C-NW, 22.3 ± 2.6 years, 21.7 ± 1.4 kg/m²; 17 C-OW, 25.0 ± 4.7 years, 31.6 ± 4.7 kg/m²	Insula differentiated food and neutral stimuli across all subjects; All 4 subject groups could be separated based on brain activation patterns, particularly the insula, VS, and ACC
Schienle et al. (2009)	Passive viewing of high caloric food, disgust-inducing or neutral items	17 BED, 26.4 ± 6.4 years, 32.2 ± 4.0 kg/m²; 14 BN, 23.1 ± 3.8 years, 22.1 ± 2.5 kg/m²; 19 C-NW, 22.3 ± 2.6 years, 21.7 ± 1.4 kg/m²; 17 C-OW, 25.0 ± 4.7 years, 31.6 ± 4.7 kg/m²	BED had greater activation versus BN, OW, NW to food stimuli in the medial OFC; BN had greater activation versus BED, OW, NW to food stimuli in insula and ACC; BED response to food stimuli in ACC and medial OFC correlated positively with behavioral activation scale and arousal
Wonderlich et al. (2018)	Viewing of palatable foods during neutral or stress condition	12 BN, 4 OSFED, 22.9 ± 5.4 years, 24.5 ± 3.3 kg/m²	BOLD response decreased to food cues following the stress induction in the right and left ventromedial PFC, the right ACC, and left amygdala; Greater decrease in activation when observing food cues following stress induction was associated with a greater increase in negative affect prior to binge eating

(continued)

Table 1 (continued)

References	Modality and task	Demographic information	Main findings
	Functional magnetic resonance imaging (fMRI)—taste stimuli		
Bohon and Stice (2011)	Milkshake consumption versus control solution	26, 13 of which with full or subthreshold BN 13 NW controls, 20.3 ± 1.9 years; 23.6 ± 2.6 kg/m²	Women with BN symptoms had trends for less activation in right anterior insula to anticipated chocolate milkshake receipt, and in left middle frontal gyrus, right posterior insula, precentral gyrus, and mid-dorsal insula to milkshake consumption
Filbey et al. (2012)	High-calorie taste (Pepsi, chocolate milk, or cream soda) or water	26 BE, 10 F, 32.88 ± 11.04 years, 32.72 ± 5.98 kg/m²	High activation to caloric versus water stimulus in medial OFC, VTA, insula, caudate, putamen, nucleus accumbens, precuneus. Functional connectivity higher for caloric versus water stimulus between nucleus accumbens and OFC. Elevated brain response to high calorie taste associated with increased binge eating behavior
Frank et al. (2011)	Sucrose solution tasting versus control solution in a prediction error paradigm to test response to unpredictable receipt or omission of the stimulus	20 BN, 25.2 ± 5.3 years, 22.6 ± 5.7 kg/m²; 23 HC, 27.2 ± 6.4, 21.5 ± 1.2	BN showed lower brain response to unexpected receipt and omission of taste as well as lower prediction error brain regression with computational model-based dopamine neuron response, in insula, ventral putamen, amygdala, and OFC
Frank et al. (2016)	Sucrose solution tasting	25 BN, 24.6 ± 4.2 years, 23.6 ± 5.8 kg/m²; 26 AN, 23.2 ± 5.3 years, 16.2 ± 1.1 kg/m²; 26 NW, 24.4 ± 3.5 years, 21.6 ± 1.2 kg/m²	In HC, the hypothalamus drove ventral striatal activity, but in BN effective connectivity was directed from ventral striatum to hypothalamus
Monteleone et al. (2018)	Tasting of sweet (sucrose), bitter (quinine) and neutral (water) solution	20 NW, 27.1 ± 4.7 years, 21.0 ± 1.5 kg/m²; 20 BN, 27.7 ± 8.0 years, 21.8 ± 2.6 kg/m²; 20 AN, 25.5 ± 7.8 years, 17.3 ± 1.0 kg/m²	BN had reduced responses to bitter taste in the right amygdala and left insula. AN had lower responses to bitter taste in the right amygdala and left ACC
Setsu et al. (2017)	Tasting of umami taste plus salt	18 BN, 25.0 ± 5.6 years, 20.2 ± 1.6 kg/m²; 18 NW, 27.1 ± 5.7 years, 20.5 ± 1.4 kg/m²	Greater right insula activation in BN versus NW
	Functional magnetic resonance imaging (fMRI)—monetary stimuli		
Balodis et al. (2013)	Monetary incentive delay task	19 BED, 14 F, 43.7 ± 12.7 years, 36.7 ± 4.05 kg/m²; 19 OB, 10 F, 38.3 ± 7.5 years, 34.6 ± 3.5 kg/m²; 19 NW 10 F, 34.8 ± 10.7 years, 23.3 ± 1.1 kg/m²	OB versus NW had greater ventral striatal and ventromedial PFC activity during anticipation of monetary gain. OB versus BED had greater bilateral ventral striatal activity during anticipation of monetary gain

Study	Task/stimuli	Sample	Results
Balodis et al. (2014)	Monetary incentive delay task	19 BED, 14 F, 43.7 ± 12.7 years, 36.7 ± 4.05 kg/m² Follow up study after discharge	Reward anticipation: Post-treatment non-binge eating versus binge eating patients had greater left superior frontal gyrus, bilateral IFG, right VS activation. Loss anticipation: post-treatment binge eating versus non-binge eating patients had higher right thalamus activation; post-treatment non-binge eating versus binge eating had higher left IFG, bilateral medial frontal gyrus activation. Winning: Post-treatment non-binge eating versus binge eating had higher activation in medial PFC, right postcentral gyrus, right superior frontal gyrus, cingulate gyrus, medial frontal gyrus. Losing: Post-treatment non-binge eating versus binge eating had higher activation in middle frontal gyrus, and left precentral gyrus
Bodell et al. (2018)	Monetary guessing game (win, lose)	122 adolescent girls, 28 of which endorsed binge eating	Greater ventromedial prefrontal cortex and caudate responses to winning money were correlated with greater binge eating severity
Simon et al. (2016)	Monetary incentive delay task	27 BED, 38.26 ± 13.75 years, 32.31 ± 4.55 kg/m²	BED and BN versus HC had greater medial OFC response to food reward receipt
	Food incentive delay task (using "abstract" snack points)	28 CW (BED), 38.0 ± 10.85 years, 34.02 ± 4.5 kg/m² 29 BN, 27.45 ± 10.55 years, 21.33 ± 2.99 kg/m² 27 CW (BN), 25.74 ± 5.25 years, 21.85 ± 1.85 kg/m²	BED and BN versus HC had greater food craving and external eating scores correlated with increased medial OFC activity
	Functional magnetic resonance imaging (fMRI)—visual food or monetary stimuli plus drug challenge		No differences in monetary task
Cambridge et al. (2013)	Viewing high-calorie foods, low-calorie foods, rewarding non-food, less rewarding non-food. Mu-opioid receptor antagonist, GSK1521498 or placebo	14 placebo OB BE, 8 F, 40.6 ± 8.5 years 16 drug OB BE, 8 F, 39.8 ± 10.2 years	Mu-opioid receptor antagonist led to a reduction in pallidum/putamen responses to high-calorie food pictures and reduced motivation to view high-calorie food pictures
Dodds et al. (2012)	Viewing high-calorie or low-calorie foods, or non-food items. D3 receptor antagonist GSK598809 or placebo	26 OW/OB BE, 11 F, 35.1 ± 7.1 years, 32.7 ± 3.7 kg/m²	Brain activation to food images was not modulated by the dopamine D3 antagonist

(continued)

Table 1 (continued)

References	Modality and task	Demographic information	Main findings
Mueller et al. (2018)	Monetary incentive delay task	17 rec BN, 29.6 ± 8.9 years, 21.6 ± 2.3 kg/m^2	AMPT affected reward effort in HC but not BN
	Catecholamine depletion agent alpha-methyl-paratyrosine (AMPT)	21 HC, 27.3 ± 9.4 years, 24.2 ± 3.3 kg/m^2	AMPT affected anteroventral striatum activation in HC but not in the BN group, suggesting a hypo-sensitive dopamine system
	Functional magnetic resonance imaging (fMRI)—other		
Cyr et al. (2016)	Reward-based spatial learning task	27 BN, 16.6 ± 1.5 years, 22.0 ± 2.0 kg/m^2	BN activated right anterior hippocampus during receipt of unexpected rewards
		27 HC, 16.3 ± 2.1 years, 21.9 ± 1.9 kg/m^2	BN deactivated left superior frontal gyrus and right anterior hippocampus during expected reward receipt, which was the opposite compared to HC
Reiter et al. (2017)	Dynamic choice task reinforcement learning model	22 BED, 16 F, 29.0 ± 9.4 years, 28.27 ± 6.58 kg/m^2	HC had higher response versus BED in the anterior insula, ventrolateral PFC for exploratory decisions
		22 HC, 15 F, 27.8 ± 4.54 years, 26.06 ± 4.35 kg/m^2	HC had higher response versus BED for ventromedial PFC learning signatures (prediction error)

ACC Anterior cingulate cortex, *PFC* Prefrontal cortex, *OFC* Orbitofrontal cortex, *HC* Healthy control, *NW* normal weight control participants viewing high-calorie foods, disgust-inducing items, and neutral items. Compared to BN, normal weight, and overweight groups, BED showed greater medial orbitofrontal cortex response to food stimuli (Schienle et al. 2009). In contrast, although the BN and BED groups scored similarly in the degree of binge eating, the BN group showed greater insula and anterior cingulate cortex response to food stimuli than the other three groups. These results implicate distinct neural mechanisms in BED, BN, and obesity, where altered orbitofrontal cortex processing of food cues may underlie binge eating in BED. However, a decoding analysis of this data set later showed that BED could be distinguished from normal weight controls, not by the orbitofrontal cortex, but by insula response to food cues and from obese controls with ventral striatum response (Weygandt et al. 2012). The same data set was further studied for brain volume across BED, BN, and controls in a voxel-based morphometry (VBM) analysis (Schafer et al. 2010). Groups showed no differences in global gray matter, white matter, or cerebrospinal fluid volume. However, regional gray matter volume differences in regions involved in food reward processing were detected: both the obese BED and BN groups showed greater orbitofrontal cortex volume compared to controls, while BED had larger anterior cingulate and BN larger ventral striatal volumes compared to controls. The BN group showed also higher orbitofrontal cortex and striatal volumes compared to BED participants. This same sample of individuals with BED displayed greater medial orbitofrontal cortex activation in response to food stimuli compared to BN, overweight and normal weight control groups (Schienle et al. 2009). The authors suggest that the structural and functional abnormalities of the orbitofrontal cortex could underlie the altered self-regulation and habit learning in BED. Brain volume could be different across groups due to developmental differences in growth and neuron pruning, and neuronal mass could explain altered brain function. While these are compelling findings, the combination of structural and functional imaging data will need to be applied in future analyses to better understand

brain structure–function relationships (Frank et al. 2018b).

In a reward incentive delay task that included both visual food cues and monetary stimuli, brain response was compared between 27 individuals with BED, 28 with obesity, 29 with BN, and 27 normal weight controls (Simon et al. 2016). That task tests brain activation during expectation of rewards, while rewards vary in magnitude and how long a person has to wait for a larger or smaller reward. No group differences for brain response in the *monetary* task were found. In response to the notification that a *food* reward had been earned, both BED and BN groups exhibited heightened medial orbitofrontal cortex activity compared to their respective control groups. This hyper-responsiveness was further correlated with increased food craving and external eating scores, but not binge eating behavior, in the BED and BN groups. That study suggested that food reward circuitry was specifically hyper-responsive in the binge eating study groups and related to the drive to approach food.

Another study investigated the effects of food image interference in a Stroop-Match-to-Sample task to investigate whether potentially reward system activating food stimuli would interfere with cognitive performance (Lee et al. 2017). Compared to healthy controls, the BED group showed greater ventral striatum response to food images. Behaviorally, there was a nonsignificant trend of impaired cognitive control over food image interference in BED. These results suggest the reward salience processing function of the ventral striatum could interfere with top-down attentional control in BED, but the groups were small and the study likely underpowered.

One study investigated brain response during viewing of palatable foods in full and subthreshold BN and compared a neutral versus a stress condition (Wonderlich et al. 2018). That study, which did not include a control group, showed that brain response decreased to food cues following the stress induction in the right and left ventromedial prefrontal cortex, the right anterior cingulate cortex, and left amygdala. Interestingly, less change in brain response was associated with greater negative and less positive affect prior to binge eating. That study suggests a direct relationship between negative affect and brain response, which in turn could control the drive to binge eat.

A recent review of the use of food images and functional brain imaging in anorexia nervosa found that using this approach resulted in more consistent findings when targeting reward pathways than other cognitive or emotional processes (Lloyd and Steinglass 2018). Thus, this approach may also have value to study reward circuit activation in binge eating. Nevertheless, presentation of food images frequently leads to results that are difficult to relate to the underlying neurobiology or neurochemistry that drives pathological eating behavior and more sophisticated studies are needed to move this field forward.

2.1.2   Taste Cue Studies

Individuals with a BED diagnosis are yet to be studied with fMRI while consuming real taste stimuli. One study has investigated binge eating behavior with taste stimuli in a single group of 26 individuals with a BMI over 25 kg/m^2 and a moderate binge eating score on a self-report measure (Filbey et al. 2012). Reward regions in the brain (medial orbitofrontal cortex, ventral tegmental area, insula, caudate, putamen, nucleus accumbens, and precuneus) were more responsive to high-calorie taste stimuli compared to water in this group. Higher binge eating symptoms, but not BMI, were associated with higher high-calorie brain taste responses. These findings provide some evidence of taste reward sensitivity underlying the increased motivation to eat in BED. A study that applied milkshake or control solution to women with full or subthreshold BN found that the BN group tended to have less activation when expecting milkshake in the right anterior cingulate, and in response to consumption of milkshake lower activation in the left middle frontal gyrus, posterior insula, precentral gyrus, and mid-dorsal insula (Bohon and Stice 2011). However, the results were not statistically significant, making the implications less clear. One recent study compared sweet taste (sucrose) with the aversive bitter stimulus quinine in BN (Monteleone et al. 2018). In that study, the BN

group had a lower response to the bitter stimulus in insula and amygdala, but sweet taste response was normal. Sweet sucrose taste was also used in a study from our group in BN that assessed brain prediction error response, a model that has been associated with brain dopamine function and tests brain response to unexpected receipt or omission of stimuli (Frank et al. 2011). The BN group showed lower activation as well as prediction error regression in insula, ventral putamen, amygdala, and orbitofrontal cortex compared to controls, suggesting a downregulation of the sensitivity of the dopamine circuitry. Importantly, higher binge eating frequency was negatively correlated with brain prediction error response, suggesting either a direct impact from eating disorder behavior on brain function or a premorbid lower brain response that could drive binge eating episodes. Any causal relationships need further study. Brain imaging can also be used to test the direction of activation between brain regions using so-called effective connectivity analyses. We used this approach in the above-described sample of individuals with BN and controls that we studied for prediction error response and extracted data for the expected sucrose solution taste condition (Frank et al. 2016). Controls showed effective connectivity from the hypothalamus to the ventral striatum, while the BN group had the opposite result; the direction of activation was from cortical structures to ventral striatum and hypothalamus. We hypothesized that sugar as a fear-inducing stimulus may stimulate a pathway that interrupts the eating drive via the ventral striatal—hypothalamic circuitry (Frank et al. 2018a).

Umami taste has been described as a savory flavor, it is different than the other taste qualities, sweet, sour, bitter or salty, and specific tongue taste receptors exist (de Araujo et al. 2003). A study in BN using this taste stimulus found that the BN group had a stronger response to the umami taste in the right insula compared to the control group (Setsu et al. 2017). In addition, in the control group, there was a significant inverse relationship between insula response and subjective umami pleasantness ratings, while there was no significant brain–behavior response in the BN group.

In summary, studies report higher or lower brain responses to taste stimuli across study designs that vary in taste quality and method of application. The key to identifying biological targets for treatment development will be that studies use basic science models that can be tested in humans and that allow identifying neurotransmitters directly involved in specific eating disorder behaviors. The prediction error model is such a model that is based on dopamine neuronal function and that can be studied using specific dopamine receptor agonists and antagonists. In addition, studies that test neurobiology cross cutting the various eating disorder groups may help identify circuitry that drives binge eating behaviors and identify for instance circuit alterations specific to binge eating versus purging or high BMI. The NIMH Research Domain Criteria provide a framework for such studies (Cuthbert 2014).

2.2 Nonfood-Specific Reward Paradigms

The first study to explore nonfood-specific reward processing in BED compared brain activity during a monetary incentive delay task in 19 individuals with obesity and BED (OB BED), 19 with obesity and without BED (OB), and 19 healthy weight controls (Balodis et al. 2013). The BED group displayed reduced ventrostriatal activity during reward anticipation and reduced prefrontal cortex and insula activity during reward outcome. The BED group then went on to complete 4 months of treatment with sibutramine (an appetite suppressant) and/or cognitive behavioral therapy. Compared to individuals in the group who successfully responded to this treatment, individuals who did not respond to treatment (i.e. continued to binge eat) showed decreased ventral striatal and inferior frontal gyrus response to reward anticipation and decreased medial prefrontal cortex response to reward outcome pretreatment (Balodis et al. 2014). These findings implicate hyporesponsivity to nonfood reward stimuli in BED.

One study in BED-linked responses on a monetary reward learning task with brain structure

(but not functional brain response) in subjects who were obese with BED and subjects who were obese without BED (Voon et al. 2015). The obese BED group displayed a greater tendency toward habit-based learning and perseveration and reduced left ventral striatal, bilateral caudate, and orbitofrontal cortex gray matter volume compared to the OB group. Interestingly, these regional volume differences were no longer significant when researchers accounted for the model-based parameter for the subjects' habit-based learning behavior. Furthermore, higher binge eating scores were associated with a stronger bias toward using a habit-based or "model-free" ("trial and error") strategy. These results point to model-free reward learning as a neurocomputational mechanism contributing to the maladaptive habit formation involved in binge eating behaviors. Importantly, the regional volumes were different in the OB groups depending on the presence of BED. Unfortunately, the study did not include normal weight controls, preventing comparison of its results with the previous MRI study (Schafer et al. 2010).

A more recent study found that BED compared to healthy controls exhibited reduced insula and ventrolateral prefrontal cortex activation during exploratory decisions using a monetary dynamic choice task that tested a reinforcement learning model (Reiter et al. 2017). The BED group further displayed reduced ventromedial prefrontal cortex activation associated with the prediction error learning signature that incorporated alternative choices. Although the study did not relate behavioral or brain activation data to binge eating symptoms, the findings suggest a neurocognitive phenotype of BED, where deficient prefrontal cortex activation during decision-making represents a neural correlate of maladaptive switching behavior. A study that investigated brain circuitry for both neurocognition and reward recruited adolescents with BN and controls and applied a spatial orientation task in a virtual maze where subjects could earn monetary rewards (Cyr et al. 2016). The BN group showed opposite responses compared to controls. The BN participants activated the right anterior hippocampus during the receipt of unexpected rewards (control condition), and deactivated the left superior frontal gyrus and right anterior hippocampus during expected reward receipt (learning condition). Furthermore, hippocampal activation in the BN during the unexpected rewards condition was significantly related to BN behavior scores. Anxiety and impulsivity are behavioral constructs that have been associated with BN (Chase et al. 2017; Xia et al. 2017; Vitousek and Manke 1994; Wagner et al. 2006). The hippocampus has traditionally been found to be involved in memory function but recent optogenetic research implicated hippocampal serotonin neurons in the modulation of anxiety and impulsivity (Ohmura et al. 2019). It is possible that for instance, hippocampal hyper-responsiveness during an unexpectancy task condition can be a neurobiological correlate that is related to impulsivity and tendency to binge eat in a person's natural environment. A different approach was taken in a study that recruited adolescent girls from a community sample and compared binge eating participants with those without that behavior (Bodell et al. 2018). In that study, the severity of binge eating correlated positively with activation in ventromedial prefrontal cortex and caudate during winning money. This study further points toward reward circuit abnormalities.

Taken together, nonfood reward tasks have led to altered brain responses in groups with binge eating behaviors. BED and BN tended to show lower brain response in frontal or subcortical brain response, although the described community sample study indicated higher brain activation with more severe binge eating severity. Lower response to unexpected stimuli could point again to reduced dopamine circuit sensitivity as found in a sweet taste paradigm in BN (Frank et al. 2011).

2.2.1 Drug Challenge Studies

The reward system involves the interaction of cortical and subcortical brain regions and associated connecting pathways to process desire, action to approach and consume reward stimuli, and learning from those experiences (Haber and Knutson 2010; Kelley et al. 2005). The neurotransmitters dopamine and opioids code key aspects of neural reward processing. Dopamine neurons code motivation ("wanting"), reward approach and learning, and the opioid system codes pleasurable experience from rewards ("liking") (Berridge 2009a; Kelley and Berridge 2002). Functional magnetic resonance brain imaging (fMRI) tests brain activation across brain regions and circuits, such as reward or anxiety pathways. Those studies usually do not test brain neurotransmitters directly, but the response during tasks that test specific behaviors might help in understanding neurotransmitters involved in the brain response (Frank 2011). Although fMRI does not allow for direct measurement of neurotransmitter levels, drugs can be administered to pharmacologically manipulate neurotransmitter systems. Several fMRI studies have used drugs to challenge specific neurotransmitter systems in BED. The first measured brain response to high- and low-calorie food images after placebo or a dopamine D3 receptor antagonist in 26 individuals who were overweight/obese and had binge eating behaviors (Dodds et al. 2012). Binge eating behavior was assessed with a self-report questionnaire and subjects fasted 15 h prior to scanning. While high-calorie food images did elicit stronger responses than low-calorie food images in reward processing regions (caudate, insula, nucleus accumbens, putamen, amygdala), this was unaffected by the dopamine D3 receptor antagonist GSK598809. These results do not support a direct role of D3 receptor function on the processing of food reward images in individuals with binge eating. However, as noted by the authors, the task only assessed reward-cue responsivity and did not require reward learning processes which would rely more on the dopamine system. The second drug challenge study also measured fasted brain response to high- and low-calorie food images, but targeted the opioid system (Cambridge et al. 2013) and included individuals with BED rather than just binge eating behavior with no prior history of eating disorders. Compared to placebo, the mu-opioid receptor antagonist, GSK1521498, reduced both behavioral motivation to view high-calorie food images as well as right pallidum and putamen response to high-calorie food images. This provides evidence of the opioid system's involvement in food-related motivational processes in BED. If the opioid system were to be targeted for treatment purposes as suggested by the authors, future studies should measure more direct effects on binge eating behaviors. A third study used the catecholamine-depleting agent, alpha-methyl-paratyrosine (AMPT), together with the monetary incentive delay (MID) task during fMRI in healthy controls and women with BN (Mueller et al. 2018). The results suggested that the BN group was less sensitive in terms of dopamine-dependent brain response and supported the notion of a dopamine downregulation in BN (Frank et al. 2011).

3 Positron Emission Tomography (PET) Imaging

Positron emission tomography (PET) imaging allows researchers to more directly probe the involvement of neurotransmitter systems, which are central to the rewarding effects of eating and have been shown to be altered in eating disorders (Frank and Kaye 2005; Bailer et al. 2013). A caveat is that those studies can tell about neurotransmitter receptor distribution and thus about up- or downregulation in numbers, but connecting those to brain response during tasks that test illness-specific pathophysiology has been more challenging. To date, only a few studies have utilized PET imaging in individuals diagnosed with BN or BED. Earlier studies showed elevated serotonin 1A receptor binding in BN when ill and after recovery, but lower serotonin 2A receptors compared to controls when recovered, and receptor binding was frequently associated with anxiety (Frank 2015). Serotonin receptors may have a specific impact on reward processing across psychiatric disorders and further study on how this pervasive neurotransmitter system is involved in binge eating warrants further study (Hayes and Greenshaw 2011). A study that investigated dopamine receptor binding did not show significant dopamine D2 receptor group differences between BN and controls, but striatal dopamine release was lower in BN, which was inversely associated with binge eating frequency (Broft et al. 2012).

One study investigated striatal dopamine changes in 10 subjects who were obese with BED (OB BED) and eight subjects who were obese without BED (OB) (Wang et al. 2011). This crossover design study included 2 days of scanning. On the first day, participants completed a neutral condition scan with placebo and then a food-stimulation scan with oral methylphenidate (MPH) to block dopamine reuptake (i.e., enhance dopamine signaling). On the second day, they completed a food-stimulation scan with a placebo and then a neutral condition scan with MPH. Subjects fasted overnight before both scan days. In the food stimulation condition, subjects viewed and smelled fresh, warm food (selected based on prior subject preference ratings), and then tasted the food indirectly via cotton swabs. Only the OB BED group displayed significant increases in caudate and putamen in response to the food stimulation condition. Across both groups, greater binge eating scores, but not BMI, were associated with caudate dopamine increases in response to food stimulation. This provides further evidence of the importance of dopamine in BED reward processing and self-reported severity of binge eating behavior. It is likely that neurotransmitters and receptors are in part trait alterations that could contribute to the development of eating disorder behaviors including binge eating, but also adapt to the effects of behaviors and hinder recovery (Frank 2016). With the same food stimulation task, another PET study explored striatal dopamine changes and attitudes toward food, such as restraint and emotionality, in a small sample of ten healthy, nonobese, and non-BED subjects (Tomasi and Volkow 2013). The group found that increased dorsal striatum DA responsivity to food stimulation correlated with higher restraint scores. No significant correlations were found between striatal DA responsivity to food stimulation and BMI. The authors' interpretation of these findings is that the increased DA changes signal greater saliency of the food stimuli and that those subjects utilize restrained eating as a compensatory strategy.

One other study measured regional cerebral blood flow (rCBF) with single photon emission tomography (SPECT) in three groups of adult women: eight subjects who were obese with BED (OB BED), 11 who were obese without BED (OB non-BED), and 12 healthy normal weight controls (Karhunen et al. 2000). Participants were scanned after an overnight fast once while viewing a control image of a landscape and on a second day while viewing a portion of real food which they selected. In the food exposure condition, the OB BED group exhibited a significantly greater increase in rCBF in the left frontal and prefrontal cortices compared to the OB non-BED and healthy controls. Furthermore, only the OB-BED group's increase in hunger ratings, but not desire to eat, during food exposure correlated with greater left frontal and prefrontal rCBF. The prefrontal cortex's role in reward expectancy, specifically the orbitofrontal cortex's response to food reward value estimation, is therefore suggested to be potentially involved in BED. The inclusion of the OB non-BED comparison group is a strength of this study, although these particular control subjects had completed an active weight reduction program before scanning. Weight stability may be an important variable to control for in such a group (Frank et al. 2018c).

4 Limitations and Future Directions

When interpreting the rather small brain imaging literature on reward processing and binge eating behaviors, it is important to take several considerations into account. First, these studies used several different self-report questionnaires and combinations of scores, while others used a structured clinical interview to identify BED. Some applied DSM-IV criteria and other DSM-5. Because the DSM-5 requires reduced frequency (1 day a week instead of 2) and duration (3 months instead of 6) of binge eating to meet BN and BED criteria, earlier studies using DSM-IV criteria may include subjects with slightly more severe symptoms. Future studies should further clarify the potential relationships between binge eating symptom severity and

neurobiological measures. Moreover, BMI was not correlated with brain function in several studies reviewed here, supporting the inclusion of nonobese individuals with BED in future work. Second, menstrual cycle phase was not consistently controlled for in the studies described here. Future investigations should take sex hormones into account given the evidence of structural and functional effects in areas of the brain that mediate reward processing, appetite, emotion, and cognition (Frank et al. 2018b). Third, not all the studies reviewed here included males with BED and none included men with BN. Significant effects of sex were not reported in these studies and sample sizes were too small to compare males and females within BED groups. Nevertheless, some evidence suggests sex differences in cortical response to food images and therefore emphasizes the need for further examination (Michaelides et al. 2012).

5 Conclusion

In summary, the still sparse literature on reward system function and binge eating yield some themes that deserve further exploration. Heightened response to visual food cues may indicate hyper-arousal to those stimuli and it could be tested in the laboratory and natural environment whether for instance mindfulness techniques could normalize such a response. Lower brain activation to taste stimuli, and especially to tasks where reward cues were received unexpectedly, point toward altered dopamine brain circuit function, which could become an important target for pharmacological intervention for binge eating. PET imaging and neurotransmitter receptor-specific drugs before fMRI can be used to study neurotransmitter circuits directly. The ideal solution may be multimodal imaging approaches that combine techniques and study binge eating pathophysiology across patient populations that exhibit that behavior to be able to identify specific brain circuit function that drives this behavior.

References

American Psychiatric Association (2013) Desk reference to the diagnostic criteria from DSM-5. American Psychiatric Publishing, Washington, DC

Avena NM (2013) Animal models of eating disorders. In: Avena NM (ed) Neuromethods, vol 74. Humana Press, Totowa, NJ, pp 281–290

Bailer UF, Frank GK, Price JC, Meltzer CC, Becker C, Mathis CA, Wagner A, Barbarich-Marsteller NC, Bloss CS, Putnam K, Schork NJ, Gamst A, Kaye WH (2013) Interaction between serotonin transporter and dopamine D2/D3 receptor radioligand measures is associated with harm avoidant symptoms in anorexia and bulimia nervosa. Psychiatry Res 211(2):160–168. https://doi.org/10.1016/j.pscychresns.2012.06.010

Balodis IM, Kober H, Worhunsky PD, White MA, Stevens MC, Pearlson GD, Sinha R, Grilo CM, Potenza MN (2013) Monetary reward processing in obese individuals with and without binge eating disorder. Biol Psychiatry 73(9):877–886. https://doi.org/10.1016/j.biopsych.2013.01.014

Balodis IM, Grilo CM, Kober H, Worhunsky PD, White MA, Stevens MC, Pearlson GD, Potenza MN (2014) A pilot study linking reduced fronto-striatal recruitment during reward processing to persistent bingeing following treatment for binge-eating disorder. Int J Eat Disord 47(4):376–384. https://doi.org/10.1002/eat.22204

Berridge KC (2009a) 'Liking' and 'wanting' food rewards: brain substrates and roles in eating disorders. Physiol Behav 97(5):537–550. https://doi.org/10.1016/j.physbeh.2009.02.044

Berridge KC (2009b) Wanting and liking: observations from the neuroscience and psychology laboratory. Inquiry (Oslo) 52(4):378. https://doi.org/10.1080/00201740903087359

Bodell LP, Wildes JE, Goldschmidt AB, Lepage R, Keenan KE, Guyer AE, Hipwell AE, Stepp SD, Forbes EE (2018) Associations between neural reward processing and binge eating among adolescent girls. J Adolesc Health 62(1):107–113. https://doi.org/10.1016/j.jadohealth.2017.08.006

Bohon C, Stice E (2011) Reward abnormalities among women with full and subthreshold bulimia nervosa: a functional magnetic resonance imaging study. Int J Eat Disord 44(7):585–595. https://doi.org/10.1002/eat.20869

Boswell RG, Kober H (2016) Food cue reactivity and craving predict eating and weight gain: a meta-analytic review. Obes Rev 17(2):159–177. https://doi.org/10.1111/obr.12354

Broft A, Shingleton R, Kaufman J, Liu F, Kumar D, Slifstein M, Abi-Dargham A, Schebendach J, Van Heertum R, Attia E, Martinez D, Walsh BT (2012) Striatal dopamine in bulimia nervosa: a PET imaging study. Int J Eat Disord 45(5):648–656. https://doi.org/10.1002/eat.20984

Cambridge VC, Ziauddeen H, Nathan PJ, Subramaniam N, Dodds C, Chamberlain SR, Koch A, Maltby K, Skeggs AL, Napolitano A, Farooqi IS, Bullmore ET, Fletcher PC (2013) Neural and behavioral effects of a novel mu opioid receptor antagonist in binge-eating obese people. Biol Psychiatry 73(9):887–894. https://doi.org/10.1016/j.biopsych.2012.10.022

Chase HW, Fournier JC, Bertocci MA, Greenberg T, Aslam H, Stiffler R, Lockovich J, Graur S, Bebko G, Forbes EE, Phillips ML (2017) A pathway linking reward circuitry, impulsive sensation-seeking and risky decision-making in young adults: identifying neural markers for new interventions. Transl Psychiatry 7(4):e1096. https://doi.org/10.1038/tp.2017.60

Craig AD (2009) How do you feel—now? The anterior insula and human awareness. Nat Rev Neurosci 10 (1):59–70. https://doi.org/10.1038/nrn2555

Cuthbert BN (2014) Translating intermediate phenotypes to psychopathology: the NIMH research domain criteria. Psychophysiology 51(12):1205–1206. https://doi.org/10.1111/psyp.12342

Cyr M, Wang Z, Tau GZ, Zhao G, Friedl E, Stefan M, Terranova K, Marsh R (2016) Reward-based spatial learning in teens with bulimia nervosa. J Am Acad Child Adolesc Psychiatry 55(11):962–971.e3. https://doi.org/10.1016/j.jaac.2016.07.778

de Araujo IE, Kringelbach ML, Rolls ET, Hobden P (2003) Representation of umami taste in the human brain. J Neurophysiol 90(1):313–319. https://doi.org/10.1152/jn.00669.2002

Dodds CM, O'Neill B, Beaver J, Makwana A, Bani M, Merlo-Pich E, Fletcher PC, Koch A, Bullmore ET, Nathan PJ (2012) Effect of the dopamine D3 receptor antagonist GSK598809 on brain responses to rewarding food images in overweight and obese binge eaters. Appetite 59(1):27–33. https://doi.org/10.1016/j.appet.2012.03.007

Filbey FM, Myers US, Dewitt S (2012) Reward circuit function in high BMI individuals with compulsive overeating: similarities with addiction. NeuroImage 63 (4):1800–1806. https://doi.org/10.1016/j.neuroimage.2012.08.073

Flagel SB, Clark JJ, Robinson TE, Mayo L, Czuj A, Willuhn I, Akers CA, Clinton SM, Phillips PE, Akil H (2011) A selective role for dopamine in stimulus-reward learning. Nature 469(7328):53–57. https://doi.org/10.1038/nature09588

Frank GK (2011) Reward and neurocomputational processes. Curr Top Behav Neurosci 6:95–110. https://doi.org/10.1007/7854_2010_81

Frank GK (2013) Altered brain reward circuits in eating disorders: chicken or egg? Curr Psychiatry Rep 15 (10):396. https://doi.org/10.1007/s11920-013-0396-x

Frank GK (2015) Advances from neuroimaging studies in eating disorders. CNS Spectr 20:1–10. https://doi.org/10.1017/S1092852915000012

Frank GK (2016) The perfect storm—a bio-psycho-social risk model for developing and maintaining eating disorders. Front Behav Neurosci 10:44. https://doi.org/10.3389/fnbeh.2016.00044

Frank GK, Kaye WH (2005) Positron emission tomography studies in eating disorders: multireceptor brain imaging, correlates with behavior and implications for pharmacotherapy. Nucl Med Biol 32(7):755–761. https://doi.org/10.1016/j.nucmedbio.2005.06.011

Frank GK, Reynolds JR, Shott ME, O'Reilly RC (2011) Altered temporal difference learning in bulimia nervosa. Biol Psychiatry 70(8):728–735. https://doi.org/10.1016/j.biopsych.2011.05.011

Frank GK, Shott ME, Riederer J, Pryor TL (2016) Altered structural and effective connectivity in anorexia and bulimia nervosa in circuits that regulate energy and reward homeostasis. Transl Psychiatry 6(11):e932. https://doi.org/10.1038/tp.2016.199

Frank GKW, DeGuzman MC, Shott ME, Laudenslager ML, Rossi B, Pryor T (2018a) Association of brain reward learning response with harm avoidance, weight gain, and hypothalamic effective connectivity in adolescent anorexia nervosa. JAMA Psychiatry 75(10):1071–1080. https://doi.org/10.1001/jamapsychiatry.2018.2151

Frank GKW, Favaro A, Marsh R, Ehrlich S, Lawson EA (2018b) Toward valid and reliable brain imaging results in eating disorders. Int J Eat Disord 51 (3):250–261. https://doi.org/10.1002/eat.22829

Frank GKW, Shott ME, DeGuzman MC, Smolen A (2018c) Dopamine D2 -141C Ins/Del and Taq1A polymorphisms, body mass index, and prediction error brain response. Transl Psychiatry 8(1):102. https://doi.org/10.1038/s41398-018-0147-1

Geliebter A, Ladell T, Logan M, Schneider T, Sharafi M, Hirsch J (2006) Responsivity to food stimuli in obese and lean binge eaters using functional MRI. Appetite 46(1):31–35. https://doi.org/10.1016/j.appet.2005.09.002

Geliebter A, Benson L, Pantazatos SP, Hirsch J, Carnell S (2016) Greater anterior cingulate activation and connectivity in response to visual and auditory high-calorie food cues in binge eating: preliminary findings. Appetite 96:195–202. https://doi.org/10.1016/j.appet.2015.08.009

Haber SN, Knutson B (2010) The reward circuit: linking primate anatomy and human imaging. Neuropsychopharmacology 35(1):4–26. https://doi.org/10.1038/npp.2009.129

Hayes DJ, Greenshaw AJ (2011) 5-HT receptors and reward-related behaviour: a review. Neurosci Biobehav Rev 35(6):1419–1449. https://doi.org/10.1016/j.neubiorev.2011.03.005

Karhunen LJ, Vanninen EJ, Kuikka JT, Lappalainen RI, Tiihonen J, Uusitupa MI (2000) Regional cerebral blood flow during exposure to food in obese binge eating women. Psychiatry Res 99(1):29–42

Kelley AE, Berridge KC (2002) The neuroscience of natural rewards: relevance to addictive drugs. J Neurosci 22 (9):3306–3311. https://doi.org/10.1523/JNEUROSCI.22-09-03306.2002

Kelley AE, Baldo BA, Pratt WE, Will MJ (2005) Corticostriatal-hypothalamic circuitry and food motivation: integration of energy, action and reward. Physiol Behav 86(5):773–795. https://doi.org/10.1016/j.physbeh.2005.08.066

Kessler RM, Hutson PH, Herman BK, Potenza MN (2016) The neurobiological basis of binge-eating disorder. Neurosci Biobehav Rev 63:223–238. https://doi.org/10.1016/j.neubiorev.2016.01.013

Lee JE, Namkoong K, Jung YC (2017) Impaired prefrontal cognitive control over interference by food images in binge-eating disorder and bulimia nervosa. Neurosci Lett 651:95–101. https://doi.org/10.1016/j.neulet.2017.04.054

Lloyd EC, Steinglass JE (2018) What can food-image tasks teach us about anorexia nervosa? A systematic review. J Eat Disord 6:31. https://doi.org/10.1186/s40337-018-0217-z

Michaelides M, Thanos PK, Volkow ND, Wang GJ (2012) Dopamine-related frontostriatal abnormalities in obesity and binge-eating disorder: emerging evidence for developmental psychopathology. Int Rev Psychiatry 24(3):211–218. https://doi.org/10.3109/09540261.2012.679918

Monteleone AM, Castellini G, Volpe U, Ricca V, Lelli L, Monteleone P, Maj M (2018) Neuroendocrinology and brain imaging of reward in eating disorders: a possible key to the treatment of anorexia nervosa and bulimia nervosa. Prog Neuropsychopharmacol Biol Psychiatry 80(Pt B):132–142. https://doi.org/10.1016/j.pnpbp.2017.02.020. Epub 2017 Mar 1

Mueller SV, Morishima Y, Schwab S, Wiest R, Federspiel A, Hasler G (2018) Neural correlates of impaired reward-effort integration in remitted bulimia nervosa. Neuropsychopharmacology 43(4):868–876. https://doi.org/10.1038/npp.2017.277

Ohmura Y, Tsutsui-Kimura I, Sasamori H, Nebuka M, Nishitani N, Tanaka KF, Yamanaka A, Yoshioka M (2019) Different roles of distinct serotonergic pathways in anxiety-like behavior, antidepressant-like, and anti-impulsive effects. Neuropharmacology:107703. https://doi.org/10.1016/j.neuropharm.2019.107703

Reiter AM, Heinze HJ, Schlagenhauf F, Deserno L (2017) Impaired flexible reward-based decision-making in binge eating disorder: evidence from computational modeling and functional neuroimaging. Neuropsychopharmacology 42(3):628–637. https://doi.org/10.1038/npp.2016.95

Rolls ET (2016) Functions of the anterior insula in taste, autonomic, and related functions. Brain Cogn 110:4–19. https://doi.org/10.1016/j.bandc.2015.07.002

Rolls BJ, Rolls ET, Rowe EA, Sweeney K (1981) Sensory specific satiety in man. Physiol Behav 27(1):137–142

Schafer A, Vaitl D, Schienle A (2010) Regional grey matter volume abnormalities in bulimia nervosa and binge-eating disorder. NeuroImage 50(2):639–643. https://doi.org/10.1016/j.neuroimage.2009.12.063

Schag K, Schonleber J, Teufel M, Zipfel S, Giel KE (2013) Food-related impulsivity in obesity and binge eating disorder—a systematic review. Obes Rev 14 (6):477–495. https://doi.org/10.1111/obr.12017

Schienle A, Schafer A, Hermann A, Vaitl D (2009) Binge-eating disorder: reward sensitivity and brain activation to images of food. Biol Psychiatry 65(8):654–661. https://doi.org/10.1016/j.biopsych.2008.09.028

Setsu R, Hirano Y, Tokunaga M, Takahashi T, Numata N, Matsumoto K, Masuda Y, Matsuzawa D, Iyo M, Shimizu E, Nakazato M (2017) Increased subjective distaste and altered insula activity to umami Tastant in patients with bulimia nervosa. Front Psychiatry 8:172. https://doi.org/10.3389/fpsyt.2017.00172

Simon JJ, Skunde M, Walther S, Bendszus M, Herzog W, Friederich HC (2016) Neural signature of food reward processing in bulimic-type eating disorders. Soc Cogn Affect Neurosci 11(9):1393–1401. https://doi.org/10.1093/scan/nsw049

Tomasi D, Volkow ND (2013) Striatocortical pathway dysfunction in addiction and obesity: differences and similarities. Crit Rev Biochem Mol Biol 48(1):1–19. https://doi.org/10.3109/10409238.2012.735642

Uddin LQ, Nomi JS, Hebert-Seropian B, Ghaziri J, Boucher O (2017) Structure and function of the human insula. J Clin Neurophysiol 34(4):300–306. https://doi.org/10.1097/WNP.0000000000000377

Vitousek K, Manke F (1994) Personality variables and disorders in anorexia nervosa and bulimia nervosa. J Abnorm Psychol 103(1):137–147

Voon V, Morris LS, Irvine MA, Ruck C, Worbe Y, Derbyshire K, Rankov V, Schreiber LR, Odlaug BL, Harrison NA, Wood J, Robbins TW, Bullmore ET, Grant JE (2015) Risk-taking in disorders of natural and drug rewards: neural correlates and effects of probability, valence, and magnitude. Neuropsychopharmacology 40(4):804–812. https://doi.org/10.1038/npp.2014.242

Wagner A, Barbarich-Marsteller NC, Frank GK, Bailer UF, Wonderlich SA, Crosby RD, Henry SE, Vogel V, Plotnicov K, McConaha C, Kaye WH (2006) Personality traits after recovery from eating disorders: do subtypes differ? Int J Eat Disord 39 (4):276–284. https://doi.org/10.1002/eat.20251

Wang GJ, Geliebter A, Volkow ND, Telang FW, Logan J, Jayne MC, Galanti K, Selig PA, Han H, Zhu W, Wong CT, Fowler JS (2011) Enhanced striatal dopamine release during food stimulation in binge eating disorder. Obesity (Silver Spring) 19(8):1601–1608. https://doi.org/10.1038/oby.2011.27

Weygandt M, Schaefer A, Schienle A, Haynes JD (2012) Diagnosing different binge-eating disorders based on reward-related brain activation patterns. Hum Brain Mapp 33(9):2135–2146. https://doi.org/10.1002/hbm.21345

Wierenga CE, Ely A, Bischoff-Grethe A, Bailer UF, Simmons AN, Kaye WH (2014) Are extremes of consumption in eating disorders related to an altered balance between reward and inhibition? Front Behav Neurosci 8:410. https://doi.org/10.3389/fnbeh.2014.00410

Wonderlich JA, Breithaupt L, Thompson JC, Crosby RD, Engel SG, Fischer S (2018) The impact of neural responses to food cues following stress on trajectories of negative and positive affect and binge eating in daily life. J Psychiatr Res 102:14–22. https://doi.org/10.1016/j.jpsychires.2018.03.005. Epub 2018 Mar 14. PMID: 29558632

Wu M, Brockmeyer T, Hartmann M, Skunde M, Herzog W, Friederich HC (2016) Reward-related decision making in eating and weight disorders: a systematic review and meta-analysis of the evidence from neuropsychological studies. Neurosci Biobehav Rev 61:177–196. https://doi.org/10.1016/j.neubiorev.2015.11.017

Xia L, Gu R, Zhang D, Luo Y (2017) Anxious individuals are impulsive decision-makers in the delay discounting task: an ERP study. Front Behav Neurosci 11:5. https://doi.org/10.3389/fnbeh.2017.00005

The Neurobiological Basis of Executive Function Alterations in Binge Eating Populations

Trevor Steward and Laura A. Berner

Abstract

Binge eating episodes are defined by the consumption of large amounts of food over a short period of time accompanied by feelings of loss of control, and in many cases, remorse, embarrassment, and disgust. Binge eating is a key feature of bulimia nervosa, binge eating disorder, and anorexia nervosa binge eating/purging subtype, yet our understanding of the neurobiological substrates of this maladaptive behavior is still in nascent stages. The current chapter provides an overview of behavioral and functional magnetic resonance imaging (fMRI) studies of executive dysfunction in eating disorder populations with binge eating. Overall, evidence suggests that alterations in cognitive interference control, response inhibition, delay discounting, set-shifting, and decision-making may make it difficult for individuals to control and adjust their food intake thereby contributing to the etiology and/or maintenance of binge eating. We conclude by covering limitations in the existing literature on this topic and propose new lines of research to address gaps in our knowledge of executive functions in individuals who binge eat.

Keywords

Binge eating · Bulimia nervosa · Anorexia nervosa · Executive function · Neuropsychology · Neuroimaging · fMRI · Cognitive interference · Response inhibition · Delay discounting · Set-shifting · Decision-making

Learning Objectives
In this chapter, you will:

- Understand the relevance of executive function to binge eating.
- Learn about behavioral and fMRI studies focused on the domains of cognitive interference control, response inhibition, delay discounting, set-shifting, and decision-making.
- Identify limitations in the existing literature and consider future lines of research focused on executive function in binge eating populations.

T. Steward
Melbourne School of Psychological Sciences, University of Melbourne, Parkville, VIC, Australia

L. A. Berner (✉)
Department of Psychiatry, Center of Excellence in Eating and Weight Disorders, Icahn School of Medicine at Mount Sinai, New York, New York, USA
e-mail: laura.berner@mssm.edu

© Springer Nature Switzerland AG 2020
G. K.W. Frank, L. A. Berner (eds.), *Binge Eating*, https://doi.org/10.1007/978-3-030-43562-2_10

1   Introduction

Binge eating is characterized by the consumption of an objectively large quantity of food in a discrete period of time and feelings of loss of control (APA 2013). Binge eating episodes are associated with marked distress and are a key feature of binge-eating disorder (BED), bulimia nervosa (BN), and anorexia nervosa, binge eating/purging subtype (AN-BP). Each of these disorders is associated with significant medical complications and high rates of comorbid psychopathology (Telch and Stice 1998; Swinbourne et al. 2012; Kessler et al. 2013), and data suggest that 50–70% of those who receive first-line eating disorder treatments do not achieve full symptom remission (Khalsa et al. 2017; Linardon et al. 2017; Linardon 2018). Improved understanding of the neurobiological mechanisms that promote and maintain binge eating could lead to the development of more effective, brain-based interventions.

A growing body of research suggests that alterations in neurocognitive processes contribute to the symptomology of eating disorders (Steinglass et al. 2019) and, more specifically, to the onset and maintenance of binge eating behaviors (Voon 2015). Given that binge eating is defined in part by a sense of loss of control over eating, much of this neurocognitive research in BED, BN, and AN-BP has focused on executive function, also called executive control or cognitive control. This broad term includes holding information in working memory and maintaining focus, inhibiting impulsive responses, resisting temptations, and shifting the focus of one's attention according to personal goals. Recent integrative theories of binge eating have posited that vulnerabilities in executive function could predispose individuals to exhibit aberrant responses, perhaps particularly in the context of food-related stimuli, which in turn could trigger binge eating behaviors (Pearson et al. 2015; Kessler et al. 2016).

Animal and human studies that pair brain-based and behavioral measures have identified distinguishable neurobiological correlates of executive function. Task-based functional magnetic resonance imaging (fMRI), in particular, has elucidated the neural underpinnings of distinct executive domains in individuals with binge eating and researchers have developed food-specific adaptations of tasks that measure executive function to probe potential eating-related alterations. For example, food choice tasks have been developed that involve participants choosing between more immediate, smaller, and delayed, larger chocolate rewards during fMRI scanning (Dong et al. 2016). However, using tasks with nonfood-related stimuli (e.g., monetary rewards) to compare the behavior and neural activation of individuals with and without binge eating provides insight into more generalized impairments that may span multiple disorders (Lempert et al. 2019). By characterizing deficits in cognitive function in contexts beyond eating behavior, this approach may ultimately identify targets to improve treatment outcomes (Insel et al. 2010).

In this chapter, we provide an overview of existing behavioral and neuroimaging data on executive function in populations with binge eating behavior, namely individuals with BED, BN, and AN-BP. We highlight studies including non-binge eating groups with overweight or obesity to control for the possible confounding effects of body mass index (BMI), and focus on subconstructs of executive function that have been most studied in these populations—cognitive interference, response inhibition, delay discounting, attentional control, set-shifting, and decision-making. We conclude with an outline of limitations of the existing literature and proposals for new lines of research.

2   Cognitive Interference Control

Attentional control refers to the ability to select and direct attention to stimuli that are goal relevant. This complex process has been divided into three sub-processes—alerting, orienting, and cognitive interference control, also called executive control of attention (Petersen and Posner 2012). Alerting refers to the ability to focus attention on upcoming events, and involves responding to a warning cue with an alert state that is then maintained. Orienting and reorienting relate to shifting attention to a stimulus outside one's current focus. Cognitive interference control involves ignoring or suppressing irrelevant information or a competing stimulus in order to maintain a goal. Cognitive interference control is, to date, the most investigated executive function in AN-BP, BN, and BED, perhaps because deficits in this domain are posited to contribute to difficulty inhibiting attention to urges or cravings that may potentiate binge episodes (Voon 2015).

Classic tests of cognitive interference control that have been most often used to study individuals with eating disorders include the Stroop and Simon Spatial Incompatibility Tasks. The Stroop Task requires that an individual ignore the color a word is printed in to be able to correctly read the word. The Simon Task requires that an individual ignore the side of the screen on which an arrow or a word is presented to be able to correctly identify the direction in which the arrow is pointing or the direction of the printed word ("left" or "right"). Accuracy and reaction time on conflict trials (i.e., when the word is different than its color or the direction of the arrow is different than the side on which it is presented), index cognitive interference control. To successfully exert cognitive interference control, two networks of the attentional control system are required—a frontoparietal control network, which maintains moment-to-moment attention on the task at hand, and a cingulo-opercular network, which maintains overall task set (e.g., only press a button to indicate arrow directionality). Data from fMRI studies indicate that during correct responses to conflict trials on both the Stroop and Simon Spatial Incompatibility Tasks healthy individuals show activation in the anterior cingulate cortex (ACC), supplementary motor area (SMA), inferior temporal, parietal, and frontal cortices, dorsolateral prefrontal cortex (DLPFC), and dorsal caudate (Peterson et al. 2002; Rubia et al. 2006).

A recent meta-analysis of individuals with any bulimic-type eating disorder examined behavioral and cognitive control on tasks that used disorder-specific (i.e., food or body-related words or images) and non-disorder-specific stimuli (Wu et al. 2013). Mostly cognitive interference control tasks were included (15 out of the 21 total studies in the meta-analysis that used non-disorder-specific stimuli). Poor performance on these tasks, as indexed by a larger number of errors and pronounced slowing in reaction time on incongruent trials, is thought to reflect deficits in interference control. Results indicated that individuals with AN-BP show large deficits, individuals with BN show small deficits, and individuals with BED do not show deficits relative to healthy individuals on tasks with general stimuli. Studies including disorder-specific stimuli only focused on BN and almost exclusively (11 out of 12 studies) included cognitive interference control tasks; deficits on these disorder-specific tasks were associated with a medium effect size and were more pronounced than for tasks using general stimuli in BN. Taken together, these findings suggest that AN-BP and BN are associated with cognitive interference control deficits, whereas BED is not. However, studies comparing overweight individuals with loss of control over eating (not necessarily BED) to weight-matched controls have documented deficits in color word interference (Manasse et al. 2014). Additional research is needed to determine how weight status, episode size and frequency, and compensatory behaviors may relate to cognitive interference control across these groups.

In one study, despite intact behavioral performance on the Stroop Task, individuals with BED showed reduced activation in the ventromedial PFC (VMPFC), inferior frontal gyrus, and insula during Stroop Task performance relative to obese and lean controls with no binge eating (Balodis et al. 2013). PFC activation in the BED group was inversely correlated with restraint scores, suggesting that those with the most abnormally reduced activation in these areas during interference control made the most attempts to limit or control their intake (via, e.g., fasting, attempting to exclude foods, attempting to follow eating rules). Similarly, data from adolescents (Marsh et al. 2011) and adults (Marsh et al. 2009) with BN suggest reduced activation in frontostriatal circuits during the engagement of cognitive interference control on the Simon task. Adults with BN performed worse than controls, and showed reduced activation in the bilateral ventrolateral PFC (VLPFC), caudate, and ACC on conflict trials, whereas adolescents performed as well as controls on an abbreviated version of the task, but also showed reduced activation in the VLPFC, DLPFC, and putamen on conflict trials. Binge eating frequency was associated with reduced frontostriatal activation in both groups, and these altered patterns of activation on the Simon Task have been shown to reliably distinguish full and subthreshold BN cases from healthy controls (Cyr et al. 2018). Longitudinal studies would be beneficial in determining whether disturbances in frontostriatal circuits ensue with the onset of disordered eating, or instead precede the onset of binge eating behaviors as a result of vulnerable self-regulatory systems.

In contrast, results of a pilot study indicate that adults with BN show increased frontostriatal activation during a food-specific Stroop paradigm, despite poorer performance on the task, and adults with BED show increased activation in reward-related regions during the paradigm (Lee et al. 2017). Therefore, neural underpinnings of food and nonfood-specific cognitive interference control problems in binge eating populations may importantly differ. To our knowledge, no neuroimaging studies have examined the neural correlates of cognitive interference control in AN-BP. Overall, existing data suggest reduced cognitive interference control in binge eating populations; however, the directionality of altered neural activation that may drive impaired interference control and, ultimately, dysregulated eating behavior, requires further investigation. Drawing distinctions among cognitive interference processing stages (e.g., stimulus encoding, response selection, and response execution) could provide possible explanations for the aforementioned differences in the directionality of regional activations and aid in dissociating the neural interference resolution mechanisms most related to clinical symptoms.

3 Response Inhibition

Response inhibition, also called behavioral inhibition or motor inhibitory control, broadly refers to the ability to inhibit or otherwise control responses or actions. Given that a sense of difficulty controlling one's eating behavior defines "binges," this domain of executive function may be particularly relevant for the study of binge eating populations. Classic computerized neurocognitive measures of response inhibition typically necessitate that individuals press a button in response to one stimulus but withhold the response when some other, usually infrequent, stimulus occurs. For example, in go/no-go tasks, individuals must inhibit a response when no-go stimuli are presented on the screen in lieu of go stimuli. In stop signal tasks (SST), individuals must inhibit a response when a stop signal sounds or appears with some delay (the "stop signal delay") after the go stimulus was presented. Of note, these two types of tasks measure slightly different aspects of inhibition—action restraint, or the withholding of a prepotent response, and action cancelation of an already initiated response (Eagle et al. 2008). However, on both tasks, the number of commission errors, or responses during no-go or stop trials, are thought to index deficits in behavioral inhibition. In addition, stop signal reaction time on the SST, or the point at which half of the stop trials are successfully inhibited relative to an individual's mean reaction time, is another index of inhibitory ability, with longer stop-signal reaction times indicating poorer inhibition (Logan et al. 1997).

In healthy individuals, inhibition on both go/no-go and SSTs involves the left and right inferior and middle frontal gyri, right superior frontal gyrus, ACC, anterior insula, subthalamic nucleus, pre-SMA, and dorsal aspects of the striatum (Buchsbaum et al. 2005; Aron 2006, 2011; Simmonds et al. 2008). However, the go/no-go task engages the frontoparietal control network mediating adaptive online control to a greater extent than the SST, whereas SST inhibition engages the cingulo-opercular control network, which maintains task set and responses to salient stimuli (Swick et al. 2011). Go/no-go task inhibition predominantly activates the right middle frontal gyrus and the right inferior parietal lobule, and the go/no-go task is more robustly associated with PFC neural activation than the SST (Swick et al. 2011). In addition, the SST more prominently activates the bilateral thalamus and left anterior insula compared to the go/no-go task. At a molecular level, go/no-go task inhibition seems to strongly involve serotonin, whereas SST inhibition involves noradrenaline (Eagle et al. 2008). Therefore, withholding a response and stopping a response involve overlapping, but also distinct neural and neuromolecular mechanisms.

Effect sizes for behavioral deficits across BED, BN, and AN-BP compared with controls are larger on tasks of response inhibition than on previously discussed tasks that measure cognitive interference control, with the largest effect size on the SST (Wu et al. 2013). These meta-analytic results support that bulimic-type eating disorders are characterized by deficits in response inhibition. However, behavioral and neuroimaging data focused on response inhibition in BED alone have been somewhat mixed. Some evidence suggests that overweight individuals with binge eating show deficits in SST performance across stimulus types compared to weight-matched controls (Manasse et al. 2016b), and most studies using food-specific go/no-go tasks show deficits in BED relative to normal weight and obese controls (Schag et al. 2013). In contrast, results of a recent meta-analysis indicate that individuals with BED do not show behavioral deficits on the SST relative to controls (Lavagnino et al. 2016). Data from individuals with BN and AN-BP more consistently suggest impairments in response inhibition (Wu et al. 2013).

Although BED relative to overeating may not translate consistently to impairments in response inhibition, the neural substrates of response inhibition in individuals who binge eat may be meaningfully altered. Similar to findings in the cognitive interference control domain, obese individuals with BED show reduced PFC activation when engaging control on a food-specific go/no-go task, even when performance is

comparable to that of obese controls (Hege et al. 2015). Somewhat similarly, results of one study indicate that a subgroup of women with BN who had more frequent binge eating showed reduced activation relative to controls in the dorsal striatum and sensorimotor cortex during the inhibition of responses, but only to neutral images, not to food images (Skunde et al. 2016). The only other study of BN to use a go/no-go task with concurrent fMRI included emotional visual stimuli and reported age-dependent reductions in PFC activation relative to controls during response inhibition (Dreyfuss et al. 2017). In sum, research to date suggests that reduced activation during action restraint may play a role in BED and BN, possibly reflecting difficulties engaging frontostriatal control networks.

Considerably less neuroimaging research has focused on response inhibition in AN, particularly AN-BP. One study that administered a go/no-go task during fMRI reported increased precentral, ACC, temporal, and DLPFC activation during response inhibition in a mixed "binge/purge" group of adolescents compared with controls (Lock et al. 2011). Because only 30% of these participants were underweight, and the healthy weight majority may not have met the criteria for a diagnosis of BN, results of this study may not be reflective of neural alterations in AN-BP and are difficult to compare to those in BN. One study administered the SST outside the scanner to a large sample of adults with current and past AN and collected resting state fMRI data. Results indicated that the underweight AN group, 21% of whom had AN-BP, showed increased stop signal reaction times relative to controls and to individuals remitted from AN, and increased positive and negative connectivity of the right inferior frontal gyrus relative to controls. However, this altered connectivity was unrelated to SST performance or to eating disorder symptoms (Collantoni et al. 2016).

Thus, fMRI findings suggest aberrant PFC functioning associated with action restraint across the weight spectrum, but the directionality of these differences is unclear. One study has shown increased PFC activation during food-specific SST inhibition in a normal weight group with binge eating (but not a diagnosis of BED) relative to controls (Oliva et al. 2019), but to our knowledge, no published studies of BED, BN, or AN-BP have used the SST with concurrent fMRI. Thus, the role of neural circuits supporting action cancelation in disorders characterized by binge eating is unclear. In addition, more neuroimaging research is needed to examine how the neural mechanisms that support response inhibition could contribute to binge eating with and without compensatory behaviors.

4 Delay Discounting

Delay discounting refers to an individual's tendency to devalue a reward as the delay of its receipt increases and is considered a proxy for an individual's willingness to delay gratification. A tendency to overvalue the immediate rewards of eating and discount the long-term benefits of not consuming excess calories has been hypothesized to promote the sense of loss of control over eating that defines a binge episode. Personal discounting rates can be empirically measured using delay discounting tasks, in which individuals choose between receiving smaller, sooner rewards and larger, later rewards (e.g., "Would you rather receive $10 in 7 days or $40 in 28 days?"). Most people discount the value of future rewards, although individuals widely differ in deciding when to shift their preferences from immediate to delayed rewards (Anokhin et al. 2011; Peters and Büchel 2011; Sanchez-Roige et al. 2018). Discounting rates are defined as high (steep) if a delayed reward possesses little value, and are defined as low (shallow) if delayed rewards are given high value. Delay discounting paradigms can be carried out in hypothetical, experimental or "incentive-compatible" contexts. Most studies employ hypothetical paradigms and do not reward participants based on their responses. In contrast, "incentive-compatible" designs randomly select a participant's choice from one trial to determine a reward they receive. Experiential delay discounting designs are most analogous to paradigms found in animal research and have participants wait through a delay of

seconds or minutes before receiving the reward they had chosen (Hayden 2016).

The processing of delayed reward has been demonstrated to generate an increase in activity in a wide range of brain regions including the basal ganglia and parietal, prefrontal, motor, and insular cortices (Wesley and Bickel 2014; Volkow and Baler 2015; Smith et al. 2018a). When facing a situation requiring a choice between immediate or delayed rewards, processed sensory input from the junction between temporal and posterior cortex is relayed to the PFC, along with valuation input from the striatum (Louie and Glimcher 2010). PFC regions associated with goal-directed control, such as the orbitofrontal cortex (OFC), are then involved in weighing the trade-offs of the presented options and send efferent signals to the motor cortex in order to generate or inhibit a response (Eagle and Baunez 2010). Importantly, motivational/emotional information related to the stimulus increases activation in the amygdala, the hippocampus, the temporal lobe, and the insula to modulate the salient value of the stimulus (Wilson et al. 2018). The subsequent selection and timing of delay discounting choices are regulated by cortico–striato–thalamo–cortical loops, which are implicated in making necessary adjustments related to motivation and memory (Vertes 2006; Frost and McNaughton 2017).

There is a strong evidence suggesting that individuals with BED discount rewards more steeply than nonobese healthy controls (Davis et al. 2010; Manwaring et al. 2011; Manasse et al. 2015; Bartholdy et al. 2017; Steward et al. 2017), though it is not clear whether delay discounting rates can distinguish individuals with obesity who binge eat from those who do not (Stojek and MacKillop 2017). Similarly, it remains unknown whether individual discounting rates can serve as a predictor of outcome in treatment for BED given that higher negative urgency levels (i.e., the tendency to act rashly when experiencing intense emotions), but not steeper individual discounting rates, have been linked to less pronounced benefits from treatment (Manasse et al. 2016a). Relative to studies focused on BED, fewer empirical studies have been carried out specifically examining delay discounting in individuals with BN, but existing results are consistent with those in BED. Kekic et al. (2016) found that individuals with BN showed a stronger preference for smaller, sooner rewards and endorsed a reduced capacity to delay gratification relative to controls. Likewise, Bartholdy et al. (2017) identified a negative association between binge eating episode frequency and self-report tendencies to delay gratification in patients with BN.

In stark contrast to BED and BN, combined samples of AN subtypes consistently show shallower discounting rates in comparison to controls (Steinglass et al. 2012, 2017; Decker et al. 2015; Steward et al. 2017). This propensity to forgo immediately available rewards is understood to contribute to the restrictive eating behaviors that are characteristic of the disorder, and longitudinal data indicate that discounting rates normalize after weight recovery (Decker et al. 2015). However, studies that separately study AN-R and AN-BP suggest that the restrictive subtype is associated with significantly shallower discounting rates than controls, whereas individuals with AN-BP either discount rewards to the same extent as (Ritschel et al. 2015), or more steeply than (Steward et al. 2017), controls.

A limited number of fMRI studies have been carried out to date examining the neural underpinnings of altered delayed discounting in individuals with binge eating. One study using a monetary delayed discounting task identified reduced activation during larger, later choices compared to smaller, sooner choices in the dorsal anterior cingulate cortex (dACC), and the striatum in a sample of patients with AN, over half of whom met criteria for AN-BP (Decker et al. 2015). As cingulo–striatal circuits have been found to play a role in the modulation of basic reward signals (Peters and Büchel 2011), reduced dACC and striatum activation during delayed choices in patients with AN-BP could be indicative of a deficit in complex decision-making during delay discounting. However, as no fMRI studies have focused specifically on delay discounting in AN-BP alone, or in BN or BED, the relative associations of weight, restriction, and

bulimic symptoms with steeper or shallower discounting rates and related neural alterations is not known.

5 Set-Shifting

Set-shifting, an aspect of cognitive flexibility, refers to the ability to divert attentional resources or take action according to changing situational demands (Lezak 1995). A general deficit in the domain of set-shifting is thought to contribute to the stereotypic behaviors (e.g., compulsive food seeking) and perseverative thought patterns (e.g., rumination about food and eating; Levinson et al. 2019) that are characteristic of individuals with binge eating (Kakoschke et al. 2019). Set-shifting impairment and a lack of mental flexibility may also interfere with treatment for binge eating, as a rigid cognitive style could impede the testing of new behaviors and setting realistic goals (Kakoschke et al. 2019). Tasks that assess set-shifting commonly require participants to switch from previously learned strategies to a new set of rules with different contingencies. The majority of studies examining binge eating populations assess set-shifting performance and task switching using the Wisconsin Card Sorting Test (WCST) (Heaton 1993) or the Trail Making Task (TMT) (Tombaugh 2004), respectively. Of note, successful performance on the WCST requires additional key cognitive operations apart from the target component of set-shifting, such as associative learning, performance monitoring (the ability to monitor and interpret cues and feedback signals to adaptively guide behavior), and working memory (Miyake et al. 2000; Ridderinkhof et al. 2002), which may also be impaired in binge eating populations.

Animal and human research highlight a role for the VLPFC in selecting responses after shifts in feedback and shifts in reward contingencies (Robbins 2007). For example, lesion studies support that VLPFC function is integral to the new learning required to render previously irrelevant abstract dimensions relevant during tasks involving switching reinforcement contingencies.

Convergent evidence from fMRI studies indicates that reduced VLPFC activation is linked to increased rates of perseverative errors, the primary outcome variables of the WCST, in populations with addictive behaviors (Verdejo-Garcia et al. 2015), as well as in patients with obsessive-compulsive disorder (Vaghi et al. 2017). In addition, activation in medial prefrontal regions, the caudate nucleus, and the mediodorsal thalamus increases following the receipt of negative feedback during set-shifting tasks, and this activation is thought to signal the need to shift to a new response set (Monchi et al. 2001).

Multiple meta-analyses and systematic reviews support poorer set-shifting performance in individuals with BED and BN compared to healthy controls (Wu et al. 2014; Voon 2015; Hirst et al. 2017; Smith et al. 2018b). However, the extent to which set-shifting difficulties are related to disorder severity remains unclear. Tchanturia et al. (2011), for example, did not identify an association between illness duration and set-shifting deficits in a large sample of patients with BN, though other researchers have linked set-shifting deficits to disorder-related rituals in BN (Roberts et al. 2010) and a reduced ability to engage in effective coping styles in BED (Dingemans et al. 2015). Some data suggest that the combination of depressive symptoms and deficits in set-shifting abilities, specifically, may increase the risk of loss-of-control eating (Dingemans et al. 2015). Results of studies including individuals with AN-BP are less consistent (Steinglass et al. 2006; Galimberti et al. 2012; Herbrich et al. 2018) and the most comprehensive meta-analysis of set-shifting to date did not find evidence to support prominent set-shifting impairment in either adults or adolescents with AN-BP (Wu et al. 2014).

Few neuroimaging studies have examined set-shifting in individuals who binge eat, and results are difficult to compare across studies because of heterogeneous task designs and sample characteristics. Van Autreve et al. (2016) separately compared AN-R patients and AN-BP patients to controls during a task in which participants had to switch between two simple number tasks, using shape cues to track which

task to perform. Groups did not differ in task performance, and aberrant activation in the insula and the precuneus during switch-specific blocks was found only in the AN-R group, not the AN-BP group (Van Autreve et al. 2016). In contrast, another study using a smaller, combined sample of patients with AN-R or AN-BP examined set-shifting via a fMRI-compatible version of the WCST and identified decreased activation compared to controls in the VLPFC in the AN-BP group, but not in the AN-R group, during shifting trials (Sato et al. 2013). One longitudinal study of patients with AN (81% with AN-BP) found that greater set-shifting ability was correlated with greater DLPFC and VLPFC activation, and improvements in set-shifting performance over time were predicted by a combination of low VLPFC and high frontopolar PFC activation at the start of treatment. These results suggest that patients with set-shifting difficulties may benefit the most from cognitive therapy that directly addresses this skill (Garrett et al. 2014). To date, no fMRI studies have examined set-shifting in patients with BN or BED, highlighting the need for more neuroimaging research to determine whether the neural mechanisms that underlie set-shifting are commonly altered across disorders featuring binge eating.

6 Decision-Making

The ability to make advantageous decisions based on past experiences and present circumstances is a core cognitive function of our daily lives. Failures in the decision-making process can lead to harmful consequences if, for instance, one is unable to detect the risks that a course of action entails or, in situations of uncertainty, to properly incorporate bodily signals which mark options as being advantageous or disadvantageous (Damasio 1994). In the case of binge eating, alterations in decision-making processes may promote or perpetuate binge eating episodes that persist despite the risks of significant distress and negative physical consequences (Horstmann 2017). Indeed, individuals who binge eat often report engaging in behavior that is incongruent with their explicitly stated goals, and some evidence suggests that these populations are more susceptible to the influence of environmental cues (Neveu et al. 2016) and negative affective states (Danner et al. 2013) during decision-making.

The Iowa Gambling Task (IGT), or variations of it, are the most commonly used instruments to measure decision-making capability under conditions of uncertainty. Originally developed for functional assessment in patients with VMPFC lesions, this task also considers factors such as response to uncertainty and sensitivity to reward and punishment (Bechara et al. 1994). The IGT is based on the principles of the Somatic Marker Hypothesis, a theory that posits decision-making is a dual combination of logical cost–benefit examination and "somatic marker" biasing signals that guide responses according to past experience and the likelihood of punishment or reward (Damasio 1994). On the IGT, participants must discern decks of cards that are advantageous (i.e., they provide greater rewards) from decks that are disadvantageous (i.e., they provide more losses than overall gains).

Weighing the uncertain positive or negative outcomes of options and selecting a final decision requires the coordinated activity of multiple prefrontal regions (Hare et al. 2009). The OFC/VMPFC has been repeatedly demonstrated to display heightened activation while regulating the affective and motivational aspects of decision-making (Bechara et al. 2000; Bechara 2005). Other regions, such as the VLPFC and the DLPFC, are understood to underpin the leveraging of selective attention and working memory, respectively, to make rational evaluations of risk and benefit (Heekeren et al. 2008). The insula and posterior cingulate cortex are implicated in generating representations of emotional states that inform decisions, whereas the ventral striatum, ACC, and the SMA show greater activation during the implementation of decisions (Li et al. 2010).

Results of a meta-analysis indicate that BED, BN, and AN-BP are all associated with significantly impaired IGT performance relative to controls (Guillaume et al. 2015). In addition, some studies in this area have included context-specific tests of decision-making. For example, numerous models have proposed that binge eating is preceded by the experience of negative affect, and that binge eating is often used as a way to cope with or to avoid negative emotions (Burton and Abbott 2019). As such, recent studies have opted to examine decision-making in binge eating populations under "hot" (i.e., affect-influenced) conditions. Danner et al. (2013) found that, in the context of experimentally increased negative affect, post-punishment trials were associated with disadvantageous choice behavior in both BN and BED women, but not in healthy controls. Thus, BN and BED may be characterized by decision-making impairments that are particularly exaggerated in the context of negative emotion. Whether decision-making impairments in AN-BP are similarly dependent on the contexts of negative affect or current low weight remains unclear. One longitudinal study found that IGT performance deficits in patients with AN-BP disappeared with weight restoration after 1 year (Steward et al. 2016), though other studies including recovered AN-BP and AN-R participants have found evidence for poorer (Danner et al. 2012), similar (Tchanturia et al. 2007), or even better (Lindner et al. 2012) decision-making performance when compared to controls.

To date, task-based fMRI has been used to study nonfood-specific decision-making in BED, but not in BN or AN-BP. Reiter et al. utilized computational modeling of choice behavior during a probabilistic reward learning task in order to classify specific signatures of altered decision-making in BED (Reiter et al. 2016). Behaviorally, participants with BED were more likely than controls to make unnecessary adjustments between choices, thereby indicating a possible bias toward exploratory decisions when adapting behavior. At a neural level, the BED group displayed reduced activation compared to controls in the anterior insula and VLPFC during exploratory decisions, which could reflect an inability to incorporate potential warning or uncertainty signals that deter excessive exploration. Additional studies are required to determine whether dysfunctional insula and PFC activation underlies disadvantageous eating choices in binge eating populations.

In individuals with AN-BP, IGT performance has been examined in relation only to brain structure. In accordance with decision-making neuro-imaging research in other psychiatric populations (Norman et al. 2018), poorer performance on the IGT was associated with reduced OFC/VMPFC volume in a mixed sample of patients with AN-R and AN-BP (Bodell et al. 2014). However, this correlation between brain volume and IGT performance was no longer significant after weight recovery.

7 Conclusions

Overall, behavioral and neuroimaging data suggest that eating disorders featuring binge eating episodes are characterized by executive dysfunction spanning multiple domains. To date, cognitive interference control and response inhibition have been most investigated in these populations, and converging evidence suggests that deficits in these domains may play a role in binge eating across disorders. Similarly, there is strong evidence indicating that individuals with BED and BN show impairments in set-shifting and decision-making, though additional research is needed to confirm whether the same holds true for individuals with AN-BP. Additional studies in larger, more diverse samples are required to confirm initial findings indicating that binge eating populations present steeper delay discounting rates (i.e., a greater tendency to prefer immediate rewards) in comparison to healthy controls and non-binge eating groups with overweight or obesity.

7.1 Considerations and Limitations of Existing Research

Heterogeneous and small samples and the use of disparate tasks to measure the same subdomains of executive function have limited the generalizability of existing findings. Study designs to date have also prevented systematic examination

of the relative associations of binge eating, weight status, and compensatory behaviors with executive dysfunction. In addition, few neurocognitive studies of binge eating have employed open-science practices such as preregistration (Gonzales and Cunningham 2015) and data sharing (Ross 2016). Future research focused on executive function in binge eating groups should strive to address these concerns and follow guidelines for brain imaging research in eating disorders to acquire more reliable and clinically useful results (Frank et al. 2018).

7.2 Future Directions in the Study of Executive Function and Binge Eating

Most of the studies reviewed herein employed tasks with non-disorder-specific stimuli. The development of tasks that examine food-specific responsivity across multiple domains could improve our understanding of the neurocognitive processes and neural circuits that contribute directly to binge eating (Berner et al. 2017). For instance, women with BN show reduced activation in control regions during passive exposure to food stimuli (Neveu et al. 2018). However, future studies using food-specific adaptations of well-validated tasks that include an active, control-related demand (e.g., the downregulation of craving, the control of a behavioral response) are needed to explicitly link activation in executive control circuits to eating-related control processes. Such findings could help inform the development of more effective psychological treatments that do not solely focus on symptoms, but that specifically target basic executive control functions (Lock et al. 2013; Tchanturia et al. 2014).

More research linking executive dysfunction to affective, interpersonal, and biological risk factors for binge eating is also needed. For example, combining neurocognitive assessments with ecological momentary assessment may provide much needed insight into how changes in mood modulate executive functioning and vice versa, possibly thereby promoting cycles of binge eating. One recent study using an ecological momentary assessment protocol found that steeper delay discounting strengthened momentary associations between positive and negative affect and binge eating, whereas lower delay discounting was linked to a negative association between momentary positive affect and binge eating (Smith et al. 2019). Delay discounting task performance could, therefore, help to identify patients whose binge eating may be most responsive to interventions targeted at negative affect (Berg et al. 2015).

In addition, although the modulating role of neuroendocrine factors in the processing of food reward has been extensively investigated (Monteleone et al. 2018), very few studies have explored the potential role of neuroendocrines in executive dysfunction in binge eating populations. Recent evidence suggests that alterations in neuroendocrine levels are associated with cognitive deficits in patients with eating disorders (Steward et al. 2019; Wollenhaupt et al. 2019), and investigating their potential specific role in executive dysfunction in binge eating could prove fruitful.

Finally, additional studies focused on attentional control and its sub-processes are needed. Findings consistently suggest that attention-deficit/hyperactivity disorder (ADHD) symptoms are associated with more frequent overeating episodes in overweight youth, ADHD is a risk factor for the development of BN (Mikami et al. 2008), children with ADHD are at 12 times higher risk for loss-of-control eating (Reinblatt et al. 2015), the prevalence of lifetime ADHD in the BN and AN-BP populations is elevated (Blinder et al. 2006; Yilmaz et al. 2011; Svedlund et al. 2017), and inattention symptoms explain most of the variance in eating disorder pathology severity in adults with BN (Seitz et al. 2013). Considerable research has focused on cognitive interference control, but very little has focused on alerting and orienting elements of attentional control (Berner et al. 2018; Kollei et al. 2018). In particular, only one study of a binge eating group has examined the performance and neural activation on a task that measures all three sub-processes of attentional control—the Attention Network Task (Fan et al. 2002). Despite behavioral performance comparable to that of controls, adults with BN showed increased activation in occipital, ACC, and parietal regions during attention alerting that correlated with more severe global eating disorder symptoms, and decreased activation in the temporoparietal junction and the ACC during reorienting that correlated with ADHD scores (Seitz et al. 2016). Thus, future studies should focus on behavioral and neural alterations associated with sub-processes of attentional control in binge eating populations and examine how those alterations may relate to symptoms.

Conflict of Interests The authors have no conflicts of interest to declare.

References

Anokhin AP, Golosheykin S, Grant JD, Heath AC (2011) Heritability of delay discounting in adolescence: a longitudinal twin study. Behav Genet 41:175–183. https://doi.org/10.1007/s10519-010-9384-7

APA (2013) Diagnostic and statistical manual of mental disorders: DSM-5. American Psychiatric Assocation, Washington, DC

Aron AR (2006) Cortical and subcortical contributions to stop signal response inhibition: role of the subthalamic nucleus. J Neurosci 26:2424–2433. https://doi.org/10.1523/jneurosci.4682-05.2006

Aron AR (2011) From reactive to proactive and selective control: developing a richer model for stopping inappropriate responses. Biol Psychiatry 69:e55–e68

Balodis IM, Molina ND, Kober H et al (2013) Divergent neural substrates of inhibitory control in binge eating disorder relative to other manifestations of obesity. Obesity 21:367–377. https://doi.org/10.1002/oby.20068

Bartholdy S, Rennalls S, Danby H et al (2017) Temporal discounting and the tendency to delay gratification across the eating disorder spectrum. Eur Eat Disord Rev 25:344–350. https://doi.org/10.1002/erv.2513

Bechara A (2005) Decision making, impulse control and loss of willpower to resist drugs: a neurocognitive perspective. Nat Neurosci 8:1458–1463

Bechara A, Damasio AR, Damasio H, Anderson SW (1994) Insensitivity to future consequences following damage to human prefrontal cortex. Cognition 50:7–15. https://doi.org/10.1016/0010-0277(94)90018-3

Bechara A, Damasio H, Damasio AR (2000) Emotion, decision making and the orbitofrontal cortex. Cereb Cortex 10:295–307

Berg KC, Crosby RD, Cao L et al (2015) Negative affect prior to and following overeating-only, loss of control eating-only, and binge eating episodes in obese adults. Int J Eat Disord 48:641–653. https://doi.org/10.1002/eat.22401

Berner LA, Winter SR, Matheson BE et al (2017) Behind binge eating: a review of food-specific adaptations of neurocognitive and neuroimaging tasks. Physiol Behav 176:59–70

Berner LA, Stefan M, Lee S et al (2018) Altered cortical thickness and attentional deficits in adolescent girls and women with bulimia nervosa. J Psychiatry Neurosci 43:151–160. https://doi.org/10.1503/jpn.170070

Blinder BJ, Cumella EJ, Sanathara VA (2006) Psychiatric comorbidities of female inpatients with eating disorders. Psychosom Med 68:454–462. https://doi.org/10.1097/01.psy.0000221254.77675.f5

Bodell LP, Keel PK, Brumm MC et al (2014) Longitudinal examination of decision-making performance in anorexia nervosa: before and after weight restoration. J Psychiatr Res 56:150–157. https://doi.org/10.1016/j.jpsychires.2014.05.015

Buchsbaum BR, Greer S, Chang WL, Berman KF (2005) Meta-analysis of neuroimaging studies of the Wisconsin card-sorting task and component processes. Hum Brain Mapp 25:35–45

Burton AL, Abbott MJ (2019) Processes and pathways to binge eating: development of an integrated cognitive and behavioural model of binge eating. J Eat Disord 7:18. https://doi.org/10.1186/s40337-019-0248-0

Collantoni E, Michelon S, Tenconi E et al (2016) Functional connectivity correlates of response inhibition impairment in anorexia nervosa. Psychiatry Res Neuroimaging 247:9–16. https://doi.org/10.1016/j.pscychresns.2015.11.008

Cyr M, Yang X, Horga G, Marsh R (2018) Abnormal fronto-striatal activation as a marker of threshold and subthreshold bulimia nervosa. Hum Brain Mapp 39:1796–1804. https://doi.org/10.1002/hbm.23955

Damasio A (1994) Descartes' error: emotion, rationality and the human brain. Putnam, New York

Danner UN, Sanders N, Smeets PAM et al (2012) Neuropsychological weaknesses in anorexia nervosa: set-shifting, central coherence, and decision making in currently ill and recovered women. Int J Eat Disord 45:685–694. https://doi.org/10.1002/eat.22007

Danner UN, Evers C, Sternheim L et al (2013) Influence of negative affect on choice behavior in individuals with binge eating pathology. Psychiatry Res 207:100–106. https://doi.org/10.1016/j.psychres.2012.10.016

Davis C, Patte K, Curtis C, Reid C (2010) Immediate pleasures and future consequences. A neuropsychological study of binge eating and obesity. Appetite 54:208–213. https://doi.org/10.1016/j.appet.2009.11.002

Decker JH, Figner B, Steinglass JE (2015) On weight and waiting: delay discounting in anorexia nervosa pretreatment and posttreatment. Biol Psychiatry 78(9):606–614

Dingemans AE, Visser H, Paul L, Van Furth EF (2015) Set-shifting abilities, mood and loss of control over eating in binge eating disorder: an experimental study. Psychiatry Res 230:242–248. https://doi.org/10.1016/j.psychres.2015.09.001

Dong D et al (2016) Impulse control and restrained eating among young women: evidence for compensatory cortical activation during a chocolate-specific delayed discounting task. Appetite 105:477–486. https://doi.org/10.1016/j.appet.2016.05.017

Dreyfuss MFW, Riegel ML, Pedersen GA et al (2017) Patients with bulimia nervosa do not show typical neurodevelopment of cognitive control under emotional influences. Psychiatry Res Neuroimaging 266:59–65. https://doi.org/10.1016/j.pscychresns.2017.05.001

Eagle DM, Baunez C (2010) Is there an inhibitory-response-control system in the rat? Evidence from anatomical and pharmacological studies of behavioral inhibition. Neurosci Biobehav Rev 34:50–72

Eagle DM, Bari A, Robbins TW (2008) The neuropsychopharmacology of action inhibition: cross-species translation of the stop-signal and go/no-go tasks. Psychopharmacology 199:439–456

Fan J, McCandliss BD, Sommer T et al (2002) Testing the efficiency and independence of attentional networks. J Cogn Neurosci 14:340–347. https://doi.org/10.1162/089892902317361886

Frank GKW, Favaro A, Marsh R et al (2018) Toward valid and reliable brain imaging results in eating disorders. Int J Eat Disord 51:250–261. https://doi.org/10.1002/eat.22829

Frost R, McNaughton N (2017) The neural basis of delay discounting: a review and preliminary model. Neurosci Biobehav Rev 79:48–65

Galimberti E, Martoni RM, Cavallini MC et al (2012) Motor inhibition and cognitive flexibility in eating disorder subtypes. Prog Neuropsychopharmacol Biol Psychiatry 36:307–312. https://doi.org/10.1016/j.pnpbp.2011.10.017

Garrett AS, Lock J, Datta N et al (2014) Predicting clinical outcome using brain activation associated with set-shifting and central coherence skills in anorexia nervosa. J Psychiatr Res 57:26–33. https://doi.org/10.1016/j.jpsychires.2014.06.013

Gonzales JE, Cunningham CA (2015) The promise of pre-registration in psychological research. Psychol Sci Agenda 29(8)

Guillaume S, Gorwood P, Jollant F et al (2015) Impaired decision-making in symptomatic anorexia and bulimia nervosa patients: a meta-analysis. Psychol Med 45:3377–3391

Hare TA, Camerer CF, Rangel A (2009) Self-control in decision-making involves modulation of the vmPFC valuation system. Science. https://doi.org/10.1126/science.1168450

Hayden BY (2016) Time discounting and time preference in animals: a critical review. Psychon Bull Rev 23:39–53. https://doi.org/10.3758/s13423-015-0879-3

Heaton R (1993) Wisconsin card sorting test: computer version 4. Psychol Assess Resour 4:1–4

Heekeren HR, Marrett S, Ungerleider LG (2008) The neural systems that mediate human perceptual decision making. Nat Rev Neurosci 9:467–479

Hege MA, Stingl KT, Kullmann S et al (2015) Attentional impulsivity in binge eating disorder modulates response inhibition performance and frontal brain networks. Int J Obes 39:353–360. https://doi.org/10.1038/ijo.2014.99

Herbrich L, Kappel V, van Noort BM, Winter S (2018) Differences in set-shifting and central coherence across anorexia nervosa subtypes in children and adolescents. Eur Eat Disord Rev 26:499–507. https://doi.org/10.1002/erv.2605

Hirst RB, Beard CL, Colby KA et al (2017) Anorexia nervosa and bulimia nervosa: a meta-analysis of executive functioning. Neurosci Biobehav Rev 83:678–690

Horstmann A (2017) It wasn't me; it was my brain—obesity-associated characteristics of brain circuits governing decision-making. Physiol Behav 176:125–133

Insel T, Cuthbert B, Garvey M et al (2010) Research domain criteria (RDoC): toward a new classification framework for research on mental disorders. Am J Psychiatry 167:748–751. https://doi.org/10.1176/appi.ajp.2010.09091379

Kakoschke N, Aarts E, Verdejo-García A (2019) The cognitive drivers of compulsive eating behavior. Front Behav Neurosci 12:338. https://doi.org/10.3389/fnbeh.2018.00338

Kekic M, Bartholdy S, Cheng J et al (2016) Increased temporal discounting in bulimia nervosa. Int J Eat Disord 49:1077–1081. https://doi.org/10.1002/eat.22571

Kessler RC, Berglund PA, Chiu WT et al (2013) The prevalence and correlates of binge eating disorder in the World Health Organization world mental health surveys. Biol Psychiatry 73:904–914. https://doi.org/10.1016/j.biopsych.2012.11.020

Kessler RM, Hutson PH, Herman BK, Potenza MN (2016) The neurobiological basis of binge-eating disorder. Neurosci Biobehav Rev 63:223–238

Khalsa SS, Portnoff LC, McCurdy-McKinnon D, Feusner JD (2017) What happens after treatment? A systematic review of relapse, remission, and recovery in anorexia nervosa. J Eat Disord 5:20

Kollei I, Rustemeier M, Schroeder S et al (2018) Cognitive control functions in individuals with obesity with and without binge-eating disorder. Int J Eat Disord 51:233–240. https://doi.org/10.1002/eat.22824

Lavagnino L, Arnone D, Cao B et al (2016) Inhibitory control in obesity and binge eating disorder: a systematic review and meta-analysis of neurocognitive and neuroimaging studies. Neurosci Biobehav Rev 68:714–726. https://doi.org/10.1016/j.neubiorev.2016.06.041

Lee JE, Namkoong K, Jung YC (2017) Impaired prefrontal cognitive control over interference by food images in binge-eating disorder and bulimia nervosa. Neurosci Lett 651:95–101. https://doi.org/10.1016/j.neulet.2017.04.054

Lempert KM, Steinglass JE, Pinto A et al (2019) Can delay discounting deliver on the promise of RDoC? Psychol Med 49:190–199. https://doi.org/10.1017/s0033291718001770

Levinson CA, Zerwas SC, Brosof LC et al (2019) Associations between dimensions of anorexia nervosa and obsessive-compulsive disorder: an examination of personality and psychological factors in patients with anorexia nervosa. Eur Eat Disord Rev 27:161–172. https://doi.org/10.1002/erv.2635

Lezak MD (1995) Neuropsychological assessment, 3rd edn. Oxford University Press, New York

Li X, Lu ZL, D'Argembeau A et al (2010) The Iowa Gambling Task in fMRI images. Hum Brain Mapp 31(3):410–423. https://doi.org/10.1002/hbm.20875

Linardon J (2018) Rates of abstinence following psychological or behavioral treatments for binge-eating disorder: meta-analysis. Int J Eat Disord 51:785–797

Linardon J, Wade TD, De La Piedad GX, Brennan L (2017) The efficacy of cognitive-behavioral therapy for eating disorders: a systematic review and meta-analysis. J Consult Clin Psychol 85:1080–1094

Lindner SE, Fichter MM, Quadflieg N (2012) Decision-making and planning in full recovery of anorexia nervosa. Int J Eat Disord 45:866–875. https://doi.org/10.1002/eat.22025

Lock J, Garrett A, Beenhakker J, Reiss AL (2011) Aberrant brain activation during a response inhibition task in adolescent eating disorder subtypes. Am J Psychiatry 168:55–64. https://doi.org/10.1176/appi.ajp.2010.10010056

Lock J, Agras WS, Fitzpatrick KK et al (2013) Is outpatient cognitive remediation therapy feasible to use in randomized clinical trials for anorexia nervosa? Int J Eat Disord 46:567–575. https://doi.org/10.1002/eat.22134

Logan GD, Schachar RJ, Tannock R (1997) Impulsivity and inhibitory control. Psychol Sci 8:60–64. https://doi.org/10.1111/j.1467-9280.1997.tb00545.x

Louie K, Glimcher PW (2010) Separating value from choice: delay discounting activity in the lateral intraparietal area. J Neurosci 30:5498–5507. https://doi.org/10.1523/jneurosci.5742-09.2010

Manasse SM, Juarascio AS, Forman EM et al (2014) Executive functioning in overweight individuals with and without loss-of-control eating. Eur Eat Disord Rev 22:373–377. https://doi.org/10.1002/erv.2304

Manasse SM, Forman EM, Ruocco AC et al (2015) Do executive functioning deficits underpin binge eating disorder? A comparison of overweight women with and without binge eating pathology. Int J Eat Disord 48:677–683. https://doi.org/10.1002/eat.22383

Manasse SM, Espel HM, Schumacher LM et al (2016a) Does impulsivity predict outcome in treatment for binge eating disorder? A multimodal investigation.

Appetite 105:172–179. https://doi.org/10.1016/j.appet. 2016.05.026

Manasse SM, Goldstein SP, Wyckoff E et al (2016b) Slowing down and taking a second look: inhibitory deficits associated with binge eating are not food-specific. Appetite 96:555–559. https://doi.org/10. 1016/j.appet.2015.10.025

Manwaring JL, Green L, Myerson J et al (2011) Discounting of various types of rewards by women with and without binge eating disorder: evidence for general rather than specific differences. Psychol Rec 61:561–582

Marsh R, Steinglass JE, Gerber AJ et al (2009) Deficient activity in the neural systems that mediate self-regulatory control in bulimia nervosa. Arch Gen Psychiatry 66:51–63. https://doi.org/10.1001/archgenpsychiatry. 2008.504

Marsh R, Horga G, Wang Z et al (2011) An fMRI study of self-regulatory control and conflict resolution in adolescents with bulimia nervosa. Am J Psychiatry 168:1210–1220. https://doi.org/10.1176/appi.ajp. 2011.11010094

Mikami AY, Hinshaw SP, Patterson KA, Lee JC (2008) Eating pathology among adolescent girls with attention-deficit/hyperactivity disorder. J Abnorm Psychol 117:225–235. https://doi.org/10.1037/0021-843X.117.1.225

Miyake A, Friedman NP, Emerson MJ et al (2000) The unity and diversity of executive functions and their contributions to complex "frontal lobe" tasks: a latent variable analysis. Cogn Psychol 41:49–100. https://doi.org/10.1006/cogp.1999.0734

Monchi O, Petrides M, Petre V et al (2001) Wisconsin card sorting revisited: distinct neural circuits participating in different stages of the task. J Neurosci 21:7733–7741

Monteleone AM, Castellini G, Volpe U et al (2018) Neuroendocrinology and brain imaging of reward in eating disorders: a possible key to the treatment of anorexia nervosa and bulimia nervosa. Prog Neuropsychopharmacol Biol Psychiatry 80(Pt B):132–142

Neveu R, Fouragnan E, Barsumian F et al (2016) Preference for safe over risky options in binge eating. Front Behav Neurosci 10:65. https://doi.org/10.3389/fnbeh. 2016.00065

Neveu R, Neveu D, Carrier E et al (2018) Goal directed and self-control systems in bulimia nervosa: an fMRI study. EBioMedicine 34, 214–222 https://doi.org/10. 1016/j.ebiom.2018.07.012

Norman LJ, Carlisi CO, Christakou A et al (2018) Frontostriatal dysfunction during decision making in attention-deficit/hyperactivity disorder and obsessive-compulsive disorder. Biol Psychiatry Cogn Neurosci Neuroimaging 3:694–703. https://doi.org/10.1016/j. bpsc.2018.03.009

Oliva R et al (2019) The impulsive brain: neural underpinnings of binge eating behavior in normal-weight adults. Appetite 136:33–49. https://doi.org/10. 1016/j.appet.2018.12.043

Pearson CM, Wonderlich SA, Smith GT (2015) A risk and maintenance model for bulimia nervosa: from impulsive action to compulsive behavior. Psychol Rev 122:516–535. https://doi.org/10.1037/a0039268

Peters J, Büchel C (2011) The neural mechanisms of inter-temporal decision-making: understanding variability. Trends Cogn Sci 15:227–239

Petersen SE, Posner MI (2012) The attention system of the human brain: 20 years after. Annu Rev Neurosci 35:73–89. https://doi.org/10.1146/annurev-neuro-062111-150525

Peterson BS, Kane MJ, Alexander GM et al (2002) An event-related functional MRI study comparing interference effects in the Simon and Stroop tasks. Cogn Brain Res 13:427–440. https://doi.org/10.1016/S0926-6410(02)00054-X

Reinblatt SP, Mahone EM, Tanofsky-Kraff M et al (2015) Pediatric loss of control eating syndrome: association with attention-deficit/hyperactivity disorder and impulsivity. Int J Eat Disord 48:580–588. https://doi.org/10. 1002/eat.22404

Reiter AM, Heinze H-J, Schlagenhauf F, Deserno L (2016) Impaired flexible reward-based decision-making in binge eating disorder: evidence from computational modeling and functional neuroimaging. Neuropsychopharmacology 42:1–10. https://doi.org/10.1038/npp. 2016.95

Ridderinkhof KR, Span MM, Van Der Molen MW (2002) Perseverative behavior and adaptive control in older adults: performance monitoring, rule induction, and set shifting. Brain Cogn 49:382–401. https://doi.org/10. 1006/brcg.2001.1506

Ritschel F, King JA, Geisler D et al (2015) Temporal delay discounting in acutely ill and weight-recovered patients with anorexia nervosa. Psychol Med 45:1229–1239. https://doi.org/10.1017/S0033291714002311

Robbins TW (2007) Shifting and stopping: Fronto-striatal substrates, neurochemical modulation and clinical implications. Philos Trans R Soc Lond B Biol Sci 362(1481):917–932

Roberts ME, Tchanturia K, Treasure JL (2010) Exploring the neurocognitive signature of poor set-shifting in anorexia and bulimia nervosa. J Psychiatr Res 44:964–970. https://doi.org/10.1016/j.jpsychires. 2010.03.001

Ross JS (2016) Clinical research data sharing: what an open science world means for researchers involved in evidence synthesis. Syst Rev 5:159. https://doi.org/10. 1186/s13643-016-0334-1

Rubia K, Smith AB, Woolley J et al (2006) Progressive increase of frontostriatal brain activation from childhood to adulthood during event-related tasks of cognitive control. Hum Brain Mapp 27:973–993. https://doi. org/10.1002/hbm.20237

Sanchez-Roige S, Fontanillas P, Elson SL, Pandit A, Schmidt EM et al (2018) Genome-wide association study of delay discounting in 23,217 adult research participants of European ancestry. Nat Neurosci. 21:16–18. https://doi.org/10.1038/s41593-017-0032-x

Sato Y, Saito N, Utsumi A et al (2013) Neural basis of impaired cognitive flexibility in patients with anorexia nervosa. PLoS One 8:e61108. https://doi.org/10.1371/journal.pone.0061108

Schag K, Schönleber J, Teufel M et al (2013) Food-related impulsivity in obesity and binge eating disorder—a systematic review. Obes Rev 14:477–495. https://doi.org/10.1111/obr.12017

Seitz J, Kahraman-Lanzerath B, Legenbauer T et al (2013) The role of impulsivity, inattention and comorbid ADHD in patients with bulimia nervosa. PLoS One 8(5):e63891. https://doi.org/10.1371/journal.pone.0063891

Seitz J, Hueck M, Dahmen B et al (2016) Attention network dysfunction in bulimia nervosa—an fMRI study. PLoS One 11:e0161329. https://doi.org/10.1371/journal.pone.0161329

Simmonds DJ, Pekar JJ, Mostofsky SH (2008) Meta-analysis of Go/No-go tasks demonstrating that fMRI activation associated with response inhibition is task-dependent. Neuropsychologia 46:224–232. https://doi.org/10.1016/j.neuropsychologia.2007.07.015

Skunde M, Walther S, Simon JJ et al (2016) Neural signature of behavioural inhibition in women with bulimia nervosa. J Psychiatry Neurosci 41:E69–E78. https://doi.org/10.1503/jpn.150335

Smith BJ, Monterosso JR, Wakslak CJ et al (2018a) A meta-analytical review of brain activity associated with intertemporal decisions: evidence for an anterior-posterior tangibility axis. Neurosci Biobehav Rev 86:85–98

Smith KE, Mason TB, Johnson JS et al (2018b) A systematic review of reviews of neurocognitive functioning in eating disorders: the state-of-the-literature and future directions. Int J Eat Disord 51:798–821

Smith KE, Mason TB, Crosby RD et al (2019) A multimodal, naturalistic investigation of relationships between behavioral impulsivity, affect, and binge eating. Appetite 136:50–57. https://doi.org/10.1016/j.appet.2019.01.014

Steinglass JE, Walsh BT, Stern Y (2006) Set shifting deficit in anorexia nervosa. J Int Neuropsychol Soc 12(3):431–435. https://doi.org/10.1017/S1355617706060528

Steinglass JE, Figner B, Berkowitz S et al (2012) Increased capacity to delay reward in anorexia nervosa. J Int Neuropsychol Soc 18:773–780. https://doi.org/10.1017/s1355617712000446

Steinglass JE, Lempert KM, Choo TH et al (2017) Temporal discounting across three psychiatric disorders: anorexia nervosa, obsessive compulsive disorder, and social anxiety disorder. Depress Anxiety 34:463–470. https://doi.org/10.1002/da.22586

Steinglass JE, Berner LA, Attia E (2019) Cognitive neuroscience of eating disorders. Psychiatr Clin North Am 42:75–91. https://doi.org/10.1016/j.psc.2018.10.008

Steward T, Mestre-Bach G, Agüera Z et al (2016) Enduring changes in decision making in patients with full remission from anorexia nervosa. Eur Eat Disord Rev 24:523–527. https://doi.org/10.1002/erv.2472

Steward T, Mestre-Bach G, Vintró-Alcaraz C et al (2017) Delay discounting of reward and impulsivity in eating disorders: from anorexia nervosa to binge eating disorder. Eur Eat Disord Rev 25:601–606. https://doi.org/10.1002/erv.2543

Steward T, Mestre-Bach G, Granero R et al (2019) Reduced plasma orexin-A concentrations are associated with cognitive deficits in anorexia nervosa. Sci Rep 9:7910. https://doi.org/10.1038/s41598-019-44450-6

Stojek MMK, MacKillop J (2017) Relative reinforcing value of food and delayed reward discounting in obesity and disordered eating: a systematic review. Clin Psychol Rev 55:1–11

Svedlund NE, Norring C, Ginsberg Y, von Hausswolff-Juhlin Y (2017) Symptoms of attention deficit hyperactivity disorder (ADHD) among adult eating disorder patients. BMC Psychiatry 17:19. https://doi.org/10.1186/s12888-016-1093-1

Swick D, Ashley V, Turken U (2011) Are the neural correlates of stopping and not going identical? Quantitative meta-analysis of two response inhibition tasks. Neuroimage 56:1655–1665. https://doi.org/10.1016/j.neuroimage.2011.02.070

Swinbourne J, Hunt C, Abbott M et al (2012) The comorbidity between eating disorders and anxiety disorders: prevalence in an eating disorder sample and anxiety disorder sample. Aust N Z J Psychiatry 46:118–131. https://doi.org/10.1177/0004867411432071

Tchanturia K, Liao PC, Uher R et al (2007) An investigation of decision making in anorexia nervosa using the Iowa Gambling Task and skin conductance measurements. J Int Neuropsychol Soc 13(4):635–641. https://doi.org/10.1017/S1355617707070798

Tchanturia K, Harrison A, Davies H et al (2011) Cognitive flexibility and clinical severity in eating disorders. PLoS One 6:e20462. https://doi.org/10.1371/journal.pone.0020462

Tchanturia K, Lounes N, Holttum S (2014) Cognitive remediation in anorexia nervosa and related conditions: a systematic review. Eur Eat Disord Rev 22:454–462. https://doi.org/10.1002/erv.2326

Telch CF, Stice E (1998) Psychiatric comorbidity in women with binge eating disorder: prevalence rates from a non-treatment-seeking sample. J Consult Clin Psychol 66:768–776. https://doi.org/10.1037/0022-006X.66.5.768

Tombaugh TN (2004) Trail making test A and B: normative data stratified by age and education. Arch Clin Neuropsychol 19:203–214. https://doi.org/10.1016/S0887-6177(03)00039-8

Vaghi MM, Vértes PE, Kitzbichler MG et al (2017) Specific frontostriatal circuits for impaired cognitive flexibility and goal-directed planning in obsessive-compulsive disorder: evidence from resting-state functional connectivity. Biol Psychiatry 81:708–717. https://doi.org/10.1016/j.biopsych.2016.08.009

Van Autreve S, De Baene W, Baeken C et al (2016) Differential neural correlates of set-shifting in the bingeing–purging and restrictive subtypes of anorexia nervosa: an fMRI study. Eur Eat Disord Rev 24:277–285. https://doi.org/10.1002/erv.2437

Verdejo-Garcia A, Clark L, Verdejo-Román J et al (2015) Neural substrates of cognitive flexibility in cocaine and gambling addictions. Br J Psychiatry 207:158–164. https://doi.org/10.1192/bjp.bp.114.152223

Vertes RP (2006) Interactions among the medial prefrontal cortex, hippocampus and midline thalamus in emotional and cognitive processing in the rat. Neuroscience 142:1–20

Volkow ND, Baler RD (2015) NOW vs LATER brain circuits: implications for obesity and addiction. Trends Neurosci 38:345–352

Voon V (2015) Cognitive biases in binge eating disorder: the hijacking of decision making. CNS Spectr 20:566–573. https://doi.org/10.1017/s1092852915000681

Wesley MJ, Bickel WK (2014) Remember the future II: meta-analyses and functional overlap of working memory and delay discounting. Biol Psychiatry 75:435–448

Wilson RP, Colizzi M, Bossong MG et al (2018) The neural substrate of reward anticipation in health: a meta-analysis of fMRI findings in the monetary incentive delay task. Neuropsychol Rev 28:496–506. https://doi.org/10.1007/s11065-018-9385-5

Wollenhaupt C, Wilke L, Erim Y et al (2019) The association of leptin secretion with cognitive performance in patients with eating disorders. Psychiatry Res 276:269–277. https://doi.org/10.1016/j.psychres.2019.05.001

Wu M, Hartmann M, Skunde M et al (2013) Inhibitory control in bulimic-type eating disorders: a systematic review and meta-analysis. PLoS One 8:e83412. https://doi.org/10.1371/journal.pone.0083412

Wu M, Brockmeyer T, Hartmann M et al (2014) Set-shifting ability across the spectrum of eating disorders and in overweight and obesity: a systematic review and meta-analysis. Psychol Med 44:3365–3385

Yilmaz Z, Kaplan AS, Zai CC et al (2011) COMT Val158Met variant and functional haplotypes associated with childhood ADHD history in women with bulimia nervosa. Prog Neuropsychopharmacol Biol Psychiatry 35:948–952. https://doi.org/10.1016/j.pnpbp.2011.01.012

Influence of Genetics and Sex Hormones on Binge Eating

Ya-Ke Wu, Courtney E. Breiner, and Jessica H. Baker

Abstract

Binge eating is a complex condition influenced by biological and environmental factors. The heritability of binge eating has been estimated at 38–57% in populations of European ancestry. Studies of epigenetic mechanisms such as deoxyribonucleic acid (DNA) methylation and gene expression revealed a potential gene by environment effect on binge eating. Sex hormones also play a significant role in binge eating such that the combined effects of estrogen and progesterone may contribute to the vulnerability for binge eating. Genetic influences on binge eating might increase during the pubertal period, suggesting an interaction between genetic factors and sex hormones on the risk for binge eating.

Keywords

Heritability · Genetics · Sex hormones · Binge eating

Y.-K. Wu
School of Nursing, The University of North Carolina at Chapel Hill, Chapel Hill, NC, USA

C. E. Breiner
Department of Psychology, University at Albany, State University of New York, Albany, NY, USA

J. H. Baker (✉)
Department of Psychiatry, The University of North Carolina at Chapel Hill, Chapel Hill, NC, USA
e-mail: jessica_baker@med.unc.edu

Learning Objectives
In this chapter, you will:

- Obtain an overview of the heritability of binge eating.
- Get an introduction to contemporary genomic studies of binge eating.
- Recognize the influences of estradiol, progesterone, and testosterone on binge eating.
- Identify the potential interplay between genetics and sex hormones in binge eating.

1 Introduction

Binge eating is a complex condition that arises from a combination of genetic, psychosocial, and environmental factors. Studies examining the heritability of binge eating suggest that genetic factors play an important role in the pathogenesis of binge eating. However, to date, psychiatric genomic studies of eating disorders have focused on anorexia nervosa and not included eating disorders primarily characterized by binge eating (e.g., binge-eating disorder, bulimia nervosa) or symptom-level data. Thus, research aimed at identifying the genetic risk for binge eating is

© Springer Nature Switzerland AG 2020
G. K.W. Frank, L. A. Berner (eds.), *Binge Eating*, https://doi.org/10.1007/978-3-030-43562-2_11

limited and needs more attention to assist in future prevention, treatment, and recovery efforts. Additionally, there is growing evidence illustrating the importance of sex hormones (i.e., estrogen, progesterone, and testosterone) in the etiology of binge eating pathology. Genetic factors and sex hormones may interact to increase risk for binge eating.

In this chapter, we review two specific biological contributions to binge eating: genetic factors and sex hormone influences from both preclinical and clinical research. Because studies in this area to date have mainly focused on female populations, the human research discussed in this chapter will focus on women. Preclinical studies will be grouped within their respective topic areas. First, we provide an overview of the heritability of binge eating and highlight the limited molecular genetic studies of binge eating that have been conducted to date. We review the influence of the sex hormones estradiol and progesterone on binge eating and provide a brief introduction to testosterone's role in binge eating. Next, we describe the potential interactions between genetics and sex hormones on binge eating. We conclude by discussing future research directions.

2 The Heritability of Binge Eating

2.1 Family Studies

Family studies are typically the first step in determining whether a trait may have a genetic component by investigating whether a specific trait or disease of interest aggregates within families. Specifically, family studies compare the risk for a trait of interest (e.g., binge eating) in relatives of probands with and without the trait of interest. Notably, family studies are only able to determine whether a disease or trait of interest "runs" in families. These studies are unable to address whether any observed familial aggregation is due to genetic or family environmental factors. Two family studies of binge eating have been conducted to date (Hudson et al. 2006; Javaras et al. 2008). Cumulatively, results indicate that binge eating aggregates strongly in families. Family members of probands with binge-eating disorder are two times more likely to have binge-eating disorder than family members of individuals without binge-eating disorder (Hudson et al. 2006; Javaras et al. 2008). Specifically, odds ratio for the risk of binge-eating disorder in a relative of a proband with binge-eating disorder compared with relatives of controls have been estimated between 1.9 and 2.2 across studies (Thornton et al. 2011).

Although family studies offered initial evidence that having a relative with binge eating increases an individual's risk of developing binge eating, as previously mentioned, family studies are unable to address whether this increased familial risk is due to genetic or family environmental factors. Thus, the results of family studies cannot tease apart the influence of genes and the environment.

2.2 Twin Studies

Unlike family studies, twin studies are able to tease apart the contribution of genetic and environmental factors on a trait of interest. Specifically, twin studies provide information regarding the heritability and shared and nonshared environmental effects for a trait. Heritability is the amount of variance in a trait that is due to genetic variation. Shared environmental factors are factors shared by each twin, such as childhood socioeconomic status. On the other hand, nonshared environmental factors are factors that are unique to one twin. Twin studies compare the concordance rates of diseases between monozygotic and dizygotic twins. Monozygotic twins develop from one zygote and therefore share approximately 100% of their genes, whereas dizygotic twins develop from a separate egg that is fertilized by its own sperm cell, only sharing 50% of their genes. In contrast, both monozygotic and dizygotic twin pairs share 100% of their shared environment (unless they were adopted separately at birth). Specifically, twin studies measure how often both twins exhibit a trait of interest (i.e., concordant) or only one is affected by the trait (i.e., discordant). If the concordance rate of a disease in monozygotic twins is approximately twice that observed in dizygotic twins, genetic factors are said to have a role in the disease. On the other hand, if the concordance rate of a disease is similar between monozygotic and dizygotic twins, this indicates that shared environmental factors are important. Nonshared environmental factors are implicated if the concordance rate between monozygotic twins is less than 1.0.

Overall, twin studies confirm family studies showing that binge eating aggregates in families and that this familial aggregation is largely due to genetic factors. Heritability estimates for binge eating have been estimated from 38 to 70% (Bulik et al. 1998; Javaras et al. 2008; Munn-Chernoff et al. 2013; Reichborn-Kjennerud et al. 2004; Root et al. 2010; Thornton et al. 2011). For example, Mitchell et al. (2010) investigated binge-eating behaviors among 1024 female twins and found that the heritability of binge eating was estimated at 45% with shared and nonshared environmental factors accounting for 13 and 42% of the variance, respectfully. In contrast, in a study of Swedish twins' heritability was estimated at 70% for binge eating (Root et al. 2010). Root et al. (2010) hypothesized that using an anonymous online binge eating assessment (i.e., online Structured Clinical Interview for DSM-IV-based instrument) might be one reason for the higher heritability estimate observed in their study compared with other studies because the online anonymity may provide advantages for more precise results when assessing sensitive information compared with a face-to-face interview.

Twin studies can also be used to examine the genetic and environmental overlap between traits of interest; for example, between binge eating and other phenotypes such as alcohol dependence, major depressive disorder, and attention-deficit hyperactivity disorder (ADHD). A significant overlap in both genetic and nonshared environmental influences has been observed between alcohol dependence and binge eating (Munn-Chernoff et al. 2013). Approximately 38% of the genetic and 7% of nonshared environmental variance between major depressive disorder and an overeating–binge eating dimension were

shared in a combined sample of European- and African-American female twins (Munn-Chernoff et al. 2015). Also, the genetic and nonshared environmental factors that influenced the variance of the overeating–binge eating dimension did not differ across European- and African-American twin women (Munn-Chernoff et al. 2015). Finally, shared genetic factors have been observed between binge eating and ADHD such that genetic factors explained almost all (91%) of the phenotypic association between the two (Capusan et al. 2017). This indicates possible shared genetic underpinnings between binge eating and alcohol dependence, major depression, and ADHD.

Twin studies paved the way for understanding the genetic architecture of binge eating. Like family studies, twin studies also have limitations in addressing the genetic contribution to a trait of interest. Specifically, twin studies are able to tell us whether a genetic component exists, but they are unable to identify the specific genes involved in risk.

3   Contemporary Genomic Studies of Binge Eating

Here, we provide the reader with an overview of contemporary genomic studies of binge eating. Although candidate gene studies exist, we do not include a discussion of these studies. Candidate gene studies are considered an out-of-date genetic methodology given that such studies depend on choosing a specific gene or set of genes based on a priori knowledge about the role of the selected gene on a disease. However, binge eating is a complex polygenic condition, meaning that many genes of small effect contribute to the development of binge eating. As such, results from candidate gene studies are fraught with limitations and have been limitedly replicated. With the advancement of genomic technology, the genetics field has moved toward more contemporary methodologies such as genome-wide association studies (GWASs). GWASs represent a promising method to examine polygenic diseases by comparing individuals with and without a trait of interest on 300,000–1,000,000 genetic markers across the genome simultaneously. However, such contemporary genomic methods have been limitedly applied to binge eating. In the following section, we review the existing studies that apply contemporary genomic methods to binge eating, including GWASs, polygenic risk scores, deoxyribonucleic acid (DNA) methylation, and gene expression.

3.1 Genome-Wide Association Studies

The National Human Genome Research Institute defines a GWAS as "an approach that involves rapidly scanning markers across the complete set of DNAs, or genomes, of many people to find genetic variations associated with a particular disease." (National Human Genome Research Institute 2015). With the advent of GWASs, we now have the ability to identify multiple genetic factors contributing to complex phenotypes, such as binge eating. This is done by comparing genetic variation between individuals with a specific disease to individuals without the disease. If the frequency of particular genetic variants is significantly higher in individuals with the disease compared with individuals without the disease, the variants are significantly associated with the disease. Because GWASs are able to span across the genome, a correction for multiple testing is applied to determine significance. Significance is based upon a Bonferroni correction, assuming $p < 0.05$ and a million independent variants across the genome, making genome-wide significance: $p < 5 \times 10^{-8}$. To date, no published GWASs exist for binge eating. However, two GWASs have been conducted in bipolar disorder patients with comorbid binge eating (McElroy et al. 2018; Winham et al. 2014). The goal of these GWASs was to identify genetic factors underlying bipolar disorder with comorbid binge eating.

The first GWAS included 729,454 single nucleotide polymorphisms (SNPs) genotyped from 930 European American bipolar disorder patients (206 with binge eating and 723 without

binge eating) and 1034 healthy controls (Winham et al. 2014). No genome-wide significant results emerged (Winham et al. 2014). The second GWAS included data from the Mayo Clinic Bipolar Disorder Biobank with 969 bipolar disorder patients and 777 controls (McElroy et al. 2018). A case-only meta-analysis of the combined Mayo Clinic data and the first published GWAS showed a significant association between a group of SNPs in the *PRR5-ARHGAP8* gene and binge eating in bipolar patients such that the common allele was associated with reduced risk for binge eating among bipolar patients (top SNPs rs726170 and rs8139558; OR = 0.52, $p = 3.05$ E-08) (McElroy et al. 2018). *PRR5-ARHGAP8* is a protein-coding gene for PRR5 (PRoline-Rich 5) and ARHGAP8 (Rho GTPase-activating protein 8), which are involved in food intake regulation (McElroy et al. 2018). Comparing bipolar cases with comorbid binge eating to non-bipolar controls showed a significant association with a variant in an intergenic region of chromosome 2q12.3— with the nearest gene encoding protein phosphatase 1 regulatory inhibitor subunit 2 (PPP1R2) (McElroy et al. 2018). PPP1R2 (Protein Phosphatase 1 Regulatory Inhibitor Subunit 2) is expressed in GABAergic neurons in the hippocampus, striatum, and cortex (McElroy et al. 2018). One final GWAS for anorexia nervosa spectrum, bulimia spectrum (includes binge eating), purging, and disordered eating behaviors among 2564 female twins also found no genome-wide significant results (Wade et al. 2013).

3.2 Polygenic Risk Scores

Given the advancement of genomic technology, our knowledge of the genetic liability of complex diseases has improved. As previously described, the available evidence suggests that an individual's risk of developing a complex disease such as binge eating is polygenic in nature, determined by variations occurring in many genes that individually contribute small effects and may act over long periods (Sugrue and Desikan 2019; Torkamani et al. 2018). A GWAS can identify variations occurring in many genes that are associated with the risk to a complex disease, but the effect of each risk variant is generally small. A polygenic risk score uses results from a GWAS to create a genetic risk profile for predicting each individual's genetic risk for the disease (Wray et al. 2008). Polygenic risk scores represent the cumulative effect of risk alleles for a complex disease and can be used to predict the likelihood of disease development (Richardson et al. 2019). No polygenic risk scores of binge eating exist. However, one study examined whether a polygenic risk score for schizophrenia was associated with binge eating among children. Results showed that a higher schizophrenia polygenic risk score was significantly associated with greater odds of binge eating such that a 1 standard deviation increase in polygenic risk for schizophrenia was associated with 1.36 increased odds of binge eating (95% CI = 1.16–1.60, $p < 0.0001$) (Solmi et al. 2019). This suggests that those with a higher genetic propensity for schizophrenia have a greater likelihood of binge eating, and these two conditions may share some genetic risk in children, although the mechanism underlying this association remains unclear.

3.3 Preclinical Studies of DNA Methylation

DNA methylation is an important epigenetic mechanism. Epigenetics is broadly defined as changes in gene function or expression that occur in the absence of change in the DNA sequence. Varying environmental factors, such as exposure to nutrients or stress, can affect the epigenome, effectively turning genes on or off. DNA methylation occurs when a methyl group is added to DNA, typically resulting in turning "off" the gene. Such changes in DNA methylation can influence observable outcomes such as binge eating. To date, human studies have limitedly addressed DNA methylation patterns in women who binge eat. However, one study examined methylation with the *SLC1A2* promoter in individuals with bipolar disorder with or without comorbidity, including binge eating (referred to as food addiction by the study authors) (Jia et al.

2017). *SLC1A2* promoter encodes the excitatory amino acid transporter 2, which has been implicated in bipolar disorder. Results showed the *SLC1A2* promoter region was hypomethylated in individuals with bipolar disorder and comorbid binge eating compared with healthy controls.

In a preclinical study, Schroeder et al. (2018) examined the link between chronic prenatal stress (i.e., stress over the whole period of pregnancy) and a later predisposition to binge eating in mice offspring using a prenatal stress protocol. DNA methylation patterns of the offspring were examined within the placenta, and binge-eating behaviors were determined by the amount of high-fat food intake over a 3-week period. Prenatal stress did not increase predisposition to binge eating for female offspring compared with controls; in fact, prenatal stress appeared to decrease the risk for binge eating. In contrast, male offspring exposed to prenatal stress showed an increased propensity to binge eat compared with controls (Schroeder et al. 2018). Prenatal stress increased global DNA methylation and reduced RNA methylation in females compared with control females. Given these same changes were not significant in male mice, these methylation patterns in female mice exposed to prenatal stress may, in part, explain the decreased risk for later binge eating.

3.4 Preclinical Studies of Gene Expression

DNA is a nucleic acid made out of nucleotides as major building blocks that consist of a sugar molecule, a phosphate group and one of four nitrogenous bases, adenine, guanine, cytosine, or thymine. A gene is a section of DNA nucleotides that carry the message for making a specific protein for a cell. The process of copying (transcription) the message from a gene to ribonucleic acid (RNA) for making a particular protein (translation) is referred to as gene expression. The expression of a gene can change with time and can be influenced by external factors, such as environmental insults (e.g., stress) and exposures (e.g., binge eating). Thus, gene expression has the potential to identify expression profiles that are related to disease risk or specific relevant states such as acute illness or recovery. We are aware of no human studies that have examined gene expression as it relates to binge eating; however, a handful of preclinical studies report links between gene expression in brain regions that are involved in food consumption and binge-eating behavior. For example, Albertz et al. (2018) tested whether running wheel activity impacts binge eating and alters gene expression in brain regions in a rat model. To determine binge-eating behavior, rats were provided high-fat food access, and the total amount of high-fat food intake per day was observed. Compared with rats with no access to a running wheel, rats with access to the wheel showed changes in gene expression in brain regions that involve appetite and energy balance (e.g., significantly decreased expression of the dopamine D2 receptor) and presented significantly less high-fat food intake (Albertz et al. 2018). Similarly, Pucci et al. (2018) found that palatable food (i.e., high-fat, high-sugar) intake was significantly increased in rats exposed to food restriction and a stress procedure in comparison with rats not exposed to food restriction and stress. After stress and intermittent food access, selective downregulation of fatty acid amide hydrolase (FAAH) gene expression was observed in the hypothalamus of rats showing binge-eating behavior (Pucci et al. 2018). FAAH is distributed in the central nervous system and peripheral tissues and involved in terminating the signaling of the endocannabinoids and oleamide as well as feeding and metabolic processes (Balsevich et al. 2018). Deficiency of FAAH may increase food intake; thus, downregulation of FAAH gene expression might drive binge eating (Pucci et al. 2018). Finally, an interaction between history of food restriction and frustration stress also changed A_{2A} Adenosine Receptor ($A_{2A}AR$) and dopaminergic D2 receptor (D2R) gene expression in binge eating female rats in the amygdala (Micioni Di Bonaventura et al. 2019). $A_{2A}AR$ and D2R are both involved in the intake of palatable food and are associated with binge eating pathology in humans and rats, while

the amygdala is a temporal lobe structure that is involved in fear processing and learning.

4 Sex Hormones Influence on Binge Eating

The primary sex hormones include estrogen, progesterone, and testosterone with estrogen and progesterone being the primary biologically female hormones and testosterone the primary biologically male sex hormone. For both females and males, these hormones are higher post-puberty. For females, estrogen and progesterone also vary cyclically across the phases of a menstrual cycle (i.e., menstrual phase, follicular phase, and luteal phase). During the follicular phase, estrogen levels are rising in preparation for ovulation and peak at ovulation. The luteal phase begins after ovulation, during which progesterone levels peak approximately halfway through, and estrogen has a secondary peak. If pregnancy does not occur, estrogen and progesterone levels fall, leading to menstruation. Preclinical and human studies have addressed the role of estrogen and progesterone on food intake and eating behaviors, including binge eating. Here, we focus this section on estrogen and progesterone, with a brief introduction to testosterone, given only a small number of studies examined the relationship between testosterone and binge eating. Additionally, while periods of reproductive hormone change such as puberty and the menopause transition have been suggested as risk periods for dysregulated eating, there is limited work that has focused on binge eating specifically.

4.1 Estradiol, Progesterone, and the Menstrual Cycle

There is culminating evidence suggesting that female sex hormones play a significant role in eating behaviors. Early animal studies established that estrogen reduces food intake and body weight, whereas progesterone may increase food intake by antagonizing the effects of estrogen (Baker et al. 2012; Leeners et al. 2017). This same effect of estrogen and progesterone on normal food intake is also present for dysregulated eating behaviors such as binge eating. Rats that have been ovariectomized engage in increased binge-eating behaviors (i.e., palatable food consumption) (Klump et al. 2011) whereas estrogen replacement in ovariectomized rats reverses these effects such that estrogen inhibits food consumption (Yu et al. 2008) and decreases binge-eating behavior (Yu et al. 2011). In contrast with estrogen replacement, progesterone replacement appears to have no direct effect on binge eating in ovariectomized rats (Butera 2010), but it may antagonize the beneficial effects of estrogen such that the presence of higher progesterone levels inhibits the effects of estrogen on binge eating (Baker et al. 2012).

Comparatively, in humans, food intake and dysregulated eating behaviors also change across the menstrual cycle: food intake significantly decreased on the days approaching ovulation when estradiol levels were low, and food intake increased after ovulation when progesterone levels increased (Roney and Simmons 2017). Similar patterns have emerged for binge eating during the menstrual cycle in women: during the mid-luteal and premenstrual phases, binge eating and emotional eating are increased compared with the follicular and ovulatory phases (Klump et al. 2008; Edler et al. 2007). Corroborating preclinical studies, direct associations between estradiol and progesterone, and binge eating have also been observed. In a sample of women with bulimia nervosa, increases in estradiol were associated with decreased binge eating frequency, whereas increases in progesterone were associated with increases in binge eating frequency (Edler et al. 2007). The same association has been observed for emotional eating in community samples (Klump et al. 2008). Taken together, binge eating in women fluctuates in a predictable pattern over the course of the menstrual cycle, and this fluctuation seems to be the direct result of changing estrogen and progesterone levels.

Indeed, binge eating may be exacerbated during the mid-luteal phase of the menstrual cycle in part due to the antagonizing effects of rising progesterone levels on estrogen. In a community sample of adult women, a significant interaction between estrogen and progesterone has been observed, indicating that binge eating is higher when both estrogen and progesterone levels are high, which corresponds to the mid-luteal phase of the menstrual cycle—corroborating the hypothesis that progesterone antagonizes the beneficial effect of estrogen (Klump et al. 2013). In a follow-up study splitting this community sample of women into those who binge eat and those who do not, Klump et al. (2014) reported that emotional eating was elevated in women *without* binge episodes when progesterone and estradiol levels were high. In contrast, emotional eating and binge eating frequencies were elevated in women *with* binge episodes when progesterone and estradiol were both low (Klump et al. 2014), corresponding to the premenstrual phase of the menstrual cycle. The differential effect of hormone levels on women with and without binge episodes provides insight into how sex hormones may influence binge eating at differing levels of pathology. Low or decreasing levels of estradiol may be the catalyst for dysregulated eating in women with high levels of pathology, whereas, for women with lower levels of pathology, the antagonizing effect of progesterone may play a more important role in symptom exacerbation. This could also suggest that individual differences exist between those with high versus low levels of pathology that moderate the impact of sex hormones on binge eating such as differences in genetic risk, a history of adverse life events (Gordon et al. 2016), or even that the presence of binge eating is modifying the effects of estrogen and progesterone.

4.2 Testosterone

Research focusing on the role of testosterone in binge eating is in its infancy. In comparison with estrogen, testosterone is associated with higher serum leptin levels (i.e., the "hunger hormone") and functions to increase food consumption (Iwasa et al. 2016). Contrary to these results, however, studies in both animals and humans examining dysregulated eating behaviors such as binge eating suggest that higher levels of testosterone may be protective against binge eating or eating disorder symptoms in general. For example, in a preclinical study of male and female rats, perinatal testosterone exposure reduced the later risk of binge eating during mid-to-post puberty (Culbert et al. 2018). Some human studies have found similar results showing that higher testosterone levels were associated with lower levels of disordered eating symptoms (including binge eating) in boys after pubertal onset (Culbert et al. 2014, 2015). Despite evidence suggesting that testosterone is related to increased food consumption, circulating testosterone may be protective against disordered eating, particularly in males (Mikhail et al. 2019). However, some human studies suggest no protective effect of testosterone on disordered eating (Baker et al. 2012). Much more preclinical and clinical research are needed to clarify the role of testosterone in the etiology of disordered eating behaviors, such as binge eating, and to explicate why testosterone may have differential effects on normative versus dysregulated food intake. Currently, it is difficult to draw definitive conclusions about the role of testosterone in binge eating etiology and maintenance across the sexes.

4.3 The Interaction Between Genetics and Sex Hormones in Binge Eating

Although in its early stages, research has begun to explore the combined effect of genetic factors and sex hormones in the risk for binge eating. For example, the genetic influences for binge eating were 5% in pre-puberty yet raised substantially to 42% in post-puberty for girls (Klump et al. 2017a). A single study in boys showed similar results such that genetic influences on disordered eating (which included binge eating) were practically nonexistent (0%) during pre-to-early adrenarche (early stage of maturation, typically peaks at age 10–14 years) but increased to 57% at early puberty (Culbert et al. 2017). Moreover, for females, the genetic effects on binge eating also appear to vary based on estrogen level: binge

eating exhibited little-to-no genetic influence in twins with low estradiol levels, but moderate-to-substantial genetic effects in twins with higher estradiol levels (Klump et al. 2010). Together, these findings suggest that estradiol levels may moderate genetic effects on binge eating during puberty. Given that pubertal development induces substantial changes in estradiol levels, the interaction between increasing estradiol during the pubertal period and genetic predisposition may be important in identifying those at higher risk for developing binge eating. Specifically, it has been proposed that increasing estrogen at puberty activates the genetic risk for eating disorders (Klump et al. 2017b). In contrast, however, a larger replication study reported that compared with higher estradiol levels, stronger genetic effects were observed at lower estradiol levels (Klump et al. 2018). The contrasting findings between the above two studies might be due to the much smaller sample size in the preliminary study (98 vs. 964 twin pairs (Klump et al. 2018)).

How sex hormones may influence the genetic risk for binge eating is unclear. In particular for males given the paucity of research in this area. Regarding female sex hormones and the genetic risk for binge eating, estrogen influences the development of the central nervous system (CNS) during the prenatal and pubertal periods and the CNS is an important system for appetite and body weight regulation (Farr et al. 2016). Estrogen receptors are expressed in the pituitary gland and brain regions such as the hypothalamus, the hippocampus, and the amygdala (Almey et al. 2015). Estrogen receptors are activated by estrogen, which in turn regulates gene expression in the CNS, ultimately affecting cognitive processes such as agonistic behaviors by altering transmission in neurotransmitter systems (Almey et al. 2015). The activation of genetic risk by estradiol at puberty and its influences on the CNS may contribute to the increased genetic effects on binge eating during puberty (Fig. 1; Klump et al. 2018). To date, no studies have empirically addressed this hypothesis. Future imaging genetic studies may help to unmask the link between estrogen's genomic effect on the CNS and risk for binge eating. Finally, this

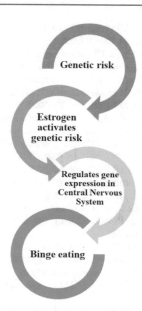

Fig. 1 A model for how estrogen may influence genetic risk for binge eating

hypothesized model may, in part, explain why eating disorders and binge eating are more prevalent in females and testosterone's protective role. For example, female rats treated with testosterone show a lower risk for binge eating post-puberty (Culbert et al. 2018).

5 Conclusion

This chapter summarized studies to date examining the genetic architecture of binge eating and its associations with sex hormones. Taken together, both genetic factors and sex hormones likely play an important role in the pathophysiology of binge eating. Based on the literature to date we conclude: (1) binge eating is a heritable condition that aggregates in families, (2) binge eating may share genetic risk with other psychiatric phenotypes such as major depression, bipolar disorder, and schizophrenia, (3) preclinical studies indicate that prenatal stress and food restriction may cause epigenetic changes which, in turn, may increase risk for binge eating, (4) estrogen and progesterone interacts to influence binge eating, (5) estrogen in girls may play an important role in the

genetic vulnerability for binge eating, and (6) more research is needed to clarify the etiological role of testosterone in binge eating.

Despite the momentum of the psychiatric genomic field, the etiology of binge eating remains to be fully elucidated. Contemporary genomic methods have been limitedly applied to eating disorders primarily characterized by binge eating or to symptom-level binge eating data. GWASs, including the creation of a polygenic risk score, are a necessary next step to further our understanding of the genetic architecture of binge eating. Moreover, it will be important for these studies to include diverse populations as a majority of the studies to date have exclusively focused on populations of European ancestry—with a limited representation of other populations, including from Asia, Oceania, and Africa. European populations represent only a certain subset of human genetic variation and cannot fully unmask the risk variants underlying binge eating in all populations. Large-scale, multiethnic samples in GWASs will provide a better understanding of the genetic risk variants for binge eating on a global scale and beyond what can be achieved with populations of European descent alone. Much more work is also needed in preclinical and clinical research to further our understanding of the role of epigenetic mechanisms in the genetic architecture of binge eating. Finally, while a plethora of studies exist showing that the sex hormones estrogen and progesterone play an important role in the etiology of binge eating for females, fewer studies have examined how genetic factors may interact with sex hormones to influence this risk. Fewer studies still have examined the role of testosterone.

Binge eating is a complex but treatable condition. The results of genomic and sex hormones studies may hold promise for the prevention and personalization of medical care. For example, polygenic risk scores could potentially be used to identify individuals who are at heightened genetic risk for binge eating so that interventions can be applied early, before the onset of binge eating. Finally, fully understanding the interplay between estrogen, progesterone, and testosterone on binge eating may also guide future interventions.

References

Albertz J, Boersma GJ, Tamashiro KL et al (2018) The effects of scheduled running wheel access on binge-like eating behavior and its consequences. Appetite 126:176–184. https://doi.org/10.1016/j.appet.2018.04.011

Almey A, Milner TA, Brake WG (2015) Estrogen receptors in the central nervous system and their implication for dopamine-dependent cognition in females. Horm Behav 74:125–138. https://doi.org/10.1016/j.yhbeh.2015.06.010

Baker JH, Girdler SS, Bulik CM (2012) The role of reproductive hormones in the development and maintenance of eating disorders. Expert Rev Obstet Gynecol 7:573–583. https://doi.org/10.1586/eog.12.54

Balsevich G, Sticht M, Bowles NP et al (2018) Role for fatty acid amide hydrolase (FAAH) in the leptin-mediated effects on feeding and energy balance. Proc Natl Acad Sci USA 115(29):7605–7610. https://doi.org/10.1073/pnas.1802251115

Bulik CM, Sullivan PF, Kendler KS (1998) Heritability of binge-eating and broadly defined bulimia nervosa. Biol Psychiatry 44(12):1210–1218

Butera PC (2010) Estradiol and the control of food intake. Physiol Behav 99(2):175–180. https://doi.org/10.1016/j.physbeh.2009.06.010

Capusan AJ, Yao S, Kuja-Halkola R et al (2017) Genetic and environmental aspects in the association between attention-deficit hyperactivity disorder symptoms and binge-eating behavior in adults: a twin study. Psychol Med 47(16):2866–2878. https://doi.org/10.1017/s0033291717001416

Culbert KM, Burt SA, Sisk CL et al (2014) The effects of circulating testosterone and pubertal maturation on risk for disordered eating symptoms in adolescent males. Psychol Med 44(11):2271–2286. https://doi.org/10.1017/s0033291713003073

Culbert KM, Breedlove SM, Sisk CL et al (2015) Age differences in prenatal testosterone's protective effects on disordered eating symptoms: developmental windows of expression? Behav Neurosci 129(1):18–36. https://doi.org/10.1037/bne0000034

Culbert KM, Burt SA, Klump KL (2017) Expanding the developmental boundaries of etiologic effects: the role of adrenarche in genetic influences on disordered eating in males. J Abnorm Psychol 126(5):593–606. https://doi.org/10.1037/abn0000226

Culbert KM, Sinclair EB, Hildebrandt BA et al (2018) Perinatal testosterone contributes to mid-to-post pubertal sex differences in risk for binge eating in male and female rats. J Abnorm Psychol 127(2):239–250. https://doi.org/10.1037/abn0000334

Edler C, Lipson SF, Keel PK (2007) Ovarian hormones and binge eating in bulimia nervosa. Psychol Med 37:131–141. https://doi.org/10.1017/s0033291706008956

Farr OM, Li CS, Mantzoros CS (2016) Central nervous system regulation of eating: insights from human brain imaging. Metabolism 65:699–713. https://doi.org/10.1016/j.metabol.2016.02.002

Gordon JL, Rubinow DR, Eisenlohr-Moul TA et al (2016) Estradiol variability, stressful life events, and the emergence of depressive symptomatology during the menopausal transition. Menopause 23(3):257–266. https://doi.org/10.1097/gme.0000000000000528

Hudson JI, Lalonde JK, Berry JM et al (2006) Binge-eating disorder as a distinct familial phenotype in obese individuals. Arch Gen Psychiatry 63(3):313–319. https://doi.org/10.1001/archpsyc.63.3.313

Iwasa T, Matsuzaki T, Tungalagsuvd A et al (2016) Effects of chronic testosterone administration on body weight and food intake differ among pre-pubertal, gonadal-intact, and ovariectomized female rats. Behav Brain Res 309:35–43. https://doi.org/10.1016/j.bbr.2016.04.048

Javaras KN, Laird NM, Reichborn-Kjennerud T et al (2008) Familiality and heritability of binge eating disorder: results of a case-control family study and a twin study. Int J Eat Disord 41(2):174–179. https://doi.org/10.1002/eat.20484

Jia YF, Choi Y, Ayers-Ringler JR et al (2017) Differential SLC1A2 promoter methylation in bipolar disorder with or without addiction. Front Cell Neurosci 11:217. https://doi.org/10.3389/fncel.2017.00217

Klump KL, Keel PK, Culbert KM et al (2008) Ovarian hormones and binge eating: exploring associations in community samples. Psychol Med 38(12):1749–1757. https://doi.org/10.1017/s0033291708002997

Klump KL, Keel PK, Sisk C et al (2010) Preliminary evidence that estradiol moderates genetic influences on disordered eating attitudes and behaviors during puberty. Psychol Med 40(10):1745–1753. https://doi.org/10.1017/s0033291709992236

Klump KL, Suisman JL, Culbert KM et al (2011) Binge eating proneness emerges during puberty in female rats: a longitudinal study. J Abnorm Psychol 120(4):948–955. https://doi.org/10.1037/a0023600

Klump KL, Keel PK, Racine SE et al (2013) The interactive effects of estrogen and progesterone on changes in emotional eating across the menstrual cycle. J Abnorm Psychol 122(1):131–137. https://doi.org/10.1037/a0029524

Klump KL, Racine SE, Hildebrandt B et al (2014) Ovarian hormone influences on dysregulated eating: a comparison of associations in women with versus without binge episodes. Clin Psychol Sci 2(4):545–559. https://doi.org/10.1177/2167702614521794

Klump KL, Culbert KM, O'Connor S et al (2017a) The significant effects of puberty on the genetic diathesis of binge eating in girls. Int J Eat Disord 50(8):984–989. https://doi.org/10.1002/eat.22727

Klump KL, Culbert KM, Sisk CL (2017b) Sex differences in binge eating: gonadal hormone effects across development. Annu Rev Clin Psychol 13:183–207. https://doi.org/10.1146/annurev-clinpsy-032816-045309

Klump KL, Fowler N, Mayhall L et al (2018) Estrogen moderates genetic influences on binge eating during puberty: disruption of normative processes? J Abnorm Psychol 127(5):458–470. https://doi.org/10.1037/abn0000352

Leeners B, Geary N, Tobler PN et al (2017) Ovarian hormones and obesity. Hum Reprod Update 23(3):300–321. https://doi.org/10.1093/humupd/dmw045

McElroy SL, Winham SJ, Cuellar-Barboza AB et al (2018) Bipolar disorder with binge eating behavior: a genome-wide association study implicates PRR5-ARHGAP8. Transl Psychiatry 8(1):40. https://doi.org/10.1038/s41398-017-0085-3

Micioni Di Bonaventura MV, Pucci M, Giusepponi ME et al (2019) Regulation of adenosine A2A receptor gene expression in a model of binge eating in the amygdaloid complex of female rats. J Psychopharmacol 33:1550–1561. https://doi.org/10.1177/0269881119845798

Mikhail ME, Culbert KM, Sisk CL et al (2019) Gonadal hormone contributions to individual differences in eating disorder risk. Curr Opin Psychiatry 32(6):484–490. https://doi.org/10.1097/yco.0000000000000543

Mitchell KS, Neale MC, Bulik CM et al (2010) Binge eating disorder: a symptom-level investigation of genetic and environmental influences on liability. Psychol Med 40(11):1899–1906. https://doi.org/10.1017/s0033291710000139

Munn-Chernoff MA, Duncan AE, Grant JD et al (2013) A twin study of alcohol dependence, binge eating, and compensatory behaviors. J Stud Alcohol Drugs 74(5):664–673

Munn-Chernoff MA, Grant JD, Agrawal A et al (2015) Are there common familial influences for major depressive disorder and an overeating-binge eating dimension in both European American and African American female twins? Int J Eat Disord 48(4):375–382. https://doi.org/10.1002/eat.22280

National Human Genome Research Institute (2015) Genome-wide association studies fact sheet. https://www.genome.gov/about-genomics/fact-sheets/Genome-Wide-Association-Studies-Fact-Sheet. Accessed 20 July 2007

Pucci M, Micioni Di Bonaventura MV, Zaplatic E et al (2018) Transcriptional regulation of the endocannabinoid system in a rat model of binge-eating behavior reveals a selective modulation of the hypothalamic fatty acid amide hydrolase gene. Int J Eat Disord 52(1):51–60. https://doi.org/10.1002/eat.22989

Reichborn-Kjennerud T, Bulik CM, Tambs K et al (2004) Genetic and environmental influences on binge eating in the absence of compensatory behaviors: a population-based twin study. Int J Eat Disord 36(3):307–314. https://doi.org/10.1002/eat.20047

Richardson TG, Harrison S, Hemani G et al (2019) An atlas of polygenic risk score associations to highlight putative causal relationships across the human phenome. Elife 8:e43657. https://doi.org/10.7554/eLife.43657

Roney JR, Simmons ZL (2017) Ovarian hormone fluctuations predict within-cycle shifts in women's food intake. Horm Behav 90:8–14. https://doi.org/10.1016/j.yhbeh.2017.01.009

Root TL, Thornton LM, Lindroos AK et al (2010) Shared and unique genetic and environmental influences on binge eating and night eating: a Swedish twin study. Eat Behav 11(2):92–98. https://doi.org/10.1016/j.eatbeh.2009.10.004

Schroeder M, Jakovcevski M, Polacheck T et al (2018) Sex dependent impact of gestational stress on predisposition to eating disorders and metabolic disease. Mol Metab 17:1–16. https://doi.org/10.1016/j.molmet.2018.08.005

Solmi F, Mascarell MC, Zammit S et al (2019) Polygenic risk for schizophrenia, disordered eating behaviours and body mass index in adolescents. Br J Psychiatry 215(1):1–6. https://doi.org/10.1192/bjp.2019.39

Sugrue LP, Desikan RS (2019) What are polygenic scores and why are they important? JAMA 321:1820–1821. https://doi.org/10.1001/jama.2019.3893

Thornton LM, Mazzeo SE, Bulik CM (2011) The heritability of eating disorders: methods and current findings. Curr Top Behav Neurosci 6:141–156. https://doi.org/10.1007/7854_2010_91

Torkamani A, Wineinger NE, Topol EJ (2018) The personal and clinical utility of polygenic risk scores. Nat Rev Genet 19(9):581–590. https://doi.org/10.1038/s41576-018-0018-x

Wade TD, Gordon S, Medland S et al (2013) Genetic variants associated with disordered eating. Int J Eat Disord 46:594–608. https://doi.org/10.1002/eat.22133

Winham SJ, Cuellar-Barboza AB, Mcelroy SL et al (2014) Bipolar disorder with comorbid binge eating history: a genome-wide association study implicates APOB. J Affect Disord 165:151–158. https://doi.org/10.1016/j.jad.2014.04.026

Wray NR, Goddard ME, Visscher PM (2008) Prediction of individual genetic risk of complex disease. Curr Opin Genet Dev 18:257–263. https://doi.org/10.1016/j.gde.2008.07.006

Yu Z, Geary N, Corwin RL (2008) Ovarian hormones inhibit fat intake under binge-type conditions in ovariectomized rats. Physiol Behav 95(3):501–507. https://doi.org/10.1016/j.physbeh.2008.07.021

Yu Z, Geary N, Corwin RL (2011) Individual effects of estradiol and progesterone on food intake and body weight in ovariectomized binge rats. Physiol Behav 104(5):687–693. https://doi.org/10.1016/j.physbeh.2011.07.017

Neuroendocrine Correlates of Binge Eating

Francesca Marciello, Alessio Maria Monteleone, Giammarco Cascino, and Palmiero Monteleone

Abstract

Neuroendocrine systems seem to play a central role in both the development and the maintenance of aberrant eating behaviors, including binge eating (BE). BE occurs in several clinical conditions such as the binge-eating/purging subtype of anorexia nervosa, bulimia nervosa, and binge-eating disorder. Because of this transdiagnostic position in the spectrum of eating disorders (EDs), it has been particularly problematic to identify hormonal alterations specifically linked to BE behavior because of the direct or indirect neuroendocrine effects of the different nutritional and psychopathological aspects of the various EDs associated with BE.

In this chapter, the literature regarding putative links between hormonal factors and BE behavior in human populations is reviewed.

We focused on those hormonal systems and substances known to be involved in the regulation of both eating behavior and biological functions directly or indirectly modulating eating behavior such as the endogenous stress response, social interaction, and reward processing. Although findings in this area are far from being conclusive, it seems that BE behavior could be connected with dysfunction in the hypothalamus–pituitary–adrenal axis as well as with alterations in substances such a leptin, ghrelin, oxytocin, BDNF and, especially, endocannabinoids, actively involved in the modulation of brain reward processes.

Keywords

Neuroendocrinology · Eating disorders · Binge eating · Cortisol · Estrogens · Gastrointestinal hormones · Oxytocin · Brain-derived neurotrophic factor · Endocannabinoids

F. Marciello · G. Cascino
Department of Psychiatry, University of Campania "Luigi Vanvitelli", Naples, Italy

Department of Medicine, Surgery and Dentistry "Scuola Medica Salernitana", University of Salerno, Baronissi, Salerno, Italy

A. M. Monteleone
Department of Psychiatry, University of Campania "Luigi Vanvitelli", Naples, Italy

P. Monteleone (✉)
Department of Medicine, Surgery and Dentistry "Scuola Medica Salernitana", University of Salerno, Baronissi, Salerno, Italy
e-mail: pmonteleone@unisa.it

Learning Objectives

After reading this chapter, you will:

- Be able to review how the central endocrine system may contribute to binge eating in humans.
- Understand how gut hormones are involved in the pathophysiology of binge eating.

(continued)

- Be able to discuss endocrine abnormalities in binge-eating-related disorders that may develop during the course of the illness or might be premorbid vulnerability factors.

1   Introduction

In the last years, scientific research has made enormous strides in trying to clarify the psycho-biological aspects of eating disorders (EDs), including their neuroendocrine correlates. Although these efforts resulted in increased knowledge about the neurobiological systems involved in these conditions, it remains still difficult to differentiate hormonal consequences of EDs from hormonal factors involved in their pathophysiology. Indeed, a large body of literature has dealt with the involvement of neuroendocrine systems in the development and maintenance of EDs and has suggested that dysregulation of neuroendocrine systems might be not only the consequence of malnutrition but also contributes to the pathophysiology of aberrant eating behaviors. These aspects remain still more debated, especially in those EDs characterized by the occurrence of binge eating (BE) behaviors, such as binge-eating disorder (BED), bulimia nervosa (BN), and the binge eating/purge subtype of anorexia nervosa (AN-BP). BE is a behavior typically characterized by the consumption of an unambiguously large amount of food in a discrete period of time with feelings of loss of control over eating. The presence of this aberrant eating behavior in such different clinical conditions has led to great heterogeneity in research findings of its pathophysiology. Therefore, at present, it is difficult to draw conclusive results on neuroendocrine alterations of BE due to the bias related to the different nutritional conditions and psychopathological aspects of the various EDs with BE behaviors.

In the present chapter, we summarize the current state of knowledge regarding putative links between hormonal factors and BE behavior in humans with the effort to try and discern, where possible, those neuroendocrine alterations that may have a direct pathogenetic role in promoting and/or maintaining BE from those that are the consequence of BE. We focused on those central and peripheral hormonal substances that have been recognized to be involved in the regulation of both eating behavior and biological functions that directly or indirectly modulate or mediate eating behavior such as endogenous stress response, social interaction, and reward processing.

2   Hypothalamic–Pituitary–Adrenal Axis

A growing amount of evidence suggests that the hypothalamus is a key region in controlling both energy homeostasis and the endogenous stress response. Furthermore, a condition of stress is the most frequently cited precursor of BE (Laessle and Schulz 2009; Levine and Marcus 1997; Pike et al. 2006) and it has been shown that cortisol released during stress promotes hunger and feeding behavior (Tataranni et al. 1996) whereas the intake of palatable food reduces the magnitude of the cortisol stress response.

Several lines of research have explored the role of hypothalamic–pituitary–adrenal (HPA) axis in EDs but, to our knowledge, only a few studies have been developed for the specific purpose of investigating HPA dysregulation in relation to BE behavior. Our knowledge in this field, therefore, remains inconclusive because of inconsistent results likely related to the heterogeneity of clinical conditions with BE. Moreover, many studies have investigated the impact of BE on HPA axis functioning in obese populations with BED, and obesity per se is associated with substantial dysregulations in the HPA axis, so it is difficult to disentangle the effects of obesity on the HPA axis from those of BE.

Studies based on a single measurement of blood cortisol levels found no significant differences in both morning (Monteleone et al. 2000a, 2003a) and evening cortisol values (Coutinho et al. 2007) between obese women with BED and those without BED. However, Coutinho et al. (2007) reported a significant correlation between the Binge Eating Scale scores and salivary cortisol levels in women without BED, suggesting possible effects of BE behavior on the functioning of the HPA axis, independently from the subject's nutritional status. Other authors found increased cortisol secretion in obese individuals with BED and interpreted these results as evidence for a hyperactive HPA system. Indeed, higher morning cortisol levels and greater cortisol response to a cold pressor stress test have been reported in obese women with BED as compared to obese women without BED (Gluck et al. 2004a, b; Gluck 2006). However, Schulz et al. (2011) found a similar increase of cortisol after the cold pressor stress test in both non-BED and BED individuals who were obese.

Several studies, instead, have suggested a downregulation of the HPA axis in obese patients with BED (Larsen et al. 2009; Lavagnino et al. 2013; Rosenberg et al. 2013). A study that performed repeated measurements of cortisol over two consecutive days and evaluated the cortisol awakening response (CAR) in obese women with and without BED after bariatric surgery showed that obese women with BED had overall cortisol levels during the day significantly lower than obese women without BED and a moderate but not significant decrease in the CAR (Larsen et al. 2009). Another study investigated the involvement of the HPA axis in the eating response to negative emotions induced by a psychosocial stressor, the Trier Social Stress Test (TSST), in obese subjects with or without BED compared with a control group (Rosenberg et al. 2013). In that study, basal cortisol levels were similar in the three groups, but obese BED participants had a blunted cortisol response to the TSST as compared to the other two groups, and their TSST-induced increases in the desire to binge and sweet craving were positively correlated to TSST-induced changes in cortisol levels. A recent study that measured the 24-h urinary free cortisol excretion rate (UFC/24) in a sample of obese individuals with BED reported an inverse correlation between Binge Eating Scale scores and UFC/24 values further supporting the hypothesis of a hypoactivity of HPA axis in obese patients with BED (Lavagnino et al. 2013).

The relationships between HPA axis functioning and metabolic aspects of BE behavior have also been explored. Gluck et al. (2004a), reported a significant positive correlation between the cortisol response to a cold pressor stress test (evaluated as the overall secretion of cortisol or the peak cortisol response to the stressor) and waist circumference in obese women with BED but not in those without BED, supporting the thesis that BE mediates the relationship between stress and metabolic abnormalities possibly representing a coping mechanism in response to stress events. Chao et al. (2016) examined the connections among BE behavior, self-reported stress, morning plasma cortisol levels, homeostatic model of insulin resistance (HOMA-IR) index, and abdominal obesity in adults. They found a significant negative relationship between self-reported stress measures and morning cortisol levels and between cortisol levels and HOMA-IR index and waist circumference with no significant evidence that BE mediated the relationship between stress and HOMA-IR index or waist circumference values. Another study, that examined cortisol levels during the day (12:00 p. m., 3:00 p.m., and 8:00 p.m.) and CAR in obese women after bariatric surgery, reported that obese women with BED had significantly lower total daytime cortisol levels and a trend toward a higher BMI but lower waist-to-hip ratio than obese women without BED (Larsen et al. 2009). Therefore, at present, the connection between HPA axis functioning and metabolic aspects of BE behavior remains elusive.

Studies carried out in clinical populations with EDs associated with BE, but other than BED, have provided mixed results concerning the functioning of the HPA axis. Indeed, both normal or higher basal cortisol levels, CAR, and cortisol response to laboratory psychosocial stress tests (e.g., TSST) have been reported in populations with BN (Koo-Loeb et al. 1998, 2000; Monteleone et al. 2015). Moreover, a decreased diurnal secretion of cortisol, with an excessive increase after daytime meals and an enhanced ACTH response to CRF, has been found in a small sample of remitted BN patients (Birketvedt et al. 2006).

Recently, an enhanced CAR has been reported in women with AN-BP as compared to both healthy controls and women with the restrictive subtype of AN (AN-R); in the AN-BP group the CAR was positively correlated with BE symptoms as assessed by the bulimia score of the Eating Disorder Inventory-2 (Monteleone et al. 2017a), supporting the idea that BE behavior, but not malnutrition, may have a specific connection with HPA axis functioning with possible pathogenetic implications.

Given the above contradictory results, no definitive conclusion can be drawn as to HPA axis functioning in individuals with BE behavior. However, some explanations can be put forward to account for such inconsistency. First, EDs with BE, similarly to those without BE, have a specific psychopathology and several psychiatric comorbidities that may affect differently HPA axis functioning. Indeed, dysregulation of the HPA axis in humans has been documented in mood and anxiety disorders, which frequently co-occur with EDs (Nemeroff et al. 1999; Vreeburg et al. 2009). Second, different methods have been employed to measure baseline cortisol levels and to explore the cortisol response to a stressor across the studies; in particular, different types of laboratory stress paradigms, based on physical, psychological, or social stressors, have been adopted and it is therefore difficult to compare results from study to study. Third, very few studies included a comparison healthy control group, so it is difficult to judge whether the levels of cortisol of the patient populations were actually "decreased" or "increased" if not compared with a healthy control population in similar conditions. Finally, most studies were based on a single measurement of cortisol that might not be the best way to characterize HPA axis function (Pruessner et al. 2003; Spencer and Deak 2017).

3 Hypothalamic–Pituitary–Gonadal Axis

There is consistent evidence for ovarian hormone effects on dysregulated eating such as BE and emotional eating. Many researchers have underlined that changes in ovarian hormones predict changes in dysregulated eating across the menstrual cycle. Initial studies were consistent in suggesting that low levels of estradiol and high levels of progesterone were associated with increased BE frequency and higher emotional eating scores both in women with BN and women from the community (Edler et al. 2007; Klump et al. 2008). However, these initial studies were conducted on very small samples and did not examine the interactive effects of hormones. Subsequent larger-scale studies did not corroborate those conclusions. Klump et al. (2013a) found that in women recruited from the community, emotional eating scores were highest during the midluteal phase when progesterone peaks and estradiol demonstrates a secondary peak. These results replicated findings from experimental studies showing that elevated levels of estrogens caused decreases in food intake when progesterone levels were low, while food intake increased when both hormones were high, probably due to the antagonizing effects of progesterone on estrogen action (Varma et al. 1999; Kemnitz et al. 1989). More recently, another study revealed stronger associations between dysregulated eating and ovarian hormone levels in women with BE as compared to women without BE and that progesterone would moderate the effects of both high and low levels of estradiol in women with BE, but not in women without BE (Klump et al. 2014). Indeed, examining changes across the menstrual cycle, emotional eating scores, and BE frequency peaked in the post-ovulatory phase in both women with BE and those without BE, and the high levels of estrogen and progesterone in this phase partially accounted for those peaks. Also, in the BE sample only, lower levels of estradiol across the menstrual cycle were found to predict higher emotional eating scores and binge frequencies. These findings are consistent with the idea that progesterone has few direct effects on BE or food intake and that it acts indirectly by antagonizing estrogen effects.

It may be concluded from the above that ovarian hormones act in concert, rather than alone, in influencing dysregulated eating across the menstrual cycle. In particular, dysregulated eating could occur both in the context of high estradiol and high progesterone, via progesterone's antagonizing effects on estradiol, and in the context of low estradiol and low progesterone, since low levels of progesterone could act as a permissive factor on the increase of food intake and BE induced by low levels of estradiol (Culbert et al. 2016). Interestingly, the predictive effects of ovarian hormonal interaction on within-person changes in dysregulated eating across the menstrual cycle do not apply to other types of eating disorder symptoms and are not affected by individual differences in body mass index (Klump et al. 2013b; Racine et al. 2013).

In addition to phenotypic effects, ovarian hormones have been shown to increase the genetic risk for eating pathology. These hormones act as gene transcription factors in several brain neurotransmitter systems that are known to be disrupted in eating disorders such as serotonin and dopamine (Ostlund et al. 2003). For instance, circulating levels of estradiol have been shown to moderate the genetic risk for disordered eating symptoms in adolescent girls, such that higher levels of estradiol during puberty affect gene transcription in the brain leading to a genetic background responsive to ovarian hormones and affecting BE risk in adulthood (Klump et al. 2010, 2013a, 2017).

4 Gastrointestinal and Adipose Tissue Hormones

The gastrointestinal tract and the adipose tissue produce substances and hormones that are known to be involved in the regulation of energy metabolism and eating behavior, including orexigenic (appetite-stimulating, e.g., ghrelin) and anorexigenic (appetite-reducing, e.g., cholecystokinin, glucagon-like peptide-1, peptide YY, leptin) peptides. Alterations in the physiology of these substances have been supposed to occur in EDs and several studies have examined basal levels (i.e., single measure of a hormone level, typically following an overnight fast) and postprandial concentrations (i.e., hormone release following administration of a standardized meal) of hormones secreted by gastrointestinal tract and adipose tissue in people with EDs with or without BE.

4.1 Ghrelin

Ghrelin, a 28-amino-acid peptide hormone mainly secreted from the stomach, is a ligand for the growth hormone secretagogue receptor and is classically known as a central appetite-stimulating hormone. Ghrelin levels in the blood rise before meals, fall following meals and show a diurnal pattern with peaks and troughs over the 24 h (Cummings et al. 2002). Several studies have examined ghrelin production in individuals with BE behavior with mixed results. Most studies indicated no significant differences in fasting levels of ghrelin in individuals with BN compared to healthy controls (Monteleone et al. 2008a), and although initial evidence pointed to elevated postprandial ghrelin levels in BN patients (Tanaka et al. 2003; Kojima et al. 2005), subsequent studies have not confirmed this finding (Sedlackova et al. 2012; Devlin et al. 2012). Studies on AN by Tanaka et al. (2003) reported that plasma ghrelin concentrations of women with AN-BP were significantly higher than those of women with AN-R and positively correlated with the frequency of binge/vomiting episodes, suggesting that binge-purging behavior may have some influence on ghrelin secretion. However, Troisi et al. (2005) reported opposite results with higher plasma ghrelin levels in AN-R than in a group with AN-BP and a negative correlation between plasma ghrelin levels and BE in AN-BP individuals. Finally, other authors found no significant difference in fasting plasma ghrelin concentrations between AN-R and AN-BP patients and no significant correlations between ghrelin concentrations and the frequency of binge/vomiting in BN subjects (Otto et al. 2005; Monteleone et al. 2005a).

Studies focusing on the pre-absorptive cephalic phase of food ingestion, that is the time period occurring immediately before the consumption of a meal and consisting mostly of vagal efferent activation and concomitant release of some gastro-entero-pancreatic hormones such as ghrelin, found increased ghrelin secretion in individuals with BN and AN-BP compared to healthy controls (Monteleone et al. 2008b, 2010), supporting the idea that an elevated ghrelin secretion during the pre-absorptive phase of food ingestion may contribute to a strong sense of hunger or urge to eat, which could increase BE propensity.

In contrast to what may be expected considering ghrelin's orexigenic effect, both normal-weight and overweight or obese individuals with BED have been found to have lower ghrelin levels than normal-weight non-BED subjects; moreover, circulating ghrelin was not significantly correlated with the frequency or severity of BE in individuals with BED (Monteleone et al. 2005a). Some authors measured various appetitive hormones before and after a meal in overweight women with BE not meeting full criteria for BED, women with BED and those without BE behavior, reporting that both women with BED and BE behavior had lower pre-meal ghrelin levels, and lower post-meal ghrelin decline, which also declined postprandially less than in controls (Geliebter et al. 2004). Similar findings (significantly lower ghrelin levels both pre-meal and post-meal with a smaller post-meal decline in the BED group as compared to the non-BED group) were also highlighted in other studies comparing overweight or obese BED women and obese women without BED (Geliebter et al. 2005, 2008a). These results are consistent with the hypothesis of a downregulation of ghrelin secretion as a consequence of BE behavior in BED. However, recent data suggest that the downregulation of ghrelin in individuals with BED may be exacerbated by higher weight status. Ghrelin levels were significantly lower in the obese compared to overweight women (Buss et al. 2014). Interestingly, in that study,

overweight women showed positive associations between ghrelin, caloric intake, and hedonic eating but this correlation was not observed in the obese group and authors hypothesized that one explanation could lie in the resistance to central signaling of ghrelin in the hypothalamus in obesity (Buss et al. 2014).

Other studies have also investigated whether ghrelin may mediate stress-induced BE. Compared to healthy controls, BN subjects showed significantly enhanced salivary ghrelin but similar cortisol responses to TSST despite no significant differences in the prestress salivary levels of both hormones and without any significant correlation between stress-induced salivary hormone changes and self-report measures of BE (Monteleone et al. 2012a). Similarly, the stress-induced urge for uncontrolled eating seemed not to be modulated by stress-related elevations in ghrelin levels in BED individuals undergoing the TSST (Rouach et al. 2007). Moreover, no significant difference was reported in ghrelin levels between obese BED and obese non-BED women after a cold pressor stress test (Gluck et al. 2014). Therefore, it seems that stress-induced BE is not connected to ghrelin secretion in BN or BED.

Finally, one of the major issues that still need to be addressed in ghrelin research is to clarify the role of acylated (active) and des-acyl (inactive) ghrelin. Results from a recent study showed that obese individuals with BE had significantly lower fasting and post-ingestive acylated ghrelin concentrations than obese subjects without BE, post-prandial decrease in acylated ghrelin concentrations was significantly smaller for the BE group than for the non-BE group while no significant differences in des-acyl ghrelin levels were observed between the two groups (Hernandez et al. 2019).

4.2   Cholecystokinin

Cholecystokinin (CCK) is a member of the gut-brain family of peptide hormones that decreases hunger and feeding in humans (Degen et al. 2001). After a meal, CCK is released into the bloodstream from endocrine I cells of the duodenum and the jejunum (Buchan et al. 1978). Studies in humans have shown that CCK inhibits food intake, acting as a satiety hormone (Kissileff et al. 1981; Ballinger et al. 1995).

Studies that examined fasting and postprandial levels of CCK in individuals with EDs have produced mixed results. Blunted postprandial CCK release has generally been found in women ill and remitted from BN (Geracioti et al. 1992; Culbert et al. 2016), and it has been suggested that this may contribute to patients' impairments in satiation as well as to their BE propensity. Further, women with BN have been found to have blunted postprandial CCK as compared to women with purging disorder, suggesting that CCK disturbances may be particularly tied to BE and not to purging behavior (Keel et al. 2007). These results raised the possibility that BED could be associated with changes in CCK release similar to BN. Yet, individuals with BED have been reported to exhibit fasting or postprandial CCK levels similar to healthy controls (Geliebter et al. 2004, 2005; Culbert et al. 2016).

4.3 Glucagon-Like Peptide 1

The glucagon-like peptide 1 (GLP-1) is synthesized by the jejunum, ileum, colon, and neurons of the brain stem nucleus tractus solitarius, and is involved in nutrient assimilation and energy homeostasis, by promoting satiety. Relatively few studies have examined whether individuals with EDs have disturbances in GLP-1.

In BN patients, plasma GLP-1 levels in response to a test meal did not differ from healthy controls, but higher than normal values were observed at 4:00 PM (Brambilla et al. 2009). Naessén et al. (2011), instead, have reported that women with BN secrete abnormally low amounts of GLP-1 in response to a test meal. This blunted postprandial GLP-1 release in patients with BN has been confirmed by a subsequent study that reported a similar finding also in patients with BED, but not in those with purging disorder (Culbert et al. 2016), thus suggesting that alternations in GLP-1

secretion may be linked to BE rather than to purging (Dossat et al. 2015). Blunted postprandial GLP-1 release may contribute to altered perceptions of satiation in patients with BN and BED, which could subsequently result in an increased BE propensity (Dossat et al. 2015). Only one study has examined GLP-1 levels in recovered patients with BED and have found normal concentrations of the peptide (Geliebter et al. 2008b), supporting the role of a state-dependent marker for this peptide.

4.4 Peptide YY

The short-term appetite regulator peptide YY (PYY) is a 36-amino-acid peptide released from the endocrine L cells of the distal ileum and colon in response to feeding. PYY in the circulation exists in two major forms: PYY_{1-36} and PYY_{3-36}.

A first study performed in women with BN reported normal CSF and plasma levels of PYY in both symptomatic and 1-year recovered patients (Gendall et al. 1999). Subsequently, Kaye et al. (1990) found that in six healthy women, who experimentally binged without vomiting, plasma PYY concentrations significantly rose after bingeing and remained elevated for the subsequent 2 h whereas in five women with BN undergoing episodes of bingeing/vomiting circulating PYY increased after the first binge and remained elevated for the duration of bingeing and vomiting. More recently, two independent research groups found a blunted PYY_{3-36} response to food ingestion in symptomatic women with BN accompanied by a decreased response of ghrelin (Kojima et al. 2005; Monteleone et al. 2005b) with a significant negative correlation between meal-induced PYY_{3-36} increase and ghrelin decrease. The authors suggested that in symptomatic BN patients, the blunted meal-induced responses of circulating ghrelin and PYY_{3-36} would denote the occurrence of an impaired suppression of the drive to eat following a meal, which might sustain an increased food consumption and BE. This idea is corroborated by the demonstration that normal-weight individuals with disinhibited eating had a blunted postprandial PYY_{3-36} and PYY_{1-36} response, which may contribute to their susceptibility to overeating (Martins et al. 2010). Recently, also a lower than normal circadian secretion of PYY_{3-36} has been reported in symptomatic BN women (Germain et al. 2010).

Higher fasting levels of PYY_{3-36} but similar postprandial PYY_{3-36} release have been detected in obese individuals with BED compared to non-BED obese subjects. The authors suggested that the higher levels of fasting PYY_{3-36} in obese BED individuals may reflect dysregulation in sensitivity to PYY's satiating signals, which could contribute to hunger and overeating in obese patients with BED.

4.5   Leptin

Leptin, secreted from the adipose tissue, acts as a satiety signal on neurons in the hypothalamus, reducing food intake. Leptin is considered to be part of a feedback loop where it informs the central nervous system about energy stores in the body adipose tissue. Levels of leptin are positively associated with body fat, and, notably, obesity has been linked to the development of leptin resistance whereby elevations in leptin fail to decrease hunger and appetite.

Leptin physiology has been studied in EDs with or without BE. Decreased, normal, or increased CSF and plasma levels of leptin have been reported in normal-weight subjects with BN, (Monteleone et al. 2008c). It has been demonstrated that women with BN exhibit plasma leptin levels ranging from low anorexic-like levels to normal values depending on the severity of BE and length of their illness, since leptin levels were found to be inversely correlated to both variables (Monteleone et al. 2002a). Moreover, it has been reported that in BN patients with basal hypoleptinaemia, acute fasting was not able to induce the physiological fall in circulating leptin and short-term normal refeeding was not sufficient to restore blood levels of the hormone similar to those of healthy controls (Monteleone et al. 2000b). Therefore, it was suggested that, since leptin behaves as a hunger-suppressant signal, lower leptin values after eating together with alterations in other key modulators of hunger and satiety occurring in people with BN, such as the impaired meal-induced ghrelin suppression (Monteleone et al. 2003b, 2005b) and the blunted postprandial CCK rise (Geracioti et al. 1992), could contribute to their BE behavior.

In individuals with BED, findings suggest that leptin disturbances are more closely linked to obesity rather than BE symptomatology. Obese individuals with BED tend to show higher levels of basal leptin compared to normal-weight controls, whereas mixed results have emerged with respect to obese non-BED controls. Regarding the leptin post-prandial response, no significant differences between obese women with BED versus obese women without BED have emerged (Geliebter et al. 2004, 2005). To note, a significant inverse correlation has been described between plasma levels of leptin and those of its soluble receptor in EDs displaying that whatever the genesis of leptin changes in the ED spectrum, they are accompanied by opposite modifications in the secretion of its receptor (Monteleone et al. 2002b).

4.6   Adiponectin

Adiponectin is a protein hormone produced exclusively by adipocytes, and that modulates insulin sensitivity. Decreased adiponectin production has been supposed to be involved in the development of obesity, diabetes, and insulin resistance (Havel 2002). Fasting adiponectin levels were found increased in the blood of symptomatic women with BN and decreased in the blood of women with BED; moreover, circulating adiponectin appeared strongly correlated with the frequency of binge/purging episodes in BN patients but not with BE frequency in BED (Monteleone et al. 2003c). Yet, another study (Tagami et al. 2004) detected decreased circulating levels of adipokine in BN subjects, whereas Housova et al. (2005) found normal concentrations of adiponectin in symptomatic women with BN. Differences in the patients' samples, the assay methods, and the time of day when blood was collected may be responsible for these discrepancies among the studies.

5   Oxytocin

The hypothalamic nine amino acid neuropeptide oxytocin has been implicated directly or indirectly in the regulation of several psychological and physiological functions in humans, including not only reproductive behavior and mother–infant interaction, but also social cognition, metabolism, and eating behavior (Skinner et al. 2019). It has been hypothesized that oxytocin could be a link between psychosocial relations and eating behavior; respective mutual impairments in these functions may stem at least in part from alterations in oxytocin pathways (Spetter and Hallschmid 2017; Giel et al. 2018). Thus far, evidence of oxytocin disturbances in subjects with EDs is largely limited to AN, where it has been put forward that oxytocin may mediate the aberrant individual's response to social cues and the relationship between insecure attachment and ED psychopathology (Connan et al. 2003). In particular, it has been proposed that adverse attachment experiences early in life might modulate the stress response system (i.e., HPA axis), leading to alterations in emotional processing and that dysfunctional eating behavior might represent an attempt to deal with altered emotion processing and stress. Although originally formulated for the pathophysiology of AN, the assumptions of this model might also explain mechanisms underlying eating disorders associated with BE behavior. Indeed, difficulties in emotional regulation have been found to trigger BE in persons with BED (Leehr et al. 2015). In spite of these suggestions, so far no study has explored oxytocin secretion in people with BED while cerebrospinal fluid (CSF) or plasma levels of the hormone have been reported to be normal in both acutely ill or recovered persons with AN-BP or BN (Demitrack et al. 1990; Chiodera et al. 1991; Frank et al. 2000; Monteleone et al. 2016a). However, in those studies, no data have been provided regarding the possible connections between CSF or plasma levels of oxytocin and BE behavior.

A few studies have focused on the potential role of genetic variants of oxytocin-related genes in EDs. Single nucleotide polymorphisms (SNPs) of oxytocin receptor gene have been found associated with dysfunctional eating patterns in a large community cohort of women. In particular, the GG rs53576 genotype was associated with BE and purging, and the rs2254298 AG/AA genotype interacted with poor maternal care to increase the odds of BE and purging (Micali et al. 2017). Another study has described associations of the G allele of the rs53576 genotype and BN (Kim et al. 2015a). These data represent a first preliminary hint that SNPs of the oxytocin receptor might be involved in the risk of developing ED with BE.

The putative role of oxytocin in the pathophysiology of EDs with or without BE has also been investigated by assessing the effects of intranasal oxytocin on food intake in subjects with BN. Kim et al. found that a single dose of intranasal oxytocin produced no significant change in appetite either immediately and 24 h after its administration in healthy controls and a group with anorexia nervosa, while it was able to decrease calorie consumption over 24 h in subjects with BN (Kim et al. 2015b). In men with obesity, instead, a single dose of intranasal oxytocin was reported to exert a potent acute inhibition of food intake greater than that found in normal-weight controls (Thienel et al. 2016). As BE is often associated with overweight or obesity, it has been suggested that oxytocin may be potentially useful in the treatment of disorders related to BE behavior (Giel et al. 2018). However, this hypothesis has not been proven yet.

6   Brain-Derived Neurotrophic Factor

The brain-derived neurotrophic factor (BDNF) is a member of the neurotrophin family and implicated not only in the processes of neuronal outgrowth and differentiation, synaptic connectivity and neuronal repair but also in energy homeostasis, acting as an inhibitor of food intake.

Several animal studies support the hypothesis that alterations in BDNF physiology could be a susceptibility factor for disordered eating (Kernie et al. 2000; Lyons et al. 1999; Rios et al. 2001). As for the possible implication of BDNF in the human BE behavior, increased plasma BDNF levels have been reported in BN patients by Mercader et al. (2007) whereas several other studies have detected circulating BDNF levels significantly lower in symptomatic BN patients (Monteleone et al. 2005c; Nakazato et al. 2003) but not in individuals with BED (Monteleone et al. 2005c). However, no statistically significant correlations have emerged between the frequency of binge vomiting episodes and serum concentrations of the neurotrophin in women with BN (Monteleone et al. 2005c). In women with BN serum BDNF concentrations were decreased, similar to underweight patients with AN, whereas in overweight BED the neurotrophin levels were similar to those of normal-weight control women; this led to the hypothesis that decreased levels of circulating BDNF were due to derangements of the patients' nutritional status and alterations in both dietary macronutrients and energy balance rather than to body weight or BE behavior (Lee et al. 1999; Wu et al. 2004; Monteleone et al. 2005c). Conversely, in a first genetic association study exploring the 196G/A (val66met) single nucleotide polymorphism (SNP) of the *BDNF* gene, this genetic marker was found significantly associated with BE behavior in women with BN or BED, even if no significant difference emerged in the frequency of the polymorphism among patients with BN or BED and healthy controls; this suggested that the 196G/A SNP of the human BDNF gene does not contribute to the genetic susceptibility to BN and BED, but likely it may predispose those patients to a more severe BE behavior pattern (Monteleone et al. 2006). In a subsequent study, a significant association between the 196G/A SNP of the BDNF gene and BE behavior was found in a sample of adolescents who made attempts to regulate their body weight by reducing their meal frequency or the total amount of daily food intake (Akkermann et al. 2011). The authors suggested that this finding may help to explain why some people develop BE in response to dieting, and others

do not. However, these genetic findings await replication in larger study samples.

7 Endocannabinoids

The endocannabinoid system, consisting of two cannabinoid receptors (CB1 and CB2) and the endogenous ligands anandamide (arachidonoylethanolamide (AEA)) and 2-arachidonoylglycerol (2-AG), has been shown to control feeding in both its homeostatic and hedonic component (Cota et al. 2003). In particular, following food deprivation, hypothalamic, and mesolimbic endocannabinoids are produced to activate CB1 receptors locally and enhance appetite by stimulating both homeostatic and rewarding aspects of food intake (Cristino et al. 2014). Indeed, it has been shown that in both normal-weight subjects and individuals with obesity, an overproduction of peripheral endocannabinoids may increase food intake (Matias et al. 2006) and modulate hedonic eating (Monteleone et al. 2012b, 2016b). Furthermore, endocannabinoid CB1 receptors have been implicated in a rat model of BE behavior (Scherma et al. 2013).

Alterations of the endocannabinoid system have been explored in the pathophysiology of EDs with or without BE behavior. In a first study (Monteleone et al. 2005d), increased plasma concentrations of AEA but no significant changes in plasma levels of 2-AG were found in overweight or obese patients with BED, but not in those with BN. No correlation with clinical characteristics of eating disorder patients emerged. Increased production of AEA in patients with BED was interpreted as an endocannabinoid-induced potentiation of the drive to eat, possibly contributing to their BE behavior. Moreover, it was suggested that in women with BED, the enhanced levels of plasma AEA could reinforce the hedonic properties of hypercaloric nutrition, thus favoring an addiction to food intake and perpetuating BE behavior. In support of this idea, a subsequent study showed distinctive responses of endocannabinoids to food-related rewards in obese BED individuals (Monteleone et al. 2017b). In fact, plasma levels of AEA decreased after eating non-favorite food and increased after eating a favorite food, while plasma levels of 2-AG did not differ significantly between the two eating conditions. Moreover, the subjects' sensations of urge to eat and of pleasantness while eating the presented food were positively correlated to meal-induced overall secretion of peripheral AEA, while the food-induced overall production of peripheral 2-AG levels was positively correlated to the subjects' sensation of pleasantness while eating the presented food and to the amount of food they would eat. These results suggest that, in obese patients with BED, food-induced changes in peripheral endocannabinoids might modulate both the "wanting" and "liking" for a food reward, and this may sustain their BE behavior.

Finally, a randomized, placebo-controlled, double-blind trial has demonstrated the efficacy of rimonabant (a blocker of G protein-coupled CB1 receptor) both in reduction of body weight and reduction in number of binge episodes per week in obese patients with BED, although this last effect did not reach statistical significance (Pataky et al. 2013).

In conclusion, it seems reasonable to hypothesize a role of endocannabinoids in the modulation of binge eating behavior, especially in light of the known rewarding aspects of food intake.

8 Conclusions

The literature reviewed above provides evidence that neuroendocrine alterations are undoubtedly associated with BE behavior, although, at present, no definitive conclusions can be drawn as to whether alterations of both central or peripheral hormones precede the appearance of BE, thus representing possible pathogenetic factors, or are the mere consequence of BE. However, it can be suggested that even if those alterations are secondary phenomena, they might contribute to the maintenance of the aberrant eating behavior, thus affecting the course and the prognosis of BE behavior (and its related syndromes) and deserving full clinical attention in the process of treatment planning (Fig. 1).

Another aspect emerging from this narrative review is that it has been problematic to identify specific hormonal features and patterns related exclusively to BE because of its particular

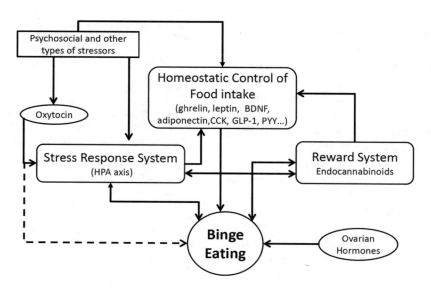

Fig. 1 Schematic representation of the possible interplay between neuroendocrine factors and binge eating behavior. Bidirectional arrows indicate reciprocal interactions. The dotted line indicates a proposed interaction

transdiagnostic position in the spectrum of EDs spanning from the underweight AN-BP disorder to the overweight BED and going through the normal-weight BN. All these disorders are characterized not only by different nutritional conditions but also by specific psychopathological features that may impact differently on endocrine physiology. Indeed, the aberrant BE behavior with its nutritional consequences represents only the surfacing tip of an iceberg whose underwater section represents the main psychopathological component of the related ED, which may have specific or unspecific effects on hormone secretions and may, in turn, be sustained by hormonal changes, although these may be secondary alterations. Indeed, a wide range of hormone abnormalities have been connected with depression, anxiety, insecure attachment, social dysregulation, and altered reward mechanisms that represent psychopathological dimensions of EDs with or without BE.

BE behavior seems unequivocally connected with food-related rewards. Indeed, it has been proposed that the ingestion of a large amount of food during the binge episode may have the aim to alleviate the patient's negative emotions by increasing food-derived feelings of pleasure. In this context, it has been found that people with BE syndromes usually report pre-binge negative emotions with a strong urge to eat, followed by a transient binge-induced reduction of negative emotional states. Therefore, BE may be conceptualized as a reward-dependent phenomenon, and most of the above reported endocrine alterations refer to substances (such as leptin, ghrelin, oxytocin, BDNF, and, especially endocannabinoids) actively involved in the modulation of brain reward processes (Monteleone and Maj 2013).

In conclusion, to clarify the role of endocrine dysregulations connected to BE behavior, further studies are required. Some help could come from other research areas as, for example, molecular genetic studies that will be helpful to elucidate the role of certain SNPs of genes coding for endocrine substances controlling eating behavior or hormone-induced changes in the expression of genes contributing to the biological vulnerability

to BE. Findings in this area are now emerging as for reported data on oxytocin receptor and BDNF gene polymorphisms and estrogen's action on genes connected to the risk for BE. Therefore, in the future, data coming from combined endocrine and genetic studies will help to clarify the mechanisms regulating BE behavior, and this knowledge will contribute to improving the prevention and/or the cure BE-related syndromes.

References

Akkermann K, Hiio K, Villa I et al (2011) Food restriction leads to binge eating dependent upon the effect of the brain-derived neurotrophic factor Val66Met polymorphism. Psychiatry Res 185(1–2):39–43

Ballinger A, McLoughlin L, Medbak S et al (1995) Cholecystokinin is a satiety hormone in humans at physiological post-prandial plasma concentrations. Clin Sci (Lond) 89(4):375–381

Birketvedt GS, Drivenes E, Agledahl I et al (2006) Bulimia nervosa—a primary defect in the hypothalamic-pituitary-adrenal axis? Appetite 46(2):164–167

Brambilla F, Monteleone P, Maj M (2009) Glucagon-like peptide-1 secretion in bulimia nervosa. Psychiatry Res 169(1):82–85

Buchan AM, Polak JM, Solcia E et al (1978) Electron immunohistochemical evidence for the human intestinal I cell as the source of CCK. Gut 19(5):403–407

Buss J, Havel PJ, Epel E et al (2014) Associations of ghrelin with eating behaviors, stress, metabolic factors, and telomere length among overweight and obese women: preliminary evidence of attenuated ghrelin effects in obesity? Appetite 76:84–94

Chao A, Grey M, Whittemore R et al (2016) Examining the mediating roles of binge eating and emotional eating in the relationships between stress and metabolic abnormalities. J Behav Med 39(2):320–332

Chiodera P, Volpi R, Capretti L et al (1991) Effect of estrogen or insulin-induced hypoglycemia on plasma oxytocin levels in bulimia and anorexia nervosa. Metabolism 40(11):1226–1230

Connan F, Campbell IC, Katzman M et al (2003) A neurodevelopmental model for anorexia nervosa. Physiol Behav 79(1):13–24

Cota D, Marsicano G, Lutz B et al (2003) Endogenous cannabinoid system as a modulator of food intake. Int J Obes Relat Metab Disord 27(3):289–301

Coutinho WF, Moreira RO, Spagnol C et al (2007) Does binge eating disorder alter cortisol secretion in obese women? Eat Behav 8:59–64

Cristino L, Becker T, Di Marzo V (2014) Endocannabinoids and energy homeostasis: an update. Biofactors 40(4):389–397

Culbert KM, Racine SE, Klump KL (2016) Hormonal factors and disturbances in eating disorders. Curr Psychiatry Rep 18(7):65

Cummings DE, Weigle DS, Frayo RS et al (2002) Plasma ghrelin levels after diet-induced weight loss or gastric bypass surgery. N Engl J Med 346(21):1623–1630

Degen L, Matzinger D, Drewe J et al (2001) The effect of cholecystokinin in controlling appetite and food intake in humans. Peptides 22(8):1265–1269

Demitrack MA, Lesem MD, Listwak SJ et al (1990) CSF oxytocin in anorexia nervosa and bulimia nervosa: clinical and pathophysiologic considerations. Am J Psychiatry 147(7):882–886

Devlin MJ, Kissileff HR, Zimmerli EJ (2012) Gastric emptying and symptoms of bulimia nervosa: effect of a prokinetic agent. Physiol Behav 106(2):238–242

Dossat AM, Bodell LP, Williams DL et al (2015) Preliminary examination of glucagon-like peptide-1 levels in women with purging disorder and bulimia nervosa. Int J Eat Disord 48(2):199–205

Edler C, Lipson SF, Keel PK (2007) Ovarian hormones and binge eating in bulimia nervosa. Psychol Med 37 (1):131–141

Frank GK, Kaye WH, Altemus M et al (2000) CSF oxytocin and vasopressin levels after recovery from bulimia nervosa and anorexia nervosa, bulimic subtype. Biol Psychiatry 48(4):315–318

Geliebter A, Yahav EK, Gluck ME et al (2004) Gastric capacity, test meal intake, and appetitive hormones in binge eating disorder. Physiol Behav 81(5):735–740

Geliebter A, Gluck ME, Hashim SA (2005) Plasma ghrelin concentrations are lower in binge-eating disorder. J Nutr 135(5):1326–1330

Geliebter A, Hashim SA, Gluck ME (2008a) Appetite-related gut peptides, ghrelin, PYY, and GLP-1 in obese women with and without binge eating disorder (BED). Physiol Behav 94(5):696–699

Geliebter A, Ochner CN, Aviram-Friedman R (2008b) Appetite-related gut peptides in obesity and binge eating disorder. Am J Lifestyle Med 2(4):305–314

Gendall KA, Kaye WH, Altemus M et al (1999) Leptin, neuropeptide Y, and peptide YY in long-term recovered eating disorder patients. Biol Psychiatry 46 (2):292–299

Geracioti TD Jr, Liddle RA, Altemus M et al (1992) Regulation of appetite and cholecystokinin secretion in anorexia nervosa. Am J Psychiatry 149(7):958–961

Germain N, Galusca B, Grouselle D et al (2010) Ghrelin and obestatin circadian levels differentiate bingeing-purging from restrictive anorexia nervosa. J Clin Endocrinol Metab 95(6):3057–3062

Giel K, Zipfel S, Hallschmid M (2018) Oxytocin and eating disorders: a narrative review on emerging findings and perspectives. Curr Neuropharmacol 16 (8):1111–1121

Gluck ME (2006) Stress response and binge eating disorder. Appetite 46(1):26–30

Gluck ME, Geliebter A et al (2004a) Cortisol, hunger, and desire to binge eat following a cold stress test in obese women with binge eating disorder. Psychosom Med 66 (6):876–881

Gluck ME, Geliebter A et al (2004b) Cortisol stress response is positively correlated with central obesity in obese women with binge eating disorder (BED) before and after cognitive-behavioral treatment. Ann N Y Acad Sci 1032:202–207

Gluck ME, Yahav E, Hashim SA (2014) Ghrelin levels after a cold pressor stress test in obese women with binge eating disorder. Psychosom Med 76(1):74–79

Havel PJ (2002) Control of energy homeostasis and insulin action by adipocyte hormones: leptin, acylation stimulating protein, and adiponectin. Curr Opin Lipidol 13(1):51–59

Hernandez D, Mehta N, Geliebter A (2019) Meal-related acyl and des-acyl ghrelin and other appetite-related hormones in people with obesity and binge eating. Obesity (Silver Spring) 27(4):629–635

Housova J, Anderlova K, Krizová J et al (2005) Serum adiponectin and resistin concentrations in patients with restrictive and binge/purge form of anorexia nervosa and bulimia nervosa. J Clin Endocrinol Metab 90 (3):1366–1370

Kaye WH, Berrettini W, Gwirtsman H et al (1990) Altered cerebrospinal fluid neuropeptide Y and peptide YY immunoreactivity in anorexia and bulimia nervosa. Arch Gen Psychiatry 47(6):548–556

Keel PK, Wolfe BE, Liddle RA et al (2007) Clinical features and physiological response to a test meal in purging disorder and bulimia nervosa. Arch Gen Psychiatry 64(9):1058–1066

Kemnitz JW, Gibber JR, Lindsay KA et al (1989) Effects of ovarian hormones on eating behaviors, body weight, and glucoregulation in rhesus monkeys. Horm Behav 23(2):235–250

Kernie SG, Liebl DJ, Parada LF (2000) BDNF regulates eating behavior and locomotor activity in mice. EMBO J 19(6):1290–1300

Kim YR, Kim JH, Kim CH et al (2015a) Association between the oxytocin receptor gene polymorphism (rs53576) and bulimia nervosa. Eur Eat Disord Rev 23(3):171–178

Kim YR, Eom JS, Yang JW et al (2015b) The impact of oxytocin on food intake and emotion recognition in patients with eating disorders: a double blind single dose within-subject cross-over design. PLoS One 10 (9):e0137514

Kissileff HR, Pi-Sunyer FX, Thornton J (1981) C-terminal octapeptide of cholecystokinin decreases food intake in man. Am J Clin Nutr 34(2):154–160

Klump KL, Keel PK, Culbert KM et al (2008) Ovarian hormones and binge eating: exploring associations in community samples. Psychol Med 38(12):1749–1757

Klump KL, Keel PK, Sisk C et al (2010) Preliminary evidence that estradiol moderates genetic influences on disordered eating attitudes and behaviors during puberty. Psychol Med 40(10):1745–1753

Klump KL, Keel PK, Racine SE et al (2013a) The interactive effects of estrogen and progesterone on changes in emotional eating across the menstrual cycle. J Abnorm Psychol 122(1):131–137

Klump KL, Keel PK, Burt SA et al (2013b) Ovarian hormones and emotional eating associations across the menstrual cycle: an examination of the potential moderating effects of body mass index and dietary restraint. Int J Eat Disord 46(3):256–263

Klump KL, Racine SE, Hildebrandt B et al (2014) Ovarian hormone influences on dysregulated eating: a comparison of associations in women with versus without binge episodes. Clin Psychol Sci 2(4):545–559

Klump KL, Culbert KM, O'Connor S et al (2017) The significant effects of puberty on the genetic diathesis of binge eating in girls. Int J Eat Disord 50(8):984–989

Kojima S, Nakahara T, Nagai N et al (2005) Altered ghrelin and peptide YY responses to meals in bulimia nervosa. Clin Endocrinol 62(1):74–78

Koo-Loeb JH, Pedersen C, Girdler SS (1998) Blunted cardiovascular and catecholamine stress reactivity in women with bulimia nervosa. Psychiatry Res 80 (1):13–27

Koo-Loeb JH, Costello N, Light KC et al (2000) Women with eating disorder tendencies display altered cardiovascular, neuroendocrine, and psychosocial profiles. Psychosom Med 62(4):539–548

Laessle RG, Schulz S (2009) Stress-induced laboratory eating behavior in obese women with binge eating disorder. Int J Eat Disord 42(6):505–510

Larsen JK, van Ramshorst B, van Doornen LJ et al (2009) Salivary cortisol and binge eating disorder in obese women after surgery for morbid obesity. Int J Behav Med 16(4):311–315

Lavagnino L, Amianto F, Parasiliti Caprino M et al (2013) Urinary cortisol and psychopathology in obese binge eating subjects. Appetite 83:112–116

Lee DR, Semba R, Kondo H et al (1999) Decrease in the levels of NGF and BDNF in brains of mice fed a tryptophan-deficient diet. Biosci Biotechnol Biochem 63(2):337–340

Leehr EJ, Krohmer K, Schag K et al (2015) Emotion regulation model in binge eating disorder and obesity—a systematic review. Neurosci Biobehav Rev 49:125–134

Levine MD, Marcus MD (1997) Eating behavior following stress in women with and without bulimic symptoms. Ann Behav Med 19(2):132–138

Lyons WE, Mamounas LA, Ricaurte GA et al (1999) Brain-derived neurotrophic factor-deficient mice develop aggressiveness and hyperphagia in conjunction with brain serotonergic abnormalities. Proc Natl Acad Sci USA 96(26):15239–15244

Martins C, Robertson MD, Morgan LM (2010) Impact of restraint and disinhibition on PYY plasma levels and subjective feelings of appetite. Appetite 55(2):208–213

Matias I, Bisogno T, Di Marzo V (2006) Endogenous cannabinoids in the brain and peripheral tissues: regulation of their levels and control of food intake. Int J Obes (Lond) 30(Suppl 1):S7–S12

Mercader JM, Fernández-Aranda F, Gratacòs M et al (2007) Blood levels of brain-derived neurotrophic factor correlate with several psychopathological symptoms in anorexia nervosa patients. Neuropsychobiology 56(4):185–190

Micali N, Crous-Bou M, Treasure J et al (2017) Association between oxytocin receptor genotype, maternal care, and eating disorder behaviours in a community sample of women. Eur Eat Disord Rev 25(1):19–25

Monteleone P, Maj M (2013) Dysfunctions of leptin, ghrelin, BDNF and endocannabinoids in eating disorders: beyond the homeostatic control of food intake. Psychoneuroendocrinology 38(3):312–330

Monteleone P, Di Lieto A, Tortorella A et al (2000a) Circulating leptin in patients with anorexia nervosa, bulimia nervosa or binge-eating disorder: relationship to body weight, eating patterns, psychopathology and endocrine changes. Psychiatry Res 94(2):121–129

Monteleone P, Bortolotti F, Fabrazzo M et al (2000b) Plamsa leptin response to acute fasting and refeeding in untreated women with bulimia nervosa. J Clin Endocrinol Metab 85(7):2499–2503

Monteleone P, Martiadis V, Colurcio B et al (2002a) Leptin secretion is related to chronicity and severity of the illness in bulimia nervosa. Psychosom Med 64 (6):874–879

Monteleone P, Fabrazzo M, Tortorella A et al (2002b) Opposite modifications in circulating leptin and soluble leptin receptor across the eating disorder spectrum. Mol Psychiatry 7(6):641–646

Monteleone P, Luisi M, De Filippis G et al (2003a) Circulating levels of neuroactive steroids in patients with binge eating disorder: a comparison with nonobese healthy controls and non-binge eating obese subjects. Int J Eat Disord 34(4):432–440

Monteleone P, Martiadis V, Fabrazzo M et al (2003b) Ghrelin and leptin responses to food ingestion in bulimia nervosa: implications for binge-eating and compensatory behaviours. Psychol Med 33 (8):1387–1394

Monteleone P, Fabrazzo M, Martiadis V et al (2003c) Opposite changes in circulating adiponectin in women with bulimia nervosa or binge eating disorder. J Clin Endocrinol Metab 88(11):5387–5391

Monteleone P, Fabrazzo M, Tortorella A (2005a) Circulating ghrelin is decreased in non-obese and obese women with binge eating disorder as well as in obese non-binge eating women, but not in patients with bulimia nervosa. Psychoneuroendocrinology 30 (3):243–250

Monteleone P, Martiadis V, Rigamonti AE et al (2005b) Investigation of peptide YY and ghrelin responses to a test meal in bulimia nervosa. Biol Psychiatry 57 (8):926–931

Monteleone P, Fabrazzo M, Martiadis V et al (2005c) Circulating brain-derived neurotrophic factor is decreased in women with anorexia and bulimia nervosa but not in women with binge-eating disorder:

relationships to co-morbid depression, psychopathology and hormonal variables. Psychol Med 35(6):897–905

Monteleone P, Matias I, Martiadis V et al (2005d) Blood levels of the endocannabinoid anandamide are increased in anorexia nervosa and in binge-eating disorder, but not in bulimia nervosa. Neuropsychopharmacology 30(6):1216–1221

Monteleone P, Zanardini R, Tortorella A et al (2006) The 196G/A (val66met) polymorphism of the BDNF gene is significantly associated with binge eating behavior in women with bulimia nervosa or binge eating disorder. Neurosci Lett 406(1–2):133–137

Monteleone P, Serritella C, Martiadis V (2008a) Plasma obestatin, ghrelin, and ghrelin/obestatin ratio are increased in underweight patients with anorexia nervosa but not in symptomatic patients with bulimia nervosa. J Clin Endocrinol Metab 93(11):4418–4421

Monteleone P, Serritella C, Martiadis V et al (2008b) Deranged secretion of ghrelin and obestatin in the cephalic phase of vagal stimulation in women with anorexia nervosa. Biol Psychiatry 64(11):1005–1008

Monteleone P, Castaldo E, Maj M (2008c) Neuroendocrine dysregulation of food intake in eating disorders. Regul Pept 149(1–3):39–50

Monteleone P, Serritella C, Scognamiglio P et al (2010) Enhanced ghrelin secretion in the cephalic phase of food ingestion in women with bulimia nervosa. Psychoneuroendocrinology 35(2):284–288

Monteleone P, Tortorella A, Scognamiglio P et al (2012a) The acute salivary ghrelin response to a psychosocial stress is enhanced in symptomatic patients with bulimia nervosa: a pilot study. Neuropsychobiology 66(4):230–236

Monteleone P, Piscitelli F, Scognamiglio P et al (2012b) Hedonic eating is associated with increased peripheral levels of ghrelin and the endocannabinoid 2-arachidonoyl-glycerol in healthy humans: a pilot study. J Clin Endocrinol Metab 97(6):E917–E924

Monteleone AM, Monteleone P, Serino I et al (2015) Childhood trauma and cortisol awakening response in symptomatic patients with anorexia nervosa and bulimia nervosa. Int J Eat Disord 48(6):615–621

Monteleone AM, Scognamiglio P, Volpe U et al (2016a) Investigation of oxytocin secretion in anorexia nervosa and bulimia nervosa: relationships to temperament personality dimensions. Eur Eat Disord Rev 24(1):52–56

Monteleone AM, Di Marzo V, Monteleone P et al (2016b) Responses of peripheral endocannabinoids and endocannabinoid-related compounds to hedonic eating in obesity. Eur J Nutr 55(4):1799–1805

Monteleone AM, Monteleone P, Marciello F et al (2017a) Differences in cortisol awakening response between binge-purging and restrictive patients with anorexia nervosa. Eur Eat Disord Rev 25(1):13–18

Monteleone AM, Piscitelli F, Dalle Grave R et al (2017b) Peripheral endocannabinoid responses to hedonic eating in binge-eating disorder. Nutrients 9(12):E1377

Naessén S, Carlström K, Holst JJ (2011) Women with bulimia nervosa exhibit attenuated secretion of glucagon-like peptide 1, pancreatic polypeptide, and insulin in response to a meal. Am J Clin Nutr 94 (4):967–972

Nakazato M, Hashimoto K, Shimizu E et al (2003) Decreased levels of serum brain-derived neurotrophic factor in female patients with eating disorders. Biol Psychiatry 54(4):485–490

Nemeroff C, Heim C, Owens M et al (1999) Neurochemical mechanisms underlying depression and anxiety disorders: the influence of early trauma. New Developments in Understanding Depression and Its Treatment. Monograph, Wyeth-Ayerst Laboratories, New York

Ostlund H, Keller E, Hurd YL (2003) Estrogen receptor gene expression in relation to neuropsychiatric disorders. Ann N Y Acad Sci 1007:54–63

Otto B, Tschöp M, Frühauf E et al (2005) Postprandial ghrelin release in anorectic patients before and after weight gain. Psychoneuroendocrinology 30(6):577–581

Pataky Z, Gasteyger C, Ziegler O et al (2013) Efficacy of rimonabant in obese patients with binge eating disorder. Exp Clin Endocrinol Diabetes 121(1):20–26

Pike KM, Wilfley D, Hilbert A et al (2006) Antecedent life events of binge-eating disorder. Psychiatry Res 142 (1):19–29

Pruessner JC, Kirschbaum C et al (2003) Two formulas for computation of the area under the curve represent measures of total hormone concentration versus time-dependent change. Psychoneuroendocrinology 28 (7):916–931

Racine SE, Keel PK, Burt SA et al (2013) Individual differences in the relationship between ovarian hormones and emotional eating across the menstrual cycle: a role for personality? Eat Behav 14(2):161–166

Rios M, Fan G, Fekete C et al (2001) Conditional deletion of brain-derived neurotrophic factor in the postnatal brain leads to obesity and hyperactivity. Mol Endocrinol 15(10):1748–1757

Rosenberg N, Bloch M, Ben Avi I et al (2013) Cortisol response and desire to binge following psychological stress: comparison between obese subjects with and without binge eating disorder. Psychiatry Res 208 (2):156–161

Rouach V, Bloch M, Rosenberg N et al (2007) The acute ghrelin response to a psychological stress challenge does not predict the post-stress urge to eat. Psychoneuroendocrinology 32(6):693–702

Scherma M, Fattore L, Satta V et al (2013) Pharmacological modulation of the endocannabinoid signalling alters binge-type eating behaviour in female rats. Br J Pharmacol 169(4):820–833

Schulz S, Laessle R et al (2011) No evidence of increased cortisol stress response in obese women with binge eating disorder. Eat Weight Disord 16(3):e209–e211

Sedlackova D, Kopeckova J, Papezova H (2012) Comparison of a high-carbohydrate and high-protein breakfast effect on plasma ghrelin, obestatin, NPY and PYY levels in women with anorexia and bulimia nervosa. Nutr Metab (Lond) 9(1):52

Skinner JA, Campbell EJ, Dayas CV et al (2019) The relationship between oxytocin, dietary intake and feeding: a systematic review and meta-analysis of studies in mice and rats. Front Neuroendocrinol 52:65–78

Spencer RL, Deak T (2017) A users guide to HPA axis research. Physiol Behav 178:43–65

Spetter MS, Hallschmid M (2017) Current findings on the role of oxytocin in the regulation of food intake. Physiol Behav 176:31–39

Tagami T, Satoh N, Usui T et al (2004) Adiponectin in anorexia nervosa and bulimia nervosa. J Clin Endocrinol Metab 89(4):1833–1837

Tanaka M, Naruo T, Nagai N et al (2003) Habitual binge/purge behavior influences circulating ghrelin levels in eating disorders. J Psychiatr Res 37(1):17–22

Tataranni PA, Larson DE, Snitker S et al (1996) Effects of glucocorticoids on energy metabolism and food intake in humans. Am J Phys 271(2 Pt 1):E317–E325

Thienel M, Fritsche A, Heinrichs M et al (2016) Oxytocin's inhibitory effect on food intake is stronger in obese than normal-weight men. Int J Obes 40 (11):1707–1714

Troisi A, Di Lorenzo G, Lega I (2005) Plasma ghrelin in anorexia, bulimia, and binge-eating disorder: relations with eating patterns and circulating concentrations of cortisol and thyroid hormones. Neuroendocrinology 81(4):259–266

Varma M, Chai JK, Meguid MM et al (1999) Effect of estradiol and progesterone on daily rhythm in food intake and feeding patterns in Fischer rats. Physiol Behav 68(1–2):99–107

Vreeburg SA, Hoogendijk WJ, van Pelt J et al (2009) Major depressive disorder and hypothalamic-pituitary-adrenal axis activity. Results from a large cohort study. Arch Gen Psychiatry 66(6):617–626

Wu A, Ying Z, Gomez-Pinilla F (2004) The interplay between oxidative stress and brain-derived neurotrophic factor modulates the outcome of a saturated fat diet on synaptic plasticity and cognition. Eur J Neurosci 19(7):1699–1707

Body Image Disturbance and Binge Eating

Andrea S. Hartmann, Merle Lewer, and Silja Vocks

Abstract

Body image disturbance is one of the diagnostic criteria for anorexia and bulimia nervosa, but not for binge-eating disorder (BED). Nevertheless, a disturbed body image is also present in patients with BED. In particular, the cognitive-affective component of body image disturbance, which encompasses overvaluation of shape and weight, body dissatisfaction, and dysfunctional body-related information processing, is more pronounced in individuals with BED than in weight-matched controls. There is no evidence regarding a perceptual body image disturbance, and findings on body-checking and avoidance behaviors are inconclusive with respect to whether these characteristics are associated with BED rather than with obesity. However, not only BED, but also other eating disorders, which feature binge eating behavior, i.e., bulimia nervosa and the binge eating/purging subtype of anorexia nervosa, are characterized by dysfunctions in these body image facets. In sum, body image disturbance might be a worthwhile treatment target, and initial research has illustrated the efficacy of body image-related interventions in patients with BED or binge eating behavior. However, future studies are warranted to investigate the add-on effects of body image-related interventions to existing state-of-the-art treatments and to identify subgroups of individuals with BED or binge eating who benefit the most from these treatments.

Keywords

Body image disturbance · Body size perception · Body image dissatisfaction · Body-related cognitive biases · Weight and shape concern · Checking and avoidance behavior

Learning Objectives

In this chapter, you will:

- Learn about the relevance of body image disturbance to binge eating-related psychopathology.
- Gain an overview of the different components of body image disturbance and their association with binge eating (disorder).
- Identify how body image disturbance can be tackled therapeutically.

A. S. Hartmann · S. Vocks (✉)
Section of Clinical Psychology and Psychotherapy, Institute of Psychology, Department of Human Sciences, Osnabrück University, Osnabrück, Germany
e-mail: andrea.hartmann@uni-osnabrueck.de; silja.vocks@uni-osnabrueck.de

M. Lewer
Mental Health Research and Treatment Center, Ruhr-University Bochum, Bochum, Germany
e-mail: merle.lewer@ruhr-uni-bochum.de

© Springer Nature Switzerland AG 2020
G. K.W. Frank, L. A. Berner (eds.), *Binge Eating*, https://doi.org/10.1007/978-3-030-43562-2_13

1   Introduction

Body image disturbance is a hallmark feature (Hrabosky et al. 2009) and a diagnostic criterion of anorexia (AN) and bulimia nervosa (BN) in the International Classification of Diseases (ICD-10) and the Diagnostic and Statistical Manual of Mental Disorders (DSM-5), and is described as a distorted perception of one's own body or undue influence of shape and weight on self-evaluation (American Psychiatric Association 2013; World Health Organization 2016; Zanetti et al. 2013). Moreover, body image disturbance has been a treatment target in interventions for these forms of eating disorders (e.g., Vocks et al. 2018). With the introduction of the DSM-5 (American Psychiatric Association 2013), binge-eating disorder (BED) has become a diagnostic entity in its own right, within the eating disorder category. However, body image disturbance has not been introduced in the diagnostic guidelines for BED (American Psychiatric Association 2013), despite discussions about its inclusion as a diagnostic specifier (Grilo 2013). Indeed, a recent review indicated that characteristic aspects of a body image disturbance might also be present in individuals with BED (e.g., Lewer et al. 2017). Its relevance as a risk factor for the etiology of binge eating in youth (Stice et al. 2002), along with the identification of overvaluation of shape and weight as unique predictors for distress and impairment in BED (Mitchison et al. 2017a, b), underscores the need for research regarding body image in BED. Such information might assist in the decision whether to introduce therapeutic techniques targeting body image disturbance in patients with BED (e.g., Lewer et al. 2018). This chapter aims to summarize the current state of research regarding the phenomenology of body image disturbance in binge eating, with a particular focus on BED, as well as evidence pertaining to the effectiveness of interventions that target it.

2   A Definition of Body Image Disturbance

Body image is a multidimensional construct (Cash and Smolak 2011), which can be defined as "... *the picture we have in our minds of the size, shape and form of our bodies; and our feelings concerning these characteristics (...)*" (Slade 1988, p. 20). Four different components of a disturbed body image have been proposed in eating disorders: First, the perceptual component involves a distorted perception of one's own body, i.e., perceiving one's body as bigger than it actually is (Vocks et al. 2007). Second, the affective component refers to negative body-related emotions such as shame, anxiety, or disgust (Hilbert et al. 2002; Kraus et al. 2015; Tuschen-Caffier et al. 2003). Third, the cognitive component describes dysfunctional body-related information processing (Williamson et al. 2004). This may consist of memory biases that present themselves as facilitated encoding and retrieval of disorder-salient information (e.g., Johansson et al. 2008), attentional biases that are defined as selective visual processing of specific body parts (Bauer et al. 2017), interpretation biases that manifest as distorted processing of ambiguous information (e.g., Korn et al. 2019), or identification biases, according to which the evaluation of bodies differs depending on whether the rater believes the body to belong to herself or to another person (Voges et al. 2019). Often, the cognitive and affective components of body image are combined into a cognitive-affective or evaluative component, consisting of attitudes toward one's body and associated emotions, body dissatisfaction, or concerns with shape and/or weight (Cash and Smolak 2011). The latter has also been termed overvaluation of shape and weight, indicating that these concerns with shape and weight might have an undue influence on self-evaluation. And, fourth, the phenomenology of a body image disturbance also includes a behavioral component. This describes avoidance behavior, e.g., hiding one's body by wearing baggy clothing, not looking in the mirror, or not visiting swimming pools (Nikodijevic et al. 2018; Williamson et al. 2004), and body-checking behavior, e.g., examining one's own body weight and shape in the mirror or on the scales, or pinching parts of the body (Nikodijevic et al. 2018).

> Body image disturbance is a multidimensional construct comprising the perception of one's body, body-related emotions, attitudes, and information processing, as well as behaviors to avoid, check, and invest in one's body.

3 Perceptual Component of Body Image Disturbance

To date, four studies have examined body image perception in BED (Legenbauer et al. 2011; Lewer et al. 2015; Nicoli and Liberatore Junior 2011; Sorbara and Geliebter 2002). In the study by Sorbara and Geliebter (2002), participants with obesity with and without a comorbid BED completed the Stunkard Figure Rating Scale (Stunkard et al. 1983). The results revealed no difference in body image distortion between the two groups. This finding was corroborated in more recent studies employing the photo distortion technique in comparable groups (Legenbauer et al. 2011; Lewer et al. 2015), which reported that individuals with and without BED tended to estimate their body size rather realistically. Again using a figure rating scale, the fourth study found that individuals with BED showed a higher self-image inadequacy than did individuals without BED (Nicoli and Liberatore Junior 2011). This latter study, however, did not assess BMI, and diagnoses were not derived from expert clinical interviews, thus limiting the generalizability of the findings. In contrast, studies on eating disorders other than BED, which are characterized by binge eating behavior, i.e., BN and AN, show evidence of a strong

overestimation of body size (for an overview: Mölbert et al. 2017). Notably, it was not possible to determine the association between binge eating behavior and body image disturbance, as the binge-purge subtype of AN (AN-BP) was not examined separately from the restrictive subtype. Interestingly, this review highlights that distorted body size estimations are caused by body-related attitudes, which influence body perception, rather than by visual perception deficits. In sum, there is no evidence of a body image disturbance on the perceptual level in BED or other disorders characterized by binge eating behavior, and specific studies in youth are lacking.

> There is no evidence of a body image disturbance on the perceptual level in individuals with BED.

4 Cognitive-Affective Component of Body Image Disturbance

Most of the research on body image disturbance in BED has focused on the affective component, namely body dissatisfaction. For the most part, this has been assessed using the Body Dissatisfaction subscale of the Eating Disorder Inventory (EDI; Garner 1991). These studies have compared individuals with BED to individuals with BN and/or nonclinical individuals with obesity. Some studies revealed no differences between individuals with BED and those with obesity who were not suffering from an eating disorder (Kuehnel and Wadden 1994; de Zwaan et al. 1994; Lewer et al. 2015; Raymond et al. 1995), which might be interpreted as indicating an association between overweight/obesity and body dissatisfaction, irrespective of the binge eating pathology per se. However, other studies demonstrated differences between individuals with overweight or obesity with and without a comorbid BED, indicating greater body dissatisfaction in those with BED as compared to those without BED (Hilbert et al. 2002; Legenbauer et al. 2011; Lloyd-Richardson et al. 2000; Mussell et al. 1996; Sorbara and Geliebter 2002; Vinai et al. 2014). Further hints toward an association between binge eating and body dissatisfaction stem from a study that reported equal levels of body dissatisfaction in patients with BED and BN (Barry et al. 2003). This latter finding was corroborated by the only study to date to have examined body dissatisfaction using a method other than classic self-report with default response options, namely the think-aloud task during mirror exposure (Hilbert and Tuschen-Caffier 2005). The study found that individuals with BED and BN did not differ in their negative body-related thoughts about their own bodies, but both groups showed a higher degree of negative thoughts about their bodies than did normal-weight healthy controls. Another experimental study (Naumann et al. 2018) used stress induction in combination with mirror exposure to examine the impact of psychological distress on body dissatisfaction in women with overweight/obesity and BED as compared to a weight-matched healthy control group. The authors reported that body dissatisfaction was significantly higher in the BED group than in the control group. Additionally, while body dissatisfaction increased after mirror exposure in both the BED and the control group, this change was greater in the BED group.

In the two eating disorders also characterized by binge eating, namely AN-BP and BN, there is also abundant evidence of higher body dissatisfaction compared to healthy controls. In fact, patients with AN and BN show comparable degrees of body dissatisfaction (e.g., Hrabosky et al. 2009). However, as studies have not distinguished between AN subtypes, it has not been possible to identify the binge eating-associated part of body dissatisfaction in AN. Furthermore, both AN and BN are characterized by other core symptoms that are representative for the disorders, besides binge eating. Another study, which focused more closely on binge eating, examined a group of adults with loss of control (LOC) eating, and found that body dissatisfaction positively predicted LOC eating (Berner et al. 2015). Body dissatisfaction may, therefore, have

relevance to binge eating, both transdiagnostically and at subthreshold levels.

Overvaluation of shape and weight is a broader concept than body dissatisfaction, as it encompasses not only cognitions and emotions regarding one's own weight and shape, but also their impact on self-esteem and self-worth (Grilo et al. 2010a; Mond and Hay 2011). A multitude of studies have looked at this concept in individuals with BED, most of which employed the Weight concern and Shape concern subscales from the Eating Disorder Examination interview (EDE; Fairburn et al. 2014) and the self-report Eating Disorder Examination Questionnaire (EDE-Q; Berg et al. 2012). Overall, these studies demonstrate that individuals with BED show a greater overvaluation of shape and weight than individuals without BED in the same age range (Allison et al. 2005; Eldredge and Agras 1996; Grilo et al. 2010b; Hilbert and Tuschen-Caffier 2005; Spitzer et al. 1993; Lewer et al. 2015; Naumann et al. 2013; Nauta et al. 2000; Wilfley et al. 2000; Yiu et al. 2017). Furthermore, it was shown that individuals with BED do not differ from those with BN regarding their increased overvaluation of shape and weight (Grilo et al. 2010b; Hilbert and Tuschen-Caffier 2005; Wilfley et al. 2000). Additionally, Goldschmidt et al. (2011) illustrated that the overvaluation of shape and weight is independent of the individual's actual weight, as participants with BED of normal weight and those with overweight did not differ in their scores on these subscales.

Body dissatisfaction and overvaluation of shape and weight are the only components of body image disturbance that have been assessed in studies that examined overeating, loss of control eating, or binge eating in youth who do not meet the diagnostic criteria for BED. Four studies in different age groups found elevated body dissatisfaction (Goldschmidt et al. 2015) and overvaluation of shape and weight (Goossens et al. 2007; Tanofsky-Kraff et al. 2004; Vannucci et al. 2013) in individuals with overeating or loss of control eating compared to those without these symptoms. Three studies corroborated these findings on a dimensional level, showing that body dissatisfaction was significantly related to binge eating in girls (Mitchison et al. 2017a; Sonneville et al. 2015). In boys, the two concepts were related, but only through a mediation by overvaluation of weight and shape (Mitchison et al. 2017a). Moreover, overvaluation of shape and weight was significantly associated with the frequency of binges in youth of both genders (Goldschmidt et al. 2010; Hilbert and Czaja 2009; Mitchison et al. 2017a).

Only three studies, all by the same research group, have targeted a further concept of the cognitive component of body image disturbance in BED, namely dysfunctional information processing (Svaldi et al. 2010, 2011, 2012). Two of these studies used eye tracking to assess attentional bias. In the first study, BMI-matched individuals with obesity with and without BED were presented with pictures of themselves and another person in random order. While both groups exhibited an attentional bias toward subjectively unattractive body areas, individuals with BED looked longer and more often at such body areas, especially when looking at their own body (Svaldi et al. 2011). In the second study (Svaldi et al. 2012), individuals with BED and weight-matched controls were concurrently presented with a picture of themselves and another person. The trials were presented in two conditions: In the cue condition, participants were informed where the picture of their own body would appear on the screen, and in the no-cue condition they received no such information. In the cue condition, the group with BED showed a higher gaze frequency for own-body pictures and a lower gaze frequency for other body pictures. This effect was not found in the no-cue condition, suggesting that the cue activated body-related schemata, resulting in an attentional bias shown by individuals with BED. In addition to these two studies using eye tracking to assess an attentional bias, in the third study, memory bias was investigated using a recall task (Svaldi et al. 2010). Women with overweight with and without BED were presented with positively and negatively valenced body-related or neutral words, and were asked to recall them after a five-minute break. While both groups retrieved negative body-related words more easily than neutral words, the BED group recalled

significantly fewer positively valenced body-related words. This finding was supported by a study in a nonclinical sample of young women, which found a positive association between memory biases and binge eating episodes (Cooper and Wade 2015).

These experimental results provide first indications of attentional and memory bias in BED. There is already plenty of evidence regarding body-related attentional biases in AN and BN (for a review: Kerr-Gaffney et al. 2018). Notably, only one of these studies analyzed eating disorder subtypes separately (Bauer et al. 2017), thus enabling a distinction not only between AN and BN but also between AN subtypes. The study found that the AN-restrictive group showed a more pronounced attentional bias toward subjectively unattractive body parts than the AN-BP and BN groups. These findings were supported by a study indicating that attention to disorder-specific threat words differs between the AN subtypes (Gilon Mann et al. 2018). Dimensionally speaking, a recent review concluded that the combination of attentional biases toward disorder-specific cues (i.e., food and weight/shape) and attentional biases toward threat may be a potent contributor to binge eating behavior (Stojek et al. 2018). In terms of memory bias, studies show evidence for body-related memory biases in AN (both subtypes) and BN (Legenbauer et al. 2010; Tekcan et al. 2008; Hermans et al. 1998; Johansson et al. 2008). To our knowledge, the only study to have looked into this finding from a dimensional perspective, as mentioned above (Cooper and Wade 2015), indicated that memory biases might be positively related to the extent of binge eating behavior.

In sum, individuals with BED seem to be less satisfied with their bodies than weight-matched individuals without BED, and show comparable levels of body dissatisfaction to individuals with BN. These latter findings also apply to the overvaluation of shape and weight. Both individuals with AN (irrespective of subtype) and BN are more dissatisfied with their bodies than healthy controls. Moreover, there is evidence that elevated levels of body dissatisfaction and overvaluation of shape and weight are also present and associated with symptom severity in youth with overeating or loss of control eating. Furthermore, research on body-related information processing suggests a more pronounced attentional bias toward pictures of one's own body, and toward subjectively unattractive areas thereof, in females with obesity and BED compared to females with obesity only, and in other eating disorders characterized by binge eating behavior compared to healthy controls. Furthermore, findings hint at a memory bias for negative body-related information and for less positive body-related information in BED, AN, and BN.

> There is initial evidence that individuals with BED might show dysfunctional body-related information processing in terms of attention and memory as well as evidence for elevated body dissatisfaction and overconcern with shape and weight.

5   Behavioral Component of Body Image Disturbance

Three studies have examined the behavioral component of body image disturbance in individuals with BED. In the first, Reas et al. (2005) found that the majority of participants with BED showed body-checking and avoidance behavior as measured by the Body Shape Questionnaire (Cooper et al. 1987). As the study did not include a control group, however, no conclusions can be drawn regarding whether these individuals displayed such behavior to a higher extent than healthy individuals. A more recent study, by Lewer et al. (2015), indicated that individuals with obesity with and without BED did not differ in terms of body-checking or avoidance behaviors assessed via self-report in the Body Checking Questionnaire (BCQ; Reas et al. 2002) and the Body Image Avoidance Questionnaire (BIAQ; Rosen et al. 1991). Taken together, these two studies might hint at an attribution of these behaviors to obesity rather than to BED per se. However, this is in contrast to a recent study

by Legenbauer et al. (2017), who found that individuals with BED reported more body avoidance behaviors than those with other eating disorders or healthy controls, but almost no body-related checking behavior. In sum, evidence regarding the behavioral component of body image disturbance in BED is conflicting. Potentially, however, characteristic behaviors in this population might be attributable to obesity rather than to BED. Plenty of studies have concluded that both body checking and avoidance are higher in patients with eating disorders in general compared to healthy controls, without differences between eating disorder diagnoses (Nikodijevic et al. 2018). While this meta-analysis confirmed that there are strong associations between the behaviors and core eating pathology, the latter centered around body image, restraint, and total eating disorder symptom scores, no conclusion regarding binge eating is justified. Studies investigating the behavioral component focusing on youth with loss of control eating or binge eating are lacking.

> The evidence regarding body image disturbance on the behavioral level is conflicting. Associations of respective behaviors with obesity or BED cannot yet be conclusively determined.

6   Interventions Related to Body Image Disturbance

While psychological treatment in general, and cognitive-behavioral and interpersonal therapy in particular, is promising for BED, there is still room for improvement (see Hilbert et al. 2019 for an overview). Given the existence of body image disturbance, especially with regard to the cognitive-affective component, it might be a worthwhile target for expanding existing effective therapies for BED (Hilbert et al. 2019). This suggestion is underscored by the strong impact of body image-focused interventions on recovery and maintenance (Delinsky and Wilson 2006;

Vocks et al. 2008, 2011), although none of these studies specifically looked at the reduction of binge eating behavior as a consequence of body image therapy.

So far, only four studies have evaluated body image-related interventions for BED (Hilbert et al. 2002; Hilbert and Tuschen-Caffier 2004; Hildebrandt et al. 2012; Lewer et al. 2018). In a study by Hilbert et al. (2002), a two-session guided exposure led to a reduction of negative body-related cognitions and emotions as well as an increase in appearance self-esteem. The same work group also showed not only that body exposure seems to be successful in BED, but also that body-related cognitive interventions achieve the same goal (Hilbert and Tuschen-Caffier 2004). In a third study, 2 five-session treatments, namely acceptance-based mirror exposure and nondirective body image therapy, were compared in a mixed sample of patients with anorexia and bulimia nervosa, BED and eating disorders not otherwise specified (Hildebrandt et al. 2012). While the findings suggested a superiority of mirror exposure with regard to the reduction of body image disturbance and eating disorder symptoms, it is important to note that only about 6% of the participants suffered from BED, which leads to a limited generalizability of the findings to this patient group. In the fourth and most recent study, Lewer et al. (2018) compared the effects of a 10-session standardized group body image therapy program consisting of psychoeducation, cognitive restructuring of negative body-related thoughts, in-session exposure, and guidance regarding exposure outside of sessions, with a waiting-list control condition. The study revealed significantly larger effects in the intervention group than in the control group regarding the cognitive-affective component of body image disturbance and body checking as a behavioral facet. In sum, targeting body image disturbance in the treatment of BED seems to be effective, especially with respect to body image disturbance, while evidence for eating disorder pathology is more limited. However, it remains open which of the body image-related treatment components (e.g., exposure-based or cognitive techniques) is the most promising for BED, to what extent

such an intervention would add to a usual cognitive-behavioral treatment protocol in terms of reducing binge eating (Hilbert et al. 2019), how the diverse facets of body image disturbance are impacted by such an intervention, and which patients would particularly benefit from the incorporation of a body image-related intervention.

First studies hint at the effectiveness of body image-related interventions in individuals with BED. Add-on effects to state-of-the-art treatments, selective indication, and differential effectiveness depending on diverse facets need to be examined.

7   Conclusion

To summarize, as in AN and BN, body image disturbance does exist in BED, even though it is not explicitly mentioned as a diagnostic criterion or specifier in the DSM-5 (American Psychiatric Association 2013) or the upcoming ICD-11 (World Health Organization 2019). In particular, BED may be characterized by cognitive-affective body image disturbance. For instance, with regard to body dissatisfaction and weight and shape concerns, individuals with BED showed higher scores than individuals with obesity without BED, and comparable scores to individuals with BN. Moreover, dysfunctional information processing has been demonstrated in BED, especially with respect to attention and memory processes. In contrast to individuals with BN and AN-BP, individuals with BED seem to rate their own body size accurately, and it remains inconclusive whether they show body-checking and avoidance behaviors to a higher degree than weight-matched individuals.

As such, body image disturbance in BED, and particularly body dissatisfaction and overvaluation of weight and shape, might indeed be comparable to body image disturbance in AN and BN, potentially suggesting the utility of its inclusion in diagnostic guidelines. However, in view of preliminary evidence that only a subgroup of individuals with BED might manifest a body image disturbance (Goldschmidt et al. 2010; Grilo et al. 2009, 2010a, b; Mond et al. 2006), it was advised that it should be introduced as a specifier rather than as a diagnostic criterion or guideline (Grilo 2013). Additionally, body dissatisfaction and overvaluation of shape and weight have been shown to be risk factors for the onset of binge eating in adolescence and young adulthood (Goldschmidt et al. 2014, 2015; Neumark-Sztainer et al. 2009; Stice et al. 2002), and already manifest in childhood and adolescence (Sonneville et al. 2012; Stice et al. 2002; Tanofsky-Kraff et al. 2004). Moreover, overvaluation of shape and weight has been found to predict non-remission from binge eating after treatment (Grilo et al. 2013). Thus, body image disturbance might be etiologically relevant in BED, as it is in AN and BN (Ahrberg et al. 2011; Allison et al. 2005; Grilo et al. 2010b; Lewer et al. 2015), and it might be worthwhile to target body image disturbance in BED through specific body image-related interventions. The studies outlined above indicate preliminary support for the effectiveness of such treatments. Next, studies are warranted that examine the impact of these interventions as add-on treatments to existing protocols (e.g., cognitive-behavior therapy) on the different components of body image disturbance (e.g., the perceptual component), associated pathology (e.g., depression), and binge eating, and that analyze which patient groups benefit the most.

As is evident from the reported literature, further studies are needed to examine the behavioral component of body image disturbance, such as body checking and avoidance. Furthermore, the associations between the different components of body image disturbance remain unclear, as only a small number of studies have hinted at an association of overvaluation of shape and weight with body checking and avoidance behavior in BED (Legenbauer et al. 2011), and with other aspects of the cognitive-affective, behavioral, and perceptual components (Lewer et al. 2015). Thus, in order to elucidate these aspects, a theory-driven study based on specific models (e.g., Williamson et al. 2004) which can be adapted to BED is needed. Furthermore, studies on the associations of the single body image components with eating disorder pathology, such as the relation between body dissatisfaction and LOC eating (e.g., Berner et al. 2015) would determine the extent of the relevance of body image disturbance for BED.

References

Ahrberg M, Trojca D, Nasrawi N, Vocks S (2011) Body image disturbance in binge eating disorder: a review. Eur Eat Disord Rev 19(5):375–381. https://doi.org/10.1002/erv.1100

Allison KC, Grilo CM, Masheb RM, Stunkard AJ (2005) Binge eating disorder and night eating syndrome: a comparative study of disordered eating. J Consult Clin Psychol 73(6):1107–1115. https://doi.org/10.1037/0022-006X.73.6.1107

American Psychiatric Association (2013) Diagnostic and statistical manual of mental disorders (DSM-V), 5th edn. American Psychiatric Association, Arlington, VA

Barry DT, Grilo CM, Masheb RM (2003) Comparison of patients with bulimia nervosa, obese patients with binge eating disorder, and nonobese patients with binge eating disorder. J Nerv Ment Dis 191 (9):589–594. https://doi.org/10.1097/01.nmd.0000087185.95446.65

Bauer A, Schneider S, Waldorf M, Braks K, Huber TJ, Adolph D, Vocks S (2017) Selective visual attention towards oneself and associated state body satisfaction: an eye-tracking study in adolescents with different types of eating disorders. J Abnorm Child Psychol 45 (8):1647–1661. https://doi.org/10.1007/s10802-017-0263-z

Berg KC, Peterson CB, Frazier P, Crow SJ (2012) Psychometric evaluation of the eating disorder examination and eating disorder examination-questionnaire: a systematic review of the literature. Int J Eat Disord 45 (3):428–438. https://doi.org/10.1002/eat.20931

Berner LA, Arigo D, Mayer LE, Sarwer DB, Lowe MR (2015) Examination of central body fat deposition as a risk factor for loss-of-control eating. Am J Clin Nutr 102(4):736–744. https://doi.org/10.3945/ajcn.115.107128

Cash TF, Smolak L (2011) Body image: a handbook of science, practice, and prevention. Guilford Press, New York

Cooper JL, Wade TD (2015) The relationship between memory and interpretation biases, difficulties with emotion regulation, and disordered eating in young women. Cogn Ther Res 39(6):853–862. https://doi.org/10.1007/s10608-015-9709-1

Cooper P, Taylor M, Cooper Z, Fairburn CG (1987) The development and validation of the body shape questionnaire. Int J Eat Disord 6(4):485–494. https://doi.org/10.1002/1098-108X(198707)6:4<485::AID-EAT2260060405>3.0.CO;2-O

de Zwaan M, Mitchell JE, Seim HC, Specker SM, Pyle RL, Raymond NC, Crosby RB (1994) Eating related and general psychopathology in obese females with binge eating disorders. Int J Eat Disord 15(1):43–52. https://doi.org/10.1002/1098-108x(199401)15:1<43::aid-eat2260150106>3.0.co;2-6

Delinsky SS, Wilson GT (2006) Mirror exposure for the treatment of body image disturbance. Int J Eat Disord 39(2):108–116. https://doi.org/10.1002/eat.20207

Eldredge KL, Agras WS (1996) Weight and shape over-concern and emotional eating in binge eating disorder. Int J Eat Disord 19(1):73–82. https://doi.org/10.1002/(SICI)1098-108X(199601)19:1<73::AID-EAT9>3.0.CO;2-T

Fairburn CG, Cooper Z, O'Connor ME (2014) Eating Disorder Examination (Edition 17.0D). The Centre of Research on Eating Disorders at Oxford, Oxford, UK. https://www.credo-oxford.com/pdfs/EDE_17.0D.pdf. Accessed 18 November 2019

Garner DM (1991) Eating Disorder Inventory-2. Professional manual. Psychological Assessment Resources, Odessa, FL

Gilon Mann T, Hamdan S, Bar-Haim Y, Lazarov A, Enoch-Levy A, Dubnov-Raz G, Treasure J, Stein D (2018) Different attention bias patterns in anorexia nervosa restricting and binge/purge types. Eur Eat Disord Rev 26(4):293–301. https://doi.org/10.1002/erv.2593

Goldschmidt AB, Hilbert A, Manwaring JL, Wilfley DE, Pike KM, Fairburn CG, Striegel-Moore RH (2010) The significance of overvaluation of shape and weight in binge eating disorder. Behav Res Ther 48(3):187–193. https://doi.org/10.1016/j.brat.2009.10.008

Goldschmidt AB, Le Grange D, Powers P, Crow SJ, Hill LL, Peterson CB, Crosby RD, Mitchell JE (2011) Eating disorder symptomatology in normal-weight-vs. obese individuals with binge eating disorder. Obesity 19(7):1515–1518. https://doi.org/10.1038/oby.2011.24

Goldschmidt AB, Wall MM, Loth KA, Bucchianeri MM, Neumark-Sztainer D (2014) The course of binge eating from adolescence to young adulthood. Health Psychol 33(5):457–460. https://doi.org/10.1037/a0033508

Goldschmidt AB, Loth KA, MacLehose RF, Pisetsky EM, Berge JM, Neumark-Sztainer D (2015) Overeating with and without loss of control: associations with weight status, weight-related characteristics, and psychosocial health. Int J Eat Disord 48(8):1150–1157. https://doi.org/10.1002/eat.22465

Goossens L, Braet C, Decaluwé V (2007) Loss of control over eating in obese youngsters. Behav Res Ther 45(1):1–9. https://doi.org/10.1016/j.brat.2006.01.006

Grilo CM (2013) Why no cognitive body image feature such as overvaluation of shape/weight in the binge eating disorder diagnosis? Int J Eat Disord 46(3):208–211. https://doi.org/10.1002/eat.22082

Grilo CM, Crosby RD, Masheb RM, White MA, Peterson CB, Wonderlich SA, Engel SG, Crow SJ, Mitchell JE (2009) Overvaluation of shape and weight in binge eating disorder, bulimia nervosa, and subthreshold bulimia nervosa. Behav Res Ther 47(8):692–696. https://doi.org/10.1016/j.brat.2009.05.001

Grilo CM, Crosby RD, Peterson CB, Masheb RM, White MA, Crow SJ, Wonderlich SA, Mitchell JE (2010a) Factor structure of the eating disorder examination interview in patients with binge-eating disorder. Obes Res 18(5):977–981. https://doi.org/10.1038/oby.2009.321

Grilo CM, Masheb RM, White MA (2010b) Significance of overvaluation of shape/weight in binge-eating disorder: comparative study with overweight and bulimia nervosa. Obes Res 18(3):499–504. https://doi.org/10.1038/oby.2009.280

Grilo CM, White MA, Gueorguieva R, Wilson GT, Masheb RM (2013) Predictive significance of the overvaluation of shape/weight in obese patients with binge eating disorder: findings from a randomized controlled trial with 12-month follow-up. Psychol Med 43(6):1335–1344. https://doi.org/10.1017/S0033291712002097

Hermans D, Pieters G, Eelen P (1998) Implicit and explicit memory for shape, body weight, and food-related words in patients with anorexia nervosa and nondieting controls. J Abnorm Psychol 107(2):193–202. https://doi.org/10.1037//0021-843x.107.2.193

Hilbert A, Czaja J (2009) Binge eating in primary school children: towards a definition of clinical significance. Int J Eat Disord 42(3):235–243. https://doi.org/10.1002/eat.20622

Hilbert A, Tuschen-Caffier B (2004) Body image interventions in cognitive-behavioural therapy of binge-eating disorder: a component analysis. Behav Res Ther 42(11):1325–1339. https://doi.org/10.1016/j.brat.2003.09.001

Hilbert A, Tuschen-Caffier B (2005) Body-related cognitions in binge-eating disorder and bulimia nervosa. J Soc Clin Psychol 24(4):561–579. https://doi.org/10.1521/jscp.2005.24.4.561

Hilbert A, Tuschen-Caffier B, Vögele C (2002) Effects of prolonged and repeated body image exposure in binge-eating disorder. J Psychosom Res 52(3):137–144. https://doi.org/10.1016/S0022-3999(01)00314-2

Hilbert A, Petroff D, Herpertz S, Pietrowsky R, Tuschen-Caffier B, Vocks S, Schmidt R (2019) Meta-analysis of the efficacy of psychological and medical treatments for binge-eating disorder. J Consult Clin Psychol 87(1):91–105. https://doi.org/10.1037/ccp0000358

Hildebrandt T, Loeb K, Troupe S, Delinsky S (2012) Adjunctive mirror exposure for eating disorders: a randomized controlled pilot study. Behav Res Ther 50(12):797–804. https://doi.org/10.1016/j.brat.2012.09.004

Hrabosky JI, Cash TF, Veale D, Neziroglu F, Soll EA, Garner DM, Strachan-Kinser M, Bakke B, Clauss LJ, Phillips KA (2009) Multidimensional body image comparisons among patients with eating disorders, body dysmorphic disorder, and clinical controls: a multisite study. Body Image 6(3):155–163. https://doi.org/10.1016/j.bodyim.2009.03.001

Johansson L, Ghaderi A, Hällgren M, Andersson G (2008) Implicit memory bias for eating- and body appearance-related sentences in eating disorders: an application of Jacoby's white noise task. Cogn Behav Ther 37(3):135–145. https://doi.org/10.1080/16506070701664821

Kerr-Gaffney J, Harrison A, Tchanturia K (2018) Eye-tracking research in eating disorders: a systematic review. Int J Eat Disord. https://doi.org/10.1002/eat.22998

Korn J, Dietel F, Hartmann AS (2019) Testing the specificity of interpretation biases in women with eating disorder symptoms: an online experimental assessment. Int J Eat Disord. https://doi.org/10.1002/eat.23201

Kraus N, Lindenberg J, Zeeck A, Kosfelder J, Vocks S (2015) Immediate effects of body checking behavior on negative and positive emotions in women with eating disorders: an ecological momentary assessment approach. Eur Eat Disord Rev 23:399–407. https://doi.org/10.1002/erv.2380

Kuehnel RH, Wadden TA (1994) Binge eating disorder, weight cycling, and psychopathology. Int J Eat Disord 15(4):321–329. https://doi.org/10.1002/eat.2260150403

Legenbauer T, Maul B, Rühl I, Kleinstäuber M, Hiller W (2010) Memory bias for schema-related stimuli in individuals with bulimia nervosa. J Clin Psychol 66:302–316. https://doi.org/10.1002/jclp.20651

Legenbauer T, Vocks S, Betz S, Báguena Puigcerver MJ, Benecke A, Troje NF, Rüddel H (2011) Differences in the nature of body image disturbances between female obese individuals with versus without a comorbid binge eating disorder: an exploratory study including static and dynamic aspects of body image. Behav Modif 35(2):162–186. https://doi.org/10.1177/0145445510393478

Legenbauer T, Martin F, Blaschke A, Schwenzfeier A, Blechert J, Schnicker KBI (2017) Two sides of the same coin? A new instrument to assess body checking and avoidance behaviors in eating disorders. Body Image 21:39–46. https://doi.org/10.1016/j.bodyim.2017.02.004

Lewer M, Nasrawi N, Schroeder D, Vocks S (2015) Body image disturbance in binge eating disorder: a comparison of obese patients with and without binge eating disorder regarding the cognitive, behavioral and perceptual component of body image. Eat Weight Disord 21(1):115–125. https://doi.org/10.1007/s40519-015-0200-5

Lewer M, Bauer A, Hartmann AS, Vocks S (2017) Different facets of body image disturbance in binge eating disorder: a review. Nutrients 9(12):1294. https://doi.org/10.3390/nu9121294

Lewer M, Kosfelder J, Michalak J, Schroeder D, Nasrawi N, Vocks S (2018) Effects of a cognitive-behavioral exposure-based body image therapy for overweight females with binge eating disorder: a pilot study. Int J Eat Disord 5:43. https://doi.org/10.1186/s40337-017-0174-y

Lloyd-Richardson EE, King TK, Forsyth LH, Clark MM (2000) Body image evaluations in obese females with binge eating disorder. Eat Behav 1(2):161–171. https://doi.org/10.1016/S1471-0153(00)00016-7

Mitchison D, Hay P, Griffiths S, Murray SB, Bentley C, Gratwick-Sarll K, Harrison C, Mond J (2017a) Disentangling body image: the relative associations of overvaluation, dissatisfaction, and preoccupation with psychological distress and eating disorder behaviors in male and female adolescents. Int J Eat Disord 50:118–126. https://doi.org/10.1002/eat.22592

Mitchison D, Rieger E, Harrison C, Murray SB, Griffiths S, Mond J (2017b) Indicators of clinical significance among women in the community with binge-eating disorder symptoms: delineating the roles of binge frequency, body mass index, and overvaluation. Int J Eat Disord 51(2):165–169. https://doi.org/10.1002/eat.22812

Mölbert SC, Klein L, Thaler A, Mohler BJ, Brozzo C, Martus P, Karnath HO, Zipfel S, Giel KE (2017) Depictive and metric body size estimation in anorexia nervosa and bulimia nervosa: a systematic review and meta-analysis. Clin Psychol Rev 57:21–31. https://doi.org/10.1016/j.cpr.2017.08.005

Mond JM, Hay PJ (2011) Dissatisfaction versus overevaluation in a general population sample of women. Int J Eat Disord 44(8):721–726. https://doi.org/10.1002/eat.20878

Mond JM, Hay PJ, Rodgers B, Owen C (2006) Recurrent binge eating with and without the "undue influence of weight or shape on self-evaluation": implications for the diagnosis of binge eating disorder. Behav Res Ther 45(5):929–938. https://doi.org/10.1016/j.brat.2006.08.011

Mussell MP, Peterson CB, Weller CL, Crosby RD, de Zwaan M, Mitchell JE (1996) Differences in body image and depression among obese women with and without binge eating disorder. Obes Res 4(5):431–439. https://doi.org/10.1002/j.1550-8528.1996.tb00251.x

Naumann E, Trentowska M, Svaldi J (2013) Increased salivation to mirror exposure in women with binge eating disorder. Appetite 65:103–110. https://doi.org/10.1016/j.appet.2013.01.021

Naumann E, Svaldi J, Wyschka T, Heinrichs M, von Dawans B (2018) Stress-induced body dissatisfaction in women with binge eating disorder. J Abnorm Psychol 127(6):548–558. https://doi.org/10.1037/abn0000371

Nauta H, Hospers H, Jansen A, Kok G (2000) Cognitions in obese binge eaters and obese non-binge eaters. Cogn Ther Res 24(5):521–531. https://doi.org/10.1023/A:1005510027890

Neumark-Sztainer D, Wall M, Story M, Sherwood NE (2009) Five-year longitudinal predictive factors for disordered eating in a population-based sample of overweight adolescents: implications for prevention and treatment. Int J Eat Disord 42(7):664–672. https://doi.org/10.1002/eat.20733

Nicoli MG, Liberatore Junior RD (2011) Binge eating disorder and body image perception among university students. Eat Behav 12(4):284–288. https://doi.org/10.1016/j.eatbeh.2011.07.004

Nikodijevic A, Buck K, Fuller-Tyszkiewicz M, de Paoli T, Krug I (2018) Body checking and body avoidance in eating disorders: systematic review and meta-analysis. Eur Eat Disord Rev 26(3):159–185. https://doi.org/10.1002/erv.2585

Raymond NC, Mussell MP, Mitchell JE, de Zwaan M, Crosby RD (1995) An age-matched comparison of subjects with binge eating disorder and bulimia nervosa. Int J Eat Disord 18(2):135–143. https://doi.org/10.1002/1098-108X(199509)18:2<135::AID-EAT2260180205>3.0.CO;2-M

Reas D, Whisenhunt B, Netemeyer R, Williamson D (2002) Development of the body checking questionnaire: a self-report measure of body checking behaviors. Int J Eat Disord 31:324–333. https://doi.org/10.1002/eat.10012

Reas DL, Grilo CM, Masheb RM, Wilson GT (2005) Body checking and avoidance in overweight patients with binge eating disorder. Int J Eat Disord 37 (4):342–346. https://doi.org/10.1002/eat.20092

Rosen J, Srebnik D, Saltzberg E, Wendt S (1991) Development of a body image avoidance questionnaire. J Consult Clin Psychol 3:32–37. https://doi.org/10.1037/1040-3590.3.1.32

Slade PD (1988) Body image in anorexia nervosa. Br J Psychiatry 153(S2):20–22. https://doi.org/10.1192/S0007125000298930

Sonneville KR, Calzo JP, Horton NJ, Haines J, Austin SB, Field AE (2012) Body satisfaction, weight gain, and binge eating among overweight adolescent girls. Int J Obes 36(7):944–949. https://doi.org/10.1038/ijo.2012.68

Sonneville KR, Grilo CM, Richmond TK, Thurston IB, Jernigan M, Gianini L, Field AE (2015) Prospective association between overvaluation of weight and binge eating among overweight adolescent girls. J Adolesc Health 56(1):25–29. https://doi.org/10.1016/j.jadohealth.2014.08.017

Sorbara M, Geliebter A (2002) Body image disturbance in obese outpatients before and after weight loss in relation to race, gender, binge eating, and age of onset of obesity. Int J Eat Disord 31(4):416–423. https://doi.org/10.1002/eat.10046

Spitzer RL, Yanovski S, Wadden T, Wing R, Marcus MD, Stunkard A, Devlin M, Mitchell J, Hasin D, Horne RL (1993) Binge eating disorder: its further validation in a multisite study. Int J Eat Disord 13(2):137–153. https://doi.org/10.1002/1098-108X(199303)13:2<137::AID-EAT2260130202>3.0.CO;2-%23

Stice E, Presnell K, Spangler D (2002) Risk factors for binge eating onset in adolescent girls: a 2-year prospective investigation. Health Psychol 21(2):131–138. https://doi.org/10.1037/0278-6133.21.2.131

Stojek M, Shank LM, Vannucci A, Bongiorno DM, Nelson EE, Waters AJ, Engel SG, Boutelle KN, Pine DS, Yanovski JA, Tanofsky-Kraff M (2018) A systematic review of attentional biases in disorders involving binge eating. Appetite 123:367–389. https://doi.org/10.1016/j.appet.2018.01.019

Stunkard A, Sorensen T, Schulsinger F, Kety S (1983) Use of the Danish Adoption Register for the study of obesity and thinness. In: Kety S, Rowland LP, Sidman RL, Matthysse SW (eds) The genetics of neurological and psychiatric disorders. Raven Press, New York, pp 115–120

Svaldi J, Bender C, Tuschen-Caffier B (2010) Explicit memory bias for positively valenced body-related cues in women with binge eating disorder. J Behav Ther Exp Psychiatry 41(3):251–257. https://doi.org/10.1016/j.jbtep.2010.02.002

Svaldi J, Caffier D, Tuschen-Caffier B (2011) Attention to ugly body parts is increased in women with binge eating disorder. Psychother Psychosom 80 (3):186–188. https://doi.org/10.1159/000317538

Svaldi J, Caffier D, Tuschen-Caffier B (2012) Automatic and intentional processing of body pictures in binge eating disorder. Psychother Psychosom 81(1):52–53. https://doi.org/10.1159/000329110

Tanofsky-Kraff M, Yanovski SZ, Wilfley DE, Marmarosh C, Morgan CM, Yanovski JA (2004) Eating-disordered behaviors, body fat, and psychopathology in overweight and normal-weight children. J Consult Clin Psychol 72(1):53–61. https://doi.org/10.1037/0022-006X.72.1.53

Tekcan AI, Caglar Tas A, Topcuoglu V, Yucel B (2008) Memory bias in anorexia nervosa: evidence from directed forgetting. J Behav Ther Exp Psychiatry 39 (3):369–380. https://doi.org/10.1016/j.jbtep.2007.09.005

Tuschen-Caffier B, Vögele C, Bracht S, Hilbert A (2003) Psychological responses to body shape exposure in patients with bulimia nervosa. Behav Res Ther 41 (5):573–586. https://doi.org/10.1016/S0005-7967(02)00030-X

Vannucci A, Shomaker LB, Field SE, Sbrocco T, Stephens M, Kozlosky M, Reynolds JC, Yanovski JA, Tanofsky-Kraff M (2013) History of weight control attempts among adolescent girls with loss of control eating. Health Psychol 33(5):419–423. https://doi.org/10.1037/a0033184

Vinai P, Da Ros A, Speciale M, Gentile N, Tagliabue A, Vinai P, Bruno C, Vinai L, Studt S, Cardetti S (2014) Psychopathological characteristics of patients seeking for bariatric surgery, either affected or not by binge eating disorder following the criteria of the DSM IV TR and of the DSM 5. Eat Behav 16:1–4. https://doi.org/10.1016/j.eatbeh.2014.10.004

Vocks S, Legenbauer T, Ruddel H, Troje NF (2007) Static and dynamic body image in bulimia nervosa: mental representation of body dimensions and biological motion patterns. Int J Eat Disord 40(1):59–66. https://doi.org/10.1002/eat.20336

Vocks S, Wächter A, Wucherer M, Kosfelder J (2008) Look at yourself: can body image therapy affect the cognitive and emotional response to seeing oneself in the mirror in eating disorders? Eur Eat Disord Rev 16 (2):147–154. https://doi.org/10.1002/erv.825

Vocks S, Schulte D, Busch M, Grönemeyer D, Herpertz S, Suchan B (2011) Changes in neuronal correlates of body image processing by means of cognitive-behavioural body image therapy for eating disorders: a randomized controlled fMRI study. Psychol Med 41(8):1651–1663. https://doi.org/10.1017/S0033291710002382

Vocks S, Bauer A, Legenbauer T (2018) Körperbildtherapie bei Anorexia und Bulimia nervosa: Ein kognitiv-verhaltenstherapeutisches Behandlungsprogramm. Hogrefe, Göttingen

Voges MM, Giabbiconi CM, Schöne B, Braks K, Huber TJ, Waldorf M, Hartmann AS, Vocks S (2019) Double standards in body evaluation? How identifying with a body stimulus influences ratings in women with anorexia nervosa and bulimia nervosa. Int J Eat Disord 51(11):1223–1232. https://doi.org/10.1002/eat.22967

Wilfley DE, Schwartz MB, Spurrell EB, Fairburn CG (2000) Using the eating disorder examination to identify the specific psychopathology of binge eating disorder. Int J Eat Disord 27(3):259–269. https://doi.org/10.1002/(SICI)1098-108X(200004)27:3<259::AID-EAT2>3.0.CO;2-G

Williamson DA, White MA, York-Crowe E, Stewart TM (2004) Cognitive-behavioral theories of eating disorders. Behav Modif 28(6):711–738. https://doi.org/10.1177/0145445503259853

World Health Organization (2016) International statistical classification of diseases and related health problems - 10th Revision (ICD-10), 5th edn. World Health Organization, Geneva

World Health Organization (2019) International statistical classification of diseases and related health problems - 11th Revision (ICD-11), 5th edn. World Health Organization, Geneva

Yiu A, Murray SM, Arlt JM, Eneva KT, Chen EY (2017) The importance of body image concerns in overweight and normal weight individuals with binge eating disorder. Body Image 22:6–12. https://doi.org/10.1016/j.bodyim.2017.04.005

Zanetti T, Santonastaso P, Sgaravatti E, Degortes D, Favaro A (2013) Clinical and temperamental correlates of body image disturbance in eating disorders. Eur Eat Disord Rev 21(1):32–37. https://doi.org/10.1002/erv.2190

Food Addiction, Binge Eating, and the Role of Dietary Restraint: Converging Evidence from Animal and Human Studies

David A. Wiss and Nicole M. Avena

Abstract

With emerging evidence of a biological basis to binge eating, questions about the role of food addiction (FA) have stimulated scholarly debate. A major criticism of the FA construct is its failure to account for dietary restraint and weight suppression, known contributors to binge eating. In this chapter, we examine animal and human models of addiction-like eating in the context of binge eating. Overlapping mechanisms such as reward dysfunction, craving, impulsivity, and attentional bias from animal and human studies are discussed. Directionality of the binge eating cascade is explored across different theoretical models with empirical support for multiple pathways. We offer a "Diet Drives the Binge" theory of food addiction, and a "Food Environment Drives Addiction" theory of binge eating. While FA research highlights the neurobiological vulnerability of certain people, there is less consensus about effective interventions at the individual level. We discuss current controversies surrounding FA and important findings that may have public health implications.

Keywords

Binge-eating disorder · Bulimia nervosa · Food addiction · Obesity · Dietary restraint · Sugar · Preclinical · Neuroimaging · Eating disorders · Substance use disorders

Learning Objectives

After reading this chapter, you will be able to:

1. Explore overlapping characteristics between substance use disorders and binge-like eating behaviors.
2. Better understand the historical basis of sugar addiction as shown by animal models.
3. Review human neuroimaging data that identify various brain regions implicated in both drug and food addiction.
4. Examine the links between highly palatable food consumption and food addiction, binge eating, and obesity.
5. Discuss divergent explanatory theories to conceptualize directionality between dietary restraint, binge eating, and addiction-like eating.

D. A. Wiss
Fielding School of Public Health, University of California Los Angeles, Los Angeles, CA, USA

N. M. Avena (✉)
Department of Neuroscience, Icahn School of Medicine at Mount Sinai, New York, NY, USA

Department of Psychology, Princeton University, Princeton, NJ, USA
e-mail: nicole.avena@mssm.edu

© Springer Nature Switzerland AG 2020
G. K.W. Frank, L. A. Berner (eds.), *Binge Eating*, https://doi.org/10.1007/978-3-030-43562-2_14

1   Introduction

The field of eating disorders (EDs) has evolved in recent years, in large part due to a broader understanding of their natural history. We now know there is a biological basis to disordered eating, yet the neurochemical pathways are not fully understood. Considerable controversy remains regarding treatment and policy implications of recent neuroscience advances in both animal and human research. Binge-eating disorder (BED) is recognized as a distinct clinical entity included in the Diagnostic and Statistical Manual of Mental Disorders (DSM-5) since 2013. This has stimulated several novel research questions in the fields of neuroscience, psychology, nutrition, and public health. Accepted criteria for BED include recurrent episodes of eating abnormally large amounts of food while experiencing a loss of control with accompanying emotional distress, absent the compensatory behaviors observed in bulimia nervosa (BN).

Prevalence estimates suggest BED is more common than anorexia nervosa (AN) and BN combined (Hudson et al. 2007). Nationally representative data have shown that all EDs are more common in women than men (Udo and Grilo 2018) and occur more frequently in higher-weight individuals (Duncan et al. 2017). Recent data suggest that while BED has a similar prevalence across genders, men are less likely to seek help (Thapliyal et al. 2018). Early research on EDs has shown that BED has a later age onset than other EDs (Fairburn and Harrison 2003). Current demographic data suggest that binge eating has increased in the last two decades in both genders, and at higher rates in lower socioeconomic classes, as well as younger ages (Mitchison et al. 2014). BED has been strongly correlated with increased BMI although it can occur in the absence of obesity (Kessler et al. 2013). With obesity impacting a third of the US population (Ogden et al. 2007), aberrant overeating is a major public health concern. Subclinical binge eating affects over 10% of adolescent females (Touchette et al. 2011). Furthermore, binge eating

is associated with an increased frequency of body weight fluctuation, depression, anxiety, and substance abuse (Ramacciotti et al. 2005; Grucza et al. 2007; Galanti et al. 2007). Approximately one-fourth of patients with BED have a co-occurring substance use disorder (SUD) (Becker and Grilo 2015).

Aside from the known medical complications associated with excess body weight, binge eating can lead to psychological distress (Mustelin et al. 2017) and impairment in work productivity (Pawaskar et al. 2017) even without obesity (Striegel et al. 2012). The vast majority of individuals with BED do not receive treatment (Kazdin et al. 2017) probably due to a lack of physician awareness and shortage of screening tools (Dorflinger et al. 2017). According to a recent study, BED is perceived by the public as less severe, less impairing, easier to treat compared to other EDs, with common attitudes reflecting blameworthiness and lack of self-discipline (Reas 2017). BED has been associated with treatment dissatisfaction and early discontinuation of care (Amianto et al. 2015). It has also been shown that BED is often associated with delayed diagnosis and that earlier interventions might improve long-term health outcomes and reduce costs (Watson et al. 2018).

Taken together, these studies suggest that binge eating behavior affects a significant proportion of our society and has deleterious consequences, making it important to study from a public health perspective. In this review, we discuss the controversies surrounding the conceptualization of binge eating as food addiction (FA) that may have policy implications in the field of public health nutrition.

1.1   Defining Addiction

At least 2 of 11 criteria in the DSM-5 must be met for the diagnosis of a SUD (see Table 1). In the brain, there are patterns of activation that are similar across addictions, whether it be to alcohol, drugs, or gambling. In recent years it has

Table 1 Four broader categories for 11 criteria used for substance use disorder

A. Impaired control	1. Use of larger amount and for longer than intended
	2. Craving
	3. Much time spent using
	4. Repeated attempts to quit and/or control use
B. Social impairment	1. Social/interpersonal problems related to use
	2. Neglected major role to use
	3. Activities given up for use
C. Continued use despite risk	1. Hazardous use
	2. Physical/psychological problems related to use
D. Pharmacological criteria	1. Tolerance
	2. Withdrawal

Table adapted from Wiss et al. (2018)

been suggested that overeating, and perhaps binge eating, may result in a state that resembles an "addiction" to food (Avena et al. 2012), stemming from the idea that binge eating shares similarities with conventional drug addiction (Davis and Carter 2009). This has been supported by clinical research demonstrating that food craving, or the urge to consume palatable food in the absence of homeostatic hunger, in normal weight and obese adults activates areas of the brain similar to those activated by drug craving in individuals with SUD (Pelchat et al. 2004; Wang et al. 2004). While the DSM-5 does not officially recognize food addiction, the American Society of Addiction Medicine (ASAM) does, defining addiction broadly as a "primary, chronic disease of brain reward, motivation, memory, and related circuitry." Clinical accounts of FA have been reported (Ifland et al. 2009) and structured interviews have been used to identify addictive eating patterns in people with BED (Cassin and von Ranson 2007). The debate about whether FA should be recognized as a substance-related or behavioral addiction has been covered extensively elsewhere (Hebebrand et al. 2014; Schulte et al. 2017). Areas of convergence within addiction literature such as reward dysfunction, impulsivity, and attentional bias will be reviewed as they relate to food consumption. We will also discuss criticism of the FA construct including the argument that this concept can be counterproductive for BED and BN recovery at the individual level.

2 Food Addiction Overview

The most widely used clinical measure and "diagnostic tool" of addictive eating is the Yale Food Addiction Scale (YFAS) (Gearhardt et al. 2009), which corresponds to the criteria for substance dependence in DSM-IV-TR. In 2016, the YFAS 2.0 was created to maintain consistency with the current diagnostic criteria of SUD in the DSM-5 which includes severity indicators (Gearhardt et al. 2016). A total of 35 questions related to loss of control, cravings, and failed attempts to stop are measured across eight frequency measures ranging from never to every day. In 2017, a modified form (mYFAS 2.0) was reduced to 13 items, with eleven for each of the SUD criteria, and two questions for clinical significance (impairment and distress) which are not counted (but considered) in the overall score ranging from 0 to 11 (Schulte and Gearhardt 2017). Foods that have been determined to have the most self-reported addictive qualities include chocolate, ice cream, French fries, pizza, and cookies (Schulte et al. 2015) all of which are common convenience foods high in fat and refined carbohydrates.

National prevalence estimates of FA, according to YFAS research, are approximately 15% with higher rates in those who are obese (Schulte and Gearhardt 2018). In other research specifically focused on obese adults, prevalence estimates range from 6.7 to 16.5% (Brunault et al. 2016; Chao et al. 2017) which are close to national prevalence rates for SUDs including alcohol (Substance Abuse and Mental Health Services Administration 2017). Common correlates of FA are negative urgency (the tendency to act rashly when distressed) (Wolz et al. 2016) and low distress tolerance (Kozak et al. 2017). In obese patients seeking bariatric surgery, FA is associated with a higher prevalence of mood and anxiety disorders, and loss of control over the consumption of foods high in fat, sugar, and salt (Benzerouk et al. 2018). It is well-established that negative mood increases palatable food consumption (Becker et al. 2016) but what is less clear is how FA may over time induce negative mood, perhaps similar to withdrawal in drug addiction. In a sample of undergraduate women and men ($n = 998$), FA was associated with emotion dysregulation and depressive symptoms (Racine et al. 2019). Six months post-sleeve gastrectomy, patients meeting criteria for FA had greater ED and depression scores (Ivezaj et al. 2019).

2.1 Historical Overview of Animal Models of Binge Eating and Addiction

Animal models of binge eating have proven important in exploring the physiological, behavioral, and neurochemical aspects of binge eating observed in humans. Here we discuss an animal model of binge eating that specifically has been used to study addictive overeating. This allows for the discernment of behavioral and neurochemical indices related to addiction in the context of overeating highly palatable foods. Because studies report people binge eat most frequently on foods that are rich in sugar and fat (Hadigan et al. 1989; Kales 1990; Raymond et al. 2007), this chapter reviews binge eating models focused on sugar alone as well as in combination (i.e., processed foods).

In the sugar alone model, we impose an intermittent (limited) food access schedule, in which animals have access to a sugar solution (e.g., 25% glucose or 10% sucrose) and chow 12 h daily, followed by 12 h of deprivation for

approximately 1 month (Avena et al. 2006a). We impose a 4-h delay between the onset of the dark cycle and the onset of food access, as rats typically feed at the onset of the dark cycle and thus will be hungry when food is presented. After just a few days of this feeding schedule, rats develop binge-like eating behavior, defined in the model as distinct, large bouts of intake in discrete periods of time. The most salient binge-like episode is during the first hour of access, during which animals have been shown to consume approximately 20% of their total daily sugar intake. However, meal analysis throughout the access period reveals that these rats spontaneously engage in binge-like episodes, in contrast with the continuous consumption of smaller meals seen in control animals (Avena et al. 2008a). Further, rats maintained on this limited access schedule of sucrose or glucose gradually increase their total daily intake of sugar, eventually drinking as much in the 12-h access period as ad libitum-fed rats do in 24 h (70 mL/day).

What is most interesting and unique about this model of sugar overeating is that it results in signs of dependence. This model has identified both behavioral and neurochemical commonalities between binge-like eating and drug use. Rats maintained on this paradigm for 3 weeks show a series of behaviors similar to the effects of drugs of abuse, including the escalation of daily sugar intake and increase in sugar intake during the first hour of daily access. Further, when administered the opioid antagonist, naloxone, somatic signs of withdrawal, such as teeth chattering, forepaw tremor, and head shakes are observed, as well as anxiety as measured by reduced time spent on the exposed arm of an elevated plus-maze (Colantuoni et al. 2002). Similarly, these signs of opiate-like withdrawal emerge when access to the sugar solution is removed (Colantuoni et al. 2002; Avena et al. 2008b).

In models of SUD, animals will self-administer more of a drug after an abstinence period, when the drug is made available again. In the sugar-binge model, after 2 weeks of forced abstinence from sugar, rats will lever-press to obtain access to sugar more than before the period of abstinence, suggesting an increase in the reinforcing value of sugar after abstinence (Avena et al. 2005). Further, in the drug literature, sensitization and cross-sensitization play a role in drug self-administration, and both are typically measured in terms of increased locomotion in response to a drug. In the sugar binge model, rats show locomotor cross-sensitization to a low dose of amphetamine (Avena and Hoebel 2003). In addition to its effects on locomotor activity, drug sensitization can lead to subsequent increased intake of another drug or substance. Using this model, we find that rats with a history of binge-like consumption of sugar drink 9% more alcohol compared to controls (Avena et al. 2004).

Concomitant with these behaviors that are similar to those seen in drug dependence, rats maintained on the sugar binge-like feeding schedule show neurochemical changes similar to those seen in models of addiction. One of the strongest neurochemical commonalities between binge-like consumption of sugar and drugs of abuse is their effect on extracellular dopamine (DA) in the nucleus accumbens (NAc). Rodents show nonabating DA release during binge-like consumption of sugar, which is similar to the DA response seen with drugs of abuse (Rada et al. 2005). This unabated release of DA can be elicited by the taste of sucrose (Avena et al. 2006b) and is enhanced when rats are at a reduced body weight (Avena et al. 2008c). We have also shown alterations in DA receptor binding and gene expression in the binge model (Colantuoni et al. 2001; Spangler et al. 2004). Again, similar to what is seen in response to drugs of abuse, mu-opioid receptor binding is significantly enhanced in the NAc shell after 3 weeks of binge-like sugar consumption (Colantuoni et al. 2001). These animals also have a significant decrease in enkephalin mRNA in the NAc (Spangler et al. 2004), which is similarly observed in chronic drug use (Noble et al. 2015).

Withdrawal from several drugs of abuse can be accompanied by alterations in DA/acetylcholine (ACh) balance in the NAc, with ACh increasing while DA is suppressed (Hoebel et al. 2007). Using our model of sugar binge eating, we have shown that these rats show the same neurochemical imbalance in DA/ACh during withdrawal precipitated by naloxone (Colantuoni et al. 2002) or after 36 h of total food deprivation (Avena et al. 2008b). A review of the role of ACh interacting and counteracting with the DA system in the context of sugar addiction has been recently published (Wiss et al. 2018). Thus, multiple addiction-like neurochemical changes can result from drinking a sugar solution in a binge-like manner.

2.2 Human Data on Binge Eating and Addiction

In a recent review of studies using the YFAS 2.0, it was found that the highest prevalence of FA is found in individuals with BN (Meule and Gearhardt 2019), which may be linked to higher levels of dietary restraint and compensatory behaviors (e.g. purging) (Meule 2012). One study has suggested that FA criteria is met by all women with BN (Meule et al. 2014) and other research has shown than FA symptoms decrease following BN treatment (Hilker et al. 2016). What remains unclear is if the aberrant eating is a cause or a consequence of the disorder (Kaye et al. 2013). Estimates of FA in BED populations range from 42% to 57% (Gearhardt et al. 2012, 2013) suggesting that these conditions do not overlap entirely. A recent study found that 92% of a community BED sample ($n = 157$) met criteria for at least mild food addiction (Carter et al. 2018), and comorbid BED and FA have been associated with increased levels of depression (Blume et al. 2018). Therefore, this combination may represent a subgroup with more severe eating pathology than either condition alone (Carter et al. 2018; Burrows et al. 2018). Recently, studies have shown that FA explains more unique variance in psychological distress and psychosocial impairment than other established features of BED (Linardon and Messer 2019).

A recent comprehensive review discusses brain regions that are altered in both BED and

SUDs: (1) ventral striatum (goal-seeking behaviors, motivation, reward sensitivity), (2) dorsal striatum (habitual and compulsive behaviors), (3) prefrontal cortex (PFC) (executive functioning), and (4) insula (interoception, decision-making, taste perception, and feeding regulation) (Kessler et al. 2016). Reduced mu-opioid receptor availability has been observed in comorbid BED and obesity (Joutsa et al. 2018) and across SUDs (Belzeaux et al. 2018) which strongly suggest neurochemical overlap in these conditions that can persist despite weight loss or periods of drug abstinence. Individuals with FA, eating disorders characterized by binge eating, and SUDs share several neurocognitive alterations and comorbid symptoms. These include reward dysfunction, craving in response to relevant cues and negative mood states, emotion dysregulation, and impulsivity and executive dysfunction (Hutson et al. 2018). Here, we focus on reward, impulsivity, and attentional bias.

2.3 Reward

Recently it has been proposed that "reward" is a vague term (Salamone and Correa 2012) that is better conceptualized with at least three separate components: hedonics (referred to as "liking" and includes the opioid system), reinforcement (learning), and motivation (incentive salience, referred to as "wanting" primarily governed by DA) (Berridge and Robinson 2016). In addictions, the "liking" is stable or even decreasing, whereas the "wanting" increases over time (Polk et al. 2017). Just as individuals with SUDs show increased expectancy of reward from substances (Volkow et al. 2011), FA populations show increased expectancy of reward from food intake relative to controls (Wolz et al. 2016). Similarly, both BN and BED have been associated with altered self-reported expectancies of reward from eating and altered neural response during food reward anticipation (Frank 2013; Simon et al. 2016; Smith et al. 2018). In addition, both BN and SUDs have been associated with altered reward-based learning (Cyr et al. 2016; Myers et al. 2016), and regardless of weight,

women with BED have shown working memory problems (Eneva et al. 2017) which may be linked to the dysfunction in the reinforcement/learning cascade. However, in contrast to the decreased sensitivity to reward documented in SUDs (Volkow et al. 2010), there are data to support the notion of increased sensitivity to food reward in BED, in part via phasic release of DA in the ventral striatum motivating food seeking (Hutson et al. 2018). A similar increased sensitivity to food reward has been documented in FA: women with FA show elevated responses in the superior frontal gyrus to images of highly processed food (Schulte et al. 2019). According to these authors, this region has been previously associated with cue-induced craving in SUD and several processes likely influenced by dopaminergic responses to reward appraisal. Furthermore, another study showed additive striatal responses to images of fat and carbohydrate foods in combination during the valuation process (DiFeliceantonio et al. 2018).

2.4 Impulsivity

Binge eating may be conceptualized as impulsive/compulsive, given the maladaptation of the corticostriatal circuitry regulating motivation and impulse control (Kessler et al. 2016). Compulsivity is defined as repetitive and persistent actions despite negative consequences. Substance-dependent rats and rats that develop binge-like eating will endure foot-shock in order to obtain a drug or palatable food (Oswald et al. 2011; Shaham et al. 2000), implicating compulsivity across conditions. Impulsivity can be defined as decision-making with limited forethought (rash-spontaneous behavior). Both impulsivity and compulsivity are widely associated with altered dopaminergic neurotransmission in the ventral midbrain and NAc. Alterations in DA-D2 receptor binding and DA release observed in neuroimaging studies of addiction are thought to constitute a neurobiological marker of impulsivity (Trifilieff and Martinez 2014) but it remains unclear if low DA-D2 is genetically determined or merely a consequence of prolonged substance/food abuse.

Genetic research has identified a locus within the HTR2A gene (which encodes a serotonin receptor) that may be a candidate for the heritability of impulsive personality traits (Gray et al. 2018).

Impulsivity is a multifaceted process and can be divided by impulsive choice and impulsive action. Impulsive choice, or delay discounting, can be defined as a propensity to choose a lower but more immediate reward over a delayed but more valuable outcome (Tang et al. 2019). Impulsive choice is a known predictor of drug use (Jentsch et al. 2017) as well as binge episodes (Smith et al. 2019). Impulsive action includes difficulty inhibiting or controlling behaviors. Impulsive action associated with increased intake of food (Nederkoorn et al. 2009) and drugs (Trifilieff and Martinez 2014). A systematic review found that BED is a distinct phenotype within the obesity spectrum characterized by increased impulsivity (Giel et al. 2017), and an fMRI study using response inhibition tasks found that impulsivity was a stable characteristic of BED even in the absence of excess weight (Oliva et al. 2019). The features of BED associated with impulsivity include: (1) inability to stop eating, (2) sense of loss control, and (3) eating more rapidly than normal (Ural et al. 2017). Although women with FA symptoms did not show behavioral disinhibition in response to food stimuli (Meule et al. 2012), in a study of 133 obese bariatric candidates, impulsivity predicted FA diagnosis (Meule et al. 2017).

Impulsive action in particular contexts may be especially relevant to FA. Impulsive action in the context of strong negative emotion, or negative urgency, is a known predictor of externalizing behavior and SUD severity (Settles et al. 2012). Negative urgency has been strongly implicated in the maintenance of BN (Fischer et al. 2013), and in a community sample of adults with BED versus controls ($n = 151$), the BED group had significantly higher levels of negative urgency (Kenny et al. 2018). However, a recent study found that negative urgency predicted FA independent of eating disorder symptomatology in a large sample of individuals with diagnoses of binge-eating disorders (Wolz et al. 2017). It has been suggested that negative urgency has a

biological basis associated with low serotonin and high DA (Wolz et al. 2017). Positron emission tomography (PET) data has shown that both BN and BED are associated with abnormalities in the serotonergic system (Majuri et al. 2017; Kaye et al. 1998). While it is known that serotonin plays a role in impulsivity (Vaz-Leal et al. 2017) more research is needed to determine its role in binge eating and addiction. Taken together, it appears impulsivity may be an important target in the assessment and treatment of co-occurring binge eating/FA.

2.5 Attentional Bias

Attentional bias toward substance-related stimuli has been shown to be involved in the development and maintenance of drug dependence (Díaz-Batanero et al. 2018). In the addiction framework, attentional biases are thought to trigger a sequence of events including craving that leads to substance use. In one study, individuals with BED had higher vigilance for food stimuli than - weight-matched controls, pointing toward differences in attentional biases between BED and obesity (Schmitz et al. 2014). In a study of adolescents with BED and weight-matched controls using eye tracking and a visual search task, the BED group showed greater detection bias for attractive food targets (Schmidt et al. 2016). Adults with BN show similar attentional biases for food and body-related cues (Mai et al. 2015; Albery et al. 2016). Functional MRI studies have shown that food images elicit neural activity in the cuneate and posterior cingulate cortex indicating attentional bias in the obese with BED phenotype (Aviram-Friedman et al. 2018). A recent systematic review of various methods to measure brain activation in response to a stimulus found an interaction between negative affect and attentional bias to disorder-specific cues (Stojek et al. 2018). The authors recommend using content-specific cues most salient to each individual to further study intensity of response. Taken together, attentional bias may prove to be an important factor in the assessment, prevention, and intervention of addiction and binge eating behaviors. Neurocognitive therapeutics (e.g. neurofeedback) may prove beneficial (Schnyer et al. 2015). Studies examining the association between FA and attentional bias are warranted.

3 Debate 1: We Need Food But Not Drugs to Survive, Therefore Food Addiction Cannot Exist

Many have argued against the notion of FA by noting that food is a ubiquitous part of our everyday lives and its consumption is required for our survival, whereas addictive substances, in most cases, must be explicitly sought out and are not required for survival. However, there may be unique characteristics of refined sugars that more closely mimic the neural and behavioral effects of substances of abuse than other foods. From a behavioral perspective, the nutritional content of processed foods is not accurately conveyed to the brain (Small and DiFeliceantonio 2019). Rodent models of sugar addiction, specifically, meet five of the eleven criteria for SUD: the use of larger amounts for longer than intended, craving, hazardous use, tolerance, and withdrawal (Wiss et al. 2018). A recent review of mostly rodent data suggests that sugar triggers hedonic mechanisms that override neuroendocrine signals (e.g. insulin, leptin, ghrelin) that preserve homeostasis (Olszewski et al. 2019). As a result, a shift in the hunger-satiety continuum propels an individual to crave sugar despite a lack of energy need. Refined sugar has no fiber, which is designed by nature to stimulate sensations of fullness and barriers to excess consumption. Sugars in fresh fruits are absorbed slowly and calorie-constant thereby preserving glucose homeostasis, whereas concentrated sugar is absorbed rapidly and can disrupt glucose homeostasis (Baschetti 2019). Along with concentration and rate of absorption, dose also appears to matter (Schulte et al. 2015) but has not been extensively studied in humans. Meanwhile, it has been shown that people who binge eat are more likely to do so on foods high in sugar and fat (Gendall et al. 1997;

Hetherington et al. 1994). Taken together, it can be argued that while we need food to survive, the addition of refined sugars to processed food does not promote survival, but rather can activate hedonic hunger in the absence of homeostatic need (Lowe and Butryn 2007).

3.1 Debate 2: Food Addiction Could Be as Relevant to Eating Disorders as It Is to Obesity

The overlap and distinctions of disorders characterized by binge eating, FA, and obesity are important and still under investigation. While the current review does not focus on AN, and AN–BP has not been studied with the YFAS, we have included AN, binge eating/purging subtype (AN–BP) in our conceptual model (see Fig. 1). The transdiagnostic theory suggests that behaviors can evolve across ED diagnostic criteria, whether in the short term or across the life course. It remains unclear to what extent FA severity contributes to overeating and/or undereating across ED domains. Future research on FA in ED populations should adjust for dietary restraint (see Debate 3).

Meanwhile, binge-like eating and overeating of palatable foods have distinct impacts on the brain, and other findings have emphasized that overeating of palatable foods, regardless of intake schedule, and obesity are associated with brain alterations similar to addiction. For example, a landmark study by Johnson and Kenny showed that striatal DA-D2 receptors were downregulated in obese rats, similar to what has been reported in humans addicted to drugs (Johnson and Kenny 2010). The investigators were able to show that the consumption triggers the neuroadaptation. More recently it was documented that animals given extended access to high-sugar and high-fat foods led to neuroadaptations in the DA system and consequential reward deficits as well as compulsive responses to these foods (DiFeliceantonio 2019). Evidence of decreased striatal transmission via DA-D2 receptors in binge eating rats has been repeated recently (Heal et al. 2017). Hyper-neural activation to reward in the early

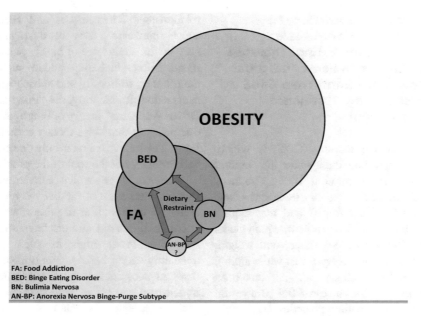

FA: Food Addiction
BED: Binge Eating Disorder
BN: Bulimia Nervosa
AN-BP: Anorexia Nervosa Binge-Purge Subtype

Fig. 1 Conceptual model for overlap between obesity, food addiction, and eating disorders

stages of binge eating and decreased activation in later stages (Hildebrandt et al. 2018) strongly support the presence of an addiction cascade. In animal models, it is without question that highly palatable foods have rewarding properties that can promote addiction-like behaviors as a consequence rather than a cause of obesity. However, it is worth emphasizing that several of the 11 criteria for addiction (e.g. those associated with social impairment) cannot be assessed in animal models, nor can the subjective sense of loss of control over eating that defines a binge eating episode, limiting the conclusive distinctions that can be drawn between FA, overeating, binge eating, and elevated weight status in animal models.

In humans, fMRI studies have suggested that aberrant ventral and frontostriatal activation in response to food images is associated with increased reward sensitivity and habitual binge eating (Lee et al. 2017). Frontostriatal circuits support self-regulatory self-control (executive function). A recent systematic review of neuroimaging studies concluded that the frequency of binge eating, across BED and BN, is related to neural changes, including diminished activity in

frontostriatal circuits (Donnelly et al. 2018). A study of overweight and obese women suggests that the interplay between increased appetite and decreased cognitive function may distinguish binge eating from overweight/obesity (Manasse et al. 2015). Interestingly, another systematic review suggested that lower prefrontal cortical activity affects inhibitory control and BMI in obesity independent of BED (Lavagnino et al. 2016), suggesting that neuroadaptations can occur in the absence of binge eating, and raising the possibilities that FA, elevated weight status itself, and/or overeating promotes prefrontal dysfunction. Data from a large, general population sample ($n = 625$) suggest that FA does not account for a major part of structural brain differences (cortical thickness) associated with BMI (Beyer et al. 2019). However, it was found that FA symptoms might explain additional variance in cortical thickness of right lateral OFC, a hub area of the reward network. Overall, disentangling causes and consequences of elevated weight status and the regular consumption of palatable foods versus binge-like consumption of the foods remains an important direction for future research.

3.2 Debate 3: Restriction, Not Food Addiction, Causes Dysregulated Eating; Therefore the Notion of Food Addiction Does More Harm Than Good by Encouraging Abstinence (Restriction)

Cognitive behavioral therapy (CBT) is widely accepted as a first-line therapeutic intervention for binge eating (Fairburn et al. 2003). The theoretical model underpinning this treatment posits that there are "no bad foods" and that dieting behavior is the root of the problem, which stems from overevaluating eating, shape, and weight. Perhaps counter to this theory is that while dieting precedes and drives binge eating, it has been shown that in many instances of BN (Brewerton et al. 2000), binge eating precedes and drives dieting behavior. Recent literature on FA reviewed herein provides preliminary support for this concept. Figure 2 identifies two potential alternative pathways that conceptualize directional

relationships between FA and BED/BN. One causal upstream factor of dysregulated eating may be the abundance of highly palatable convenience foods including added sugars strongly contributing to obesity and subsequent body dissatisfaction. Importantly, the "Food Environment Drives Addiction" theory of binge eating includes one pathway that skips dietary restraint, suggesting that restrictive eating is only one factor in BED and certainly not the root of the problem.

The extent to which restriction of binge foods or restriction of "addictive" foods may precipitate a state similar to withdrawal in humans is debated. Several authors have suggested that restricting foods high in sugar may foster unhealthy eating behaviors and have unintended consequences, including but not limited to social impairment and binge eating. This assertion has been supported by data showing that the reinforcing value of foods high in sugar increased after consuming a low-sugar diet for 1 week in both obese and normal weight participants (Flack et al. 2019), similar to our animal models. Many

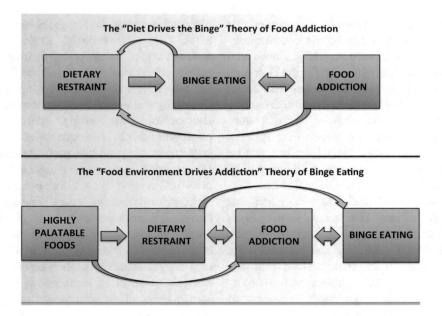

Fig. 2 Divergent models of food addiction and binge eating

researchers and clinicians believe that the FA construct does more harm than good, by encouraging an "abstinence" approach to foods that ultimately promote binge eating (Kirschenbaum and Krawczyk 2018). Thus, dietary restraint associated with EDs and FA remains at the core of our conceptual model (Fig. 1).

However, as noted, the overlap and distinctions of disorders characterized by binge eating, FA, and obesity are important. Unique aspects of EDs include restraint/rules, shape, and weight concerns, whereas unique aspects of addiction include the importance of the particular substance, withdrawal, and tolerance (Schulte et al. 2016). Four processes that differentiate BED from obesity include: (1) emotion reactivity (related to emotional vulnerability), (2) food-cue reactivity (cue-induced craving), (3) food craving, and (4) cognitive control (ability to regulate thought/action in accordance with goals) (Kober and Boswell 2018). All four of these processes are observed in FA, highlighting phenotypic overlap between FA and BED. However, as noted above, only roughly half of individuals with BED also meet "criteria" for FA, with slightly higher rates in BN. Therefore, the implications of restriction and methods for its implementation across these three conditions are likely distinct.

If the neuroadaptations associated with addiction persist over time, targeted nutrition counseling may be critical to overriding the neurobiology of FA. Viewing food as a potent biological substrate with profound neurobiological and behavioral implications in FA may be particularly helpful in guiding individualized nutrition interventions for FA. The FA model has been criticized because the ubiquity of high-sugar foods in the environment preclude an abstinence-based approach and may in fact promote dysregulated eating. Good vs. bad cognitive distortions about food (i.e., "black-and-white thinking") identified by CBT models as a threat to long-term positive outcomes are commonly endorsed by 12-Step programs that depend on clearly defined food-based abstinence. However, the usefulness of abstinence-based models for other addictions is still debated (Meule 2019), and has not been tested for FA. Meule suggests that "dietary restraint does not have to be dysfunctional as long as flexible elements are added" (Meule 2019). The optimal degree of nutritional flexibility in complex presentations of co-occurring FA/binge eating is not yet known.

4 Treatment Implications

People who identify as being addicted to food report a desire to have their condition formally recognized in order to receive appropriate care (Edwards et al. 2019). What that care entails remains to be determined. Both FA and binge eating may benefit from interventions that address impulsive behavior (Giel et al. 2017). For example, manualized group CBT treatment with a focus on reducing impulsivity led to reduced eating pathology and depression (Schag et al. 2019). In addition, intervention studies have successfully employed mindfulness meditation to decrease binge eating (Katterman et al. 2014) which has also been widely used in SUD treatment. Both of these interventions may be effective for FA. However, despite convergence of psychological and neuroscience research demonstrating overlap between food and drug misuse and some comorbidity of binge eating with FA, intervention strategies for addiction-like eating remain less clear at this time.

Potential treatment strategies based on the FA approach might include (1) psychoeducation, including new information on addiction (which may resonate with some patients as well as distract others), (2) psychotherapeutic strategies to reduce cravings (used in addiction treatment), and (3) abstinence (whether it be substance-related or behavioral) (Meule 2019). For example, as research shows that obesity causes changes in opioid and DA signaling which alter reward processing (Matikainen-Ankney and Kravitz 2018), patients with obesity scoring high in FA could be taught about the cyclic nature of weight gain and increased propensity to engage in more overeating. In addition to using the YFAS 2.0 to assess addiction-like eating, other constructs such as delay discounting, attentional bias, novelty seeking, and incentive salience can contribute to targeted behavioral interventions to reduce cravings, particularly in the face of cues and negative affect. Psychoeducation using these addiction-related constructs may prove effective for FA, offering novel explanatory opportunities for problematic eating. Alternative options could include trusting some cravings as "intuition" and enjoying the foods without guilt in order to decrease distress associated with cognitive control. This approach could be viewed as a "harm reduction" strategy. Of note, animal models have suggested addiction transfer from mother to offspring following exposure to highly palatable foods during pregnancy and lactation, suggesting that tangible progress in reducing FA may take generations (Wiss et al. 2017).

5 Conclusion and Future Directions

Failure to account for individual differences in the way people respond to food may contribute to poor outcomes in patients with binge eating. EDs are heterogenous, and reconceptualizing the binge eating cascade may prove important for the development of interventions at the individual and policy levels, as well as for efforts to further investigate overlapping mechanisms between FA and binge eating. It is also important to see where these constructs diverge to improve nutritional as well as therapeutic treatments. The message of "a calorie is a calorie" may be effective in the cognitive restructuring process of restrictive EDs, but less helpful for individuals with FA. More research on FA and FA-informed treatment is needed. For example, qualitative interviews of individuals with addiction-like eating have revealed that social and situational cues contribute to addiction-like eating; themes not captured by the YFAS (Paterson et al. 2019). Investigations of individualized treatments for individuals with and without binge eating and overweight/obesity comorbid with FA may prove fruitful, but ultimately, targeting the food environment through public health interventions may be more effective than individual nutrition treatments.

References

Albery IP, Wilcockson T, Frings D et al (2016) Examining the relationship between selective attentional bias for food- and body-related stimuli and purging behaviour in bulimia nervosa. Appetite 107:208–212. https://doi.org/10.1016/j.appet.2016.08.006

Amianto F, Ottone L, Daga G, Fassino S (2015) Binge-eating disorder diagnosis and treatment: a recap in front of DSM-5. BMC Psychiatry 15:70. https://doi.org/10.1186/s12888-015-0445-6

Avena NM, Hoebel BG (2003) A diet promoting sugar dependency causes behavioral cross-sensitization to a low dose of amphetamine. Neuroscience 122:17–20. https://doi.org/10.1016/s0306-4522(03)00502-5

Avena NM, Carrillo CA, Needham L et al (2004) Sugar-dependent rats show enhanced intake of unsweetened ethanol. Alcohol 34:203–209. https://doi.org/10.1016/j.alcohol.2004.09.006

Avena NM, Long KA, Hoebel BG (2005) Sugar-dependent rats show enhanced responding for sugar after abstinence: evidence of a sugar deprivation effect. Physiol Behav 84:359–362. https://doi.org/10.1016/j.physbeh.2004.12.016

Avena NM, Rada P, Hoebel BG (2006a) Current protocols in neuroscience. Curr Protoc Neurosci Éditor Board Jacqueline N Crawley et al Chapter 9:9.23C.1–9.23C.6. doi: https://doi.org/10.1002/0471142301.ns0923cs36

Avena NM, Rada P, Moise N, Hoebel BG (2006b) Sucrose sham feeding on a binge schedule releases accumbens dopamine repeatedly and eliminates the acetylcholine satiety response. Neuroscience 139:813–820. https://doi.org/10.1016/j.neuroscience.2005.12.037

Avena NM, Rada P, Hoebel BG (2008a) Evidence for sugar addiction: behavioral and neurochemical effects of intermittent, excessive sugar intake. Neurosci Biobehav Rev 32:20–39. https://doi.org/10.1016/j.neubiorev.2007.04.019

Avena NM, Bocarsly ME, Rada P et al (2008b) After daily bingeing on a sucrose solution, food deprivation induces anxiety and accumbens dopamine/acetylcholine imbalance. Physiol Behav 94:309–315. https://doi.org/10.1016/j.physbeh.2008.01.008

Avena NM, Rada P, Hoebel BG (2008c) Underweight rats have enhanced dopamine release and blunted acetylcholine response in the nucleus accumbens while bingeing on sucrose. Neuroscience 156:865–871. https://doi.org/10.1016/j.neuroscience.2008.08.017

Avena NM, Gold JA, Kroll C, Gold MS (2012) Further developments in the neurobiology of food and addiction: update on the state of the science. Nutrition 28:341–343. https://doi.org/10.1016/j.nut.2011.11.002

Aviram-Friedman R, Astbury N, Ochner CN et al (2018) Neurobiological evidence for attention bias to food, emotional dysregulation, disinhibition and deficient somatosensory awareness in obesity with binge eating disorder. Physiol Behav 184:122–128. https://doi.org/10.1016/j.physbeh.2017.11.003

Baschetti R (2019) Evolutionary physiology shows the need for an unprecedented study on sugar. Clin Nutr ESPEN. https://doi.org/10.1016/j.clnesp.2019.03.012

Becker DF, Grilo CM (2015) Comorbidity of mood and substance use disorders in patients with binge-eating disorder: associations with personality disorder and eating disorder pathology. J Psychosom Res 79:159–164. https://doi.org/10.1016/j.jpsychores.2015.01.016

Becker K, Fischer S, Smith GT, Miller JD (2016) The influence of negative urgency, attentional bias, and emotional dimensions on palatable food consumption. Appetite 100:236–243. https://doi.org/10.1016/j.appet.2016.02.019

Belzeaux R, Lalanne L, Kieffer BL, Lutz P-E (2018) Focusing on the opioid system for addiction biomarker discovery. Trends Mol Med. https://doi.org/10.1016/j.molmed.2017.12.004

Benzerouk F, Gierski F, Ducluzeau P-H et al (2018) Food addiction, in obese patients seeking bariatric surgery, is associated with higher prevalence of current mood and anxiety disorders and past mood disorders. Psychiatry Res 267:473–479. https://doi.org/10.1016/j.psychres.2018.05.087

Berridge KC, Robinson TE (2016) Liking, wanting, and the incentive-sensitization theory of addiction. Am Psychol 71:670. https://doi.org/10.1037/amp0000059

Beyer F, García-García I, Heinrich M et al (2019) Neuroanatomical correlates of food addiction symptoms and body mass index in the general population. Hum Brain Mapp 40:2747–2758. https://doi.org/10.1002/hbm.24557

Blume M, Schmidt R, Hilbert A (2018) Executive functioning in obesity, food addiction, and binge-eating disorder. Nutrients 11. https://doi.org/10.3390/nu11010054

Brewerton TD, Dansky BS, Kilpatrick DG, O'Neil PM (2000) Which comes first in the pathogenesis of bulimia nervosa: dieting or bingeing? Int J Eat Disord 28:259–264. https://doi.org/10.1002/1098-108x(200011)28:3<259::aid-eat2>3.0.co;2-d

Brunault P, Ducluzeau P-H, Bourbao-Tournois C et al (2016) Food addiction in bariatric surgery candidates: prevalence and risk factors. Obes Surg 26:1650–1653. https://doi.org/10.1007/s11695-016-2189-x

Burrows T, Kay-Lambkin F, Pursey K et al (2018) Food addiction and associations with mental health symptoms: a systematic review with meta-analysis. J Hum Nutr Diet. https://doi.org/10.1111/jhn.12532

Carter JC, Wijk M, Rowsell M (2018) Symptoms of 'food addiction' in binge eating disorder using the Yale Food Addiction Scale version 2.0. Appetite 133:362–369. https://doi.org/10.1016/j.appet.2018.11.032

Cassin SE, von Ranson KM (2007) Is binge eating experienced as an addiction? Appetite 49:687–690. https://doi.org/10.1016/j.appet.2007.06.012

Chao A, Shaw J, Pearl R et al (2017) Prevalence and psychosocial correlates of food addiction in persons with obesity seeking weight reduction. Compr Psychiatry 73:97–104. https://doi.org/10.1016/j.comppsych.2016.11.009

Colantuoni C, Schwenker J, McCarthy J et al (2001) Excessive sugar intake alters binding to dopamine and mu-opioid receptors in the brain. Neuroreport 12:3549–3552. https://doi.org/10.1097/00001756-200111160-00035

Colantuoni C, Rada P, McCarthy J et al (2002) Evidence that intermittent, excessive sugar intake causes endogenous opioid dependence. Obes Res 10:478–488. https://doi.org/10.1038/oby.2002.66

Cyr M, Wang Z, Tau GZ et al (2016) Reward-based spatial learning in teens with bulimia nervosa. J Am Acad Child Adolesc Psychiatry 55:962–971.e3. https://doi.org/10.1016/j.jaac.2016.07.778

Davis C, Carter JC (2009) Compulsive overeating as an addiction disorder. A review of theory and evidence. Appetite 53:1–8. https://doi.org/10.1016/j.appet.2009.05.018

Díaz-Batanero C, Domínguez-Salas S, Moraleda E et al (2018) Attentional bias toward alcohol stimuli as a predictor of treatment retention in cocaine dependence and alcohol user patients. Drug Alcohol Depend 182:40–47. https://doi.org/10.1016/j.drugalcdep.2017.10.005

DiFeliceantonio AG, all D (2019) Dopamine and diet-induced obesity. Nat Neurosci 22:1–2. doi: https://doi.org/10.1038/s41593-018-0304-0

DiFeliceantonio AG, Coppin G, Rigoux L et al (2018) Supra-additive effects of combining fat and carbohydrate on food reward. Cell Metab 28:33–44.e3. https://doi.org/10.1016/j.cmet.2018.05.018

Donnelly B, Touyz S, Hay P et al (2018) Neuroimaging in bulimia nervosa and binge eating disorder: a systematic review. J Eat Disord 6:3. https://doi.org/10.1186/s40337-018-0187-1

Dorflinger LM, Ruser CB, Masheb RM (2017) A brief screening measure for binge eating in primary care. Eat Behav 26:163–166. https://doi.org/10.1016/j.eatbeh.2017.03.009

Duncan AE, Ziobrowski HN, Nicol G (2017) The prevalence of past 12-month and lifetime DSM-IV eating disorders by BMI category in US men and women. Eur Eat Disord Rev 25:165–171. https://doi.org/10.1002/erv.2503

Edwards S, Lusher J, Murray E (2019) The lived experience of obese people who feel that they are addicted to food. Int J Psychol Cogn Sci 5:79–87

Eneva KT, Arlt JM, Yiu A et al (2017) Assessment of executive functioning in binge-eating disorder independent of weight status. Int J Eat Disord 50:942–951. https://doi.org/10.1002/eat.22738

Fairburn CG, Harrison PJ (2003) Eating disorders. Lancet 361:407–416. https://doi.org/10.1016/S0140-6736(03)12378-1

Fairburn CG, Cooper Z, Shafran R (2003) Cognitive behaviour therapy for eating disorders: a "transdiagnostic" theory and treatment. Behav Res Ther 41:509–528. https://doi.org/10.1016/S0005-7967(02)00088-8

Fischer S, Peterson CM, McCarthy D (2013) A prospective test of the influence of negative urgency and expectancies on binge eating and purging. Psychol Addict Behav 27:294. https://doi.org/10.1037/a0029323

Flack KD, Ufholz K, Casperson S et al (2019) Decreasing the consumption of foods with sugar increases their reinforcing value: a potential barrier for dietary behavior change. J Acad Nutr Diet. https://doi.org/10.1016/j.jand.2018.12.016

Frank GK (2013) Altered brain reward circuits in eating disorders: chicken or egg? Curr Psychiatry Rep 15:396. https://doi.org/10.1007/s11920-013-0396-x

Galanti K, Gluck ME, Geliebter A (2007) Test meal intake in obese binge eaters in relation to impulsivity and compulsivity. Int J Eat Disord 40:727–732. https://doi.org/10.1002/eat.20441

Gearhardt A, Corbin W, Brownell K (2009) Preliminary validation of the Yale Food Addiction Scale. Appetite 52:430–436. https://doi.org/10.1016/j.appet.2008.12.003

Gearhardt AN, White MA, Masheb RM et al (2012) An examination of the food addiction construct in obese patients with binge eating disorder. Int J Eat Disord 45:657–663. https://doi.org/10.1002/eat.20957

Gearhardt AN, White MA, Masheb RM, Grilo CM (2013) An examination of food addiction in a racially diverse sample of obese patients with binge eating disorder in primary care settings. Compr Psychiatry 54:500–505. https://doi.org/10.1016/j.comppsych.2012.12.009

Gearhardt A, Corbin W, Brownell K (2016) Development of the Yale Food Addiction Scale Version 2.0. Psychol Addict Behav 30:113–121. https://doi.org/10.1037/adb0000136

Gendall KA, Sullivan PE, Joyce PR et al (1997) The nutrient intake of women with bulimia nervosa. Int J Eat Disord 21:115–127. https://doi.org/10.1002/(sici)1098-108x(199703)21:2<115::aid-eat2>3.0.co;2-o

Giel KE, Teufel M, Junne F et al (2017) Food-related impulsivity in obesity and binge eating disorder—a systematic update of the evidence. Nutrients 9:1170. https://doi.org/10.3390/nu9111170

Gray JC, MacKillop J, Weafer J et al (2018) Genetic analysis of impulsive personality traits: Examination of a priori candidates and genome-wide variation. Psychiatry Res 259:398–404. https://doi.org/10.1016/j.psychres.2017.10.047

Grucza RA, Przybeck TR, Cloninger C (2007) Prevalence and correlates of binge eating disorder in a community sample. Compr Psychiatry 48:124–131. https://doi.org/10.1016/j.comppsych.2006.08.002

Hadigan C, Kissileff H, Walsh B (1989) Patterns of food selection during meals in women with bulimia. Am J Clin Nutr 50:759–766. https://doi.org/10.1093/ajcn/50.4.759

Heal DJ, Hallam M, Prow M et al (2017) Dopamine and μ-opioid receptor dysregulation in the brains of binge-eating female rats – possible relevance in the psychopathology and treatment of binge-eating disorder. J Psychopharmacol 31:770–783. https://doi.org/10.1177/0269881117699607

Hebebrand J, Albayrak O, Adan R et al (2014) "Eating addiction", rather than "food addiction", better captures addictive-like eating behavior. Neurosci Biobehav Rev 47:295–306. https://doi.org/10.1016/j.neubiorev.2014.08.016

Hetherington M, Altemus M, Nelson M et al (1994) Eating behavior in bulimia nervosa: multiple meal analyses. Am J Clin Nutr 60:864–873. https://doi.org/10.1093/ajcn/60.6.864

Hildebrandt BA, Sinclair EB, Sisk CL, Klump KL (2018) Exploring reward system responsivity in the nucleus accumbens across chronicity of binge eating in female rats. Int J Eat Disord 51:989–993. https://doi.org/10.1002/eat.22895

Hilker I, Sánchez I, Steward T et al (2016) Food addiction in bulimia nervosa: clinical correlates and association with response to a brief psychoeducational intervention. Eur Eat Disord Rev 24:482–488. https://doi.org/10.1002/erv.2473

Hoebel BG, Avena NM, Rada P (2007) Accumbens dopamine-acetylcholine balance in approach and avoidance. Curr Opin Pharmacol 7:617–627. https://doi.org/10.1016/j.coph.2007.10.014

Hudson JI, Hiripi E, Pope HG, Kessler RC (2007) The prevalence and correlates of eating disorders in the National Comorbidity Survey Replication. Biol Psychiatry 61:348–358. https://doi.org/10.1016/j.biopsych.2006.03.040

Hutson PH, Balodis IM, Potenza MN (2018) Binge-eating disorder: clinical and therapeutic advances. Pharmacol Ther 182:15–27. https://doi.org/10.1016/j.pharmthera.2017.08.002

Ifland JR, Preuss HG, Marcus MT et al (2009) Refined food addiction: a classic substance use disorder. Med Hypotheses 72:518–526. https://doi.org/10.1016/j.mehy.2008.11.035

Ivezaj V, Wiedemann AA, Lawson JL, Grilo CM (2019) Food addiction in sleeve gastrectomy patients with loss-of-control eating. Obes Surg 29:2071–2077. https://doi.org/10.1007/s11695-019-03805-8

Jentsch DJ, Ashenhurst JR, Cervantes CM et al (2017) Dissecting impulsivity and its relationships to drug addictions. Ann N Y Acad Sci 1327:1–26. https://doi.org/10.1111/nyas.12388

Johnson PM, Kenny PJ (2010) Dopamine D2 receptors in addiction-like reward dysfunction and compulsive eating in obese rats. Nat Neurosci 13:635. https://doi.org/10.1038/nn.2519

Joutsa J, Karlsson HK, Majuri J et al (2018) Binge eating disorder and morbid obesity are associated with lowered mu-opioid receptor availability in the brain. Neuroimaging, Psychiatry Res. https://doi.org/10.1016/j.pscychresns.2018.03.006

Kales E (1990) Macronutrient analysis of binge eating in bulimia. Physiol Behav 48:837–840. https://doi.org/10.1016/0031-9384(90)90236-w

Katterman SN, Kleinman BM, Hood MM et al (2014) Mindfulness meditation as an intervention for binge eating, emotional eating, and weight loss: a systematic review. Eat Behav 15:197–204. https://doi.org/10.1016/j.eatbeh.2014.01.005

Kaye WH, Greeno CG, Moss H et al (1998) Alterations in serotonin activity and psychiatric symptoms after recovery from bulimia nervosa. Arch Gen Psychiatry 55:927–935. https://doi.org/10.1001/archpsyc.55.10.927

Kaye WH, Wierenga CE, Bailer UF et al (2013) Does a shared neurobiology for foods and drugs of abuse contribute to extremes of food ingestion in anorexia and bulimia nervosa? Biol Psychiatry 73:836–842. https://doi.org/10.1016/j.biopsych.2013.01.002

Kazdin AE, Fitzsimmons-Craft EE, Wilfley DE (2017) Addressing critical gaps in the treatment of eating disorders. Int J Eat Disord 50:170–189. https://doi.org/10.1002/eat.22670

Kenny TE, Singleton C, Carter JC (2018) An examination of emotion-related facets of impulsivity in binge eating disorder. Eat Behav 32:74–77. https://doi.org/10.1016/j.eatbeh.2018.12.006

Kessler RC, Berglund PA, Chiu W et al (2013) The prevalence and correlates of binge eating disorder in the World Health Organization World Mental Health Surveys. Biol Psychiatry 73:904–914. https://doi.org/10.1016/j.biopsych.2012.11.020

Kessler RM, Hutson PH, Herman BK, Potenza MN (2016) The neurobiological basis of binge-eating disorder. Neurosci Biobehav Rev 63:223–238. https://doi.org/10.1016/j.neubiorev.2016.01.013

Kirschenbaum DS, Krawczyk R (2018) The food addiction construct may do more harm than good: weight controllers are athletes, not addicts. Child Obes 14:227–236. https://doi.org/10.1089/chi.2018.0100

Kober H, Boswell RG (2018) Potential psychological & neural mechanisms in binge eating disorder: implications for treatment. Clin Psychol Rev 60:32–44. https://doi.org/10.1016/j.cpr.2017.12.004

Kozak AT, Davis J, Brown R, Grabowski M (2017) Are overeating and food addiction related to distress tolerance? An examination of residents with obesity from a U.S. metropolitan area. Obes Res Clin Pract 11:287–298. https://doi.org/10.1016/j.orcp.2016.09.010

Lavagnino L, Arnone D, Cao B et al (2016) Inhibitory control in obesity and binge eating disorder: a systematic review and meta-analysis of neurocognitive and neuroimaging studies. Neurosci Biobehav Rev 68:714–726. https://doi.org/10.1016/j.neubiorev.2016.06.041

Lee J, Namkoong K, Jung Y-C (2017) Impaired prefrontal cognitive control over interference by food images in binge-eating disorder and bulimia nervosa. Neurosci Lett 651:95–101. https://doi.org/10.1016/j.neulet.2017.04.054

Linardon J, Messer M (2019) Assessment of food addiction using the Yale Food Addiction Scale 2.0 in individuals with binge-eating disorder symptomatology: factor structure, psychometric properties, and clinical significance. Psychiatry Res 279:216–221. https://doi.org/10.1016/j.psychres.2019.03.003

Lowe MR, Butryn ML (2007) Hedonic hunger: a new dimension of appetite? Physiol Behav 91:432–439. https://doi.org/10.1016/j.physbeh.2007.04.006

Mai S, Gramann K, Herbert BM et al (2015) Electrophysiological evidence for an attentional bias in processing body stimuli in bulimia nervosa. Biol Psychol 108:105–114. https://doi.org/10.1016/j.biopsycho.2015.03.013

Majuri J, Joutsa J, Johansson J et al (2017) Serotonin transporter density in binge eating disorder and pathological gambling: a PET study with [^{11}C]MADAM. Eur Neuropsychopharmacol 27:1281–1288. https://doi.org/10.1016/j.euroneuro.2017.09.007

Manasse SM, Espel HM, Forman EM et al (2015) The independent and interacting effects of hedonic hunger and executive function on binge eating. Appetite 89:16–21. https://doi.org/10.1016/j.appet.2015.01.013

Matikainen-Ankney BA, Kravitz AV (2018) Persistent effects of obesity: a neuroplasticity hypothesis. Ann N Y Acad Sci 1428:221–239. https://doi.org/10.1111/nyas.13665

Meule A (2012) Food addiction and body-mass-index: a non-linear relationship. Med Hypotheses 79:508–511. https://doi.org/10.1016/j.mehy.2012.07.005

Meule A (2019) A critical examination of the practical implications derived from the food addiction concept. Curr Obes Rep 8:11–17. https://doi.org/10.1007/s13679-019-0326-2

Meule A, Gearhardt AN (2019) Ten years of the Yale Food Addiction Scale: a review of version 2.0. Curr Addict Rep 1–11 doi: https://doi.org/10.1007/s40429-019-00261-3

Meule A, Lutz A, Vögele C, Kübler A (2012) Women with elevated food addiction symptoms show accelerated reactions, but no impaired inhibitory control, in response to pictures of high-calorie food-cues. Eat Behav 13:423–428. https://doi.org/10.1016/j.eatbeh.2012.08.001

Meule A, Rezori V, Blechert J (2014) Food addiction and bulimia nervosa. Eur Eat Disord Rev 22:331–337. https://doi.org/10.1002/erv.2306

Meule A, de Zwaan M, Müller A (2017) Attentional and motor impulsivity interactively predict 'food addiction' in obese individuals. Compr Psychiatry 72:83–87. https://doi.org/10.1016/j.comppsych.2016.10.001

Mitchison D, Hay P, Slewa-Younan S, Mond J (2014) The changing demographic profile of eating disorder behaviors in the community. BMC Public Health 14:943. https://doi.org/10.1186/1471-2458-14-943

Mustelin L, Bulik CM, Kaprio J, Keski-Rahkonen A (2017) Prevalence and correlates of binge eating disorder related features in the community. Appetite 109:165–171. https://doi.org/10.1016/j.appet.2016.11.032

Myers CE, Sheynin J, Balsdon T et al (2016) Probabilistic reward- and punishment-based learning in opioid addiction: experimental and computational data. Behav Brain Res 296:240–248. https://doi.org/10.1016/j.bbr.2015.09.018

Nederkoorn C, Guerrieri R, Havermans R et al (2009) The interactive effect of hunger and impulsivity on food intake and purchase in a virtual supermarket. Int J Obes 33:905. https://doi.org/10.1038/ijo.2009.98

Noble F, Lenoir M, Marie N (2015) The opioid receptors as targets for drug abuse medication. Br J Pharmacol 172:3964–3979. https://doi.org/10.1111/bph.13190

Ogden CL, Yanovski SZ, Carroll MD, Flegal KM (2007) The epidemiology of obesity. Gastroenterology 132:2087–2102. https://doi.org/10.1053/j.gastro.2007.03.052

Oliva R, Morys F, Horstmann A et al (2019) The impulsive brain: neural underpinnings of binge eating behavior in normal-weight adults. Appetite. https://doi.org/10.1016/j.appet.2018.12.043

Olszewski PK, Wood EL, Klockars A, Levine AS (2019) Excessive consumption of sugar: an insatiable drive for reward. Curr Nutr Rep 8:120–128. https://doi.org/10.1007/s13668-019-0270-5

Oswald KD, Murdaugh DL, King VL, Boggiano MM (2011) Motivation for palatable food despite consequences in an animal model of binge eating. Int J Eat Disord 44:203–211. https://doi.org/10.1002/eat.20808

Paterson C, Lacroix E, von Ranson KM (2019) Conceptualizing addictive-like eating: a qualitative analysis. Appetite 141:104326. https://doi.org/10.1016/j.appet.2019.104326

Pawaskar M, Witt EA, Supina D et al (2017) Impact of binge eating disorder on functional impairment and work productivity in an adult community sample in the United States. Int J Clin Pract 71:e12970. https://doi.org/10.1111/ijcp.12970

Pelchat M, Johnson A, Chan R et al (2004) Images of desire: food-craving activation during fMRI. Neuroimage 23:1486–1493. https://doi.org/10.1016/j.neuroimage.2004.08.023

Polk SE, Schulte E, Furman CR, Gearhardt AN (2017) Wanting and liking: separable components in problematic eating behavior? Appetite 115:45–53. https://doi.org/10.1016/j.appet.2016.11.015

Racine SE, Hagan KE, Schell SE (2019) Is all nonhomeostatic eating the same? Examining the latent structure of nonhomeostatic eating processes in women and men. Psychol Assess. https://doi.org/10.1037/pas0000749

Rada P, Avena NM, Hoebel BG (2005) Daily bingeing on sugar repeatedly releases dopamine in the accumbens shell. Neuroscience 134:737–744. https://doi.org/10.1016/j.neuroscience.2005.04.043

Ramacciotti CE, Coli E, Paoli R et al (2005) The relationship between binge eating disorder and non-purging bulimia nervosa. Eat Weight Disord 10:8–12. https://doi.org/10.1007/bf03353413

Raymond NC, Bartholome LT, Lee SS et al (2007) A comparison of energy intake and food selection during laboratory binge eating episodes in obese women with and without a binge eating disorder diagnosis. Int J Eat Disord 40:67–71. https://doi.org/10.1002/eat.20312

Reas D (2017) Public and healthcare professionals' knowledge and attitudes toward binge eating disorder: a narrative review. Nutrients 9:1267. https://doi.org/10.3390/nu9111267

Salamone JD, Correa M (2012) The mysterious motivational functions of mesolimbic dopamine. Neuron 76:470–485. https://doi.org/10.1016/j.neuron.2012.10.021

Schag K, Rennhak SK, Leehr EJ et al (2019) IMPULS: impulsivity-focused group intervention to reduce binge eating episodes in patients with binge eating disorder – a randomised controlled trial. Psychother Psychosom 88:141–153. https://doi.org/10.1159/000499696

Schmidt R, Lüthold P, Kittel R et al (2016) Visual attentional bias for food in adolescents with binge-eating disorder. J Psychiatr Res 80:22–29. https://doi.org/10.1016/j.jpsychires.2016.05.016

Schmitz F, Naumann E, Trentowska M, Svaldi J (2014) Attentional bias for food cues in binge eating disorder. Appetite 80:70–80. https://doi.org/10.1016/j.appet.2014.04.023

Schnyer D, Beevers CG, deBettencourt MT et al (2015) Neurocognitive therapeutics: from concept to application in the treatment of negative attention bias. Biol Mood Anxiety Disord 5:1. https://doi.org/10.1186/s13587-015-0016-y

Schulte EM, Gearhardt AN (2017) Development of the modified Yale Food Addiction Scale Version 2.0. Eur Eat Disord Rev 25:302–308. https://doi.org/10.1002/erv.2515

Schulte EM, Gearhardt AN (2018) Associations of food addiction in a sample recruited to be nationally representative of the United States. Eur Eat Disord Rev 26:112–119. https://doi.org/10.1002/erv.2575

Schulte E, Avena N, Gearhardt A (2015) Which foods may be addictive? The roles of processing, fat content, and glycemic load. PLoS One 10:e0117959. https://doi.org/10.1371/journal.pone.0117959

Schulte EM, Grilo CM, Gearhardt AN (2016) Shared and unique mechanisms underlying binge eating disorder and addictive disorders. Clin Psychol Rev 44:125–139. https://doi.org/10.1016/j.cpr.2016.02.001

Schulte EM, Potenza MN, Gearhardt AN (2017) A commentary on the "eating addiction" versus "food addiction" perspectives on addictive-like food consumption.

Appetite 115:9–15. https://doi.org/10.1016/j.appet.2016.10.033

Schulte E, Yokum S, Jahn A, Gearhardt AN (2019) Food cue reactivity in food addiction: a functional magnetic resonance imaging study. Physiol Behav 112574. https://doi.org/10.1016/j.physbeh.2019.112574

Settles RE, Fischer S, Cyders MA et al (2012) Negative urgency: a personality predictor of externalizing behavior characterized by neuroticism, low conscientiousness, and disagreeableness. J Abnorm Psychol 121:160. https://doi.org/10.1037/a0024948

Shaham Y, Erb S, Stewart J (2000) Stress-induced relapse to heroin and cocaine seeking in rats: a review. Brain Res Rev 33:13–33. https://doi.org/10.1016/s0165-0173(00)00024-2

Simon JJ, Skunde M, Walther S et al (2016) Neural signature of food reward processing in bulimic-type eating disorders. Soc Cogn Affect Neurosci 11:1393–1401. https://doi.org/10.1093/scan/nsw049

Small D, DiFeliceantonio AG (2019) Processed foods and food reward. Science 363:346–347. https://doi.org/10.1126/science.aav0556

Smith KE, Mason TB, Peterson CB, Pearson CM (2018) Relationships between eating disorder-specific and transdiagnostic risk factors for binge eating: an integrative moderated mediation model of emotion regulation, anticipatory reward, and expectancy. Eat Behav 31:131–136. https://doi.org/10.1016/j.eatbeh.2018.10.001

Smith KE, Mason TB, Crosby RD et al (2019) A multimodal, naturalistic investigation of relationships between behavioral impulsivity, affect, and binge eating. Appetite 136:50–57. https://doi.org/10.1016/j.appet.2019.01.014

Spangler R, Wittkowski KM, Goddard NL et al (2004) Opiate-like effects of sugar on gene expression in reward areas of the rat brain. Mol Brain Res 124:134–142. https://doi.org/10.1016/j.molbrainres.2004.02.013

Stojek M, Shank LM, Vannucci A et al (2018) A systematic review of attentional biases in disorders involving binge eating. Appetite 123:367–389. https://doi.org/10.1016/j.appet.2018.01.019

Striegel RH, Bedrosian R, Wang C (2012) Comparing work productivity in obesity and binge eating. Int J Eat Disord 45:995–998. https://doi.org/10.1002/eat.22069

Substance Abuse and Mental Health Services Administration (2017) Key substance use and mental health indicators in the United States: results from the 2016 National Survey on Drug Use and Health

Tang J, Chrzanowski-Smith OJ, Hutchinson G et al (2019) Relationship between monetary delay discounting and obesity: a systematic review and meta-regression. Int J Obes 43:1135–1146. https://doi.org/10.1038/s41366-018-0265-0

Thapliyal P, Mitchison D, Mond J, Hay P (2018) Gender and help-seeking for an eating disorder: findings from a general population sample. Eat Weight Disord 1–6. doi: https://doi.org/10.1007/s40519-018-0555-5

Touchette E, Henegar A, Godart NT et al (2011) Subclinical eating disorders and their comorbidity with mood and anxiety disorders in adolescent girls. Psychiatry Res 185:185–192. https://doi.org/10.1016/j.psychres.2010.04.005

Trifilieff P, Martinez D (2014) Imaging addiction: D2 receptors and dopamine signaling in the striatum as biomarkers for impulsivity. Neuropharmacology 76:498–509. https://doi.org/10.1016/j.neuropharm.2013.06.031

Udo T, Grilo CM (2018) Prevalence and correlates of DSM-5-defined eating disorders in a Nationally Representative Sample of U.S. Adults. Biol Psychiatry 84:345–354. https://doi.org/10.1016/j.biopsych.2018.03.014

Ural C, Belli H, Akbudak M et al (2017) Relation of binge eating disorder with impulsiveness in obese individuals. World J Psychiatry 7:114–120. https://doi.org/10.5498/wjp.v7.i2.114

Vaz-Leal FJ, Ramos-Fuentes MI, Rodríguez-Santos L, Álvarez-Mateos CM (2017) Eating disorders - a paradigm of the biopsychosocial model of illness. Intech. https://doi.org/10.5772/65248

Volkow ND, Wang G, Fowler JS et al (2010) Addiction: decreased reward sensitivity and increased expectation sensitivity conspire to overwhelm the brain's control circuit. Bioessays 32:748–755. https://doi.org/10.1002/bies.201000042

Volkow ND, Wang G-J, Fowler JS et al (2011) Addiction: beyond dopamine reward circuitry. Proc Natl Acad Sci USA 108:15037–15042. https://doi.org/10.1073/pnas.1010654108

Wang G-J, Volkow ND, Thanos PK, Fowler JS (2004) Similarity between obesity and drug addiction as assessed by neurofunctional imaging. J Addict Dis 23:39–53. https://doi.org/10.1300/j069v23n03_04

Watson HJ, Jangmo A, Smith T et al (2018) A register-based case-control study of health care utilization and costs in binge-eating disorder. J Psychosom Res 108:47–53. https://doi.org/10.1016/j.jpsychores.2018.02.011

Wiss DA, Criscitelli K, Gold M, Avena N (2017) Preclinical evidence for the addiction potential of highly palatable foods: current developments related to maternal influence. Appetite 115:19–27. https://doi.org/10.1016/j.appet.2016.12.019

Wiss DA, Avena N, Rada P (2018) Sugar addiction: from evolution to revolution. Front Psych 9:545. https://doi.org/10.3389/fpsyt.2018.00545

Wolz I, Hilker I, Granero R et al (2016) "Food addiction" in patients with eating disorders is associated with negative urgency and difficulties to focus on long-term goals. Front Psychol 7:61. https://doi.org/10.3389/fpsyg.2016.00061

Wolz I, Granero R, Fernández-Aranda F (2017) A comprehensive model of food addiction in patients with binge-eating symptomatology: the essential role of negative urgency. Compr Psychiatry 74:118–124. https://doi.org/10.1016/j.comppsych.2017.01.012

Part IV

Treatment Development

Prevention of Binge Eating

Taylor Perry and Tiffany A. Brown

Abstract

Given the public health burden of binge eating, and eating disorders more generally, preventative interventions that target risk factors prior to binge eating onset have become increasingly popular in recent decades. Research to date has tested prevention programs delivered as early as elementary school and as late as mid-life, using in-person and internet-based delivery modalities. In this chapter, we provide an overview of risk factors for binge eating and then review the evidence supporting programs targeting these risk factors using: (1) in-person school-based programs, (2) in-person programs for young adults, and (3) internet-based prevention programs for adolescents and adults. While many programs have helped reduce risk factors for binge eating, only a few programs have demonstrated the ability to truly prevent binge eating. We also highlight new and developing research on preventative interventions and models to help disseminate effective programs to a larger and more diverse population of people at risk for binge eating and other eating disorders.

Keywords

Prevention · Binge eating · Adolescents · Adults · School-based · Dissonance-based · Internet-based · Interventions

Learning Objectives

After reading this chapter, you will be able to:
1. Describe the evidence for school-based prevention interventions for binge eating in adolescents.
2. Describe the evidence for in-person prevention interventions for binge eating in young adults.
3. Describe the evidence for internet-based prevention interventions for binge eating in adolescents and young adults.

1 Introduction

Given the serious psychological and medical consequences associated with binge eating, the high cost of eating disorder treatment, and barriers to treatment access, developing interventions that target risk factors prior to binge eating onset have the potential to reduce the public health burden associated with these disorders. Broadly, prevention programs can be applied either universally

T. Perry · T. A. Brown (✉)
Department of Psychiatry, University of California San Diego, San Diego, CA, USA
e-mail: tiffanybrown@ucsd.edu

(i.e., to the general population), or selectively (i.e., to populations specifically deemed at risk for eating disorders, such as only girls in a school). In recent decades, both universal and selected prevention efforts have made strides in helping reduce problematic eating behavior. However, it is important to note that few studies have actually examined true binge eating "prevention," that is, demonstrating that a program reduces the onset of binge eating compared to a control condition. Thus, most of the preventative intervention efforts described in this chapter are more accurately described as efforts to reduce risk factors for eating disorders, rather than prevention per se. It is also important to note that most prevention efforts have focused on risk for disordered eating or eating disorders more broadly, without explicitly targeting or focusing on binge eating. Relatedly, many of these programs have examined continuous measures of bulimic symptoms as outcome variables (i.e., binge eating, vomiting, laxative misuse, and overexercise). However, most studies do not distinguish between these behaviors, which makes it difficult to determine whether programs reduce binge eating versus tendency toward compensatory behaviors. Thus, throughout the chapter, we have sought to distinguish specific binge eating outcomes, when available. The following chapter briefly outlines risk factors for binge eating, and subsequently summarizes the literature on (1) in-person school-based prevention programs, (2) in-person prevention programs for young adults, and (3) internet-based prevention programs for adolescents and adults that target these risk factors.

2   Risk Factors for Binge Eating

While most eating disorders onset during adolescence, binge eating has a relatively later age of onset, often starting in early adulthood (Udo and Grilo 2018). As such, programs that aim to help reduce the risk for binge eating are applicable across both adolescents, prior to the time of peak risk, and college students/young adults. While eating disorders are more common in females as compared to males, rates of binge-eating disorder are more equivalent across genders (Hudson et al. 2007). Research has shown that dieting, peer influences, pressure to be thin, body dissatisfaction, stress, negative affect and depressive symptoms, emotional eating, body mass, and low self-esteem are all associated with increased risk for binge eating in adolescent and young adult females and, to some extent, males (Neumark-Sztainer et al. 2006b; Stice 2002; Stice et al. 2002; Keel et al. 2013; Napolitano and Himes 2011; Vincent and McCabe 2000). Thus, programs for both adolescents and adults have largely focused on targeting these risk factors in order to prevent the development of binge eating and other unhealthy eating behaviors in both adolescents and adults.

3   In-Person School-Based Prevention Programs

For adolescents, school-based settings provide an ideal environment to disseminate binge eating prevention programs (Neumark-Sztainer et al. 2006a). School-based programs have largely been implemented universally in middle schools with both boys and girls simultaneously; however, many programs have selectively focused on girls only, given increased risk for eating disorders more broadly in this group (Udo and Grilo 2018). School-based programs have typically focused on targeting risk factors such as self-esteem, peer influences, media pressures, body image, and body mass.

3.1   Self-Esteem and Peer-Based Programs

Several school-based programs have focused on improving self-esteem in female and male adolescents. These programs have demonstrated improvements in body image and esteem (O'Dea and Abraham 2000; McVey et al. 2003; Norwood et al. 2011; Richardson et al. 2009), weight satisfaction (Norwood et al. 2011), and reduced dieting behavior (McVey et al. 2003).

Given the influence of peers on binge eating in adolescence (Vincent and McCabe 2000), several school-based programs have incorporated peer influences into prevention efforts. *Happy Being Me* (Richardson and Paxton 2010) is an interactive, 3-lesson, school-based program for 7th-grade girls, that explores risk factors with and among peers. *Happy Being Me* has demonstrated greater improvements in body image, self-esteem, thin-ideal internalization, appearance comparisons and conversations, and lower dietary restraint compared to class as usual at post-intervention and 3-month follow-up. *Happy Being Me* has also been expanded to eight lessons to include more content regarding eating concerns, although this program does not appear to improve eating disorder risk factors to a greater extent than alternative interventions (i.e., *Media Smart*; Wilksch et al. 2015). *Happy Being Me* has also been adapted for younger girls and boys (10–11 years old) to address peer influences across genders (Bird et al. 2013). This mixed gender program has demonstrated improvements in body dissatisfaction, appearance conversations, and disordered eating for girls; improvements in body-ideal internalization for boys; and appearance comparisons for both boys and girls at post-intervention. Of note, body dissatisfaction improvements for girls were maintained at 3-month follow-up.

Additionally, two peer-focused programs also target improving self-esteem to reduce unhealthy eating behaviors. *Girl's Circle,* aims to increase self-efficacy/esteem, social support, and body image and has demonstrated improvements on these variables in an open trial for females aged 10–17 years (Steese et al. 2006). *Girl Talk/Every BODY is a Somebody* is a program for 7th to 8th graders, which aims to improve body esteem and has demonstrated increased weight-related esteem and decreased dieting compared to assessment-only controls at post-intervention and 3-month follow-up (McVey et al. 2003). *Making Choices* (Weiss and Wertheim 2005) is a peer-based program that aims to decrease "fat-talk" between friends/peers. For high-risk girls, *Making Choices* has demonstrated greater reductions in the Eating Disorder Inventory

Drive for Thinness, Bulimia (a measure largely assessing tendency toward binge eating), and Body Dissatisfaction subscales compared to an assessment-only condition at post-intervention; however, effects were not maintained at 3-month follow-up. *Healthy Buddies* (Stock et al. 2007) is a peer-led program in which 4th to 7th grade girls and boys are paired with a student in K-3rd grade to promote a healthy body image, physical activity, and eating healthy foods. *Healthy Buddies* has demonstrated improvements in nutritional and physical knowledge, healthy eating, and physical activity compared to individuals who attended school as usual. Thus, there is some evidence supporting the use of peer-based programs in reducing risk factors that may contribute to binge eating for school-aged girls, and to a lesser extent, boys.

3.2 School-Based Media Literacy Programs

Media literacy prevention programs may reduce binge eating through promoting education and activism around binge eating risk factors of pressures to be thin and attractive from the media. Students are educated about gender stereotypes that are promoted through media, how these contribute to Western body ideals, and techniques used to create media images (i.e., airbrushing and digital manipulation). Students are also encouraged to write letters to industries and advertisers about their feelings about the images used. One of the most successful media literacy programs to date has been *Media Smart* (Wilksch and Wade 2009), an 8-week program for middle school boys and girls. For boys, *Media Smart* has demonstrated reductions in weight-related peer teasing, perfectionism, and body dissatisfaction at post-intervention (Wilksch et al. 2015), dieting, body dissatisfaction, and weight/shape concerns at post-intervention and 6-month follow-up (Wilksch and Wade 2009) and depression and media internalization at 6- and 12-month follow-up (Wilksch et al. 2015) compared to class as usual. For girls, *Media Smart* has demonstrated reductions in ineffectiveness at post-intervention and weight/shape concerns at 30-month follow-up compared to class as usual (Wilksch and Wade 2009). Only 8% of girls in the intervention developed clinical weight/shape concerns compared to 19% of controls at 12-month follow-up (Wilksch et al. 2015; Wilksch and Wade 2009). Thus, although media literacy programs have not directly targeted binge eating, such programs may have potential for reducing weight and shape concerns, a relevant risk factor for future binge eating behavior.

3.3 School-Based Behavior Modification-Based Programs Targeting Weight and Eating

In recent years, school-based programs have also aimed to prevent both disordered eating behaviors and obesity in adolescents. One of these programs, *New Moves* (Neumark-Sztainer et al. 2003, 2010), targets girls who are "overweight" or "obese" according to body mass index or those at risk of being in these weight categories. *New Moves* includes physical education classes, nutritional and social support/self-empowerment sessions, as well as individual counseling, which involves motivational interviewing and lunch get-togethers. The intervention has demonstrated improvements in physical activity, body image, self-worth, portion control, and a decrease in unhealthy weight control behaviors compared to physical education class as usual at 9-month follow-up, but no significant differences in weight or binge eating.

Behavioral modification programs in school-aged children have also been tested outside of the classroom. Some of these interventions have targeted children and adolescents who exhibit eating patterns that may increase risk for binge eating, including eating in the absence of hunger and loss of control eating. In particular, one study randomized 8–12-year-old boys and girls with elevated eating in the absence of hunger to either an 8-week appetite awareness training (*CAAT*) or an 8-week food-focused cue exposure intervention (*Volcravo*) (Boutelle et al. 2011). The latter condition included skills to help cope with urges to eat in the absence of hunger and cued exposures to high craving foods. Both interventions resulted in significant decreases in binge eating through 1-year follow-up, with no changes in caloric intake; however, only children in *Volcravo* showed significant decreases in eating in the absence of hunger at post-intervention and 6-month follow-up. Another study by Tanofsky-Kraff et al. (2014) randomized adolescent girls with loss of control eating (a risk factor for binge eating later in life) in the 75th–97th weight percentile to either an interpersonal therapy (IPT)-based intervention or a health education program. Notably, IPT has demonstrated efficacy as a treatment for reducing binge eating in clinical samples of patients with eating disorders. Both conditions demonstrated significant decreases in weight, body fat percentile, depression, anxiety, and the frequency of loss-of-control eating over 1-year follow-up, with no significant differences between groups. IPT was more efficacious than health education at reducing objective binge eating by 1-year follow-up.

3.4 Summary of School-Based Programs

In sum, in-person school-based programs have generally demonstrated effectiveness in reducing risk factors for unhealthy eating behaviors in adolescents. While many school-based prevention programs have included both boys and girls, given potential gender-based differences in the expression of binge eating and body image (Murray et al. 2017), future research may benefit from examining how gender may moderate intervention effects. Further, while including control groups can be challenging in school-based research, future research in this area would benefit from comparing results to active control groups (versus class as usual) to help determine which programs have the best potential to reduce binge eating. As many of these programs have not examined effects at longer-term follow-up or examined whether these programs actually reduce the onset of binge eating, future research in this area would be helpful.

4 In-Person Prevention Programs for Young Adults

In-person prevention programs for young adults have largely used cognitive-behavioral strategies to target risk factors for disordered eating by reinforcing and shaping attitudes and behaviors to foster a positive relationship with eating and body image. These programs have largely targeted thin-ideal internalization, body dissatisfaction, and healthy eating/physical activity. Critically, two of these programs, the *Body Project* and *Healthy Weight*, are among the few programs that have demonstrated the ability to prevent eating disorder onset in adolescent and adult females.

4.1 Cognitive Dissonance-Based Programs in Adolescents and Adults

The *Body Project* is a cognitive dissonance-based eating disorder prevention program that targets thin-ideal internalization and body dissatisfaction by having females actively argue against the appearance ideal that is perpetuated by Western media and culture (Stice et al. 2006). The *Body Project* consists of 4 h of programming, delivered over two to six in-person sessions by either professional and/or peer facilitators. The program originally was developed for adolescent girls, and since has been administered to college-aged women (Becker et al. 2005, 2006, 2008, 2010; Stice et al. 2006, 2015a). Most college versions of the program rely on peer facilitators (i.e., other college students) co-leading the program to maximize peer influences on eating behaviors (Forney et al. 2012; Keel et al. 2013).

The *Body Project* has demonstrated reductions in thin-ideal internalization, body dissatisfaction, dieting, negative affect, and bulimic symptoms at post-intervention, with effects largely maintained at 6-month, 1-year (Becker et al. 2005, 2010; Stice et al. 2006, 2008, 2009), and 2–3 year follow-up (Stice et al. 2008, 2015a) compared to assessment-only or alternative intervention conditions. The continuous bulimic symptoms measure used in the *Body Project* reflects a symptom composite using the diagnostic items from the Eating Disorder Examination or Eating Disorder Examination-Questionnaire for bulimia nervosa (e.g., binge eating, vomiting, laxative misuse, overexercise, weight, and shape concerns) and does not distinguish binge eating from compensatory behaviors. However, it is important to note that the *Body Project* has been successful at reducing bulimia nervosa or binge-eating disorder onset by 60% through 3-year follow-up (Stice et al. 2008). Supporting the hypothesized mechanisms of intervention effects, reductions in thin-ideal internalization mediate the *Body Project's* effects on body dissatisfaction, dieting, negative affect, and bulimic symptoms (Stice et al. 2007). Further, supporting that the *Body Project* helps reduce thin-ideal internalization, participants in the *Body Project* were less likely to activate brain regions involved in reward at post-intervention when looking at thin models compared to an educational control group (Stice et al. 2015b).

The *Body Project* has also been adapted for and/or delivered in a variety of different female populations, including sororities (Becker et al. 2005, 2006), adolescent girls in Western and non-Western countries (http://www.free-being-me.com; Amaral et al. 2019), and female athletes (Becker et al. 2012). Recently, versions of the *Body Project* have also been adapted to target mixed-gender groups (*Body Project 4 All;* Kilpela et al. 2016), groups of at-risk males (Brown and Keel 2015; Brown et al. 2017) and efforts to develop more inclusive and racially-, sexual orientation-, and gender-diverse groups are currently underway (*EVERYbody Project*; Ciao et al. 2018). In terms of efficacy, results from the *Body Project 4 All* (Kilpela et al. 2016), which included both males and females in groups together, found significant improvements in body fat and muscularity dissatisfaction for males compared to waitlist control through 6-month follow-up. However, results for females were less encouraging; suggesting that mixed gender groups may decrease the well-documented impact of the program for females.

Two programs have focused on male-only adaptations: the *PRIDE Body Project* (Brown and Keel 2015) for sexual minority males and the *Body Project: More than Muscles* (Brown et al. 2017) for body dissatisfied men, regardless of sexual orientation. Both programs have demonstrated excellent acceptability and significant improvements in risk factors for both traditional (restraint, body-ideal internalization, continuously measured bulimic symptoms) and more male-specific eating pathology (body fat/muscularity, dissatisfaction, drive for muscularity) compared to waitlist controls post-intervention, with effects being largely maintained through 1-month follow-up. A large-scale trial of the *PRIDE Body Project* is currently underway and will test whether the program actually reduces the likelihood of developing an eating disorder compared to an active control condition (R01MD012698, PI: Aaron Blashill). Thus, the *Body Project* appears to be successful at reducing eating disorder risk factors and preventing bulimic-type eating disorders in females, and appears to hold promise in more diverse populations.

4.2 Behavior Modification-Based Programs Targeting Weight and Eating in Adolescents and Adults

Initially developed as an active control comparison group for the *Body Project*, *Healthy Weight* (Stice et al. 2001) uses behavioral modification strategies and psychoeducation to promote healthy eating and weight behaviors. Participants monitor their food intake and exercise and follow individualized behavioral plans. Compared to assessment-only controls, *Healthy Weight* has demonstrated greater decreases in thin-ideal internalization, body dissatisfaction, negative affect, eating disorder symptoms, psychosocial impairment, and lower risk for eating disorder and obesity onset through 3-year follow-up (Stice et al. 2006, 2008). Supporting the hypothesized mechanisms of treatment, healthy eating and physical activity mediated the relationship between intervention condition and body dissatisfaction and negative affect (Stice et al. 2007). *Healthy Weight* has

been modified for peer-facilitation, and has demonstrated reductions in negative affect, body dissatisfaction, internalization of the thin ideal, and continuously measured bulimic symptoms (binge eating, compensatory behaviors, weight/shape concerns) through 14-month follow-up (Becker et al. 2010).

Healthy Weight has also been delivered to college female athletes and has demonstrated reductions in negative affect, shape concerns, and bulimic symptoms at post-intervention, 6-week follow-up, and 1-year follow-up; however, results did not differ from the athlete modified *Body Project* group (Becker et al. 2012). Coaches and athlete peer-leaders reported that they preferred the *Healthy Weight* intervention compared to the *Body Project*, due to greater acceptability of the nutrition-oriented focus versus the body image focus in an athlete population. This program is now known as the *Female Athlete Body Project*, and has demonstrated greater reductions in dietary restraint and objective binge episodes compared to waitlist controls at 18-month follow-up (Stewart et al. 2019). In summary, similar to the *Body Project*, *Healthy Weight* has demonstrated efficacy in reducing eating disorder risk factors in female adolescents, college students, and female athletes; however, the overall amount of research support is lower for *Healthy Weight*.

4.3 Summary of In-Person Prevention Programs for Young Adults

In sum, in-person prevention programs for adults have been effective in reducing eating disorder risk factors and, to some degree, binge eating. Critically, the *Body Project* and *Healthy Weight* have demonstrated the ability to actually reduce the onset of bulimic syndromes. Most of the adult prevention research has been conducted in female populations, with some recent encouraging expansions to more diverse populations. Future research should continue to include more diverse samples in terms of gender, sexual orientation, and cultural/ethnic identity and examine whether these factors moderate intervention effects to help determine whether additional modification for these groups is needed.

5 Internet-Based Programs

In recent years, prevention programs have also been delivered in online formats. Internet-based prevention programs can help minimize barriers to treatment, reduce in-person clinical demands, increase access to services, and reduce costs compared to in-person delivery methods. Much like in-person programs, internet-based eating disorder prevention programs have typically targeted weight and shape concerns, peer influences, pressures from media, thin-ideal internalization, and healthy eating/physical activity. Of note, a recent meta-analysis supports that internet-based programs as a whole are successful at reducing eating disorder risk, with small, but significant effects on reducing bulimic symptoms, including binge eating (Melioli et al. 2016).

5.1 Student Bodies

One of the most studied internet-based prevention programs in adults is *Student Bodies,* which uses traditional cognitive-behavioral prevention techniques to improve knowledge of eating disorder risk factors, body image, unhealthy weight regulation behaviors, and binge eating (Winzelberg et al. 2000). *Student Bodies* is an 8-week intervention for women who are at high risk of developing an eating disorder, with a psychoeducational reading on relevant topics each week (e.g., body dissatisfaction, weight/shape concerns, excessive exercise, and nutrition). Participants also engage in online self-monitoring journals, behavioral change exercises, and participate in online chat rooms and discussion boards. *Student Bodies* has demonstrated reductions in weight/shape concerns, drive for thinness, and overall eating pathology at post-intervention, 3-month follow-up (Winzelberg et al. 2000; Zabinski et al. 2001, 2004; Jacobi et al. 2007), and 2-year follow-up (Taylor et al.

2006, 2016) compared to waitlist control. Supporting the efficacy of the online format, Celio et al. (2000) found that intervention effects from *Student Bodies* did not differ from an alternative in-person psychoeducational intervention. While *Student Bodies* appears to reduce several risk factors for eating disorders, *Student Bodies* does not appear to significantly decrease the likelihood of developing a *clinical eating disorder* at 2-year follow-up compared to waitlist control (Taylor et al. 2006, 2016). However, moderator analyses revealed that those with elevated weight, those engaging in compensatory behaviors at the beginning of program (Taylor et al. 2006), or those with elevated weight/shape concerns (Taylor et al. 2016), did exhibit reduced incidence of eating disorder onset compared to waitlist control. *Student Bodies* has also been tested in a group of females with lifetime depression who were at high risk of developing an eating disorder and demonstrated reductions in weight concerns, drive for thinness, and overall eating disorder symptoms compared to waitlist control at post-intervention, 1-year, and 2-year follow-up (Taylor et al. 2016).

Student Bodies + or *Student Bodies-ED* is a version of *Student Bodies* for women with current symptoms of disordered eating or subclinical eating disorders (Jacobi et al. 2012; Saekow et al. 2015; Fitzsimmons-Craft et al. 2019a, b). Compared to waitlist control, this program has demonstrated reductions in weight concerns, eating disorder psychopathology, restrictive eating, loss of control eating, purging, and binge eating at post-intervention (Jacobi et al. 2012; Saekow et al. 2015; Fitzsimmons-Craft et al. 2019a, b), with reductions in restrictive eating, purging, and binge eating being maintained at 6-month follow-up (Jacobi et al. 2012). Thus, *Student Bodies* has demonstrated efficacy in reducing eating disorder risk factors and behaviors, including binge eating, for females at various levels of eating disorder risk.

Recently the *Healthy Body Image* Program was launched throughout the state of Missouri to help identify an individual's risk level for developing an eating disorder and subsequently match them with the appropriate online prevention services (Fitzsimmons-Craft et al. 2019a, b). The program uses an online screening tool and then directs the woman to an evidenced-based online program based on their level of risk: *Staying Fit* (a universal prevention program for low-risk women), *Student Bodies* (a selective prevention program for high-risk women), and *Student Bodies-ED* (a guided cognitive behavioral self-help intervention for those with subthreshold or clinical eating disorders). In the initial implementation of the *Healthy Body Image* Program, 1.9%–2.5% of the undergraduate population participated in the screening, with promising initial results in improving risk factors for binge eating across conditions (Fitzsimmons-Craft et al. 2019a, b).

5.2   Peer-Focused Internet Programs

Like in-person programs, internet-based programs have also specifically targeted peer influences on eating behavior. *Set Your Body Free* is a cognitive-behavioral program that has been delivered both in-person and online and aims to address body image concerns in women in peer groups, aged 18–30 years (Gollings and Paxton 2006). The online version consists of six 90-min online group-based cognitive-behavioral therapy sessions with elements of motivational interviewing. Similar to *Student Bodies,* participants use online chat rooms and discussion boards to help facilitate conversations between peers. Both in-person and online versions of *Set Your Body Free* have demonstrated improvements in body dissatisfaction, depression, and disordered eating, with the in-person version being superior at post-intervention; however, the online program demonstrated continued improvement post-intervention, such that there were no differences between conditions at 6-month follow-up (Paxton et al. 2007). *My Body My Life* (Heinicke et al. 2007) is an adapted version of *Set Your Body Free* for adolescents (Gollings and Paxton 2006). The program has demonstrated improvements in pressures from media to be thin, body dissatisfaction, and

extreme weight loss behaviors (crash dieting, fasting, laxative use, and diuretics) at post-intervention and 2-month follow-up compared to waitlist controls, with improvements in media pressures maintained at 6-month follow-up. Additionally, given the increasing understanding that eating disorders can also onset in midlife (Cumella and Kally 2008), *Set Your Body Free* has been expanded to women aged 30–60 years through *Set Your Body Free Midlife* (McLean et al. 2011). Program sessions focus on acceptance of age-related appearance changes, the importance of self-worth, increasing self-care, and body acceptance. *Set Your Body Free Midlife* has demonstrated improvements in body dissatisfaction, body attitudes, external eating (eating in response to food regardless of hunger), and emotional eating at post-intervention and 6-month follow-up compared to a waitlist control. Thus, *Set Your Body Free* has been adapted for multiple age ranges and has demonstrated effectiveness in reducing eating disorder risk factors from late adolescence through late midlife. Although *Set Your Body Free* has shown initial promising results in reducing disorder risk factors, more research is needed to determine if this program is effective at preventing binge eating onset.

5.3   Media Literacy-Focused Internet Programs in Adults

Media Smart-Targeted is an online implementation of *Media Smart* adapted for at-risk young women (Wilksch et al. 2018). *Media Smart-Targeted* had demonstrated greater reductions in overall eating disorder psychopathology compared to assessment-only controls at 12-month follow-up; however, binge eating was not differentiated as an outcome. Further, the program also demonstrated greater reductions in media internalization post-intervention, depression at 6-month and 12-month follow-up, and ineffectiveness and clinical impairment at 12-month follow-up, compared to an alternative online intervention (*Student Bodies*).

5.4 Dissonance-Based Internet Programs in Adults

The *Body Project* has also been modified for online delivery in the *eBody Project* (Stice et al. 2012). The program consists of six 30–40 min sessions over the course of 3 weeks. Similar to the *Body Project*, participants engage in activities critiquing the thin-ideal. Participants also earn "Body Project Bucks" for completing assignments that could be used to purchase items with the Body Project logo (e.g., water bottles, shirts, and mugs). Importantly, the *eBody Project* did not significantly differ on most outcomes compared to the in-person version at post-intervention. Thus, the *eBody Project* appears to be an acceptable and efficacious alternative when in-person versions of the *Body Project* are not available or feasible.

5.5 Behavior Modification Internet Programs Targeting Weight and Eating in Adolescents and Adults

Similar to in-person programs, several internet-based programs have also sought to reduce the risk for both eating disorders and obesity, all of which have been adapted from *Student Bodies*. One of these programs is *StudentBodies2* (Doyle et al. 2008), a 16-week online program that uses cognitive-behavioral techniques to help promote positive body image and reduced weight, for overweight adolescents. The program has demonstrated improvements in healthy eating skills and physical activity at post-intervention through 4-month follow-up, and decreases in BMI at post-intervention, but not 4-month follow-up, compared to an educational brochure control group. The program did not demonstrate significant reductions in eating disorder psychopathology from baseline to post-intervention.

Another obesity and eating disorder prevention program is *StudentBodies2-BED*, which specifically targets binge eating in overweight adolescents (Jones et al. 2008), and also aims to increase healthy eating, physical activity, movement, and help maintain weight. Participants in *StudentBodies2-BED* demonstrated significant reductions in objective and subjective binges at post-treatment and 9-month follow-up. Additionally, *StudentBodies2-BED* has demonstrated greater reductions in BMI at post-intervention and 9-month follow-up compared to waitlist controls.

Similar to the screening model in the *Healthy Body Image* Program, recent online prevention research in adolescents has used weight status to target appropriate prevention programs for specific individuals. *StayingFit* (Taylor et al. 2012; Jones et al. 2014; Bell et al. 2019) is an online 10–12 week, two-track program that emphasizes healthy habits and promotes positive body image, adapted from *Student Bodies*. Normal weight adolescents receive *StayingFit general*, a universal program that focuses on healthy habits related to nutrition and physical activity, and overweight adolescents receive *StayingFit weight maintenance*, a targeted program that incorporates cognitive-behavioral techniques, adolescent weight loss strategies, and awareness of hunger and satiety. Both programs demonstrated increases pre- to post-intervention in fruit and vegetable consumption (Taylor et al. 2012) and reductions in weight/shape concerns for women at risk for overweight (Taylor et al. 2012) or with high initial weight and shape concerns (Taylor et al. 2012; Jones et al. 2014). *StayingFit weight maintenance* also demonstrated reductions in BMI pre- to post-intervention (Taylor et al. 2012; Jones et al. 2014).

A similar screening platform is currently being tested for adult women, the *everyBody* program, which uses online weight and eating screening to determine which adapted version of *Student Bodies* should be delivered, with the goal of simultaneously reducing risk for eating disorders and increasing general health promotion (Nacke et al. 2019). Participants initially answer questions about their weight, weight/shape concerns, eating disorder symptoms, and dietary restraint. Individuals with a body mass index (BMI) of 18–25, with no eating disorder symptoms, will receive *everyBody Basic*, a 4-week program that promotes healthy eating and exercise habits. Those with a BMI of 21–25, with elevated weight/shape concerns, but no other eating disorder symptoms, will receive *everyBody Original*, an 8-week program that aims to improve body image, promote healthy eating and exercise, and improve self-esteem. Individuals with a BMI of 18.5–21, with elevated weight and shape concerns, but no eating disorder behaviors will receive *everyBody AN* a 10-week program that expands upon *everyBody Original* by also targeting dietary restraint. Those with a BMI greater than 18.5 with occasional binge eating and/or compensatory behaviors will receive *everyBody Plus*, an 8-week program that expands upon *everyBody Original* content to help reduce binge eating and/or purging. Lastly, those with a BMI of 25 or greater and no eating disorder concerns will receive *everyBody Fit,* a 12-week program focused on healthy weight regulation and improving self-esteem. Once enrolled in a program, participants will complete self-assessments at mid-intervention, post-intervention, 6-month follow-up, and 12-month follow-up. Thus, similar to the *Healthy Body Image* Program, this protocol seeks to match prevention interventions for young women based on the level of risk; however, this program also incorporates weight screening to a greater degree.

5.6 Summary of Internet-Based Interventions

In summary, internet-based interventions hold promise for helping implement binge eating prevention programs in a more cost-effective manner. It is important to note that of the current online prevention programs, only Student Bodies has demonstrated the ability to reduce eating disorder onset for subsets of individuals in a randomized controlled trial, and several variants of this program have been adapted to target individuals at different levels of risk. More broadly, online interventions may be particularly beneficial to help connect individuals who belong to diverse groups across the country/world and/or who may not have access to in-person services with a group of their peers (e.g., LGBTQ community members in rural areas). To help prevent

binge eating more strategically, future research should use online screening platforms, like the *Healthy Body Image* Program, to help determine the best prevention program for individuals based on eating disorder risk status.

6 Conclusions and Future Directions

In sum, several universal and targeted prevention programs have demonstrated the ability to reduce risk factors for binge eating for female adolescents and young adults across a variety of settings (schools, universities, and online). Overall, more support exists for programs targeting at-risk groups versus females more broadly.

Thus far, three targeted programs (the *Body Project, Healthy Weight,* and *Student Bodies*) have demonstrated the ability to prevent eating disorder onset in randomized controlled trials in young adult women. Critically, the *Body Project* is the only program that has demonstrated the ability to prevent binge-eating disorder onset (in addition to bulimia nervosa) in young women. As such, research has more recently emphasized efforts to help disseminate these programs to a larger audience and adapt the content from these programs to target more diverse populations. For example, a randomized controlled trial is currently underway comparing different implementation training models for the *Body Project* to determine the optimal amount of training that is needed for peers to ensure intervention effectiveness (R01MH112743; PI: Eric Stice). In efforts to be more inclusive and help prevent eating disorders beyond female samples, adaptations of the *Body Project* have started to expand the reach of prevention programs to more diverse groups that have been overlooked in previous intervention efforts, including athletes, males, the LGBTQ community, individuals in middle age, and individuals from non-Western countries, with many of these programs demonstrating promising results.

Across both adolescents and adults, a large body of research has begun to examine behavior modification programs focused on targeting individuals at risk for overweight and/or eating disorders. Future research in this area may also benefit from incorporating newer psychosocial risk factors, such as food insecurity and weight stigma, into binge eating prevention efforts. Given recent research linking food insecurity to binge eating (Becker et al. 2017; Rasmusson et al. 2019), low-cost and scalable interventions that can be delivered to low-income populations to help reduce binge eating are warranted. Further, given research supporting how weight stigma can contribute to binge eating (Puhl et al. 2007; Vartanian and Porter 2016) and to poor eating intervention outcomes (Wott and Carels 2010), future research could benefit from examining weight stigma in the context of these programs and/or examine how addressing weight stigma in preventative interventions may affect binge eating outcomes.

With the increasing use and accessibility of mobile technology, online interventions will continue to increase in relevancy for mental health intervention research, including binge eating. Indeed, the success of online programs like *Student Bodies*, and the ability to translate in-person programs to online formats (e.g., the *eBody Project, Media Smart, Set your Body Free*) suggests that this is a promising direction for future research and development. In particular, the use of online screening and implementation platforms, like the *Healthy Body Image* program, provide an excellent opportunity to help match prevention programs and help personalize targeted prevention efforts more precisely and efficiently. As both in-person and online programs have shown promise, future research would benefit from further examining when, or for whom, online versus in-person programs would be most appropriate and/or effective. In sum, future research for the programs that have demonstrated the ability to prevent eating disorder onset (*Body Project, Healthy Weight,* and *Student Bodies*) will likely continue to focus on identifying more efficient methods to disseminate and expand these programs to reach a broader population of individuals at risk for binge eating.

Conflict of Interest Dr. Brown is a master trainer for the Body Project Collaborative.

References

Amaral ACS, Stice E, Ferreira MEC (2019) A controlled trial of a dissonance-based eating disorders prevention program with Brazilian girls. Psicologia Reflexão e Crítica 32(1):13

Becker CB, Smith LM, Ciao AC (2005) Reducing eating disorder risk factors in sorority members: a randomized trial. Behav Ther 36(3):245–253

Becker CB, Smith LM, Ciao AC (2006) Peer-facilitated eating disorder prevention: a randomized effectiveness trial of cognitive dissonance and media advocacy. J Couns Psychol 53(4):550

Becker CB, Bull S, Schaumberg K, Cauble A, Franco A (2008) Effectiveness of peer-led eating disorders prevention: a replication trial. J Consult Clin Psychol 76 (2):347

Becker CB, Wilson C, Williams A, Kelly M, McDaniel L, Elmquist J (2010) Peer-facilitated cognitive dissonance versus healthy weight eating disorders prevention: a randomized comparison. Body Image 7(4):280–288. https://doi.org/10.1016/j.bodyim.2010.06.004

Becker CB, McDaniel L, Bull S, Powell M, McIntyre K (2012) Can we reduce eating disorder risk factors in female college athletes? A randomized exploratory investigation of two peer-led interventions. Body Image 9(1):31–42

Becker CB, Middlemass K, Taylor B, Johnson C, Gomez F (2017) Food insecurity and eating disorder pathology. Int J Eat Disord 50(9):1031–1040

Bell MJ, Zeiler M, Herrero R, Kuso S, Nitsch M, Etchemendy E, Fonseca-Baeza S, Oliver E, Adamcik T, Karwautz A (2019) Healthy Teens @ School: evaluating and disseminating transdiagnostic preventive interventions for eating disorders and obesity for adolescents in school settings. Internet Interv 16:65–75

Bird EL, Halliwell E, Diedrichs PC, Harcourt D (2013) Happy Being Me in the UK: a controlled evaluation of a school-based body image intervention with pre-adolescent children. Body Image 10(3):326–334

Boutelle KN, Zucker NL, Peterson CB, Rydell SA, Cafri G, Harnack L (2011) Two novel treatments to reduce overeating in overweight children: a randomized controlled trial. J Consult Clin Psychol 79(6):759

Brown TA, Keel PK (2015) A randomized controlled trial of a peer co-led dissonance-based eating disorder prevention program for gay men. Behav Res Ther 74:1–10. https://doi.org/10.1016/j.brat.2015.08.008

Brown TA, Forney KJ, Pinner D, Keel PK (2017) A randomized controlled trial of The Body Project: more than muscles for men with body dissatisfaction.

Int J Eat Disord 50(8):873–883. https://doi.org/10.1002/eat.22724

Celio AA, Winzelberg AJ, Wilfley DE, Eppstein-Herald D, Springer EA, Dev P, Taylor CB (2000) Reducing risk factors for eating disorders: comparison of an internet- and a classroom-delivered psychoeducational program. J Consult Clin Psychol 68(4):650–657. https://doi.org/10.1037/0022-006X.68.4.650

Ciao AC, Ohls OC, Pringle KD (2018) Should body image programs be inclusive? A focus group study of college students. Int J Eat Disord 51(1):82–86

Cumella EJ, Kally Z (2008) Profile of 50 women with midlife-onset eating disorders. Eat Disord 16 (3):193–203

Doyle AC, Goldschmidt A, Huang C, Winzelberg AJ, Taylor CB, Wilfley DE (2008) Reduction of over-weight and eating disorder symptoms via the internet in adolescents: a randomized controlled trial. J Adolesc Health 43(2):172–179

Fitzsimmons-Craft EE, Balantekin KN, Eichen DM, Graham AK, Monterubio GE, Sadeh-Sharvit S, Goel NJ, Flatt RE, Saffran K, Karam AM (2019a) Screening and offering online programs for eating disorders: reach, pathology, and differences across eating disorder status groups at 28 US universities. Int J Eat Disord. https://doi.org/10.1002/eat.23134

Fitzsimmons-Craft EE, Firebaugh M-L, Graham AK, Eichen DM, Monterubio GE, Balantekin KN, Karam AM, Seal A, Funk B, Taylor CB (2019b) State-wide university implementation of an online platform for eating disorders screening and intervention. Psychol Serv 16(2):239

Forney KJ, Holland LA, Keel PK (2012) Influence of peer context on the relationship between body dissatisfaction and eating pathology in women and men. Int J Eat Disord 45(8):982–989

Gollings EK, Paxton SJ (2006) Comparison of internet and face-to-face delivery of a group body image and disordered eating intervention for women: a pilot study. Eat Disord 14(1):1–15

Heinicke BE, Paxton SJ, McLean SA, Wertheim EH (2007) Internet-delivered targeted group intervention for body dissatisfaction and disordered eating in adolescent girls: a randomized controlled trial. J Abnorm Child Psychol 35(3):379–391

Hudson JI, Hiripi E, Pope HG Jr, Kessler RC (2007) The prevalence and correlates of eating disorders in the National Comorbidity Survey Replication. Biol Psychiatry 61(3):348–358. https://doi.org/10.1016/j.biopsych.2006.03.040

Jacobi C, Morris L, Beckers C, Bronisch-Holtze J, Winter J, Winzelberg AJ, Taylor CB (2007) Maintenance of internet-based prevention: a randomized controlled trial. Int J Eat Disord 40(2):114–119

Jacobi C, Völker U, Trockel MT, Taylor CB (2012) Effects of an internet-based intervention for subthreshold eating disorders: a randomized controlled trial. Behav Res Ther 50(2):93–99

Jones M, Luce KH, Osborne MI, Taylor K, Cunning D, Doyle AC, Wilfley DE, Taylor CB (2008) Randomized, controlled trial of an internet-facilitated intervention for reducing binge eating and overweight in adolescents. Pediatrics 121(3):453–462

Jones M, Lynch KT, Kass AE, Burrows A, Williams J, Wilfley DE, Taylor CB (2014) Healthy weight regulation and eating disorder prevention in high school students: a universal and targeted Web-based intervention. J Med Internet Res 16(2):e57

Keel PK, Forney KJ, Brown TA, Heatherton TF (2013) Influence of college peers on disordered eating in women and men at 10-year follow-up. J Abnorm Psychol 122(1):105–110. https://doi.org/10.1037/a0030081

Kilpela LS, Blomquist K, Verzijl C, Wilfred S, Beyl R, Becker CB (2016) The body project 4 all: a pilot randomized controlled trial of a mixed-gender disso-nance-based body image program. Int J Eat Disord 49 (6):591–602. https://doi.org/10.1002/eat.22562

McLean SA, Paxton SJ, Wertheim EH (2011) A body image and disordered eating intervention for women in midlife: a randomized controlled trial. J Consult Clin Psychol 79(6):751

McVey GL, Lieberman M, Voorberg N, Wardrope D, Blackmore E (2003) School-based peer support groups: a new approach to the prevention of disordered eating. Eat Disord 11(3):169–185

Melioli T, Bauer S, Franko DL, Moessner M, Ozer F, Chabrol H, Rodgers RF (2016) Reducing eating disorder symptoms and risk factors using the internet: a meta-analytic review. Int J Eat Disord 49(1):19–31

Murray SB, Nagata JM, Griffiths S, Calzo JP, Brown TA, Mitchison D, Blashill AJ, Mond JM (2017) The enigma of male eating disorders: a critical review and synthesis. Clin Psychol Rev 57:1–11

Nacke B, Beintner I, Görlich D, Vollert B, Schmidt-Hantke J, Hütter K, Taylor CB, Jacobi C (2019) everyBody–Tailored online health promotion and eating disorder prevention for women: study protocol of a dissemination trial. Internet Interv 16:20–25

Napolitano MA, Himes S (2011) Race, weight, and correlates of binge eating in female college students. Eat Behav 12(1):29–36

Neumark-Sztainer D, Story M, Hannan PJ, Rex J (2003) New Moves: a school-based obesity prevention program for adolescent girls. Prev Med 37(1):41–51

Neumark-Sztainer D, Levine MP, Paxton SJ, Smolak L, Piran N, Wertheim EH (2006a) Prevention of body dissatisfaction and disordered eating: What next? Eat Disord 14(4):265–285

Neumark-Sztainer D, Paxton SJ, Hannan PJ, Haines J, Story M (2006b) Does body satisfaction matter? Five-year longitudinal associations between body satisfaction and health behaviors in adolescent females and males. J Adolesc Health 39(2):244–251

Neumark-Sztainer DR, Friend SE, Flattum CF, Hannan PJ, Story MT, Bauer KW, Feldman SB, Petrich CA (2010)

New moves—preventing weight-related problems in adolescent girls: a group-randomized study. Am J Prev Med 39(5):421–432

Norwood SJ, Murray M, Nolan A, Bowker A (2011) Beautiful from the inside out: a school-based programme designed to increase self-esteem and positive body image among preadolescents. Can J Sch Psychol 26(4):263–282

O'Dea JA, Abraham S (2000) Improving the body image, eating attitudes, and behaviors of young male and female adolescents: a new educational approach that focuses on self-esteem. Int J Eat Disord 28(1):43–57

Paxton SJ, McLean SA, Gollings EK, Faulkner C, Wertheim EH (2007) Comparison of face-to-face and internet interventions for body image and eating problems in adult women: an RCT. Int J Eat Disord 40(8):692–704

Puhl RM, Moss-Racusin CA, Schwartz MB (2007) Internalization of weight bias: implications for binge eating and emotional well-being. Obesity 15(1):19–23

Rasmusson G, Lydecker JA, Coffino JA, White MA, Grilo CM (2019) Household food insecurity is associated with binge-eating disorder and obesity. Int J Eat Disord 52(1):28–35

Richardson SM, Paxton SJ (2010) An evaluation of a body image intervention based on risk factors for body dissatisfaction: a controlled study with adolescent girls. Int J Eat Disord 43(2):112–122

Richardson SM, Paxton SJ, Thomson JS (2009) Is BodyThink an efficacious body image and self-esteem program? A controlled evaluation with adolescents. Body Image 6(2):75–82

Saekow J, Jones M, Gibbs E, Jacobi C, Fitzsimmons-Craft EE, Wilfley D, Taylor CB (2015) StudentBodies-eating disorders: a randomized controlled trial of a coached online intervention for subclinical eating disorders. Internet Interv 2(4):419–428

Steese S, Dollette M, Phillips W, Hossfeld E, Matthews G, Taormina G (2006) Understanding girls' circle as an intervention on perceived social support, body image, self-efficacy, locus of control, and self-esteem. Adolescence 41(161):55–74

Stewart TM, Pollard T, Hildebrandt T, Wesley NY, Kilpela LS, Becker CB (2019) The Female Athlete Body project study: 18-month outcomes in eating disorder symptoms and risk factors. Int J Eat Disord. https://doi.org/10.1002/eat.23145

Stice E (2002) Risk and maintenance factors for eating pathology: a meta-analytic review. Psychol Bull 128(5):825–848

Stice E, Chase A, Stormer S, Appel A (2001) A randomized trial of a dissonance-based eating disorder prevention program. Int J Eat Disord 29(3):247–262

Stice E, Presnell K, Spangler D (2002) Risk factors for binge eating onset in adolescent girls: a 2-year prospective investigation. Health Psychol 21(2):131

Stice E, Shaw H, Burton E, Wade E (2006) Dissonance and healthy weight eating disorder prevention programs: a randomized efficacy trial. J Consult Clin Psychol 74(2):263–275. https://doi.org/10.1037/0022-006X.74.2.263

Stice E, Presnell K, Gau J, Shaw H (2007) Testing mediators of intervention effects in randomized controlled trials: an evaluation of two eating disorder prevention programs. J Consult Clin Psychol 75(1):20–32. https://doi.org/10.1037/0022-006X.75.1.20

Stice E, Marti CN, Spoor S, Presnell K, Shaw H (2008) Dissonance and healthy weight eating disorder prevention programs: long-term effects from a randomized efficacy trial. J Consult Clin Psychol 76(2):329–340. https://doi.org/10.1037/0022-006X.76.2.329

Stice E, Rohde P, Gau J, Shaw H (2009) An effectiveness trial of a dissonance-based eating disorder prevention program for high-risk adolescent girls. J Consult Clin Psychol 77(5):825–834. https://doi.org/10.1037/a0016132

Stice E, Rohde P, Durant S, Shaw H (2012) A preliminary trial of a prototype internet dissonance-based eating disorder prevention program for young women with body image concerns. J Consult Clin Psychol 80(5):907

Stice E, Rohde P, Butryn ML, Shaw H, Marti CN (2015a) Effectiveness trial of a selective dissonance-based eating disorder prevention program with female college students: effects at 2-and 3-year follow-up. Behav Res Ther 71:20–26

Stice E, Yokum S, Waters A (2015b) Dissonance-based eating disorder prevention program reduces reward region response to thin models; how actions shape valuation. PLoS One 10(12):e0144530

Stock S, Miranda C, Evans S, Plessis S, Ridley J, Yeh S, Chanoine J-P (2007) Healthy Buddies: a novel, peer-led health promotion program for the prevention of obesity and eating disorders in children in elementary school. Pediatrics 120(4):e1059–e1068

Tanofsky-Kraff M, Shomaker LB, Wilfley DE, Young JF, Sbrocco T, Stephens M, Ranzenhofer LM, Elliott C, Brady S, Radin RM (2014) Targeted prevention of excess weight gain and eating disorders in high-risk adolescent girls: a randomized controlled trial. Am J Clin Nutr 100(4):1010–1018

Taylor CB, Bryson S, Luce KH, Cunning D, Doyle AC, Abascal LB, Rockwell R, Dev P, Winzelberg AJ, Wilfley DE (2006) Prevention of eating disorders in at-risk college-age women. Arch Gen Psychiatry 63(8):881–888

Taylor C, Taylor K, Jones M, Shorter A, Yee M, Genkin B, Burrows A, Kass A, Rizk M, Redman M (2012) Obesity prevention in defined (high school) populations. Int J Obes Suppl 2(S1):S30

Taylor CB, Kass AE, Trockel M, Cunning D, Weisman H, Bailey J, Sinton M, Aspen V, Schecthman K, Jacobi C (2016) Reducing eating disorder onset in a very high risk sample with significant comorbid depression: a randomized controlled trial. J Consult Clin Psychol 84(5):402

Udo T, Grilo CM (2018) Prevalence and correlates of DSM-5–defined eating disorders in a nationally representative sample of US adults. Biol Psychiatry 84 (5):345–354

Vartanian LR, Porter AM (2016) Weight stigma and eating behavior: a review of the literature. Appetite 102:3–14

Vincent MA, McCabe MP (2000) Gender differences among adolescents in family, and peer influences on body dissatisfaction, weight loss, and binge eating behaviors. J Youth Adolesc 29(2):205–221

Weiss K, Wertheim EH (2005) An evaluation of a prevention program for disordered eating in adolescent girls: examining responses of high-and low-risk girls. Eat Disord 13(2):143–156

Wilksch SM, Wade TD (2009) Reduction of shape and weight concern in young adolescents: a 30-month controlled evaluation of a media literacy program. J Am Acad Child Adolesc Psychiatry 48(6):652–661

Wilksch S, Paxton S, Byrne S, Austin S, McLean S, Thompson K, Dorairaj K, Wade T (2015) Prevention across the spectrum: a randomized controlled trial of three programs to reduce risk factors for both eating disorders and obesity. Psychol Med 45(9):1811–1823

Wilksch SM, O'shea A, Wade TD (2018) Media Smart-Targeted: diagnostic outcomes from a two-country pragmatic online eating disorder risk reduction trial for young adults. Int J Eat Disord 51(3):270–274

Winzelberg AJ, Eppstein D, Eldredge KL, Wilfley D, Dasmahapatra R, Dev P, Taylor CB (2000) Effectiveness of an internet-based program for reducing risk factors for eating disorders. J Consult Clin Psychol 68(2):346

Wott CB, Carels RA (2010) Overt weight stigma, psychological distress and weight loss treatment outcomes. J Health Psychol 15(4):608–614

Zabinski MF, Pung MA, Wilfley DE, Eppstein DL, Winzelberg AJ, Celio A, Taylor CB (2001) Reducing risk factors for eating disorders: targeting at-risk women with a computerized psychoeducational program. Int J Eat Disord 29(4):401–408

Zabinski MF, Wilfley DE, Calfas KJ, Winzelberg AJ, Taylor CB (2004) An interactive psychoeducational intervention for women at risk of developing an eating disorder. J Consult Clin Psychol 72(5):914

Medication for Binge Eating

Susan L. McElroy, Anna I. Guerdjikova, Nicole Mori, and Francisco Romo-Nava

Abstract

This chapter reviews the use of medications in the treatment of individuals with bulimia nervosa (BN) and binge-eating disorder (BED), the two mental disorders defined by the presence of binge eating. Drug classes evaluated include antidepressants, stimulants, and other medications for attention-deficit/ hyperactivity disorder (ADHD), antiepileptic drugs, opioid antagonists, and weight loss agents, among others. The only two drugs with regulatory approval for binge eating are fluoxetine for BN and lisdexamfetamine for BED. Other available drugs with established efficacy in BN and BED include antidepressants (especially selective serotonin reuptake inhibitors) and the antiepileptic topiramate, though the efficacy of these compounds is modest at best. We found no evidence of a drug developed specifically for the treatment of individuals with BN or BED. Importantly, until drugs are developed specifically for eating disorders with binge eating, drugs developed for other conditions that are centrally acting and associated with beneficial psychotropic effects or reduced appetite or weight loss might be considered for repurposing in BN and BED.

Keywords

Binge eating · Bulimia · Medication · Pharmacotherapy

Learning Objectives

In this chapter you will:

1. Gain knowledge about the medications with regulatory approval for the treatment of bulimia nervosa and binge-eating disorder.
2. Gain knowledge about experimental medications for the treatment of binge eating.
3. Gain knowledge about situations when medications might be helpful for the treatment of binge eating.

S. L. McElroy (✉) · A. I. Guerdjikova · F. Romo-Nava
Lindner Center of HOPE, Mason, OH, USA

University of Cincinnati College of Medicine, Cincinnati, OH, USA
e-mail: Susan.McElroy@LindnerCenter.org

N. Mori
Lindner Center of HOPE, Mason, OH, USA

1 Introduction

Individuals with eating disorders (ED) often receive psychiatric medication (Garner et al. 2016; Watson et al. 2016). However, only two

© Springer Nature Switzerland AG 2020
G. K.W. Frank, L. A. Berner (eds.), *Binge Eating*, https://doi.org/10.1007/978-3-030-43562-2_16

drugs have regulatory approval for the treatment of an ED: the selective serotonin reuptake inhibitor (SSRI) antidepressant fluoxetine for bulimia nervosa (BN) and the stimulant prodrug lisdexamfetamine (LDX) for binge-eating disorder (BED). BN and BED are the only two EDs defined by the presence of recurrent binge eating. Though distinct diagnostic entities, the binge-eating episodes for BN and BED are similarly defined in modern nosologic systems (American Psychiatric Association 2013; ICD-11 2018), and there is a substantial diagnostic shift between the two disorders (Fichter et al. 2008). Although not yet definitively proven, a drug that reduces binge eating in one of these EDs might also reduce binge eating in the other (McElroy et al. 2019).

To our knowledge, no other mental (or medical) disorder requires binge eating as a defining feature. Binge eating may occur in individuals with anorexia nervosa (AN), and when it does, the binge-eating/purging type of AN may be diagnosed. However, no medication has regulatory approval for individuals with AN, including for AN with binge eating. Moreover, medication treatment studies in AN rarely report on binge eating as an outcome variable (McElroy et al. 2018a).

In this chapter, we review past and ongoing research with specific medications or medication classes in treating individuals with BN and BED. After searching PubMed and clinical trial registries, we found a broad array of drugs evaluated in the treatment of BN or BED, including antidepressants, stimulants, and other medications for attention-deficit hyperactivity disorder (ADHD), antiepileptic drugs, agents with hormonal properties, and weight loss drugs, among others. We summarize these data below and suggest future areas for research.

2   Antidepressants

As noted earlier, the SSRI antidepressant fluoxetine, which has also been approved for panic disorder and OCD, is the only medication with regulatory approval for the treatment of individuals with BN. In the largest randomized

controlled trial (RCT) of fluoxetine in BN, 387 women with BN were randomized to receive fluoxetine 60 mg/day, fluoxetine 20 mg/day, or placebo for 8 weeks at 13 sites (Fluoxetine Bulimia Nervosa Collaborative Study Group 1992). Fluoxetine 60 mg/day was superior to placebo for reducing binge-eating and vomiting episodes, while 20 mg/day had an intermediate effect (Fluoxetine Bulimia Nervosa Collaborative Study Group 1992). Fluoxetine 60 mg/day was also efficacious for decreasing depression, carbohydrate craving, and pathological eating attitudes and behaviors.

Antidepressants other than fluoxetine have been shown to be superior to placebo for reducing the frequency of both binge-eating and purging episodes in BN (Shapiro et al. 2007; Yager and Powers 2007) and guidelines have concluded that antidepressants, in general, are efficacious for BN (Aigner et al. 2011; Hay et al. 2014). This includes SSRIs other than fluoxetine, tricyclic antidepressants, monoamine oxidase inhibitors, and atypical agents such as mianserin, trazodone, and bupropion. A 2003 meta-analysis of 19 randomized, placebo-controlled trials of antidepressants in individuals with BN showed that antidepressants were modestly superior to placebo for stopping binge-eating behavior [pooled relative risk = 0.87 (95% CI 0.81–0.93, $P < 0.001$) favoring drugs] but associated with a higher dropout rate (Bacaltchuk and Hay 2003). Antidepressants were also safe for the treatment of BN with the important exception of the norepinephrine/dopamine-reuptake inhibitor bupropion. Though efficacious for reducing binge eating and purging in one randomized controlled trial in 55 women with BN, this agent was associated with an increased risk of seizures and is, therefore, contraindicated for the treatment of BN (and AN) (Horne et al. 1988).

Antidepressants may also be modestly efficacious for BED (McElroy et al. 2018a). A meta-analysis of seven RCTs (six with SSRIs and one with a tricyclic) showed significantly higher binge-eating remission rates for the antidepressant group compared with the placebo group: 40.5% versus 22.2% [relative risk = 0.77 (95% CI = 0.65, 0.92, $P = 0.003$)] (Stefano et al.

2008). Preliminary data suggest that the selective serotonin and norepinephrine reuptake inhibitor duloxetine may be effective for reducing binge-eating and depressive symptoms in individuals with BED and a co-occurring depressive syndrome (Guerdjikova et al. 2012). However, a RCT of bupropion in 61 overweight or obese women with BED failed to find a difference between drug and placebo for reducing binge eating (possibly due to the high placebo response), though bupropion recipients lost significantly more weight than placebo recipients did and bupropion was well tolerated with no seizures (White and Grilo 2013). Additionally, an investigator-initiated RCT of vortioxetine (a selective serotonin reuptake inhibitor, 5-HT3 receptor antagonist, and 5-HT1A receptor agonist) in BED failed to find any significant differences between drug and placebo on binge eating or any secondary outcome measures, including body weight (Grant et al. 2019).

Though results are mixed, randomized, placebo-controlled, maintenance of efficacy trials with antidepressants in BN suggest these agents reduce binge eating and purging over up to 3 or 12 months (Fichter et al. 1996; Romano et al. 2002). However, treatment discontinuation rates were high. Randomized, placebo-controlled maintenance of efficacy studies with antidepressants have not yet been conducted in BED.

Antidepressants have been studied both against and in combination with a variety of psychological interventions in BN and BED, including cognitive behavioral therapy (CBT) (Bacaltchuk et al. 2001; McElroy et al. 2018a; Shapiro et al. 2007). Designs and results have varied, making firm conclusions difficult. A Cochrane Review of RCTs in individuals with BN in which antidepressants were compared with psychological treatments, or the combination of antidepressants with psychological treatments, concluded that combined antidepressant–psychological treatment was superior to psychotherapy alone, but that the number of trials might be insufficient to show combination therapy or psychotherapy alone superior to antidepressants alone (Bacaltchuk et al. 2001). The authors also concluded that psychotherapy was more acceptable to BN patients and that the

addition of antidepressants to psychotherapy reduced its acceptability. However, in an important RCT, fluoxetine was superior to placebo for reducing binge eating in BN patients who had an inadequate response to CBT (Walsh et al. 2000).

3 Medications for ADHD

The only drug besides fluoxetine that has regulatory approval for the treatment of an ED is lisdexamfetamine (LDX), a stimulant pro-drug converted to d-amphetamine. LDX is approved for the treatment of moderate-to-severe BED in adults (and for ADHD in children and adults). The approval of LDX for BED was based on a development program in adults with moderate-to-severe BED that included one phase 2 proof-of-concept, dose-finding, 11-week study (McElroy et al. 2015b), two identically-designed, dose-optimization phase 3 acute 12-week trials (McElroy et al. 2016), a phase 3 12-month open-label safety and tolerability study (Gasior et al. 2017), and a phase 3 38-week placebo-controlled, randomized-withdrawal maintenance of efficacy study (Hudson et al. 2017). In the phase-2 RCT, 50 and 70 mg/day of LDX (but not LDX 30 mg/day) decreased binge eating significantly more than placebo. In the two phase-3 acute RCTs, LDX at 50 or 70 mg/day again decreased binge eating significantly more than the placebo. In all three trials, LDX also significantly reduced other aspects of ED psychopathology (including obsessive-compulsive features of binge eating, such as recurrent, persistent, and distressing thoughts to binge eat) and excess body weight. Post hoc analyses of the two phase-3 trials showed that LDX was superior to placebo for reduction of binge eating after 1 week of treatment and remained significantly better than placebo for decreasing binge eating for the duration of the trials (McElroy et al. 2017). In the phase-3 randomized-withdrawal, maintenance of efficacy study (12 weeks of open-label LDX followed by 26 weeks of double-blind treatment with LDX or placebo), LDX demonstrated superiority maintenance of efficacy over placebo for time to relapse (Hudson et al. 2017). Specifically, after 6 months,

3.7% of LDX recipients relapsed as compared to 32.1% of placebo recipients ($P < 0.001$). No new safety concerns (in addition to those known when using the drug in individuals with ADHD) occurred in any of the four trials. The most common side effects of LDX were dry mouth, headache, insomnia, and upper respiratory tract infection. As expected, LDX recipients had minor increases in pulse and blood pressure.

A recently published open-label, 12-week comparison of long-acting methylphenidate (18–72 mg/day) and CBT in the treatment of 49 women with BED found both treatments significantly and comparably reduced binge eating (the primary outcome) as well as other BED symptoms (Quilty et al. 2019). However, long-acting methylphenidate was associated with a greater decrease in body mass index (BMI) compared with CBT.

Preliminary data suggests stimulants might also be efficacious for reducing binge eating in BN. In a double-blind, randomized, crossover trial (the only RCT of a stimulant in BN we were able to locate), eight patients with BN were given methylamphetamine or placebo intravenously followed by a test meal and separated by an 1-week interval (Ong et al. 1983). Significantly fewer mean ($\pm$ SD) calories were consumed after methylamphetamine (224 $\pm$ 111 calories) than after placebo (943 $\pm$ 222 calories; $P < 0.02$). In addition, "the frequency of bulimia" was significantly lower after methylamphetamine (zero of eight patients) than after placebo (four of eight patients; $P < 0.05$). In addition, a growing number of case reports describe the successful use of methylphenidate in treating patients with BN, including those who respond inadequately to psychotherapy and antidepressants and those with comorbid cluster B personality disorders or ADHD (Drimmer 2003; Dukarm 2005; Guerdjikova and McElroy 2013; Keshen and Ivanova 2013; Schweickert et al. 1997; Sokol et al. 1999). Importantly, a feasibility study of LDX is ongoing in patients with BN in Nova Scotia, Canada (NCT03397446).

That LDX is efficacious for reducing binge eating in BED raises the question of whether non-stimulant ADHD drugs might also decrease binge eating. Though no such drugs have been evaluated in RCTs in individuals with BN, the non-stimulant ADHD drug atomoxetine, a selective norepinephrine reuptake inhibitor, was found superior to placebo for reducing binge eating and excess body weight in 40 BED patients (McElroy et al. 2007a). The mean ($\pm$ SD) atomoxetine daily dose at endpoint was 106 $\pm$ 21 mg. Dasotraline is a novel dopamine and norepinephrine reuptake inhibitor that is now undergoing evaluation for the treatment of ADHD and BED (Koblan et al. 2015). Dasotraline (4–8 mg/day) has been shown superior to placebo for reducing binge eating, obsessive-compulsive features of binge eating, and excess body weight in individuals with BED in two RCTs that have been presented at scientific meetings but not yet published (Goldman et al. 2018; Navia et al. 2017).

4 Antiepileptic Drugs

Several antiepileptic drugs have been evaluated for the treatment of BN or BED in RCTs, including topiramate, zonisamide, and lamotrigine. Topiramate has been the most extensively studied, with two positive RCTs in BN and three positive RCTs in BED. In the first BN study, 69 participants received topiramate or placebo for 10 weeks (Hedges et al. 2003; Hoopes et al. 2003). Topiramate (median dose 100 mg/day) was statistically superior to placebo in reducing the frequency of binge-eating and purge days (days during which at least one binge-eating or purging episode occurred), as well as other measures of ED psychopathology and excess body weight. Attrition rates were numerically lower for topiramate (34%) than placebo (47%), and the most common topiramate side effects were fatigue, flu-like symptoms, and paresthesias (the sensation of pins and needles). In the second BN study, 60 participants received topiramate (titrated to 250 mg/day by 6 weeks) or placebo for 10 weeks (Nickel et al. 2005). Compared with placebo, topiramate produced statistically significant decreases in the frequency of binge-eating/purging episodes and excess body weight. Five (17%) patients on topiramate and 6 (20%)

patients on placebo stopped the study prematurely. All patients tolerated topiramate well.

In the first BED study, 61 obese patients received topiramate or placebo for 14 weeks (McElroy et al. 2003). Topiramate (median dose 212 mg daily) was significantly superior to placebo in reducing binge eating, as well as obsessive-compulsive features of binge eating, global illness severity, body weight, and BMI. The dropout rate, however, was high: 14 (47%) topiramate recipients and 12 (39%) placebo recipients failed to complete the trial. The most common topiramate side effects were paresthesias, dry mouth, headache, and dyspepsia.

The second BED study was a multicenter trial in which 407 patients with BED and >3 binge-eating days/week, a BMI between 30 and 52 kg/m^2, and no current comorbid psychiatric disorders were randomized to receive topiramate or placebo for 16 weeks (McElroy et al. 2007b). Compared with placebo, topiramate (median final dose 300 mg daily) significantly decreased binge eating, body weight, and BMI (all P's < 0.001). Topiramate also significantly reduced obsessive-compulsive features of binge eating and other aspects of ED psychopathology. Discontinuation rates were 30% in each group; adverse events were the most common reason for topiramate discontinuation (16%; placebo, 8%). Paresthesias, upper respiratory tract infection, somnolence, and nausea were the most frequent topiramate side effects.

The third BED study was another multicenter trial in which 73 obese patients were randomized to receive 19 sessions of CBT in conjunction with topiramate (target dose 200 mg daily) or placebo for 21 weeks (Claudino et al. 2007). Compared with participants receiving placebo, those receiving topiramate had a significantly greater rate of reduction in body weight, the primary outcome measure ($P < 0.001$). Topiramate recipients lost a mean of 6.8 kg, while placebo recipients lost 0.9 kg. Rates of reduction of binge-eating frequencies and depression scores did not differ between the groups, but a greater percentage of topiramate-treated patients (84%) attained remission of binge eating as compared to placebo-treated patients (61%; $P = 0.03$). There was no difference between groups in completion rates. Paresthesias and taste perversion (altered taste) were more frequent with topiramate, while insomnia was more frequent with placebo.

Topiramate has also been reported to reduce binge eating or purging in BN or BED patients with treatment-resistant illness, those with comorbid mood or personality disorders, and those receiving the drug as adjunctive therapy in combination with antidepressants, mood stabilizers, or antipsychotics (Barbee 2003; Bruno et al. 2009; Felstrom and Blackshaw 2002; Schmidt do Prado-Lima and Bacaltchuck 2002). There is also a report of topiramate decreasing binge eating in a woman with BN and epilepsy (Knable 2001). However, there also are reports of ED patients misusing topiramate to promote weight loss (Chung and Reed 2004; Colom et al. 2001), as well as patients developing EDs, including AN and BN, after topiramate initiation (Lebow et al. 2015).

Regarding studies of other antiepileptic drugs, a RCT in 60 BED patients found that zonisamide produced a statistically significant reduction in binge eating and excess body weight but was associated with a high dropout rate due to adverse effects (27%) (McElroy et al. 2006). In an open-label trial in 12 BN patients, zonisamide reduced binge eating and purging but was again associated with a high dropout rate (50%) (Guerdjikova et al. 2013a). In a RCT of carbamazepine, 16 participants with BN received carbamazepine or placebo in a crossover design (Hudson and Pope 1988; Kaplan 1987; Kaplan et al. 1983). There was no significant difference in response between carbamazepine and placebo. The one patient who completely stopped binge eating also had bipolar disorder. Finally, in a 16-week RCT of lamotrigine in 51 obese patients with BED, lamotrigine and placebo were similarly effective in reducing binge eating, though lamotrigine was superior to placebo for reducing fasting levels of glucose, insulin, and triglycerides (Guerdjikova et al. 2009). There are no RCTs of lamotrigine in BN, but the drug was reported to reduce ED symptoms and mood instability in five patients with BN or eating disorder not otherwise specified (EDNOS) and co-occurring affective dysregulation (Trunko et al. 2014).

There are no RCTs of valproate in BN or BED, but the drug was effective in three hospitalized women with BN and comorbid rapid-cycling bipolar disorder who were previously inadequately unresponsive to lithium and antipsychotics (Herridge and Pope 1985, 1988; McElroy et al. 1987). In contrast, there are reports of valproate increasing binge-eating behavior in BED patients (Shapira et al. 2000).

Of note, antiepileptic drugs with novel mechanisms have recently come to market and are in development (Younus and Reddy 2018). Those agents associated with reducing food craving or hyperphagia but promoting weight loss (like topiramate), in particular, should be considered for evaluation in BN or BED (McElroy et al. 2019). This includes pharmaceutical-grade cannabidiol, which behaves as a negative allosteric modulator of cannabinoid (CB) 1 receptors and is approved for Dravet and Lennox–Gastaut syndromes (Devinsky et al. 2018a, b).

5   Opioid Antagonists and Other Drugs for Addiction

Mu-opioid receptor antagonists have been studied in RCTs in the treatment of individuals with BN or BED with largely negative results. There are three published RCTs of naltrexone in BN. In the first, 16 of 19 women with BN completed a 6-week, placebo-controlled, crossover trial of naltrexone (50 mg/day) and no significant differences in frequency of binge-eating or vomiting episodes between drug and placebo were found at endpoint (Mitchell et al. 1989). In the second study, 28 women with BN (and 41 obese patients with binge eating—see below) were randomized to receive naltrexone (100–150 mg/day), imipramine, or placebo for 8 weeks (Alger et al. 1991). Across all patients, there was no change in the frequency or duration of binge eating. Among the 22 BN patients who completed the trial,

naltrexone was associated with a significant reduction in binge-eating duration ($P = 0.02$), but not binge-eating frequency. In the third study, 13 BN patients received naltrexone up to 200 mg/day or placebo in individualized crossover 6-week trials (Marrazzi et al. 1995a). Significant reductions in binge eating, purging, urges to binge eat and urges to purge occurred with naltrexone. In a fourth study, intravenous administration of naloxone suppressed the consumption of sweet high-fat foods in 20 normal weight and obese women with BN, but not in 21 controls (Drewnowski et al. 1995). Finally, a recently completed but unpublished RCT of intranasal naloxone in 86 women with BN (EudraCT number 2016-003107-65) found no benefit for naloxone over placebo for reducing binge eating or any secondary outcome measure (Opiant Pharmaceuticals 2019).

Three small published RCTs of opioid antagonists for binge eating in patients with BED were negative (Alger et al. 1991; McElroy et al. 2013a; Ziauddeen et al. 2013). In the first (noted above), 8 weeks of naltrexone did not significantly reduce binge-eating frequency or duration in 33 obese binge eaters compared with 8 weeks of placebo or imipramine (Alger et al. 1991). In the second trial, 62 participants with BED were randomized to the novel mu-opioid receptor antagonist ALKS-33 (now called samidorphan) or placebo for 6 weeks (McElroy et al. 2013a). Both drug and placebo produced similar large reductions in binge eating, raising the possibility of a failed rather than a negative trial. However, there were also no differences between drug and placebo in other measures of ED psychopathology or body weight, suggesting a true negative trial. In the third RCT, 63 individuals with BED received the investigational mu-opioid receptor antagonist GSK1521498 at one of two doses or placebo for 4 weeks (Ziauddeen et al. 2013). GSK1521498 had no effect on binge eating or on body weight at either dose.

Nonetheless, open studies suggest some BN patients may respond when treated with doses of naltrexone up to 400 mg/day (Raingeard et al. 2004). In a comparison of standard-dose

(50–100 mg/day) versus high-dose (200–300 mg/day) naltrexone in 16 patients with antidepressant-resistant BN, participants receiving standard-dose naltrexone had no significant change in frequency of binge eating or purging after 6 weeks of treatment, whereas participants receiving high-dose naltrexone had significant reductions in both behaviors (Jonas and Gold 1988). Similarly, there are two favorable case reports of high-dose naltrexone in individuals with BED. The first was a positive on-off-on case of naltrexone monotherapy using doses of 200 and 400 mg/day (Marrazzi et al. 1995b). The second was the successful augmentation of fluoxetine with naltrexone 100 mg/day (Neumeister et al. 1999). Of note, we found no studies of other types of opioid receptor antagonists (e.g., delta or kappa) in the treatment of BN or BED.

Other anti-addiction drugs evaluated in individuals with binge eating in RCTs include acamprosate and baclofen. An RCT of acamprosate in 40 BED patients failed to show drug–placebo separation in the primary statistical (longitudinal) analyses, but showed greater improvement in binge-day frequency with drug versus placebo in a secondary (baseline to endpoint) analysis (McElroy et al. 2011b). In a small crossover RCT in 12 individuals with binge eating, participants were randomized to receive baclofen (titrated to 60 mg/day) for 48 days followed by placebo for 48 days, or the reverse (Corwin et al. 2012). Compared with placebo, baclofen produced a slight but statistically significant reduction in binge-eating frequency, but also a small but statistically significant increase in depressive symptoms. The most commonly reported side effects were tiredness, fatigue, and upset stomach. Of note, BED patients who do not respond to baclofen 60 mg/day may respond to higher doses (up to 180 mg daily) (de Beaurepaire et al. 2015) though side effects may be problematic (Kiel et al. 2015).

Finally, an open-label, 16-week trial of disulfiram 250 mg/day was conducted in 12 participants with BED (Farci et al. 2015). Disulfiram decreased binge eating in all participants but was associated with a high rate of adverse effects, including drowsiness, headaches, altered taste, tachycardia, dizziness, and nausea.

6 Serotonin 5-HT3 Receptor Antagonists

In a 4-week randomized, placebo-controlled trial of the potent and selective antagonist of the serotonin 5-HT3 receptor ondansetron in 26 women with severe BN, ondansetron, which was self-administered in 4-mg capsules up to six per day upon the urge to binge eat or vomit, was associated with a significantly greater decrease in frequency of binge-eating/vomiting episodes and significant improvement in the time spent engaging in bulimic behaviors compared with placebo. There was no difference in weight change between groups. We were unable to locate any RCTs of ondansetron or other serotonin 5-HT3 receptor antagonists in the treatment of patients with BED.

7 Weight Loss Drugs

A number of centrally active weight loss drugs have been studied in the treatment of BN and BED. Regarding drugs that have been removed from the market for safety concerns, RCTs suggest dexfenfluramine (Blouin et al. 1988; Russell et al. 1988; Stunkard et al. 1996), sibutramine (Appolinario et al. 2003; Milano et al. 2005; Wilfley et al. 2008), and the CB1 receptor agonist rimonabant (Pataky et al. 2013) may reduce binge eating in BN or BED. Several available centrally active weight loss agents are undergoing evaluation treatment of BN and BED. These include an ongoing randomized, placebo-controlled crossover study of phentermine–topiramate combination in individuals with BN or BED (NCT02553824) (Dalai et al. 2018), and RCTs of bupropion–naltrexone combination (NCT02317744; NCT03539900) and the glucagon-like peptide 1 (GLP-1) agonist liraglutide 3 mg daily (NCT03279731) in individuals with BED. Additionally, there are open-label reports of obese BED patients displaying reduction of binge-eating behavior (and weight loss) in response to treatment with the phentermine–topiramate combination (Guerdjikova et al. 2015), the bupropion–naltrexone combination (Guerdjikova et al. 2017; McElroy et al. 2013b), and liraglutide 1.8 mg daily (the dose indicated for the treatment of type 2 diabetes) (Robert et al. 2015). However, in a 24-week randomized, controlled weight loss maintenance trial, lorcaserin- and placebo-treated participants showed no difference in change of binge eating (Chao et al. 2018).

RCTs of the peripherally active lipase inhibitor orlistat in obese individuals with BED suggest this drug may produce weight loss in such persons, but its effects on binge-eating behavior have been mixed (Golay et al. 2005; Grilo et al. 2005). It is thus important to note that misuse of orlistat by ED patients, including those with BN, has been reported (Cochrane and Malcolm 2002; Deb et al. 2014; Fernandez-Aranda et al. 2001; Malhotra and McElroy 2002). These observations suggest that peripherally acting weight loss agents might not be efficacious for BN and might even be associated with the risk of misuse in BN patients.

8   Hormonal Treatments

Observations that oxytocin reduces food intake in animals have led to two experimental studies of the effects of oxytocin on eating behavior in women with eating disorders (Leslie et al. 2019). In the first, a single-dose, placebo-controlled, crossover study, 34 BN patients, along with 35 AN patients and 33 healthy control subjects, received oxytocin 40 IU intravenously followed by an emotion recognition task and an apple juice drink (Kim et al. 2015). In healthy controls, oxytocin enhanced emotional sensitivity but had no impact on calorie consumption. AN patients showed no response to the intervention on either outcome, while BN patients displayed enhanced emotion recognition and a decrease in 24-h caloric consumption. In the second study, 25 women with BN or BED and 27 healthy control subjects received intranasal oxytocin 64 IU or intranasal placebo in a randomized, crossover design (Leslie et al. 2019). Oxytocin had no effects on any measure of eating behavior,

including binge eating. The authors concluded that future studies should assess the dose-response effect of oxytocin in human population.

Findings that some women with BN have hyper-androgenic manifestations (e.g., acne, hirsutism, elevated serum androgen levels, and polycystic ovaries) have led to two published RCTs of compounds with anti-androgenic properties in women with BN (Sundblad et al. 2005). In the first study, 46 women with BN received the anti-androgenic drug flutamide, the SSRI citalopram, flutamide plus citalopram, or placebo for 3 months (Sundblad et al. 2005). Final flutamide and citalopram doses were 500 and 40 mg/day, respectively. All three active drug groups had superior improvement on a self-rated global assessment of symptom intensity as compared to the placebo group. A comparison of all flutamide recipients versus placebo recipients showed significant reductions in global ratings ($P = 0.03$) and binge eating ($P = 0.02$), but not vomiting. Compared with the placebo group, binge eating was significantly reduced only in the group receiving flutamide plus citalopram ($P = 0.04$); vomiting was not significantly decreased in any group. In the second study, spironolactone, a diuretic with mineralocorticoid, aldosterone, and androgen antagonistic properties, showed no therapeutic effect compared with placebo among 93 women with BN (von Wietersheim et al. 2008). Of note, an RCT of the anti-androgenic oral contraceptive drospirenone plus ethinyl estradiol in women with BN is ongoing in Sweden (EudraCT number 2011-006099-38).

We found no published or ongoing studies of hormonal agents in individuals with BED. However, in a prospective, 12-month, non-randomized trial, binge-eating scores, BMI, and leptin levels were similar in healthy adults in the contraceptive depot medroxyprogesterone acetate group compared to control women (Silva Dos Santos et al. 2016).

9   Prokinetic Agents

A 6-week RCT of the prokinetic agent erythromycin (up to 500 mg three times daily) in

29 individuals with BN showed no benefit for erythromycin over placebo (Devlin et al. 2012). We found no published or ongoing RCTs of prokinetic drugs in individuals with BED.

10 Lithium

In an RCT in 91 patients with BN, lithium (mean level 0.62 mEq/L) was not superior to placebo in decreasing binge-eating episodes (Hsu et al. 1991). Importantly, there were no serious adverse events. By contrast, case reports have described the successful treatment with lithium of patients with BN or BED and co-occurring bipolar disorder (McElroy et al. 2005). Also, among bipolar patients with binge-eating behavior, lithium may be less likely than the second-generation antipsychotic quetiapine to cause weight gain (Yaramala et al. 2018).

11 Antipsychotics

We found no published or ongoing RCTs of antipsychotics in the treatment of individuals with BN or BED. The successful use of aripiprazole in the treatment of BN patients has been described (Takaki and Okabe 2015; Trunko et al. 2011), but there are also reports of second-generation antipsychotics inducing or exacerbating binge-eating behavior in patients receiving the drugs for mood or psychotic disorders (Brewerton and Shannon 1992; Crockford et al. 1997; Gebhardt et al. 2007; Theisen et al. 2003).

12 Other Medications

An RCT of the alerting agent armodafinil in 60 individuals with BED failed to show drug–placebo separation for the primary outcome of binge-eating day frequency, but showed greater improvement in the secondary outcomes of binge-eating episode frequency, obsessive-compulsive features of binge eating, and BMI (McElroy et al. 2015a). A 12-week, open-label trial of N-acetyl cysteine (NAC) in patients with BN was discontinued after none of the first eight patients responded to NAC, and six (75%) patients stopped the compound prematurely (Guerdjikova et al. 2013b). Small open-label trials have reported that the glutamate-modulating agent memantine (Brennan et al. 2008; Hermanussen and Tresguerres 2005) and the narcolepsy medication sodium oxybate (McElroy et al. 2011a) reduced binge eating in patients with BED without published follow-up.

13 Conclusions

Research into the pharmacotherapy of binge eating has substantially lagged behind that into most other serious psychiatric conditions. Indeed, only two medications have regulatory approval for the treatment of binge eating—fluoxetine for BN and LDX for BED, and no drug is approved for the treatment of binge eating in AN. Moreover, many of the published pharmacotherapy studies in BN and BED have limitations with unclear generalizability of findings to real-world clinical situations. Nonetheless, some preliminary conclusions about the pharmacotherapy of binge eating in individuals with BN or BED can be made.

SSRIs and antidepressants from several other classes are modestly efficacious for reducing binge eating in individuals with BN and those with BED. Though they probably do not produce clinically relevant weight loss in obese individuals, they may also be efficacious for associated depressive symptoms. The antiepileptic drug topiramate appears efficacious for reducing binge eating in individuals with BN and those with BED. The drug also reduces other aspects of ED psychopathology as well as excessive body weight. Additionally, the stimulant prodrug LDX is efficacious for the short- and long-term treatment of BED. It reduces binge-eating behavior, obsessive-compulsive features of binge eating, and excess body weight. Stimulants might also be effective for reducing binge eating in BN, and an important feasibility study of LDX in women with BN is ongoing in Ontario, Canada (NCT03397446). Finally, growing data suggests non-stimulant ADHD drugs, such as atomoxetine and dasotraline, may be efficacious for decreasing binge eating in patients with BED (Goldman et al. 2018; McElroy et al. 2007a; Navia et al. 2017). If dasotraline successfully comes to the market for BED, it should also be evaluated in individuals with BN.

Regarding other medications evaluated in binge eating, further RCTs of serotonin 5-HT3 receptor antagonists in BN are needed to see whether the one small positive study of ondansetron (Faris et al. 2000) can be replicated. Extremely preliminary data suggest RCTs of oxytocin and anti-androgenic compounds in individuals with BN may prove fruitful. Finally, some medications appear to be ineffective for binge eating (e.g., mu-opioid receptor antagonists, lithium, prokinetics, and spironolactone) or may even exacerbate it (second-generation antipsychotics), but more data are needed.

Further pharmacotherapy research into BN and BED is greatly needed at many levels. At a clinical level, and similar to what has been done in other major mental disorders, RCTs are needed that explore strategies where available medications are optimized, switched, augmented, or combined. Thus, studies of topiramate in combination with antidepressants or psychological treatments would be important in patients with treatment-resistant or chronic forms of BN, as has been reported for BED (Brambilla et al. 2009; Claudino et al. 2007). Studies of LDX or other stimulants in combination with antidepressants, topiramate, or psychotherapy would be important for patients with BED. If stimulants prove efficacious in BN, it would be important to conduct such studies in individuals with BN. In addition, medication studies are needed in BN and BED patients who have clinically important psychiatric comorbidities, such as major depressive disorder, bipolar disorder, anxiety disorders, substance use disorders, and self-injurious behavior (Woodside and Staab 2006), as well as patients with co-occurring obesity and -obesity-related medical conditions (e.g., diabetes).

Perhaps most importantly, novel efficacious compounds need to be identified for treating individuals with binge eating. Available drugs that hold promise for BN or BED and merit

evaluation in RCTs include those associated with a decrease in appetite or weight loss, such as the antiepileptics zonisamide and cannabidiol and the long-term weight loss agents naltrexone–bupropion combination, phentermine–topiramate combination, and liraglutide 3 mg daily (Bray et al. 2016). Compounds that might be repurposed for BN or BED include some of those in development for ADHD, mood disorders, epilepsy, obesity, and possibly addiction (McElroy et al. 2019). Future compounds developed specifically for binge eating might target one or some of the many systems involved in regulating feeding behavior and body weight, especially those under central nervous system control (Faulconbridge and Hayes 2011). Thus, potential novel targets might include the cocaine- and amphetamine-regulated transcript peptide (Bharne et al. 2015), glucagon-like peptide 1 (McElroy et al. 2018b), and orexin (Barson 2018) systems, among others.

In sum, pharmacotherapy has an important role in the management of many individuals with BN or BED, especially those who refuse or are unresponsive to psychotherapy and those with chronic or intractable binge-eating behavior. However, the available pharmacotherapeutic armamentarium and its supporting evidence base for BN are BED are far from adequate, and no drug has been shown to reduce binge eating in AN patients. Further study is needed to clarify which available agents might be most useful for which patient subgroups and to identify agents with novel mechanisms of action. In the meantime, current and future medications with psychotropic benefits and/or effects on appetite and weight might be considered as potential therapeutic agents that may be repurposed for EDs with binge eating.

Funding None

Conflicts of Interest SLM is a consultant to or member of the scientific advisory boards of Allergan, Avanir, Bracket, F. Hoffmann-La Roche Ltd., Idorsia, Mitsubishi Tanabe Pharma America, Myriad, Opiant, Shire, and Sunovion. She is a principal or co-investigator on studies sponsored by the Allergan, Avanir, Brainsway,

Marriott Foundation, Myriad, Neurocrine, Novo Nordisk, Shire, and Sunovion. She is also an inventor on the United States Patent No. 6,323,236 B2, Use of Sulfamate Derivatives for Treating Impulse Control Disorders, and along with the patent's assignee, University of Cincinnati, Cincinnati, Ohio, has received payments from Johnson & Johnson in the past, which has exclusive rights under the patent.

AIG, NM, and FR-N declare that they have no conflicts of interest.

References

Aigner M, Treasure J, Kaye W, Kasper S (2011) World Federation of Societies of Biological Psychiatry (WFSBP) guidelines for the pharmacological treatment of eating disorders. World J Biol Psychiatry 12:400–443

Alger SA, Schwalberg MD, Bigaouette JM, Michalek AV, Howard LJ (1991) Effect of a tricyclic antidepressant and opiate antagonist on binge-eating behavior in normoweight bulimic and obese, binge-eating subjects. Am J Clin Nutr 53:865–871

American Psychiatric Association (2013) Diagnostic and statistical manual of mental disorders, 5th edn. American Psychiatric Association, Arlington, VA

Appolinario JC, Bacaltchuk J, Sichieri R, Claudino AM, Godoy-Matos A, Morgan C, Zanella MT, Coutinho W (2003) A randomized, double-blind, placebo-controlled study of sibutramine in the treatment of binge-eating disorder. Arch Gen Psychiatry 60:1109–1116

Bacaltchuk J, Hay P (2003) Antidepressants versus placebo for people with bulimia nervosa. Cochrane Database Syst Rev. https://doi.org/10.1002/14651858.CD003391

Bacaltchuk J, Hay P, Trefiglio R (2001) Antidepressants versus psychological treatments and their combination for bulimia nervosa. Cochrane Database Syst Rev. https://doi.org/10.1002/14651858.CD003385

Barbee JG (2003) Topiramate in the treatment of severe bulimia nervosa with comorbid mood disorders: a case series. Int J Eat Disord 33:468–472

Barson JR (2018) Orexin/hypocretin and dysregulated eating: promotion of foraging behavior. Brain Res. https://doi.org/10.1016/j.brainres.2018.08.018

Bharne AP, Borkar CD, Subhedar NK, Kokare DM (2015) Differential expression of CART in feeding and reward circuits in binge eating rat model. Behav Brain Res 291:219–231

Blouin AG, Blouin JH, Perez EL, Bushnik T, Zuro C, Mulder E (1988) Treatment of bulimia with fenfluramine and desipramine. J Clin Psychopharmacol 8:261–269

Brambilla F, Samek L, Company M, Lovo F, Cioni L, Mellado C (2009) Multivariate therapeutic approach to binge-eating disorder: combined nutritional, psychological and pharmacological treatment. Int Clin Psychopharmacol 24:312–317

Bray GA, Fruhbeck G, Ryan DH, Wilding JP (2016) Management of obesity. Lancet 387:1947–1956

Brennan BP, Roberts JL, Fogarty KV, Reynolds KA, Jonas JM, Hudson JI (2008) Memantine in the treatment of binge eating disorder: an open-label, prospective trial. Int J Eat Disord 41:520–526

Brewerton TD, Shannon M (1992) Possible clozapine exacerbation of bulimia nervosa. Am J Psychiatry 149:1408–1409

Bruno A, Riganello D, Marino A (2009) Treatment with aripiprazole and topiramate in an obese subject with borderline personality disorder, obsessive-compulsive symptoms and bulimia nervosa: a case report. Cases J 2:7288

Chao AM, Wadden TA, Pearl RL, Alamuddin N, Leonard SM, Bakizada ZM, Pinkasavage E, Gruber KA, Walsh OA, Berkowitz RI, Alfaris N, Tronieri JS (2018) A randomized controlled trial of lorcaserin and lifestyle counselling for weight loss maintenance: changes in emotion- and stress-related eating, food cravings and appetite. Clin Obes 8:383–390

Chung AM, Reed MD (2004) Intentional topiramate ingestion in an adolescent female. Ann Pharmacother 38:1439–1442

Claudino AM, de Oliveira IR, Appolinario JC, Cordas TA, Duchesne M, Sichieri R, Bacaltchuk J (2007) Double-blind, randomized, placebo-controlled trial of topiramate plus cognitive-behavior therapy in binge-eating disorder. J Clin Psychiatry 68:1324–1332

Cochrane C, Malcolm R (2002) Case report of abuse of Orlistat. Eat Behav 3:167–169

Colom F, Vieta E, Benabarre A, Martinez-Aran A, Reinares M, Corbella B, Gasto C (2001) Topiramate abuse in a bipolar patient with an eating disorder. J Clin Psychiatry 62:475–476

Corwin RL, Boan J, Peters KF, Ulbrecht JS (2012) Baclofen reduces binge eating in a double-blind, placebo-controlled, crossover study. Behav Pharmacol 23:616–625

Crockford DN, Fisher G, Barker P (1997) Risperidone, weight gain, and bulimia nervosa. Can J Psychiatr 42:326–327

Dalai SS, Adler S, Najarian T, Safer DL (2018) Study protocol and rationale for a randomized double-blinded crossover trial of phentermine-topiramate ER versus placebo to treat binge eating disorder and bulimia nervosa. Contemp Clin Trials 64:173–178

de Beaurepaire R, Joussaume B, Rapp A, Jaury P (2015) Treatment of binge eating disorder with high-dose baclofen: a case series. J Clin Psychopharmacol 35:357–359

Deb KS, Gupta R, Varshney M (2014) Orlistat abuse in a case of bulimia nervosa: the changing Indian society. Gen Hosp Psychiatry 36(549):e543–e544

Devinsky O, Patel AD, Cross JH, Villanueva V, Wirrell EC, Privitera M, Greenwood SM, Roberts C, Checketts D, VanLandingham KE, Zuberi SM, GWPCARE3 Study Group (2018a) Effect of cannabidiol on drop seizures in the Lennox-Gastaut Syndrome. N Engl J Med 378:1888–1897

Devinsky O, Patel AD, Thiele EA, Wong MH, Appleton R, Harden CL, Greenwood S, Morrison G, Sommerville K, GWPCARE1 Part A Study Group (2018b) Randomized, dose-ranging safety trial of cannabidiol in Dravet syndrome. Neurology 90: e1204–e1211

Devlin MJ, Kissileff HR, Zimmerli EJ, Samuels F, Chen BE, Brown AJ, Geliebter A, Walsh BT (2012) Gastric emptying and symptoms of bulimia nervosa: effect of a prokinetic agent. Physiol Behav 106:238–242

Drewnowski A, Krahn DD, Demitrack MA, Nairn K, Gosnell BA (1995) Naloxone, an opiate blocker, reduces the consumption of sweet high-fat foods in obese and lean female binge eaters. Am J Clin Nutr 61:1206–1212

Drimmer EJ (2003) Stimulant treatment of bulimia nervosa with and without attention-deficit disorder: three case reports. Nutrition 19:76–77

Dukarm CP (2005) Bulimia nervosa and attention deficit hyperactivity disorder: a possible role for stimulant medication. J Womens Health (Larchmt) 14:345–350

Farci AM, Piras S, Murgia M, Chessa A, Restivo A, Gessa GL, Agabio R (2015) Disulfiram for binge eating disorder: an open trail. Eat Behav 16:84–87

Faris PL, Kim SW, Meller WH, Goodale RL, Oakman SA, Hofbauer RD, Marshall AM, Daughters RS, Banerjee-Stevens D, Eckert ED, Hartman BK (2000) Effect of decreasing afferent vagal activity with ondansetron on symptoms of bulimia nervosa: a randomised, double-blind trial. Lancet 355:792–797

Faulconbridge LF, Hayes MR (2011) Regulation of energy balance and body weight by the brain: a distributed system prone to disruption. Psychiatr Clin North Am 34:733–745

Felstrom A, Blackshaw S (2002) Topiramate for bulimia nervosa with bipolar II disorder. Am J Psychiatry 159:1246–1247

Fernandez-Aranda F, Amor A, Jimenez-Murcia S, Gimenez-Martinez L, Turon-Gil V, Vallejo-Ruiloba J (2001) Bulimia nervosa and misuse of orlistat: two case reports. Int J Eat Disord 30:458–461

Fichter MM, Kruger R, Rief W, Holland R, Dohne J (1996) Fluvoxamine in prevention of relapse in bulimia nervosa: effects on eating-specific psychopathology. J Clin Psychopharmacol 16:9–18

Fichter MM, Quadflieg N, Hedlund S (2008) Long-term course of binge eating disorder and bulimia nervosa: relevance for nosology and diagnostic criteria. Int J Eat Disord 41:577–586

Fluoxetine Bulimia Nervosa Collaborative Study Group (1992) Fluoxetine in the treatment of bulimia nervosa. A multicenter, placebo-controlled, double-blind trial. Arch Gen Psychiatry 49:139–147

Garner DM, Anderson ML, Keiper CD, Whynott R, Parker L (2016) Psychotropic medications in adult and adolescent eating disorders: clinical practice versus evidence-based recommendations. Eat Weight Disord 21:395–402

Gasior M, Hudson J, Quintero J, Ferreira-Cornwell MC, Radewonuk J, McElroy SL (2017) A phase 3, multi-center, open-label, 12-month extension safety and tolerability trial of lisdexamfetamine dimesylate in adults with binge eating disorder. J Clin Psychopharmacol 37:315–322

Gebhardt S, Haberhausen M, Krieg JC, Remschmidt H, Heinzel-Gutenbrunner M, Hebebrand J, Theisen FM (2007) Clozapine/olanzapine-induced recurrence or deterioration of binge eating-related eating disorders. J Neural Transm 114:1091–1095

Golay A, Laurent-Jaccard A, Habicht F, Gachoud JP, Chabloz M, Kammer A, Schutz Y (2005) Effect of orlistat in obese patients with binge eating disorder. Obes Res 13:1701–1708

Goldman R, Hudson JI, McElroy SL, Grilo CM, Tsai J, Deng L, et al. (2018) Efficacy and safety of dastroline in adults with binge-eating disorder: a randomized, double-blind, fixed dose trial, ACNP meeting, Dec 9–13, Orlando, FL

Grant JE, Valle S, Cavic E, Redden SA, Chamberlain SR (2019) A double-blind, placebo-controlled study of vortioxetine in the treatment of binge-eating disorder. Int J Eat Disord 52:786–794

Grilo CM, Masheb RM, Salant SL (2005) Cognitive behavioral therapy guided self-help and orlistat for the treatment of binge eating disorder: a randomized, double-blind, placebo-controlled trial. Biol Psychiatry 57:1193–1201

Guerdjikova AI, McElroy SL (2013) Adjunctive methylphenidate in the treatment of bulimia nervosa co-occurring with bipolar disorder and substance dependence. Innov Clin Neurosci 10:30–33

Guerdjikova AI, McElroy SL, Welge JA, Nelson E, Keck PE, Hudson JI (2009) Lamotrigine in the treatment of binge-eating disorder with obesity: a randomized, placebo-controlled monotherapy trial. Int Clin Psychopharmacol 24:150–158

Guerdjikova AI, McElroy SL, Winstanley EL, Nelson EB, Mori N, McCoy J, Keck PE Jr, Hudson JI (2012) Duloxetine in the treatment of binge eating disorder with depressive disorders: a placebo-controlled trial. Int J Eat Disord 45:281–289

Guerdjikova AI, Blom TJ, Martens BE, Keck PE Jr, McElroy SL (2013a) Zonisamide in the treatment of bulimia nervosa: an open-label, pilot, prospective study. Int J Eat Disord 46:747–750

Guerdjikova AI, Blom TJ, Mori N, McElroy SL (2013b) N-acetylcysteine in bulimia nervosa—open-label trial. Eat Behav 14:87–89

Guerdjikova AI, Fitch A, McElroy SL (2015) Successful treatment of binge eating disorder with combination phentermine/topiramate extended release. Prim Care

Companion CNS Disord 17(2). https://doi.org/10.4088/PCC.14l01708

Guerdjikova AI, Walsh B, Shan K, Halseth AE, Dunayevich E, McElroy SL (2017) Concurrent improvement in both binge eating and depressive symptoms with naltrexone/bupropion therapy in overweight or obese subjects with major depressive disorder in an openlLabel, uncontrolled study. Adv Ther 34:2307–2315

Hay P, Chinn D, Forbes D, Madden S, Newton R, Sugenor L, Touyz S, Ward W, Royal Australian and New Zealand College of Psychiatrists (2014) Royal Australian and New Zealand College of Psychiatrists clinical practice guidelines for the treatment of eating disorders. Aust N Z J Psychiatry 48:977–1008

Hedges DW, Reimherr FW, Hoopes SP, Rosenthal NR, Kamin M, Karim R, Capece JA (2003) Treatment of bulimia nervosa with topiramate in a randomized, double-blind, placebo-controlled trial, part 2: improvement in psychiatric measures. J Clin Psychiatry 64:1449–1454

Hermanussen M, Tresguerres JA (2005) A new anti-obesity drug treatment: first clinical evidence that, antagonising glutamate-gated Ca2+ ion channels with memantine normalises binge-eating disorders. Econ Hum Biol 3:329–337

Herridge PL, Pope HG Jr (1985) Treatment of bulimia and rapid-cycling bipolar disorder with sodium valproate: a case report. J Clin Psychopharmacol 5:229–230

Hoopes SP, Reimherr FW, Hedges DW, Rosenthal NR, Kamin M, Karim R, Capece JA, Karvois D (2003) Treatment of bulimia nervosa with topiramate in a randomized, double-blind, placebo-controlled trial, part 1: improvement in binge and purge measures. J Clin Psychiatry 64:1335–1341

Horne RL, Ferguson JM, Pope HG Jr, Hudson JI, Lineberry CG, Ascher J, Cato A (1988) Treatment of bulimia with bupropion: a multicenter controlled trial. J Clin Psychiatry 49:262–266

Hsu LK, Clement L, Santhouse R, Ju ES (1991) Treatment of bulimia nervosa with lithium carbonate. A controlled study. J Nerv Ment Dis 179:351–355

Hudson JI, Pope HG Jr (1988) The role of anticonvulsants in the treatment of bulimia. In: McElroy SL, Pope HG Jr (eds) Use of anticonvulsants in psychiatry. Oxford Health Care, Clifton, NJ, pp 141–153

Hudson JI, McElroy SL, Ferreira-Cornwell MC, Radewonuk J, Gasior M (2017) Efficacy of lisdexamfetamine in adults with moderate to severe binge-eating disorder: a randomized controlled clinical trial. JAMA Psychiat 74:903–910

ICD-11 (2018) The international classification diseases, 11th Revision

Jonas JM, Gold MS (1988) The use of opiate antagonists in treating bulimia: a study of low-dose versus high-dose naltrexone. Psychiatry Res 24:195–199

Kaplan AS (1987) Anticonvulsant treatment of eating disorders. In: Garfinkel PE, Garner DM (eds) The role of drug treatments for eating disorders. Brunner/Mazel, New York, pp 96–123

Kaplan AS, Garfinkel PE, Darby PL, Garner DM (1983) Carbamazepine in the treatment of bulimia. Am J Psychiatry 140:1225–1226

Keshen A, Ivanova I (2013) Reduction of bulimia nervosa symptoms after psychostimulant initiation in patients with comorbid ADHD: five case reports. Eat Disord 21:360–369

Kiel LB, Hoegberg LC, Jansen T, Petersen JA, Dalhoff KP (2015) A nationwide register-based survey of baclofen toxicity. Basic Clin Pharmacol Toxicol 116:452–456

Kim YR, Eom JS, Yang JW, Kang J, Treasure J (2015) The impact of oxytocin on food intake and emotion recognition in patients with eating disorders: a double blind single dose within-subject cross-over design. PLoS One 10:e0137514

Knable M (2001) Topiramate for bulimia nervosa in epilepsy. Am J Psychiatry 158:322–323

Koblan KS, Hopkins SC, Sarma K, Jin F, Goldman R, Kollins SH, Loebel A (2015) Dasotraline for the treatment of attention-deficit/hyperactivity disorder: a randomized, double-blind, placebo-controlled, proof-of-concept trial in adults. Neuropsychopharmacology 40:2745–2752

Lebow J, Chuy JA, Cedermark K, Cook K, Sim LA (2015) The development or exacerbation of eating disorder symptoms after topiramate initiation. Pediatrics 135: e1312–e1316

Leslie M, Leppanen J, Paloyelis Y, Treasure J (2019) The influence of oxytocin on eating behaviours and stress in women with bulimia nervosa and binge eating disorder. Mol Cell Endocrinol 497:110354

Malhotra S, McElroy SL (2002) Orlistat misuse in bulimia nervosa. Am J Psychiatry 159:492–493

Marrazzi MA, Bacon JP, Kinzie J, Luby ED (1995a) Naltrexone use in the treatment of anorexia nervosa and bulimia nervosa. Int Clin Psychopharmacol 10:163–172

Marrazzi MA, Markham KM, Kinzie J, Luby ED (1995b) Binge eating disorder: response to naltrexone. Int J Obes Relat Metab Disord 19:143–145

McElroy SL, Keck PE Jr, Pope HG Jr (1987) Sodium valproate: its use in primary psychiatric disorders. J Clin Psychopharmacol 7:16–24

McElroy SL, Arnold LM, Shapira NA, Keck PE Jr, Rosenthal NR, Karim MR, Kamin M, Hudson JI (2003) Topiramate in the treatment of binge eating disorder associated with obesity: a randomized, placebo-controlled trial. Am J Psychiatry 160:255–261

McElroy SL, Kotwal R, Keck PE Jr, Akiskal HS (2005) Comorbidity of bipolar and eating disorders: distinct or related disorders with shared dysregulations? J Affect Disord 86:107–127

McElroy SL, Kotwal R, Guerdjikova AI, Welge JA, Nelson EB, Lake KA, D'Alessio DA, Keck PE, Hudson JI (2006) Zonisamide in the treatment of binge eating disorder with obesity: a randomized controlled trial. J Clin Psychiatry 67:1897–1906

McElroy SL, Guerdjikova A, Kotwal R, Welge JA, Nelson EB, Lake KA, Keck PE Jr, Hudson JI (2007a) Atomoxetine in the treatment of binge-eating disorder: a randomized placebo-controlled trial. J Clin Psychiatry 68:390–398

McElroy SL, Hudson JI, Capece JA, Beyers K, Fisher AC, Rosenthal NR, Topiramate Binge Eating Disorder Research Group (2007b) Topiramate for the treatment of binge eating disorder associated with obesity: a placebo-controlled study. Biol Psychiatry 61:1039–1048

McElroy SL, Guerdjikova AI, Winstanley EL, O'Melia AM, Mori N, Keck PE Jr, Hudson JI (2011a) Sodium oxybate in the treatment of binge eating disorder: an open-label, prospective study. Int J Eat Disord 44:262–268

McElroy SL, Guerdjikova AI, Winstanley EL, O'Melia AM, Mori N, McCoy J, Keck PE Jr, Hudson JI (2011b) Acamprosate in the treatment of binge eating disorder: a placebo-controlled trial. Int J Eat Disord 44:81–90

McElroy SL, Guerdjikova AI, Blom TJ, Crow SJ, Memisoglu A, Silverman BL, Ehrich EW (2013a) A placebo-controlled pilot study of the novel opioid receptor antagonist ALKS-33 in binge eating disorder. Int J Eat Disord 46:239–245

McElroy SL, Guerdjikova AI, Kim DD, Burns C, Harris-Collazo R, Landbloom R, Dunayevich E (2013b) Naltrexone/bupropion combination therapy in overweight or obese patients with major depressive disorder: results of a pilot study. Prim Care Companion CNS Disord 15. https://doi.org/10.4088/PCC.12m01494

McElroy SL, Guerdjikova AI, Mori N, Blom TJ, Williams S, Casuto LS, Keck PE Jr (2015a) Armodafinil in binge eating disorder: a randomized, placebo-controlled trial. Int Clin Psychopharmacol 30:209–215

McElroy SL, Hudson JI, Mitchell JE, Wilfley D, Ferreira-Cornwell MC, Gao J, Wang J, Whitaker T, Jonas J, Gasior M (2015b) Efficacy and safety of lisdexamfetamine for treatment of adults with moderate to severe binge-eating disorder: a randomized clinical trial. JAMA Psychiat 72:235–246

McElroy SL, Hudson J, Ferreira-Cornwell MC, Radewonuk J, Whitaker T, Gasior M (2016) Lisdexamfetamine dimesylate for adults with moderate to severe binge eating disorder: results of two pivotal phase 3 randomized controlled trials. Neuropsychopharmacology 41:1251–1260

McElroy SL, Hudson JI, Gasior M, Herman BK, Radewonuk J, Wilfley D, Busner J (2017) Time course of the effects of lisdexamfetamine dimesylate in two phase 3, randomized, double-blind, placebo-controlled trials in adults with binge-eating disorder. Int J Eat Disord 50:884–892

McElroy SL, Guerdjikova AI, Mori N, Keck PE Jr (2018a) Pharmacotherapy for eating disorders. In: Agras WS, Robinson A (eds) The Oxford handbook of eating disorders, 2nd edn. Oxford University Press, Oxford

McElroy SL, Mori N, Guerdjikova AI, Keck PE Jr (2018b) Would glucagon-like peptide-1 receptor agonists have efficacy in binge eating disorder and bulimia nervosa? A review of the current literature. Med Hypotheses 111:90–93

McElroy SL, Guerdjikova AI, Mori N, Romo-Nava F (2019) Progress in developing pharmacologic agents to treat bulimia nervosa. CNS Drugs 33:31–46

Milano W, Petrella C, Casella A, Capasso A, Carrino S, Milano L (2005) Use of sibutramine, an inhibitor of the reuptake of serotonin and noradrenaline, in the treatment of binge eating disorder: a placebo-controlled study. Adv Ther 22:25–31

Mitchell JE, Christenson G, Jennings J, Huber M, Thomas B, Pomeroy C, Morley J (1989) A placebo-controlled, double-blind crossover study of naltrexone hydrochloride in outpatients with normal weight bulimia. J Clin Psychopharmacol 9:94–97

Navia B, Hudson J, McElroy S, Guerdjikova A, Deng L, Sarma K, Hopkins S, Koblan K, Loebel A, Goldman R (2017) Dasotraline for the treatment of moderate-to-severe binge eating disorder in adults: results from a randomized, double-blind, placebo-controlled study, APA Meeting, May 20–24, San Diego, CA

Neumeister A, Winkler A, Wober-Bingol C (1999) Addition of naltrexone to fluoxetine in the treatment of binge eating disorder. Am J Psychiatry 156:797

Nickel C, Tritt K, Muehlbacher M, Pedrosa Gil F, Mitterlehner FO, Kaplan P, Lahmann C, Leiberich PK, Krawczyk J, Kettler C, Rother WK, Loew TH, Nickel MK (2005) Topiramate treatment in bulimia nervosa patients: a randomized, double-blind, placebo-controlled trial. Int J Eat Disord 38:295–300

Ong YL, Checkley SA, Russell GF (1983) Suppression of bulimic symptoms with methylamphetamine. Br J Psychiatry 143:288–293

Opiant Pharmaceuticals (2019) Opiant pharmaceuticals announces top-line results from phase 2 clinical trial of OPNT001 for treatment of Bulimia Nervosa. https://ir.opiant.com/news-releases/news-release-details/opiant-pharmaceuticals-announces-top-line-results-phase-2

Pataky Z, Gasteyger C, Ziegler O, Rissanen A, Hanotin C, Golay A (2013) Efficacy of rimonabant in obese patients with binge eating disorder. Exp Clin Endocrinol Diabetes 121:20–26

Quilty LC, Allen TA, Davis C, Knyahnytska Y, Kaplan AS (2019) A randomized comparison of long acting methylphenidate and cognitive behavioral therapy in the treatment of binge eating disorder. Psychiatry Res 273:467–474

Raingeard I, Courtet P, Renard E, Bringer J (2004) Naltrexone improves blood glucose control in type 1 diabetic women with severe and chronic eating disorders. Diabetes Care 27:847–848

Robert SA, Rohana AG, Shah SA, Chinna K, Wan Mohamud WN, Kamaruddin NA (2015) Improvement in binge eating in non-diabetic obese individuals after

3 months of treatment with liraglutide - a pilot study. Obes Res Clin Pract 9:301–304

Romano SJ, Halmi KA, Sarkar NP, Koke SC, Lee JS (2002) A placebo-controlled study of fluoxetine in continued treatment of bulimia nervosa after successful acute fluoxetine treatment. Am J Psychiatry 159:96–102

Russell GF, Checkley SA, Feldman J, Eisler I (1988) A controlled trial of d-fenfluramine in bulimia nervosa. Clin Neuropharmacol 11(Suppl 1):S146–S159

Schmidt do Prado-Lima PA, Bacaltchuck J (2002) Topiramate in treatment-resistant depression and binge-eating disorder. Bipolar Disord 4:271–273

Schweickert LA, Strober M, Moskowitz A (1997) Efficacy of methylphenidate in bulimia nervosa comorbid with attention-deficit hyperactivity disorder: a case report. Int J Eat Disord 21:299–301

Shapira NA, Goldsmith TD, McElroy SL (2000) Treatment of binge-eating disorder with topiramate: a clinical case series. J Clin Psychiatry 61:368–372

Shapiro JR, Berkman ND, Brownley KA, Sedway JA, Lohr KN, Bulik CM (2007) Bulimia nervosa treatment: a systematic review of randomized controlled trials. Int J Eat Disord 40:321–336

Silva Dos Santos PN, de Souza AL, Batista GA, Melhado-Kimura V, de Lima GA, Bahamondes L, Fernandes A (2016) Binge eating and biochemical markers of appetite in new users of the contraceptive depot medroxyprogesterone acetate. Arch Gynecol Obstet 294:1331–1336

Sokol MS, Gray NS, Goldstein A, Kaye WH (1999) Methylphenidate treatment for bulimia nervosa associated with a cluster B personality disorder. Int J Eat Disord 25:233–237

Stefano SC, Bacaltchuk J, Blay SL, Appolinario JC (2008) Antidepressants in short-term treatment of binge eating disorder: systematic review and meta-analysis. Eat Behav 9:129–136

Stunkard A, Berkowitz R, Tanrikut C, Reiss E, Young L (1996) D-fenfluramine treatment of binge eating disorder. Am J Psychiatry 153:1455–1459

Sundblad C, Landen M, Eriksson T, Bergman L, Eriksson E (2005) Effects of the androgen antagonist flutamide and the serotonin reuptake inhibitor citalopram in bulimia nervosa: a placebo-controlled pilot study. J Clin Psychopharmacol 25:85–88

Takaki M, Okabe N (2015) Aripiprazole may be effective as an add-on treatment in bulimic symptoms of eating disorders. J Clin Psychopharmacol 35:93–95

Theisen FM, Linden A, Konig IR, Martin M, Remschmidt H, Hebebrand J (2003) Spectrum of binge eating symptomatology in patients treated with clozapine and olanzapine. J Neural Transm 110:111–121

Trunko ME, Schwartz TA, Duvvuri V, Kaye WH (2011) Aripiprazole in anorexia nervosa and low-weight bulimia nervosa: case reports. Int J Eat Disord 44:269–275

Trunko ME, Schwartz TA, Marzola E, Klein AS, Kaye WH (2014) Lamotrigine use in patients with binge eating and purging, significant affect dysregulation, and poor impulse control. Int J Eat Disord 47:329–334

von Wietersheim J, Muler-Bock V, Rauh S, Danner B, Chrenko K, Buhler G (2008) No effect of spironolactone on bulimia nervosa symptoms. J Clin Psychopharmacol 28:258–260

Walsh BT, Agras WS, Devlin MJ, Fairburn CG, Wilson GT, Kahn C, Chally MK (2000) Fluoxetine for bulimia nervosa following poor response to psychotherapy. Am J Psychiatry 157:1332–1334

Watson HJ, Jangmo A, Munn-Chernoff MA, Thornton LM, Welch E, Wiklund C, von Hausswolff-Juhlin Y, Norring C, Herman BK, Larsson H, Bulik CM (2016) A register-based case-control study of prescription medication utilization in binge-eating disorder. Prim Care Companion CNS Disord 18. https://doi.org/10.4088/PCC.16m01939

White MA, Grilo CM (2013) Bupropion for overweight women with binge-eating disorder: a randomized, double-blind, placebo-controlled trial. J Clin Psychiatry 74:400–406

Wilfley DE, Crow SJ, Hudson JI, Mitchell JE, Berkowitz RI, Blakesley V, Walsh BT (2008) Efficacy of sibutramine for the treatment of binge eating disorder: a randomized multicenter placebo-controlled double-blind study. Am J Psychiatry 165:51–58

Woodside BD, Staab R (2006) Management of psychiatric comorbidity in anorexia nervosa and bulimia nervosa. CNS Drugs 20:655–663

Yager J, Powers PS (2007) Clinical manual of eating disorders. American Psychiatric Publishing, Washington, DC

Yaramala SR, McElroy SL, Geske J, Winham S, Gao K, Reilly-Harrington NA, Ketter TA, Deckersbach T, Kinrys G, Kamali M, Sylvia LG, McInnis MG, Friedman ES, Thase ME, Kocsis JH, Tohen M, Calabrese JR, Bowden CL, Shelton RC, Nierenberg AA, Bobo WV (2018) The impact of binge eating behavior on lithium- and quetiapine-associated changes in body weight, body mass index, and waist circumference during 6 months of treatment: findings from the bipolar CHOICE study. J Affect Disord. https://doi.org/10.1016/j.jad.2018.09.025

Younus I, Reddy DS (2018) A resurging boom in new drugs for epilepsy and brain disorders. Expert Rev Clin Pharmacol 11:27–45

Ziauddeen H, Chamberlain SR, Nathan PJ, Koch A, Maltby K, Bush M, Tao WX, Napolitano A, Skeggs AL, Brooke AC, Cheke L, Clayton NS, Sadaf Farooqi I, O'Rahilly S, Waterworth D, Song K, Hosking L, Richards DB, Fletcher PC, Bullmore ET (2013) Effects of the mu-opioid receptor antagonist GSK1521498 on hedonic and consummatory eating behaviour: a proof of mechanism study in binge-eating obese subjects. Mol Psychiatry 18:1287–1293

Oxytocin: Potential New Treatment for Binge Eating

Youl-Ri Kim, Soo Min Hong, and Jung-Joon Moon

Abstract

The brain neuropeptide hormone, oxytocin, is of particular interest in eating disorders, given its known role in appetite regulation. Oxytocin appears to be involved not only in the maintenance of energy homeostasis but also in the regulation of hedonic eating. Evidence points to a potential genetic risk for binge-eating disorders associated with variants of the oxytocin receptor. Moreover, there are several lines of evidence that indicate a reduction in calorie intake after oxytocin administration in bulimia nervosa, and that oxytocin has distinct hypophagic effects in individuals with obesity, which is commonly associated with binge-eating behavior. The currently available results suggest that oxytocin could become a therapeutic consideration in the treatment of binge-eating-related disorders.

Keywords

Oxytocin · Binge eating · Hedonic eating · Oxytocin receptor · Bulimia nervosa

Learning Objectives

After reading this chapter, the learner will be able to:

- Understand the role of oxytocin on metabolism in humans.
- Discuss research on the associations between oxytocin levels and eating behaviors in humans.
- Discuss how oxytocin could be developed as a useful therapeutic agent for the treatment of eating disorders associated with binge eating.

1 Introduction

Oxytocin is a polypeptide hormone consisting of nine amino acids (Lawson 2017). Oxytocin is synthesized in the magnocellular neurons and parvocellular neurons of the paraventricular hypothalamic nucleus (PVH) and magnocellular neurons of the supraoptic nucleus in the hypothalamus (Gimpl and Fahrenholz 2001). The parvocellular oxytocin neurons protrude into the forebrain amygdala and brain stem. In contrast,

Y.-R. Kim (✉)
Department of Psychiatry, Seoul Paik Hospital, Inje University, Seoul, Republic of Korea
e-mail: youlri.kim@paik.ac.kr

S. M. Hong
Department of Endocrinology and Metabolism, Seoul Paik Hospital, Inje University, Seoul, Republic of Korea

J.-J. Moon
Department of Psychiatry, Busan Paik Hospital, Inje University, Busan, Republic of Korea

the magnocellular oxytocin neurons protrude into the neurohypophysis and secrete oxytocin to the body periphery via the posterior pituitary (Poulain and Wakerley 1982). Oxytocin is also synthesized in multiple peripheral organs, such as the placenta, ovaries, uterus, testis, thymus, heart, blood vessels, kidneys, and skin (Gimpl and Fahrenholz 2001; Denda et al. 2012).

Oxytocin receptors are expressed in the myometrium and myoepithelial cells in the muscle layer of the uterus, pancreas, adipocytes, heart, blood vessels, kidneys, thymus, airway smooth muscle cells, and macrophages (Jankowski et al. 2000; Kiss and Mikkelsen 2005). Oxytocin receptors, which belong to the G-protein-coupled receptor family, are involved in the regulation of various physiologic processes, including eating behavior and metabolism (Lawson 2017). The oxytocin receptors induce uterine contraction, nitric oxide production, cardiomyogenesis, and lipolysis in adipocytes by coupling with G-protein and activating the Cβ pathway (Chaves et al. 2011; Danalache et al. 2010; Stoop 2012; Egan et al. 1990). Oxytocin acts in brain areas, such as the ventral tegmental area, nucleus accumbens, and the bed nucleus of the stria terminalis, areas that are involved in reward and eating behavior (Olszewski et al. 2016; Sabatier et al. 2013; Spetter and Hallschmid 2017).

Oxytocin is best known for its involvement in labor during childbirth as well as breastfeeding, but it also impacts social behavior, and metabolic processes (Ferguson et al. 2000). Furthermore, 30 years of accumulated research have demonstrated that oxytocin administration can reduce food intake and body weight, and increases energy expenditure. Oxytocin has been suggested as a potential new treatment option for binge eating due to those effects on food intake. In this chapter, we will describe and discuss the effects of oxytocin related to eating behavior and their implications on binge eating.

2   Metabolic Effects of Oxytocin

Oxytocin regulates appetite, reduces caloric intake, and increases fat oxidation and lipolysis. Further, it elevates thermoregulation and energy consumption while lowering body weight and visceral fat. It also improves insulin sensitivity and is involved in glucose and metabolic homeostasis (Lawson 2017; Chaves et al. 2013). It has been hypothesized that oxytocin decreases food intake by acting as a satiety hormone that induces a feeling of satiation in the brain. Conversely, oxytocin levels have been positively correlated with body mass index (weight kg/height m^2) BMI and fat mass (Schorr et al. 2017). Several studies reported that oxytocin levels increased with increasing waist circumference, suggesting that it could be a suitable marker for central and visceral fat. Moreover, oxytocin levels are higher in obese people and those with metabolic syndrome (Stock et al. 1989; Szulc et al. 2016). Importantly, variations in the gene that encodes the oxytocin receptor have been associated with obesity (Wheeler et al. 2013). Paradoxically, studies have shown that, for instance, a single dose or repeated intranasal oxytocin-induced weight loss in overweight or obese adults, indicating that oxytocin affects metabolism and suggests that it might have a role in the treatment of metabolic disorders, including obesity and diabetes mellitus (Lawson 2017). Elevated oxytocin levels signal the need to reduce caloric intake and increase energy expenditure. Thus, dysfunctional oxytocin signaling may cause weight gain and induce obesity. The following sections will describe oxytocin's effects on metabolism, but a comprehensive mechanistic framework that can be tested in humans still needs to be established.

2.1   Lipid Metabolism

Independent from its effect on food intake, oxytocin has been shown to decrease body weight and body fat, including visceral and liver fat, by inducing lipolysis and fat oxidation (Blevins et al. 2015, 2016; Maejima et al. 2011; Deblon et al. 2011). G-protein-coupled oxytocin receptors activate adenylate cyclase, thereby increasing cyclic adenosine monophosphate (cAMP) production, which stimulates lipolysis (Stoop 2012; Chaves et al. 2011; Alberi et al. 1997).

2.2 Energy Expenditure

Although the exact mechanism through which oxytocin increases energy expenditure is unknown, oxytocin causes thermogenesis after repeated exposure (Lawson et al. 2015) by activating brown fat tissue (Kasahara et al. 2007). Oxytocin not only has direct metabolic actions but also affects metabolism through downregulation of the hypothalamic–pituitary–adrenal (HPA) axis. Previous studies have reported that oxytocin administration acts on the oxytocin receptors in the anterior pituitary gland and adrenal gland to lower the release of adrenocorticotropic hormone (ACTH) and cortisol, reducing negative metabolic effects (Antoni 1986; Bathgate and Sernia 1995; Heinrichs et al. 2003; Legros et al. 1982, 1984; Linnen et al. 2012; Nieman 2015; Cardoso et al. 2014).

2.3 Glucose Homeostasis

Patients with type 1 or type 2 diabetes showed low levels of oxytocin (Qian et al. 2014; Kujath et al. 2015). In addition to low oxytocin levels, patients with type 2 diabetes showed high fasting blood glucose and insulin, high blood glucose and insulin after oral glucose tolerance test, and insulin resistance and high HbA1c levels (Qian et al. 2014). Oxytocin receptors are expressed in α and β cells of pancreatic islet cells, and it was found that oxytocin increases GLUT-4 (an insulin-dependent glucose transporter) expression (Eckertova et al. 2011). Oxytocin also directly stimulates insulin secretion in pancreatic β cells through phosphoinositide turnover and protein kinase C activation. Alternatively, it may trigger insulin secretion through vagal cholinergic neurons, which stimulate pancreatic β cells (Bjorkstrand et al. 1996; Gao et al. 1991).

3 Nonhuman Studies of Oxytocin Effects on Eating Behavior

Arletti et al. demonstrated that injections of oxytocin (intraperitoneal or intracerebroventricular) dose-dependently reduced feeding and drinking behavior in rats, regardless of their sex. These anorexigenic effects were blocked by oxytocin receptor antagonists (Arletti et al. 1989, 1990; Benelli et al. 1991). Thus, these studies have suggested the potential of oxytocin as a new therapeutic agent in various diseases involving appetite problems to control body weight (Arletti et al. 1989, 1990; Benelli et al. 1991). However, these studies were limited because they only assessed the short-term effects of oxytocin, usually within 24 h.

Longer-term infusions of oxytocin were also reported to reduce body weight gain in obese rats, whereas, chronic oxytocin infusions did not alter their total food intake or meal patterns (Maejima et al. 2011; Deblon et al. 2011). Maejima et al. (2011) injected daily oxytocin or saline into each standard diet-fed, normal-weight mouse, and high-fat diet-induced obesity in mice after 17 days of feeding. In this study, the decrease in short-term food intake was significant at 0.5 and 6 h after oxytocin injection in both groups and food intake decreased in the oxytocin-injected high-fat diet group until the sixth day, but the decrease in food intake did not differ from the control group from the seventh day on. The body weight of the mice was significantly reduced until the ninth day of oxytocin injection, and there was no rebound weight recovery until the seventeenth day. However, body weight recovery began when oxytocin administration stopped.

A meta-analysis of Leslie et al. (2018) showed that acute, single-dose oxytocin administration (central and peripheral) significantly reduced feeding, but chronic repeated oxytocin administration did not, and the effect of oxytocin on reducing the intake has decreased over time. Furthermore, the anorexigenic effect of oxytocin was greater in male rats than female rats. It needs to be clarified whether different dosing schedules might prevent their attenuation with chronic administration.

The effect of oxytocin has repeatedly been shown to be greater in obese rats than in lean rats (Morton et al. 2012; Altirriba et al. 2014). Morton et al. examined the effects of central and peripheral oxytocin administration in high-fat diet and low-fat diet-fed rats (Morton et al. 2012). The administration of oxytocin dose-dependently reduced food intake in both the low- and high-fat diet-fed rats. However, the authors concluded that obese rodents were more sensitive to oxytocin treatment than thin rodents in the reduction of food intake and weight because food intake and weight were reduced more in the obese rats than the thin rats. The findings suggest that increased oxytocin signaling is capable of inducing both acute and chronic reductions of body weight in rats both by suppressing food intake and by preventing the effect of weight loss from lowering energy expenditure. In a study by Altirriba et al., oxytocin dose-dependently reduced food intake and weight in obese, diabetic mice, and the effect was more prominent at 1 week than at 2 weeks (Altirriba et al. 2014). Thin rats showed decreases in food intake and body weight, which occurred only in the first 24 h, and after 1 day, they showed weight regain. Therefore, after 2 weeks of oxytocin administration, the weight gains were similar in the oxytocin and saline groups. The study found that higher oxytocin receptor specificity and a low oxytocin dose resulted in less weight gain, without changes in food intake, suggesting that the weight-reducing effect was not due to less food intake but the action of oxytocin on fat mass in the rats. The body weight gain–reducing effect was limited to the fat mass only, with decreased lipid uptake, lipogenesis, and inflammation, combined with increased cycling in abdominal adipose tissue.

Studies reporting that oxytocin treatment reduced food intake presented several hypotheses to explain the findings, including oxytocin reducing food intake via delaying gastric emptying (Ohlsson et al. 2006) and affecting food selection (Sclafani et al. 2007). Oxytocin knockout mice did not show enhanced intake of palatable fat but showed enhanced carbohydrate intake (Miedlar et al. 2007). Long-term (4 weeks) third ventricular infusion of oxytocin elicited a sustained reduction of fat mass in rats (Blevins et al. 2016). The absence of oxytocin may increase daily intake of palatable sweet and non-sweet solutions of carbohydrate (but not fat) by selectively blunting or masking processes that contribute to

postingestive satiety (Sclafani et al. 2007). The oxytocin pathway may, therefore, limit the intake of carbohydrates, but does not play a role in limiting the intake of fats. However, further studies are needed to understand the mechanisms of how oxytocin regulates food intake clearly.

4 Human Studies of the Effects of Oxytocin on Quality and Quantity of Food Intake

The intranasal administration route has turned out as a feasible tool to study the contribution of (neuro)peptidergic messengers to human brain function because the intranasal route of administration minimizes peripheral uptake and bypasses the blood–brain barrier (Spetter and Hallschmid 2015). Additionally, intranasal administration is easy to apply and well-tolerated (MacDonald et al. 2011). A previous study showed that plasma oxytocin levels peaked 15 min after intranasal oxytocin administration, oxytocin levels in the CSF were maximally elevated at 75 min, and the most potent brain effect of intranasal oxytocin may occur approximately 60 min after administration (Striepens et al. 2013). Thus, intranasal oxytocin application is a reliable method in the study of its effects in humans.

A variety of studies have tested food intake in response to intranasal oxytocin application.

Studies of healthy men given a single intranasal oxytocin administration (24 IU) reported conflicting outcomes. One study of 20 men reported that oxytocin administration did not change the total caloric intake but reduced chocolate cookie consumption by 25% (Ott et al. 2013). Another study of 25 men (13 normal-weight and 12 overweight or obese men) reported that oxytocin reduced the total caloric intake and fat intake, but did not decrease the carbohydrate and protein intake. Furthermore, the results were not related to body weight (Lawson et al. 2015). Another study included 38 men (20 normal-weight and 18 obese men) and reported that oxytocin treatment decreased the total food intake in fasted, obese men, but not in normal-weight men.

However, oxytocin reduced snack consumption in both groups (Thienel et al. 2016).

In two of the above-mentioned studies (Lawson et al. 2015; Ott et al. 2013), intranasal oxytocin administration decreased the consumption of palatable food, such as carbohydrates and fats, indicating that oxytocin might reduce food intake by modulating reward-related food motivation (Lawson 2017). According to a functional magnetic resonance imaging (fMRI) study of 15 fasted, normal-weight young men, intranasal oxytocin administration (24 IU) increased activity in the ventromedial prefrontal cortex, supplementary motor area, anterior cingulate, and ventrolateral prefrontal cortices in response to food image stimuli and reduced caloric intake by 12%. The study authors concluded that oxytocin stimulated the dopaminergic reward-processing circuits (Spetter et al. 2018), consistent with previous studies (Herisson et al. 2016; Mullis et al. 2013).

In a study investigating the effects of chronic oxytocin administration on obesity, intranasal oxytocin spray (24 IU each time, four times a day) was used for 8 weeks (Zhang et al. 2013). The study found that weight was decreased by 4.6 kg after 4 weeks and decreased by 8.9 kg after 8 weeks. The weight reduction was more prominent in people with higher-degree obesity.

A meta-analytic review (Leslie et al. 2018) reported that the reduction in food intake caused by oxytocin was greater in males than in females when the participants were satiated rather than fasted and in those with greater degrees of obesity. More studies are needed before oxytocin can be used for the treatment of diseases related to eating behavior, but the results of past studies have been encouraging.

5 The Role of Oxytocin in the Regulation of Binge Eating

Low cerebrospinal fluid (CSF) oxytocin levels have been reported in patients with restrictive anorexia nervosa, but no differences compared to controls were found in patients with the binge-eating/purging subtype of anorexia nervosa or individuals with bulimia nervosa (Demitrack et al. 1990). Studies have reported peripheral oxytocin levels in obese people to be higher (Stock et al. 1989), lower (Qian et al. 2014), and similar (Coiro et al. 1988) compared to normal-weight controls. One study reported that cerebrospinal fluid oxytocin levels were normal in subjects who had recovered from bulimia nervosa (Frank et al. 2000).

Intranasal oxytocin treatment has also emerged as a possible approach to manage hedonic and overeating behavior in healthy and overweight individuals (Thienel et al. 2016; Burmester et al. 2018; Ott et al. 2013; Spetter et al. 2018) and in bulimia nervosa (Kim et al. 2015a). Oxytocin effects on food intake were hypothesized to be driven, at least partially, by oxytocin's effects on homeostatic pathways, and cognitive control (Plessow et al. 2018) and the modulation of reward processes (Ott et al. 2013; Spetter et al. 2018). In addition to playing a key role in appetite, oxytocin is involved in social and emotional processing. Problems that involve social, emotional cognition may underpin some of the interpersonal difficulties in patients with eating disorders, which are thought to contribute to the maintenance of the disorders.

A single-dose, randomized, placebo-controlled, crossover study of the intranasal administration of oxytocin (40 IU) to women with anorexia nervosa, bulimia nervosa, and healthy controls showed that oxytocin decreased the 24-h caloric consumption of patients with bulimia nervosa, but not that of anorexia nervosa patients or healthy controls (Kim et al. 2015a). Two other studies found no evidence for altered caloric intake in women with anorexia nervosa in acute juice of smoothie consumption following administration of a single dose of 40 IU oxytocin (Kim et al. 2014a; Leppanen et al. 2017). Another study found no effect of a single dose of oxytocin on the amount of smoothie intake in patients with anorexia nervosa (Leppanen et al. 2017).

In the other study of Kim et al. to examine the effect of oxytocin on the processing of anger in patients with bulimia nervosa, patients with bulimia nervosa show similar increases in attentional processes to anger and similar moderation effects with oxytocin as found in the healthy comparison group (Kim et al. 2018). The findings of no differential impact of oxytocin between patients with bulimia nervosa and healthy controls in the attentional bias task are congruent with previous findings of no difference in the modulation of emotional sensitivity by oxytocin in patients with bulimia nervosa (Kim et al. 2015a).

A study investigating the influence of intranasal oxytocin on risk-taking in a balloon analog risk task (BART) among women with bulimia nervosa and binge-eating disorder, and those without a history of eating disorders (Leslie et al. 2019) found that oxytocin did not have a main effect on performance in the BART. However, there was evidence of a significant interaction for the balloon explosion variable such that a divided dose of 64 IU oxytocin enhanced initial numerical differences in risk-taking between the participant groups: increasing risk-taking in the healthy control group and decreasing risk-taking in the bulimia nervosa/binge-eating disorder participant group.

The motivational processes that regulate food intake involve the reward system and specifically the neurotransmitter dopamine. Oxytocin has been shown to interact with dopamine to regulate reward pathways, and the pathophysiology of eating disorders associated with binge eating could involve a disruption in oxytocin–dopamine pathways (Gamal-Eltrabily and Manzano-García 2018; Baskerville and Douglas 2010).

6 Genetic Effects of Oxytocin on Eating Disorders and Obesity

Oxytocin appears to be particularly involved in inhibiting the appetite for sugar and carbohydrates. Wild-type mice injected with oxytocin receptor antagonists showed a preference for sucrose rather than fat (Olszewski et al. 2010). Oxytocin receptor knockout animals consumed greater amounts of sweet solutions than wild-type animals (Amico et al. 2005) and developed late-onset obesity (Nishimori et al. 2008;

Takayanagi et al. 2008). Animals engineered not to express oxytocin overconsumed sweetened food (Miedlar et al. 2007) and carbohydrates (Sclafani et al. 2007). Oxytocin expression was downregulated by long-term intermittent exposure to sugar and may represent neuroadaptation to the high-sugar diet (Mitra et al. 2010).

The functional relevance of oxytocin in appetite and weight control has been demonstrated in animal models of obesity caused by the global or selective hypothalamic loss of oxytocin neurons (Leng et al. 2008; Wu et al. 2012). Additionally, deletions in the single-minded 1 gene caused obesity in animals, a phenotype associated with reduced hypothalamic oxytocin (Kublaoui et al. 2008). Mice with dietary-induced obesity also exhibited functional abnormalities in their oxytocin systems (Zhang et al. 2011). The relevance of oxytocin in animal models of obesity has also been reported in humans. Variants of the single-minded 1 gene in humans were also associated with obesity (Holder et al. 2000; Ramachandrappa et al. 2013). Rare variants of genes that encode for G-protein-coupled receptors (GPCRs), one of which is the oxytocin receptor gene, have been associated with childhood obesity (Wheeler et al. 2013). Patients with Prader–Willi syndrome were reported to have decreased numbers and sizes of PVN oxytocin neurons (Swaab et al. 1995).

Preliminary evidence has suggested that oxytocin receptor alterations may play a role in eating disorders. High levels of methylation of some CpG sites on the oxytocin receptor (OXTR) gene have been found in patients with anorexia nervosa and were associated with BMI (Kim et al. 2014b). Variations in sensitivity to oxytocin have been reported and are assumed to result from genetic and epigenetic variations in the *OXTR* gene (Ebstein et al. 2010; Insel 2010; Kogan et al. 2011). SNP rs53576 in the third intron of the gene is thought to be involved in the differences in oxytocinergic functioning (Meyer-Lindenberg et al. 2011). Connelly et al. (2014) found an association between the rs53576 GG allele and bulimic behavior in a study of a community cohort. Kim et al. (2015b) found a positive association between the G allele of OXTR rs53576 and bulimia nervosa, which suggests that genetic variations in the oxytocin system may modulate the vulnerability to developing this eating disorder. The genetic evidence point to a potential genetic risk for binge-eating behavior and bulimic-type eating disorders associated with variants of the oxytocin receptor gene.

7 Conclusion

Preliminary evidence has indicated a potential genetic risk for binge-eating behavior and bulimia-type eating disorders associated with variants of the oxytocin receptor gene (Micali et al. 2017; Kim et al. 2015b). Moreover, preliminary evidence has demonstrated decreased caloric intake after oxytocin administration in patients with bulimia nervosa (Kim et al. 2015a). No study has yet investigated oxytocin function in binge-eating disorder. However, studies have shown a distinct hypophagic effect of oxytocin in obese subjects, which is commonly associated with binge-eating behavior. Developing oxytocin as a robust therapeutic, however, remains challenging (Lawson et al. 2019), and well-controlled trials are needed to investigate the potential of oxytocin in the treatment of binge-eating-related disorders.

References

Alberi S, Dreifuss JJ, Raggenbass M (1997) The oxytocin-induced inward current in vagal neurons of the rat is mediated by G protein activation but not by an increase in the intracellular calcium concentration. Eur J Neurosci 9(12):2605–2612. https://doi.org/10.1111/j.1460-9568.1997.tb01690.x

Altirriba J, Poher AL, Caillon A, Arsenijevic D, Veyrat-Durebex C, Lyautey J, Dulloo A, Rohner-Jeanrenaud F (2014) Divergent effects of oxytocin treatment of obese diabetic mice on adiposity and diabetes. Endocrinology 155(11):4189–4201. https://doi.org/10.1210/en.2014-1466

Amico JA, Vollmer RR, Cai HM, Miedlar JA, Rinaman L (2005) Enhanced initial and sustained intake of sucrose solution in mice with an oxytocin gene deletion. Am J Phys Regul Integr Comp Phys 289(6):R1798–R1806. https://doi.org/10.1152/ajpregu.00558.2005

Antoni FA (1986) Oxytocin receptors in rat adenohypophysis: evidence from radioligand binding studies. Endocrinology 119(5):2393–2395. https://doi.org/10.1210/endo-119-5-2393

Arletti R, Benelli A, Bertolini A (1989) Influence of oxytocin on feeding behavior in the rat. Peptides 10(1):89–93. https://doi.org/10.1016/0196-9781(89)90082-x

Arletti R, Benelli A, Bertolini A (1990) Oxytocin inhibits food and fluid intake in rats. Physiol Behav 48 (6):825–830. https://doi.org/10.1016/0031-9384(90)90234-u

Baskerville TA, Douglas AJ (2010) Dopamine and oxytocin interactions underlying behaviors: potential contributions to behavioral disorders. CNS Neurosci Ther 16(3):e92–e123. https://doi.org/10.1111/j.1755-5949.2010.00154.x

Bathgate RA, Sernia C (1995) Characterization of vasopressin and oxytocin receptors in an Australian marsupial. J Endocrinol 144(1):19–29. https://doi.org/10.1677/joe.0.1440019

Benelli A, Bertolini A, Arletti R (1991) Oxytocin-induced inhibition of feeding and drinking: no sexual dimorphism in rats. Neuropeptides 20(1):57–62. https://doi.org/10.1016/0143-4179(91)90040-p

Bjorkstrand E, Eriksson M, UvnasMoberg K (1996) Evidence of a peripheral and a central effect of oxytocin on pancreatic hormone release in rats. Neuroendocrinology 63(4):377–383. https://doi.org/10.1159/000126978

Blevins JE, Graham JL, Morton GJ, Bales KL, Schwartz MW, Baskin DG, Havel PJ (2015) Chronic oxytocin administration inhibits food intake, increases energy expenditure, and produces weight loss in fructose-fed obese rhesus monkeys. Am J Physiol Regul Integr Comp Physiol 308(5):R431–R438. https://doi.org/10.1152/ajpregu.00441.2014

Blevins JE, Thompson BW, Anekonda VT, Ho JM, Graham JL, Roberts ZS, Hwang BH, Ogimoto K, Wolden-Hanson T, Nelson J, Kaiyala KJ, Havel PJ, Bales KL, Morton GJ, Schwartz MW, Baskin DG (2016) Chronic CNS oxytocin signaling preferentially induces fat loss in high-fat diet-fed rats by enhancing satiety responses and increasing lipid utilization. Am J Physiol Regul Integr Comp Physiol 310(7):R640–R658. https://doi.org/10.1152/ajpregu.00220.2015

Burmester V, Higgs S, Terry P (2018) Rapid-onset anorectic effects of intranasal oxytocin in young men. Appetite 130:104–109. https://doi.org/10.1016/j.appet.2018.08.003

Cardoso C, Kingdon D, Ellenbogen MA (2014) A meta-analytic review of the impact of intranasal oxytocin administration on cortisol concentrations during laboratory tasks: moderation by method and mental health. Psychoneuroendocrinology 49:161–170. https://doi.org/10.1016/j.psyneuen.2014.07.014

Chaves VE, Frasson D, Kawashita NH (2011) Several agents and pathways regulate lipolysis in adipocytes. Biochimie 93(10):1631–1640. https://doi.org/10.1016/j.biochi.2011.05.018

Chaves VE, Tilelli CQ, Britob NA, Brito MN (2013) Role of oxytocin in energy metabolism. Peptides 45:9–14. https://doi.org/10.1016/j.peptides.2013.04.010

Coiro V, Passeri M, Davoli C, d'Amato L, Gelmini G, Fagnoni F, Schianchi L, Bentivoglio M, Volpi R, Chiodera P (1988) Oxytocin response to insulin-induced hypoglycemia in obese subjects before and after weight loss. J Endocrinol Investig 11 (2):125–128. https://doi.org/10.1007/BF03350119

Connelly J, Golding J, Gregor S, Ring S, Davis J, Smith G, Harris J, Carter S, Pembrey M (2014) Personality, behavior and environmental features associated with OXTR genetic variants in British mothers. Plos One 9(3):e90465. https://doi.org/10.1371/journal.pone.0090465

Danalache BA, Gutkowska J, Slusarz MJ, Berezowska I, Jankowski M (2010) Oxytocin-Gly-Lys-Arg: a novel cardiomyogenic peptide. PLoS One 5(10):e13643. https://doi.org/10.1371/journal.pone.0013643

Deblon N, Veyrat-Durebex C, Bourgoin L, Caillon A, Bussier AL, Petrosino S, Piscitelli F, Legros JJ, Geenen V, Foti M, Wahli W, Di Marzo V, Rohner-Jeanrenaud F (2011) Mechanisms of the anti-obesity effects of oxytocin in diet-induced obese rats. PLoS One 6(9):e25565. https://doi.org/10.1371/journal.pone.0025565

Demitrack MA, Lesem MD, Listwak SJ, Brandt HA, Jimerson DC, Gold PW (1990) CSF oxytocin in anorexia nervosa and bulimia nervosa: clinical and pathophysiologic considerations. Am J Psychiatry 147(7):882–886. https://doi.org/10.1176/ajp.147.7.882

Denda S, Takei K, Kumamoto J, Goto M, Tsutsumi M, Denda M (2012) Oxytocin is expressed in epidermal keratinocytes and released upon stimulation with adenosine 5′-[gamma-thio]triphosphate in vitro. Exp Dermatol 21(7):535–537. https://doi.org/10.1111/j.1600-0625.2012.01507.x

Ebstein RP, Israel S, Chew SH, Zhong S, Knafo A (2010) Genetics of human social behavior. Neuron 65 (6):831–844. https://doi.org/10.1016/j.neuron.2010.02.020

Eckertova M, Ondrejcakova M, Krskova K, Zorad S, Jezova D (2011) Subchronic treatment of rats with oxytocin results in improved adipocyte differentiation and increased gene expression of factors involved in adipogenesis. Br J Pharmacol 162(2):452–463. https://doi.org/10.1111/j.1476-5381.2010.01037.x

Egan JJ, Saltis J, Wek SA, Simpson IA, Londos C (1990) Insulin, oxytocin, and vasopressin stimulate protein kinase C activity in adipocyte plasma membranes. Proc Natl Acad Sci U S A 87(3):1052–1056. https://doi.org/10.1073/pnas.87.3.1052

Ferguson JN, Young LJ, Hearn EF, Matzuk MM, Insel TR, Winslow JT (2000) Social amnesia in mice lacking the oxytocin gene. Nat Genet 25(3):284–288. https://doi.org/10.1038/77040

Frank GK, Kaye WH, Altemus M, Greeno CG (2000) CSF oxytocin and vasopressin levels after recovery from bulimia nervosa and anorexia nervosa, bulimic subtype. Biol Psychiatry 48(4):315–318. https://doi.org/10.1016/s0006-3223(00)00243-2

Gamal-Eltrabily M, Manzano-García A (2018) Role of central oxytocin and dopamine systems in nociception and their possible interactions: suggested hypotheses. Rev Neurosci 29(4):377–386. https://doi.org/10.1515/revneuro-2017-0068

Gao ZY, Drews G, Henquin JC (1991) Mechanisms of the stimulation of insulin release by oxytocin in normal mouse islets. Biochem J 276(Pt 1):169–174. https://doi.org/10.1042/bj2760169

Gimpl G, Fahrenholz F (2001) The oxytocin receptor system: structure, function, and regulation. Physiol Rev 81(2):629–683

Heinrichs M, Baumgartner T, Kirschbaum C, Ehlert U (2003) Social support and oxytocin interact to suppress cortisol and subjective responses to psychosocial stress. Biol Psychiatry 54(12):1389–1398. https://doi.org/10.1016/s0006-3223(03)00465-7

Herisson FM, Waas JR, Fredriksson R, Schioth HB, Levine AS, Olszewski PK (2016) Oxytocin acting in the nucleus accumbens core decreases food intake. J Neuroendocrinol 28(4). https://doi.org/10.1111/jne.12381

Holder JL, Butte NF, Zinn AR (2000) Profound obesity associated with a balanced translocation that disrupts the SIM1 gene. Hum Mol Genet 9(1):101–108

Insel TR (2010) The challenge of translation in social neuroscience: a review of oxytocin, vasopressin, and affiliative behavior. Neuron 65(6):768–779. https://doi.org/10.1016/j.neuron.2010.03.005

Jankowski M, Wang D, Hajjar F, Mukaddam-Daher S, McCann SM, Gutkowska J (2000) Oxytocin and its receptors are synthesized in the rat vasculature. Proc Natl Acad Sci U S A 97(11):6207–6211. https://doi.org/10.1073/pnas.110137497

Kasahara Y, Takayanagi Y, Kawada T, Itoi K, Nishimori K (2007) Impaired thermoregulatory ability of oxytocin-deficient mice during cold-exposure. Biosci Biotechnol Biochem 71(12):3122–3126. https://doi.org/10.1271/bbb.70498

Kim YR, Kim CH, Cardi V, Eom JS, Seong Y, Treasure J (2014a) Intranasal oxytocin attenuates attentional bias for eating and fat shape stimuli in patients with anorexia nervosa. Psychoneuroendocrinology 44:133–142

Kim YR, Kim JH, Kim MJ, Treasure J (2014b) Differential methylation of the oxytocin receptor gene in patients with anorexia nervosa: a pilot study. PLoS One 9(3):e90721

Kim YR, Eom JS, Yang JW, Kang J, Treasure J (2015a) The impact of oxytocin on food intake and emotion recognition in patients with eating disorders: a double blind single dose within-subject cross-over Design. PLoS One 10(9):e0137514. https://doi.org/10.1371/journal.pone.0137514

Kim YR, Kim JH, Kim CH, Shin JG, Treasure J (2015b) Association between the oxytocin receptor gene polymorphism (rs53576) and bulimia nervosa. Eur Eat Disord Rev 23(3):171–178. https://doi.org/10.1002/erv.2354

Kim YR, Eom JS, Leppanen J, Leslie M, Treasure J (2018) Effects of intranasal oxytocin on the attentional bias to emotional stimuli in patients with bulimia nervosa. Psychoneuroendocrinology 91:75–78. https://doi.org/10.1016/j.psyneuen.2018.02.029

Kiss A, Mikkelsen JD (2005) Oxytocin--anatomy and functional assignments: a minireview. Endocr Regul 39(3):97–105

Kogan A, Saslow L, Impett E, Oveis C, Keltner D, Rodrigues Saturn S (2011) Thin-slicing study of the oxytocin receptor (OXTR) gene and the evaluation and expression of the prosocial disposition. Proc Natl Acad Sci U S A 108:19192–19189. https://doi.org/10.1073/pnas.1112658108

Kublaoui BM, Gemelli T, Tolson KP, Wang Y, Zinn AR (2008) Oxytocin deficiency mediates hyperphagic obesity of Sim1 haploinsufficient mice. Mol Endocrinol 22 (7):1723–1734. https://doi.org/10.1210/me.2008-0067

Kujath AS, Quinn L, Elliott ME, Varady KA, LeCaire TJ, Carter CS, Danielson KK (2015) Oxytocin levels are lower in premenopausal women with type 1 diabetes mellitus compared with matched controls. Diabetes Metab Res Rev 31(1):102–112. https://doi.org/10.1002/dmrr.2577

Lawson EA (2017) The effects of oxytocin on eating behaviour and metabolism in humans. Nat Rev Endocrinol 13(12):700–709. https://doi.org/10.1038/nrendo.2017.115

Lawson EA, Marengi DA, DeSanti RL, Holmes TM, Schoenfeld DA, Tolley CJ (2015) Oxytocin reduces caloric intake in men. Obesity (Silver Spring) 23 (5):950–956. https://doi.org/10.1002/oby.21069

Lawson EA, Olszewski PK, Weller A, Blevins JE (2019 October) The role of oxytocin in regulation of appetitive behavior, body weight and glucose homeostasis. J Neuroendocrinol 28:e12805. https://doi.org/10.1111/jne.12805

Legros JJ, Chiodera P, Demey-Ponsart E (1982) Inhibitory influence of exogenous oxytocin on adrenocorticotropin secretion in normal human subjects. J Clin Endocrinol Metab 55(6):1035–1039. https://doi.org/10.1210/jcem-55-6-1035

Legros JJ, Chiodera P, Geenen V, Smitz S, von Frenckell R (1984) Dose-response relationship between plasma oxytocin and cortisol and adrenocorticotropin concentrations during oxytocin infusion in normal men. J Clin Endocrinol Metab 58(1):105–109. https://doi.org/10.1210/jcem-58-1-105

Leng G, Meddle SL, Douglas AJ (2008) Oxytocin and the maternal brain. Curr Opin Pharmacol 8(6):731–734. https://doi.org/10.1016/j.coph.2008.07.001

Leppanen J, Cardi V, Ng KW, Paloyelis Y, Stein D, Tchanturia K, Treasure J (2017) The effects of intranasal oxytocin on smoothie intake, cortisol and attentional bias in anorexia nervosa. Psychoneuroendocrinology 79:167–174. https://doi.org/10.1016/j.psyneuen.2017.01.017

Leslie M, Silva P, Paloyelis Y, Blevins J, Treasure J (2018) A systematic review and quantitative meta-analysis of oxytocin's effects on feeding. J Neuroendocrinol 30(8): e12584. https://doi.org/10.1111/jne.12584

Leslie M, Leppanen J, Paloyelis Y, Nazar BP, Treasure J (2019) The influence of oxytocin on risk-taking in the balloon analogue risk task among women with bulimia nervosa and binge eating disorder. J Neuroendocrinol 31(8):e12771. https://doi.org/10.1111/jne.12771

Linnen AM, Ellenbogen MA, Cardoso C, Joober R (2012) Intranasal oxytocin and salivary cortisol concentrations during social rejection in university students. Stress 15 (4):393–402. https://doi.org/10.3109/10253890.2011.631154

MacDonald E, Dadds MR, Brennan JL, Williams K, Levy F, Cauchi AJ (2011) A review of safety, side-effects and subjective reactions to intranasal oxytocin in human research. Psychoneuroendocrinology 36 (8):1114–1126. https://doi.org/10.1016/j.psyneuen.2011.02.015

Maejima Y, Iwasaki Y, Yamahara Y, Kodaira M, Sedbazar U, Yada T (2011) Peripheral oxytocin treatment ameliorates obesity by reducing food intake and visceral fat mass. Aging (Albany NY) 3(12):1169–1177. https://doi.org/10.18632/aging.100408

Meyer-Lindenberg A, Domes G, Kirsch P, Heinrichs M (2011) Oxytocin and vasopressin in the human brain: social neuropeptides for translational medicine. Nat Rev Neurosci 12(9):524–538. https://doi.org/10.1038/nrn3044

Micali N, Crous-Bou M, Treasure J, Lawson EA (2017) Association between oxytocin receptor genotype, maternal care, and eating disorder behaviours in a community sample of women. Eur Eat Disord Rev 25 (1):19–25. https://doi.org/10.1002/erv.2486

Miedlar JA, Rinaman L, Vollmer RR, Amico JA (2007) Oxytocin gene deletion mice overconsume palatable sucrose solution but not palatable lipid emulsions. Am J Physiol Regul Integr Comp Physiol 293(3): R1063–R1068. https://doi.org/10.1152/ajpregu.00228.2007

Mitra A, Gosnell BA, Schioth HB, Grace MK, Klockars A, Olszewski PK, Levine AS (2010) Chronic sugar intake dampens feeding-related activity of neurons synthesizing a satiety mediator, oxytocin. Peptides 31 (7):1346–1352. https://doi.org/10.1016/j.peptides.2010.04.005

Morton GJ, Thatcher BS, Reidelberger RD, Ogimoto K, Wolden-Hanson T, Baskin DG, Schwartz MW, Blevins JE (2012) Peripheral oxytocin suppresses food intake and causes weight loss in diet-induced obese rats. Am J Physiol Endocrinol Metab 302(1):E134–E144. https://doi.org/10.1152/ajpendo.00296.2011

Mullis K, Kay K, Williams DL (2013) Oxytocin action in the ventral tegmental area affects sucrose intake. Brain Res 1513:85–91. https://doi.org/10.1016/j.brainres.2013.03.026

Nieman LK (2015) Cushing's syndrome: update on signs, symptoms and biochemical screening. Eur J

Endocrinol 173(4):M33–M38. https://doi.org/10.1530/EJE-15-0464

Nishimori K, Takayanagi Y, Yoshida M, Kasahara Y, Young LJ, Kawamata M (2008) New aspects of oxytocin receptor function revealed by knockout mice: sociosexual behaviour and control of energy balance. In: Neumann ID, Landgraf R (eds) Advances in vasopressin and oxytocin: from genes to behaviour to disease, Progress in brain research, vol 170. Elsevier, Amsterdam, pp 79–90. https://doi.org/10.1016/s0079-6123(08)00408-1

Ohlsson B, Truedsson M, Djerf P, Sundler F (2006) Oxytocin is expressed throughout the human gastrointestinal tract. Regul Pept 135(1–2):7–11. https://doi.org/10.1016/j.regpep.2006.03.008

Olszewski PK, Klockars A, Olszewska AM, Fredriksson R, Schioth HB, Levine AS (2010) Molecular, Immunohistochemical, and pharmacological evidence of oxytocin's role as inhibitor of carbohydrate but not fat intake. Endocrinology 151(10):4736–4744. https://doi.org/10.1210/en.2010-0151

Olszewski PK, Klockars A, Levine AS (2016) Oxytocin: a conditional Anorexigen whose effects on appetite depend on the physiological, Behavioural and social contexts. J Neuroendocrinol 28(4). https://doi.org/10.1111/jne.12376

Ott V, Finlayson G, Lehnert H, Heitmann B, Heinrichs M, Born J, Hallschmid M (2013) Oxytocin reduces reward-driven food intake in humans. Diabetes 62 (10):3418–3425. https://doi.org/10.2337/db13-0663

Plessow F, Eddy KT, Lawson EA (2018) The neuropeptide hormone oxytocin in eating disorders. Curr Psychiatry Rep 20(10):91. https://doi.org/10.1007/s11920-018-0957-0

Poulain DA, Wakerley JB (1982) Electrophysiology of hypothalamic magnocellular neurones secreting oxytocin and vasopressin. Neuroscience 7(4):773–808. https://doi.org/10.1016/0306-4522(82)90044-6

Qian W, Zhu T, Tang B, Yu S, Hu H, Sun W, Pan R, Wang J, Wang D, Yang L, Mao C, Zhou L, Yuan G (2014) Decreased circulating levels of oxytocin in obesity and newly diagnosed type 2 diabetic patients. J Clin Endocrinol Metab 99(12):4683–4689. https://doi.org/10.1210/jc.2014-2206

Ramachandrappa S, Raimondo A, Cali AMG, Keogh JM, Henning E, Saeed S, Thompson A, Garg S, Bochukova EG, Brage S, Trowse V, Wheeler E, Sullivan AE, Dattani M, Clayton PE, Datta V, Bruning JB, Wareham NJ, O'Rahilly S, Peet DJ, Barroso I, Whitelaw ML, Farooqi IS (2013) Rare variants in single-minded 1 (SIM1) are associated with severe obesity. J Clin Investig 123(7):3042–3050. https://doi.org/10.1172/jci68016

Sabatier N, Leng G, Menzies J (2013) Oxytocin, feeding, and satiety. Front Endocrinol (Lausanne) 4:35. https://doi.org/10.3389/fendo.2013.00035

Schorr M, Marengi DA, Pulumo RL, Yu E, Eddy KT, Klibanski A, Miller KK, Lawson EA (2017) Oxytocin and its relationship to body composition, bone mineral density, and hip geometry across the weight spectrum.

J Clin Endocrinol Metab 102(8):2814–2824. https://doi.org/10.1210/jc.2016-3963

Sclafani A, Rinaman L, Vollmer RR, Amico JA (2007) Oxytocin knockout mice demonstrate enhanced intake of sweet and non-sweet carbohydrate solutions. Am J Physiol Regul Integr Comp Physiol 292(5):R1828–R1833. https://doi.org/10.1152/ajpregu.00826.2006

Spetter MS, Hallschmid M (2015) Intranasal neuropeptide administration to target the human brain in health and disease. Mol Pharm 12(8):2767–2780. https://doi.org/10.1021/acs.molpharmaceut.5b00047

Spetter MS, Hallschmid M (2017) Current findings on the role of oxytocin in the regulation of food intake. Physiol Behav 176:31–39. https://doi.org/10.1016/j.physbeh.2017.03.007

Spetter MS, Feld GB, Thienel M, Preissl H, Hege MA, Hallschmid M (2018) Oxytocin curbs calorie intake via food-specific increases in the activity of brain areas that process reward and establish cognitive control. Sci Rep 8(1):2736. https://doi.org/10.1038/s41598-018-20963-4

Stock S, Granstrom L, Backman L, Matthiesen AS, Uvnas-Moberg K (1989) Elevated plasma levels of oxytocin in obese subjects before and after gastric banding. Int J Obes 13(2):213–222

Stoop R (2012) Neuromodulation by oxytocin and vasopressin. Neuron 76(1):142–159. https://doi.org/10.1016/j.neuron.2012.09.025

Striepens N, Kendrick KM, Hanking V, Landgraf R, Wullner U, Maier W, Hurlemann R (2013) Elevated cerebrospinal fluid and blood concentrations of oxytocin following its intranasal administration in humans. Sci Rep 3:3440. https://doi.org/10.1038/srep03440

Swaab DF, Purba JS, Hofman MA (1995) Alterations in the hypothalamic paraventricular nucleus and its oxytocin neurons (putative satiety cells) in Prader-Willi-syndrome - a study of 5 cases. J Clin Endocrinol Metab 80(2):573–579. https://doi.org/10.1210/jc.80.2.573

Szulc P, Amri EZ, Varennes A, Panaia-Ferrari P, Fontas E, Goudable J, Chapurlat R, Breuil V (2016) High serum oxytocin is associated with metabolic syndrome in older men - the MINOS study. Diabetes Res Clin Pract 122:17–27. https://doi.org/10.1016/j.diabres.2016.09.022

Takayanagi Y, Kasahara Y, Onaka T, Takahashi N, Kawada T, Nishimori K (2008) Oxytocin receptor-deficient mice developed late-onset obesity. Neuroreport 19(9):951–955. https://doi.org/10.1097/WNR.0b013e3283021ca9

Thienel M, Fritsche A, Heinrichs M, Peter A, Ewers M, Lehnert H, Born J, Hallschmid M (2016) Oxytocin's inhibitory effect on food intake is stronger in obese than normal-weight men. Int J Obesity 40 (11):1707–1714. https://doi.org/10.1038/ijo.2016.149

Wheeler E, Huang N, Bochukova EG, Keogh JM, Lindsay S, Garg S, Henning E, Blackburn H, Loos RJF, Wareham NJ, O'Rahilly S, Hurles ME, Barroso I, Farooqi IS (2013) Genome-wide SNP and CNV analysis identifies common and low-frequency variants associated with severe early-onset obesity. Nat

Genet 45(5):513–U576. https://doi.org/10.1038/ng.2607

Wu ZF, Xu YZ, Zhu YM, Sutton AK, Zhao RJ, Lowell BB, Olson DP, Tong QC (2012) An obligate role of oxytocin neurons in diet induced energy expenditure. PLoS One 7(9):e45167. https://doi.org/10.1371/journal.pone.0045167

Zhang G, Bai H, Zhang H, Dean C, Wu QA, Li JX, Guariglia S, Meng QY, Cai DS (2011) Neuropeptide exocytosis involving Synaptotagmin-4 and oxytocin in hypothalamic programming of body weight and energy balance. Neuron 69(3):523–535. https://doi.org/10.1016/j.neuron.2010.12.036

Zhang H, Wu C, Chen Q, Chen X, Xu Z, Wu J, Cai D (2013) Treatment of obesity and diabetes using oxytocin or analogs in patients and mouse models. PLoS One 8(5):e61477. https://doi.org/10.1371/journal.pone.0061477

Psychotherapy for Binge Eating

Mary Katherine Ray, Anne Claire Grammer,
Genevieve Davison, Ellen E. Fitzsimmons-Craft,
and Denise E. Wilfley

Abstract

There are many types of psychotherapy used to treat binge eating in bulimia nervosa (BN) and binge-eating disorder (BED). Some of the most common psychotherapies include cognitive behavioral therapy (CBT), interpersonal psychotherapy (IPT), behavioral weight-loss treatment (BWL), third-wave therapies (e.g., dialectical behavior therapy, DBT), and family-based therapy (FBT). This chapter was designed to review the empirical evidence for each of these interventions to treat binge eating in both BN and BED. The chapter highlights the most well-supported psychotherapies and briefly discusses promising psychotherapies that are emerging in the field. Given the vast differences in binge eating treatment for adults and youth, the chapter discusses each population separately to provide the most comprehensive overview of the literature. The chapter also discusses the efficacy of these psychotherapies to address comorbidities that are often associated with binge-eating-related disorders (e.g., depression) and reviews future directions of the field.

Keywords

Binge eating · Cognitive behavioral therapy (CBT) · Interpersonal psychotherapy (IPT) · Behavioral weight-loss treatment (BWL) · Family-based therapy (FBT) · Third-wave therapy · Dialectical behavior therapy (DBT)

Learning Objectives

In this chapter, you will:

- Learn about the most evidence-based psychotherapies used to treat binge eating in adults and youth.
- Learn about promising psychotherapies emerging in the field to treat binge eating in adults and youth.
- Learn about how psychotherapies treat comorbidities associated with binge-eating-related disorders.

1 Introduction to Psychotherapy for Binge Eating

There are many types of psychotherapies used to treat binge eating with varying levels of evidence to support their use. The aim of this chapter is to introduce you to these psychotherapies and provide the most up-to-date support for their efficaciousness in treating binge eating in both bulimia

M. K. Ray · A. C. Grammer · G. Davison ·
E. E. Fitzsimmons-Craft · D. E. Wilfley (✉)
Department of Psychiatry, Washington University in
St. Louis School of Medicine, St. Louis, MO, USA
e-mail: wilfleyd@wustl.edu

nervosa (BN) and binge-eating disorder (BED). Given that the use of psychotherapy differs vastly between adults and youth, the chapter will be separated by the two populations to provide the most comprehensive review of each. Within each section, the most well-established psychotherapies will be the focus, followed by discussion of other, less-supported, psychotherapies emerging in the field. It is important to note that this chapter will not cover self-help versions of psychotherapy as they will be discussed in the next chapter. The literature will focus on the primary outcomes of treatment, defined by the National Institute for Health and Care Excellence (NICE) as remission, binge eating, and compensatory behaviors (with BN) (National Institute for Health Care Excellence 2017). However, secondary outcomes of treatment, such as psychopathological comorbidities will also be reviewed briefly. Each section will conclude with information on mediators, moderators, and predictors of psychotherapy outcomes and future directions of the field.

2 Psychotherapy for Binge Eating in Adults

To date, as shown in Table 1, the two most well-established psychotherapies recommended for binge eating in adults are cognitive behavioral therapy (CBT) and interpersonal psychotherapy (IPT). Therefore, CBT and IPT will be the focus of the adult section. Other, less supported, psychotherapies emerging in the field will also be discussed including behavioral weight-loss

Table 1 Overall evidence for psychotherapies to treat binge eating in adults

	Well-established	More evidence needed before recommendation
BN	CBT, IPT	BWL, third-wave therapies
BED	CBT, IPT	BWL, third-wave therapies

BN bulimia nervosa, *BED* binge-eating disorder, *CBT* cognitive behavioral therapy, *IPT* interpersonal psychotherapy, *BWL* behavioral weight-loss treatment

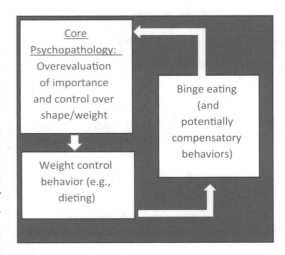

Fig. 1 Illustration of the cycle between the core psychopathology of binge-eating-related diseases and binge eating behaviors from a cognitive behavioral perspective

treatment (BWL) and third-wave therapies such as dialectical behavior therapy (DBT).

3 Cognitive Behavioral Therapy

As the name implies, CBT addresses the maladaptive "cognitive" and "behavioral" facets that contribute to psychological distress. Specifically, CBT teaches individuals to identify their own unhealthy thinking patterns and understand how those thought processes translate to behavior (Beck 1964; Fenn and Byrne 2013). Patients can then set specific, measurable, achievable, realistic, and time-limited goals to develop healthier thought patterns and behaviors (Fenn and Byrne 2013). Specific to binge eating, CBT works under the theoretical model that disordered eating behavior results from a "core psychopathology" of overevaluating the importance of weight and shape and letting the ability to control them determine self-worth. This thinking pattern sets in motion a cycle of disordered weight control behaviors (e.g., dieting) and binge eating (and potentially compensatory behaviors) as illustrated in Fig. 1 (Murphy et al. 2010; Agras 2019). Overall, CBT strives to help people break this cycle through four major treatment phases outlined in Fig. 2.

THEORETICAL RATIONALE FOR CBT:

Binge eating is driven by overvaluation of the importance of shape and weight.

PHASE 1:	PHASE 2:
• Establish factors that lead to binge eating • Develop structured eating patterns (e.g., eating regular meals vs. dieting/binging) • Self-monitor eating/binging • Weekly weigh in with the therapist	• Review progress in treatment • Change treatment focus as needed
PHASE 3:	PHASE 4:
• Introduce feared foods • Discuss concerns about shape and weight • Discuss factors that trigger binge eating	• Review progress in changing triggers • Reflect on future triggers • Develop strategies to reduce risk of relapse

Fig. 2 Phases of cognitive behavioral therapy (CBT) for binge eating (adapted from Agras 2019)

3.1 Bulimia Nervosa

Randomized controlled trials examining CBT for BN have found it to be effective at reducing symptoms of the disorder and helping patients achieve remission (Linardon et al. 2017c; Slade et al. 2018). Linardon et al. found that in a meta-analysis of randomized controlled trials (RCTs) of CBT for BN, therapist-led CBT was significantly better at producing remission, reducing binge/purge frequency, and improving cognitive symptoms in patients with BN compared to active and inactive control groups (Linardon et al. 2017c). Meanwhile, Slade et al. conducted a network meta-analysis of treatments for BN and found that both group and individual CBT were effective in achieving full remission in patients with BN by the end of treatment (Slade et al. 2018). In addition to specific eating disorder psychopathology, CBT for BN has also been found to improve self-esteem and general psychopathology (Chen et al. 2003).

Key Points: CBT for BN in Adults

- CBT effectively reduces binge eating symptoms, helps achieve remission, and improves self-esteem and general psychopathology.
- Both group and individual CBT are effective.

3.2 Binge-Eating Disorder

CBT emerged as a treatment for BED as early as 1990 (Telch et al. 1990) and subsequent research has largely supported its effectiveness (e.g., Wilfley et al. 1993, 2002). A recent meta-analysis examining treatments for BED found that, compared to a wait-list control, CBT reliably helps patients achieve remission, reduces binge eating frequency, improves eating disorder psychopathology, and improves symptoms of depression (Ghaderi et al. 2018). These findings echo those of Linardon et al. who conducted a meta-analysis of CBT for eating disorders and concluded that CBT produces greater improvements in symptoms of BED compared to both inactive and active comparison groups (Linardon et al. 2017c). Studies that included long-term follow-up assessment found that the effects of CBT for BED remain robust after 1 year (Wilfley et al. 2002) and 4 years (Hilbert et al. 2012).

Key Points: CBT for BED in Adults

- CBT reliably improves binge eating symptoms, eating disorder psychopathology, and depression.
- Effects of CBT appear to be long lasting.

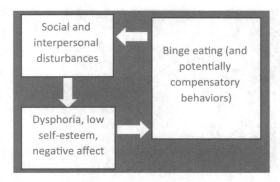

Fig. 3 Illustration of the cycle between interpersonal disturbances and binge eating behaviors through the lens of interpersonal therapy

4　Interpersonal Psychotherapy

IPT is based on the theoretical model that healthy interpersonal functioning is vital to psychological well-being; therefore, it follows that deficits in interpersonal functioning lead to psychological distress and disordered behavior. IPT does not focus on uncovering the cause of the disordered behavior, but understanding the interpersonal environment where the behavior was developed and is subsequently maintained (Karam et al. 2019).

Specific to binge eating, IPT works under the idea that disturbances in social and interpersonal functioning lead to dysphoria, low self-esteem, and negative affect which then leads to binge eating as a coping mechanism. The disordered eating then leads to more social and interpersonal disturbances, creating a propagating cycle of interpersonal disturbance and binge eating (illustrated in Fig. 3; Karam et al. 2019).

The overall aim of IPT is to break this cycle by changing the interpersonal environment where the disordered behavior developed and is maintained (Karam et al. 2019). To achieve this, IPT is broken down into three main phases, the initial phase, the intermediate phase, and the termination phase. The treatment targets for each phase are outlined in Fig. 4.

4.1　Bulimia Nervosa

In a recent meta-analysis of therapies for BN, Svaldi et al. found that IPT was associated with large reductions in self-reported eating pathology in patients with BN (2019). Further, although initial randomized controlled comparisons of IPT and CBT treatment tailored to BN treatment (CBT–BN) found that CBT–BN was more effective at reducing symptoms of BN post treatment (Fairburn et al. 1991, 1993), the two treatments were equivalent at 8- and 12-month follow-ups (Fairburn et al. 1993). Furthermore, in a 6-year follow-up study using the same patients, IPT and CBT–BN produced similar improvements in well-being and similar decreases in general psychopathology (Fairburn et al. 1995). A later, multicenter RCT comparing IPT and CBT–BN in 220 patients with BN found similar results. Agras et al. found that following 20 weeks of treatment, CBT–BN was more effective than IPT, but that when patients were assessed after 8–12 months, these differences disappeared; IPT appeared to "catch up" to CBT (2000). Fairburn et al. compared an "enhanced" CBT treatment (CBT-E) that addresses external factors associated with disordered eating to IPT in patients with BN, and although patients in both groups showed improvements post treatment, CBT-E was more effective (Fairburn et al. 2015). However, at a 60-week follow-up, the two groups were statistically equivalent (Fairburn et al. 2009). Karam et al. suggest that this difference in effect timing could be due to the potential failing of CBT to link the interpersonal context to specific eating disorder symptoms or not including techniques covered by both IPT and CBT (2019). Overall, it appears that IPT is an effective long-term treatment for BN, but more RCTs are needed before robust meta-analyses of IPT's overall effectiveness can be conducted.

Key Points: IPT for BN in Adults

- CBT–BN is initially more effective at reducing BN symptoms than IPT, but the two are equivocal at 8- and 12-month follow-up.
- IPT improves well-being, general psychopathology, and self-reported eating pathology.

THEORETICAL RATIONALE FOR IPT:

Binge eating acts as a coping mechanism to manage interpersonal distress.

PHASE 1: INITIAL	PHASE 2: INTERMEDIATE	PHASE 3: TERMINATION

• Assess symptom onset
• Give formal diagnosis and psychoeducation
• Give hope for recovery
• Establish relationship
• Conduct interpersonal inventory
  ○ Relate interpersonal environment and symptom development to establish the timeline of the disorder
• Develop interpersonal case formulation
  ○ Relate illness onset/maintenance to problem area (usually focus on 1):
    ▪ Interpersonal deficits (e.g. inability to form relationships)
    ▪ Role transitions (e.g. divorce)
    ▪ Grief (e.g. death)
    ▪ Interpersonal disputes (e.g. conflict with friend)
  ○ Determine treatment goals
	• Assess symptoms at weekly sessions
	• Link symptoms (i.e., binge eating) to problem area (e.g. grief)
	• Emphasize the link between improvements in binge eating and interpersonal functioning
	• Incorporate treatment strategies
	• Focus on treatment goals
		• Review treatment progress
		• Develop continuing work
		• Discuss relapse prevention and sign of potential relapse

Fig. 4 Rationale for interpersonal psychotherapy (IPT) and targets for each treatment phase

4.2 Binge-Eating Disorder

Research suggests that IPT is an effective treatment for BED (Karam et al. 2019). Group IPT specifically (Wilfley et al. 1998), has been shown to be effective in several randomized controlled trials. One was a comparison of group CBT, group IPT, and a wait-list control condition in 56 women with non-purging BN. Participants had to meet all binge eating diagnostic criteria for BN (i.e., recurrent binge eating episodes, lack of control over eating, having an average of two or more binges each week for 6 months, and body shape and weight concerns) with the exception of being excluded if they had a current or past history of purging. Therefore, participants had a BED-like phenotype, but at the time of study, BED had not been defined by the DSM. The study demonstrated that both group CBT and IPT led to significant reductions in binge eating while the control condition did not (Wilfley et al. 1993). In this study, the CBT and IPT conditions were also found to have reduced binge eating after 6 months and 1 year. In a larger randomized comparison of IPT and CBT, Wilfley et al. found that in 162 patients with BED, CBT and IPT produced similar recovery rates: 79% of CBT patients had recovered post treatment compared to 73% of IPT patients. After 1 year, those rates were 59% and 62%, respectively (Wilfley et al. 2002). In a follow-up, Hilbert et al. found that after 4 years, more than 64% of patients in the CBT and IPT groups had full recovery from binge eating and 80% of patients showed clinically significant improvement. Patients in the IPT group showed greater symptom stability, while patients in the CBT group were more likely to relapse. Additionally, psychopathology in the IPT group either remained stable or improved but had a tendency to worsen in the CBT group (Hilbert et al. 2012).

A randomized study comparing IPT and a behavioral weight loss treatment (BWL) in patients with BED found that IPT reduced binge eating behavior more than BWL, but BWL was associated with greater reductions in patient BMI. The study also found that IPT may be more effective in patients with very low self-esteem or high eating disorder psychopathology (Wilson et al. 2010). This finding was echoed by a latent class analysis of 205 treatment-seeking patients with BED and overweight or obesity. Sysko et al. found that when comparing patients enrolled in IPT and BWL, those with high negative affect, high shape and weight concerns, and generally high eating disorder psychopathology, benefited most from IPT (2010).

Taken together, these findings suggest that IPT is an effective treatment for BED. Additionally, while IPT may not produce results as quickly as CBT, it appears that it is equally efficacious long-term, and maybe more beneficial for patients with low self-esteem and high eating disorder pathology. Finally, as Brownley et al. note in their meta-analysis of treatments for BED in adults, more RCTs of IPT for BED are needed in order to generate pooled estimates of its effectiveness (2016).

Key Points: IPT for BED in Adults

- IPT effectively treats BED.
- The effectiveness of IPT is long lasting.
- IPT may produce greater symptom stability than CBT.
- IPT may be more beneficial in patients with low self-esteem and high eating disorder pathology.

5 Behavioral Weight-Loss Treatment

BWL is a lifestyle intervention that aims to encourage weight loss through increased energy expenditure and decreased energy intake. BWL treatment targets include (1) self-monitoring physical activity, energy consumption, and thoughts about eating; (2) implementing small, steady changes in lifestyle (e.g., food choices, physical activity); and (3) losing one pound each week. Although BWL does not target binge eating specifically, it is proposed that improvements in physical activity and eating will result in successful weight loss and discourage binge eating (Iacovino et al. 2012).

5.1 Binge-Eating Disorder

Evidence for BWL to treat binge eating is mixed. In their review, Iacovino and colleagues highlight the inconsistencies across studies but determine that, overall, BWL is not as effective as CBT or IPT in BED short- or long-term and that there is no added benefit to a sequential program (i.e., CBT program followed by BWL program). Further, they suggest that BWL could be an effective weight-loss tool for patients who respond to CBT or IPT (2012). In concert, a review conducted in 2015 highlighted the inconsistency of evidence for BWL to treat binge eating and determined that BWL could be comparable to or less effective than more established treatments (McElroy et al. 2015). Lastly, a recent meta-analysis found that BWL was inferior to CBT in reducing binge eating frequency post treatment in individuals with overweight/obesity. However, there was no difference at 1-year follow-up (Palavras et al. 2017).

Key Points: BWL for BED in Adults

- Evidence for BWL to treat binge eating is mixed.
- BWL may be comparable to or less effective than CBT and IPT.
- More research is needed before it is recommended as a first-line treatment.

Taken together, BWL could potentially help improve binge eating, but the results are mixed.

More evidence is needed before BWL could be recommended as a first-line treatment.

6 Third-Wave Therapies

Third-wave therapies focus on strategies that help patients with mindfulness, acceptance, metacognition, psychological flexibility, and experiential avoidance (Linardon et al. 2017a). The umbrella of third-wave therapies includes dialectical behavior therapy (DBT), schema therapy (ST), acceptance and commitment therapy (ACT), mindfulness-based interventions (MBI), and compassion-focused therapy (CFT). Of these treatments, DBT is the most studied and will, thus, be the focus of the section.

6.1 Bulimia Nervosa

Evidence of the effectiveness of DBT for BN is promising but sparse. Safer, Telch, and Agras found that in a randomized, wait-list controlled trial of 31 women with BN, DBT was effective at reducing binging and purging symptoms. Specifically, 28.6% of patients in the treatment group were abstinent from binge/purge behaviors by the end of treatment while an additional 35.7% showed clinically meaningful reductions in their eating disorder symptoms. By comparison, 80% of wait-list control participants remained fully symptomatic (2001). Another randomized, wait-list controlled trial examining appetite focused DBT (DBT-AF) designed specifically to target symptoms of BN, found that participants in the DBT-AF condition reported significantly fewer symptoms of BN compared to controls. At the end of treatment, 61.5% of participants in the treatment condition no longer met full or sub-threshold criteria for BN (Hill et al. 2011).

> **Key Points: DBT for BN in Adults**
>
> - Preliminary evidence shows that DBT improves binge eating.

6.2 Binge-Eating Disorder

Emerging evidence from randomized trials suggests that DBT may be an effective psychotherapeutic approach to treating BED in adults. Telch et al. tested the effectiveness of DBT (adapted for binge-eating disorder) in women diagnosed with BED. They found that 89% of women in the treatment condition stopped binging by the end of treatment compared to 12.5% of women in the control condition. However, this rate fell to 56% in 6 months after treatment (2001). Further, although Safer and colleagues found that DBT led to quicker reductions in binge eating compared to an active comparison group, differences between the groups did not persist at 3-, 6-, or 12-month follow-up (2010). A more recent RCT found that DBT improves binge eating measured by the Binge Eating Scale (BES) (Rahmani et al. 2018).

> **Key Points: DBT for BED in Adults**
>
> - DBT improves binge eating.
> - Follow-up studies suggest DBT effects may not be long lasting.
> - Effects of DBT on emotion regulation are mixed.

Rahmani and colleagues also found reductions in BMI and improvements in emotion regulation (2018). However, the effects of DBT on BED comorbidities are mixed as Safer and colleagues found no statistically significant differences in emotion regulation between the two groups (2010). Taken together, the research summarized here suggests that DBT may be effective at reducing binge eating in patients with BED in the short term. However, it is unclear if its effectiveness is due to hypothesized mechanisms of action (i.e., improved emotion regulation and distress tolerance) or nonspecific therapeutic factors.

Finally, Linardon and colleagues recently conducted a systematic review and meta-analysis of third-wave therapies for eating disorders overall. They found that all third-wave therapies, including DBT, ST, ACT, MBI, and CFT, improved eating disorder symptoms from pre-treatment to post-treatment. However, none of the third-wave therapies were better than active control or CBT. Thus, it was determined that no third-wave therapy has yet met criteria for an empirically supported treatment for eating disorders (2017a), and these treatments warrant further investigation. However, it is important to note that the recent RCT with promising results in favor of DBT conducted by Rahmani et al. 2018 was not included in the Linardon meta-analysis.

7 Mediators, Moderators, and Predictors of Psychotherapy Outcomes

In their review of mediators and moderators of CBT for eating disorders, Linardon et al. (2017b) found that early change, that is, improvement in eating disorder symptoms early in treatment, and reductions in dietary restraint consistently mediated outcomes across disorders, but other findings were inconsistent. Similarly, in their meta-analysis of predictors, moderators, and mediators of treatment outcomes across eating disorder diagnoses and treatment types, Vall and Wade found that early symptom reduction was the strongest predictor of treatment outcome at both the end of treatment and at follow-up. They also found that fewer eating disorder behaviors, high motivation to recover, less psychopathology and comorbidities, lower shape and weight concern, and better interpersonal functioning predicted better treatment outcomes (2016). As discussed previously, there is some evidence that IPT may work better in patients with BED who have high negative affect, and higher overall eating disorder psychopathology (Wilson et al. 2010; Sysko et al. 2010). However, more work is needed to explore the mechanisms underlying psychotherapies for disorders characterized by binge eating.

8 Conclusions and Future Directions for Psychotherapy in Adults

Table 1 summarizes evidence for psychotherapies to treat binge eating in adults. Overall, CBT and IPT are the most well-supported treatments for binge eating in BN and BED in adults. Although it appears that CBT works more quickly to treat disordered eating compared to IPT, IPT seems to "catch up" to CBT in the long term. IPT has also been described as a "best-buy intervention" as it addresses problems outside of the eating disorder itself (e.g., depression, anxiety) (Kazdin et al. 2017).

However, it is important to note that although CBT appears to be a superior treatment and is associated with significant reductions in bulimic symptoms, only about 30–40% of treatment completers with BN achieve symptom abstinence (Linardon and Wade 2018). This highlights the need for novel psychotherapeutic interventions. Promising treatments are emerging in the field to treat binge eating (e.g., BWL, third-wave therapies), but, to date, there is not enough evidence to support them as first-line interventions. Additional research is also needed to help tailor treatment to specific populations. A better understanding of the predictors, mediators, and moderators of treatment response could aid in this pursuit. For example, if high motivation to recover consistently predicts better treatment outcomes (Vall and Wade 2016) across studies, a target for early treatment could be to focus on increasing motivation to promote treatment success. Specialized, and more complex, adaptations of psychotherapy could also hold promise for binge-eating-related disorder treatment in specific populations. For example, "enhanced" CBT (CBT-E) treatment that addresses external factors that contribute to disordered eating (e.g., perfectionism, mood

Table 2 Overall evidence for psychotherapies to treat binge eating in youth (adapted, Lock 2015)

	Well-established	Probably efficacious	Possibly efficacious	Experimental
BN	None	None	FBT	CBT; SP
BED	None	None	None	IPT; DBT

BN bulimia nervosa, *BED* binge-eating disorder, *FBT* family-based treatment, *CBT* cognitive behavioral therapy, *SP* supportive psychotherapy, *IPT* interpersonal psychotherapy, *DBT* dialectical behavioral therapy

intolerance, low self-esteem, and interpersonal difficulties) could be especially beneficial in individuals that have marked psychopathology in these areas (Fairburn et al. 2009; Thompson-Brenner et al. 2016).

Another very important future direction for the field is to increase access to care. Preliminary research shows that technology-based platforms could be a promising tool to achieve this goal. Agras et al. (2017) recently reviewed RCTs of CD-ROM, e-mail, and internet-based CBT treatments for eating disorders, that ranged from entirely self-help oriented treatments to treatments that involved periodic contact with a therapist, and observed that several of these treatments produced greater improvements in eating disorder behavior and psychopathology compared to control groups.

9   Psychotherapy for Binge Eating in Youth

There are some similarities between psychotherapies for binge eating in adults and youth. For example, CBT, IPT, and DBT have all been tested in both populations. The theoretical frameworks for these psychotherapies are the same for adults and youth and will, therefore, not be readdressed.

Despite these theoretical similarities, evidence for each treatment in youth does not always match evidence for treatment in adults. It is also important to note that binge-eating-related disorders are difficult to assess in youth (Tanofsky-Kraff et al. 2008), therefore, most of the research has focused on loss of control (LOC) eating, the hallmark feature of binge eating episodes. Given this

distinction, the empirical evidence for youth is separated by BN, BED, and/or general binge eating/LOC eating, where applicable.

To date, there are no "well-established" or "probably efficacious" treatments for BN or BED in youth (Lock 2015). However, the "possibly efficacious" and "experimental treatments" that have been investigated, are discussed below, and summarized in Table 2.

10   Cognitive Behavioral Therapy

10.1 Bulimia Nervosa

Findings from a recent review (Hail and Le Grange 2018) indicate that only two RCTs have examined the efficacy of CBT for BN in youth.

Le Grange et al. compared the efficacy of an adolescent-adapted CBT (CBT-A) to an adapted version of family therapy specifically for adolescents with BN (FBT-BN). Compared to CBT-A, adolescents randomized to FBT-BN demonstrated higher abstinence rates from binge eating and purging post treatment and at a 6-month follow-up; however, abstinence rates were no longer statistically different at a 12-month follow-up (2015). A second RCT compared CBT to psychodynamic therapy (PDT) in a sample of adolescent females with threshold and subthreshold BN (Stefini et al. 2017). Both CBT and PDT were associated with high remission rates post treatment (33.3% and 31.0%, respectively). When examining between-group effects across types of eating pathology, adolescents randomized to CBT demonstrated greater reductions in the frequency of binge eating and purging compared to PDT.

Key Points: CBT for BN in Youth

- CBT could reduce binge eating in BN more than other treatments, but the evidence is inconsistent.

In summary, data on the efficacy of CBT in samples of youth are mixed. Novel treatment approaches for BN are needed to further improve and sustain abstinence rates in this population.

10.2 Binge-Eating Disorder

Support for CBT for BED in youth is limited. One pilot trial of CBT-A found that adolescent females randomized to CBT demonstrated greater reductions in the frequency of objective binge episodes and eating-related psychopathology (e.g., shape, weight, and eating concerns) post treatment compared to the usual/delayed treatment control group. Moreover, improvements were sustained at follow-up, such that 100% of adolescent females in CBT-A achieved abstinence from binge eating (DeBar et al. 2013). Further research is needed to examine the efficacy of CBT adapted for adolescents.

Key Points: CBT for BED in Youth

- In a preliminary study, in-person CBT improves binge eating and eating-related psychopathology (e.g., shape concerns).
- Symptom improvements are maintained.

11 Interpersonal Psychotherapy

IPT is currently considered an experimental treatment for BED in youth (Lock 2015). However, a recent review highlights the growing body of literature on the efficacy of IPT for the treatment of binge eating/LOC eating and prevention of excess weight gain in youth (Byrne et al. 2019). In the first pilot trial of group-based IPT in

adolescent girls (aged 12–17 years) at risk for excess weight, IPT promoted greater reductions in LOC eating episodes and weight gain compared to standard-of-care health education (HE) (Tanofsky-Kraff et al. 2010). In an adequately powered RCT, IPT and HE were equally effective in reducing frequency of LOC eating episodes, psychopathology, and weight and fat gain, while IPT was more efficacious than HE at reducing objective binge episodes at 1-year follow-up (Tanofsky-Kraff et al. 2014). Further follow-up analyses suggested that IPT may be more effective for particular youth subgroups. For example, racially and ethnically diverse adolescents showed greater reductions in LOC eating episodes at 1-year follow-up (Tanofsky-Kraff et al. 2014) and 3-year follow-up (Burke et al. 2017). Additionally, compared to HE, IPT promoted greater reductions in weight and fat gain among adolescent females with high levels of baseline anxiety or social problems at a 3-year follow-up (Tanofsky-Kraff et al. 2017).

IPT has also been piloted for family-based delivery (FB-IPT) in preadolescents (aged 8–13 years) with LOC eating and overweight/obesity. In an RCT that examined changes in child psychosocial functioning, LOC eating, and body weight and fat, the FB-IPT group reported greater reductions in depression, anxiety, and odds of LOC eating after treatment compared to standard of care family-based health education (FB-HE). Further, improvements in disordered eating attitudes and depressive symptoms were also reported at 6-month and 1-year follow-up, respectively (Shomaker et al. 2017).

These findings highlight the efficacy of group-based IPT for prevention of LOC eating and excess weight gain, particularly among adolescents who are ethnically/racially diverse and/or who exhibit high levels of anxiety or social problems. Preliminary data also support the acceptability and feasibility of family-based IPT for preadolescents. Given that psychosocial problems are commonly reported among youth with LOC eating and overweight/obesity (Morgan et al. 2002; Doyle et al. 2007; Goossens et al. 2007), IPT may be especially effective in addressing the underlying psychopathology that contributes to binge eating and exacerbated weight gain in youth.

> **Key Points: IPT for Binge/LOC Eating in Youth**
>
> - IPT could prevent LOC eating and excess weight gain, but evidence is mixed.
> - IPT may be more effective for certain groups (e.g., racially diverse youth and youth with anxiety).
> - Family-based IPT reduces depression, anxiety, and odds of LOC eating.
> - IPT may be helpful in addressing underlying psychopathology that contributes to binge eating.

12 Family-Based Treatment

As the name implies, family involvement is a core treatment component in family-based treatment (FBT). In short, FBT aims to (1) help patients recognize changes needed for recovery, (2) externalize the illness (i.e., highlight that the patient and the eating disorder are not the same and that the eating disorder is responsible for food, eating, shape, and weight issues, not the child), and (3) empower parents to be the child's agent for change and support system. Overall, FBT takes a very practical approach and focuses on rapid symptom reduction (Rienecke 2017).

12.1 Bulimia Nervosa

FBT is regarded as a possibly efficacious treatment for adolescent BN (Lock 2015), but it has not been tested in youth with BED. Insights from a recent review (Rienecke 2017) indicate that only two studies have compared the efficacy of FBT to other evidenced-based treatments for BN,

including CBT (see "CBT for BN" for a review of Le Grange et al. 2015) and Supportive Psychotherapy (SP; Le Grange et al. 2007). In an RCT of FBT versus SP, adolescents with BN who were randomized to FBT demonstrated significantly higher rates of abstinence from binge eating and purging compared to those in the SP condition (Le Grange et al. 2007). At a 6-month follow up, however, differences between FBT and SP in binge–purge behavior were marginally significant ($p = 0.05$). In summary, studies (Le Grange et al. 2007, 2015) indicate that FBT produces greater abstinence from binge–purge behavior post treatment compared to other evidenced-based therapies. However, differences tend to subside at follow-up periods.

13 Third-Wave Therapies

No studies to date have adapted DBT for youth with BN, but it is considered an experimental treatment for youth with BED (Lock 2015). In a sample of racially diverse adolescent girls with LOC eating or BED, one RCT found no differences between a DBT condition and a weight management condition on disordered eating cognitions, dietary restraint, or eating in response to negative affect (Mazzeo et al. 2016). Given that behavior weight management is also effective in reducing eating pathology among youth (Goldschmidt et al. 2014; Balantekin et al. 2017), the active treatment components of the two interventions may have been too similar. More research is needed to determine the efficacy of DBT in the treatment of binge eating/LOC eating in youth.

There is a lack of studies investigating the use of ST, ACT, MBI, or CFT to treat binge eating in youth. However, preliminary results indicate that dispositional mindfulness is associated with lower odds of binge eating among adolescent females at risk for type 2 diabetes (Pivarunas et al. 2015). Thus, the development of mindfulness-based interventions for youth with binge eating may alter ED attitudes and behaviors, but this remains to be tested.

> **Key Points: FBT and Third-Wave Therapies for Binge/LOC Eating in Youth**
>
> - FBT improves binge eating compared to other evidenced-based therapies post treatment, but differences tend to subside at follow-up periods.
> - DBT is not superior to a weight management program at reducing disordered eating cognitions or dietary restraint.
> - There is a lack of studies investigating third-wave therapies such as ST, ACT, MBI, or CFT.

14 Mediators, Moderators, and Predictors of Psychotherapy Outcomes

Due to the scarcity of RCTs for BN and BED in youth, few studies have examined variables that could contribute to treatment outcomes. Data suggest that IPT has more prolonged effects for adolescent females who are racially and ethnically diverse (Tanofsky-Kraff et al. 2014; Burke et al. 2017), have higher baseline levels of anxiety or social problems (Tanofsky-Kraff et al. 2017), or who are older (Burke et al. 2017). With regard to FBT, data suggest that FBT may be favorable to SP in the long term for individuals with lower baseline levels of disordered eating (Le Grange et al. 2008). Moreover, Le Grange et al. (2015) found that adolescents with higher scores on a family environment scale (e.g., high family cohesion), lower eating-related obsessions, and male gender had higher abstinence rates at treatment end. There are virtually no data on mediators and moderators of CBT or DBT for binge-eating-related disorders in youth. Further examination of the mediators and moderators of treatment outcomes is an important next step in the design of tailored interventions.

15 Conclusions and Future Directions for Psychotherapy in Youth

Overall, there are no well-established treatments for BN and BED in youth, however, there is a growing body of research on the development of possibly efficacious and experimental treatments. Preliminary evidence suggests that IPT could be especially helpful in addressing the underlying psychopathology related to binge eating, especially in certain subgroups (e.g., racially diverse adolescent females, adolescent females with social problems). FBT may be particularly effective for adolescents with lower baseline levels of eating disorder psychopathology and better family functioning. The limited research on treatment-specific predictors and moderators of CBT and DBT precludes any conclusions regarding who benefits the most from these treatments. Further research in youth is needed to compare the effectiveness of these interventions.

Although the growing body of research on psychotherapeutic treatment for BN and BED in youth is promising, there are several remaining gaps in the literature that warrant further investigation. Data on which type of psychotherapeutic treatment is best and for whom are equivocal, and short follow-up periods prevent the examination of long-term effects across treatment type. Novel treatment approaches for BN and BED in children and teens are needed to further improve and sustain abstinence rates and develop personalized interventions that target individual risk factors. Future research should also examine the efficacy of more cost-effective, scalable forms of psychotherapy to increase accessibility of treatment for individuals struggling with binge-eating-related disorders (i.e., internet facilitated interventions and mobile application platforms).

16 Summary

Overall, there are many psychotherapies being tested to treat binge eating in BN and BED in both adults and youth. While there are several well-supported treatments for binge-eating-related disorders (i.e., CBT, IPT) in adults, evidence-based support for treatment in youth is lacking. It is important to note that there are several promising treatments emerging in the field in both populations that warrant further exploration. More research is also needed to develop more disseminable techniques that will improve access to care and to develop a more tailored treatment for specific populations to promote success.

References

Agras WS (2019) Cognitive behavior therapy for the eating disorders. Psychiatr Clin N Am 42:169–179. https://doi.org/10.1016/j.psc.2019.01.001

Agras WS, Walsh T, Fairburn CG et al (2000) A multicenter comparison of cognitive-behavioral therapy and interpersonal psychotherapy for bulimia nervosa. Arch Gen Psychiatry 57:459–466. https://doi.org/10.1001/archpsyc.57.5.459

Agras WS, Fitzsimmons-Craft EE, Wilfley DE (2017) Evolution of cognitive-behavioral therapy for eating disorders. Behav Res Ther 88:26–36. https://doi.org/10.1016/j.brat.2016.09.004

Balantekin KN, Hayes JF, Sheinbein DH et al (2017) Patterns of eating disorder pathology are associated with weight change in family-based behavioral obesity treatment. Obesity (Silver Spring) 25:2115–2122. https://doi.org/10.1002/oby.22028

Beck JS (1964) Cognitive therapy: basics and beyond. Guilford Press, New York

Brownley KA, Berkman ND, Peat CM et al (2016) Binge-eating disorder in adults. Ann Intern Med 165:409–420. https://doi.org/10.7326/M15-2455

Burke NL, Shomaker LB, Brady S et al (2017) Impact of age and race on outcomes of a program to prevent excess weight gain and disordered eating in adolescent girls. Nutrients 9:947. https://doi.org/10.3390/nu9090947

Byrne ME, LeMay-Russell S, Tanofsky-Kraff M (2019) Loss-of-control eating and obesity among children and adolescents. Curr Obes Rep 8:33–42. https://doi.org/10.1007/s13679-019-0327-1

Chen E, Touyz SW, Beumont PJV et al (2003) Comparison of group and individual cognitive-behavioral therapy for patients with bulimia nervosa. Int J Eat Disord 33:241–254. https://doi.org/10.1002/eat.10137

DeBar LL, Wilson GT, Yarborough BJ et al (2013) Cognitive behavioral treatment for recurrent binge eating in adolescent girls: a pilot trial. Cogn Behav Pract 20:147–161. https://doi.org/10.1016/j.cbpra.2012.04.001

Doyle AC, le Grange D, Goldschmidt A, Wilfley DE (2007) Psychosocial and physical impairment in overweight adolescents at high risk for eating disorders. Obesity (Silver Spring) 15:145–154. https://doi.org/10.1038/oby.2007.515

Fairburn CG, Jones R, Peveler RC et al (1991) Three psychological treatments for bulimia nervosa. A comparative trial. Arch Gen Psychiatry 48:463–469. https://doi.org/10.1001/archpsyc.1991.01810290075014

Fairburn CG, Jones R, Peveler RC et al (1993) Psychotherapy and bulimia nervosa. Longer-term effects of interpersonal psychotherapy, behavior therapy, and cognitive behavior therapy. Arch Gen Psychiatry 50:419–428. https://doi.org/10.1001/archpsyc.1993.01820180009001

Fairburn CG, Norman PA, Welch SL et al (1995) A prospective study of outcome in bulimia nervosa and the long-term effects of three psychological treatments. Arch Gen Psychiatry 52:304–312. https://doi.org/10.1001/archpsyc.1995.03950160054010

Fairburn CG, Cooper Z, Doll HA et al (2009) Transdiagnostic cognitive-behavioral therapy for patients with eating disorders: a two-site trial with 60-week follow-up. Am J Psychiatry 166:311–319. https://doi.org/10.1176/appi.ajp.2008.08040608

Fairburn CG, Bailey-Straebler S, Basden S et al (2015) A transdiagnostic comparison of enhanced cognitive behaviour therapy (CBT-E) and interpersonal psychotherapy in the treatment of eating disorders. Behav Res Ther 70:64–71. https://doi.org/10.1016/j.brat.2015.04.010

Fenn K, Byrne M (2013) The key principles of cognitive behavioral therapy. InnovAiT 6:579–585. https://doi.org/10.1177/1755738012471029

Ghaderi A, Odeberg J, Gustafsson S et al (2018) Psychological, pharmacological, and combined treatments for binge eating disorder: a systematic review and meta-analysis. PeerJ 2018(6):e5113. https://doi.org/10.7717/peerj.5113

Goldschmidt AB, Best JR, Stein RI et al (2014) Predictors of child weight loss and maintenance among family-based treatment completers. J Consult Clin Psychol 82:1140–1150. https://doi.org/10.1037/a0037169

Goossens L, Braet C, Decaluwé V (2007) Loss of control over eating in obese youngsters. Behav Res Ther 45:1–9. https://doi.org/10.1016/j.brat.2006.01.006

Hail L, Le Grange D (2018) Bulimia nervosa in adolescents: prevalence and treatment challenges. Adolesc Health Med Ther 9:11–16. https://doi.org/10.2147/AHMT.S135326

Hilbert A, Bishop ME, Stein RI et al (2012) Long-term efficacy of psychological treatments for binge eating disorder. Br J Psychiatry 200:232–237. https://doi.org/10.1192/bjp.bp.110.089664

Hill DM, Craighead LW, Safer DL (2011) Appetite-focused dialectical behavior therapy for the treatment of binge eating with purging: a preliminary trial. Int J Eat Disord 44:249–261. https://doi.org/10.1002/eat. 20812

Iacovino JM, Gredysa DM, Altman M, Wilfley DE (2012) Psychological treatments for binge eating disorder. Curr Psychiatry Rep 14:432–446. https://doi.org/10. 1007/s11920-012-0277-8

Karam AM, Fitzsimmons-Craft EE, Tanofsky-Kraff M, Wilfley DE (2019) Interpersonal psychotherapy and the treatment of eating disorders. Psychiatr Clin N Am 42:205–218. https://doi.org/10.1016/j.psc.2019. 01.003

Kazdin AE, Fitzsimmons-Craft EE, Wilfley DE (2017) Addressing critical gaps in the treatment of eating disorders. Int J Eat Disord 50:170–189. https://doi. org/10.1002/eat.22670

Le Grange D, Crosby RD, Rathouz PJ, Leventhal BL (2007) A randomized controlled comparison of family-based treatment and supportive psychotherapy for adolescent bulimia nervosa. Arch Gen Psychiatry 64:1049–1056. https://doi.org/10.1001/archpsyc.64.9. 1049

Le Grange D, Crosby RD, Lock J (2008) Predictors and moderators of outcome in family-based treatment for adolescent bulimia nervosa. J Am Acad Child Adolesc Psychiatry 47:464–470. https://doi.org/10.1097/CHI. 0b013e3181640816

Le Grange D, Lock J, Agras WS et al (2015) Randomized clinical trial of family-based treatment and cognitive-behavioral therapy for adolescent bulimia nervosa. J Am Acad Child Adolesc Psychiatry 54:886–894.e2. https://doi.org/10.1016/j.jaac.2015.08.008

Linardon J, Wade TD (2018) How many individuals achieve symptom abstinence following psychological treatments for bulimia nervosa? A meta-analytic review. Int J Eat Disord 51:287–294. https://doi.org/ 10.1002/eat.22838

Linardon J, Fairburn CG, Fitzsimmons-Craft EE et al (2017a) The empirical status of the third-wave behaviour therapies for the treatment of eating disorders: a systematic review. Clin Psychol Rev 58:125–140. https://doi.org/10.1016/j.cpr.2017.10.005

Linardon J, de la Piedad Garcia X, Brennan L (2017b) Predictors, moderators, and mediators of treatment outcome following manualised cognitive-behavioural therapy for eating disorders: a systematic review. Eur Eat Disord Rev 25:3–12. https://doi.org/10.1002/erv. 2492

Linardon J, Wade TD, de la Piedad GX, Brennan L (2017c) The efficacy of cognitive-behavioral therapy for eating disorders: a systematic review and meta-analysis. J Consult Clin Psychol 85:1080–1094. https://doi.org/10.1037/ccp0000245

Lock J (2015) An update on evidence-based psychosocial treatments for eating disorders in children and adolescents. J Clin Child Adolesc Psychol 44:707–721. https://doi.org/10.1080/15374416.2014. 971458

Mazzeo SE, Lydecker J, Harney M et al (2016) Development and preliminary effectiveness of an innovative treatment for binge eating in racially diverse adolescent girls. Eat Behav 22:199–205. https://doi.org/10.1016/j. eatbeh.2016.06.014

McElroy SL, Guerdjikova AI, Mori N et al (2015) Overview of the treatment of binge eating disorder. CNS Spectr 20:546–556. https://doi.org/10.1017/ S1092852915000759

Morgan CM, Yanovski SZ, Nguyen TT et al (2002) Loss of control over eating, adiposity, and psychopathology in overweight children. Int J Eat Disord 31:430–441. https://doi.org/10.1002/eat.10038

Murphy R, Straebler S, Cooper Z, Fairburn CG (2010) Cognitive behavioral therapy for eating disorders. Psychiatr Clin North Am 33:611–627. https://doi.org/ 10.1016/j.psc.2010.04.004

National Institute for Health Care Excellence (2017) Eating disorders: recognition and treatment. NICE, London

Palavras MA, Hay P, dos Santos Filho CA, Claudino A (2017) The efficacy of psychological therapies in reducing weight and binge eating in people with bulimia nervosa and binge eating disorder who are overweight or obese—a critical synthesis and meta-analyses. Nutrients 9:E299. https://doi.org/10.3390/ nu9030299

Pivarunas B, Kelly NR, Pickworth CK, Cassidy O, Radin RM, Shank LM, Vannucci A, Courville AB, Chen KY, Tanofsky-Kraff M, Yanovski JA, Shomaker LB (2015) Mindfulness and eating behavior in adolescent girls at risk for type 2 diabetes. Int J Eat Disord 48:563–569. https://doi.org/10.1002/eat.22435

Rahmani M, Omidi A, Asemi Z, Akbari H (2018) The effect of dialectical behaviour therapy on binge eating, difficulties in emotion regulation and BMI in overweight patients with binge-eating disorder: a randomized controlled trial. Mental Health & Prevention 9:13–18. https://doi.org/10.1016/j.mhp.2017.11.002

Rienecke RD (2017) Family-based treatment of eating disorders in adolescents: current insights. Adolesc Health Med Ther 8:69–79. https://doi.org/10.2147/ AHMT.S115775

Safer DL, Telch CF, Agras WS (2001) Dialectical behavior therapy for bulimia nervosa. AJP 158:632–634. https://doi.org/10.1176/appi.ajp.158.4.632

Safer DL, Robinson AH, Jo B (2010) Outcome from a randomized controlled trial of group therapy for binge eating disorder: comparing dialectical behavior therapy adapted for binge eating to an active comparison group therapy. Behav Ther 41:106–120. https://doi.org/10. 1016/j.beth.2009.01.006

Shomaker LB, Tanofsky-Kraff M, Matherne CE et al (2017) A randomized, comparative pilot trial of family-based interpersonal psychotherapy for reducing psychosocial symptoms, disordered-eating, and excess weight gain in at-risk preadolescents with loss-of-control-eating. Int J Eat Disord 50:1084–1094. https://doi. org/10.1002/eat.22741

Slade E, Keeney E, Mavranezouli I et al (2018) Treatments for bulimia nervosa: a network meta-analysis. Psychol Med 48:2629–2636. https://doi.org/10.1017/S0033291718001071

Stefini A, Salzer S, Reich G et al (2017) Cognitive-behavioral and psychodynamic therapy in female adolescents with bulimia nervosa: a randomized controlled trial. J Am Acad Child Adolesc Psychiatry 56:329–335. https://doi.org/10.1016/j.jaac.2017.01.019

Svaldi J, Schmitz F, Baur J et al (2019) Efficacy of psychotherapies and pharmacotherapies for bulimia nervosa. Psychol Med 49:898–910. https://doi.org/10.1017/S0033291718003525

Sysko R, Hildebrandt T, Wilson GT et al (2010) Heterogeneity moderates treatment response among patients with binge eating disorder. J Consult Clin Psychol 78:681–690. https://doi.org/10.1037/a0019735

Tanofsky-Kraff M, Marcus MD, Yanovski SZ, Yanovski JA (2008) Loss of control eating disorder in children age 12 years and younger: proposed research criteria. Eat Behav 9:360–365. https://doi.org/10.1016/j.eatbeh.2008.03.002

Tanofsky-Kraff M, Wilfley DE, Young JF et al (2010) A pilot study of interpersonal psychotherapy for preventing excess weight gain in adolescent girls at-risk for obesity. Int J Eat Disord 43:701–706. https://doi.org/10.1002/eat.20773

Tanofsky-Kraff M, Shomaker LB, Wilfley DE et al (2014) Targeted prevention of excess weight gain and eating disorders in high-risk adolescent girls: a randomized controlled trial12345. Am J Clin Nutr 100:1010–1018. https://doi.org/10.3945/ajcn.114.092536

Tanofsky-Kraff M, Shomaker LB, Wilfley DE et al (2017) Excess weight gain prevention in adolescents: three-year outcome following a randomized controlled trial. J Consult Clin Psychol 85:218–227. https://doi.org/10.1037/ccp0000153

Telch CF, Agras WS, Rossiter EM et al (1990) Group cognitive-behavioral treatment for the nonpurging bulimic: an initial evaluation. J Consult Clin Psychol 58:629–635. https://doi.org/10.1037/0022-006X.58.5.629

Telch CF, Agras WS, Linehan MM (2001) Dialectical behavior therapy for binge eating disorder. J Consult Clin Psychol 69:1061–1065. https://doi.org/10.1037/0022-006X.69.6.1061

Thompson-Brenner H, Shingleton RM, Thompson DR et al (2016) Focused vs. broad enhanced cognitive behavioral therapy for bulimia nervosa with comorbid borderline personality: a randomized controlled trial. Int J Eat Disord 49:36–49. https://doi.org/10.1002/eat.22468

Vall E, Wade TD (2016) Predictors of treatment outcome in individuals with eating disorders: a systematic review and meta-analysis. Int J Eat Disord 49:432–433. https://doi.org/10.1002/eat.22518

Wilfley DE, Agras WS, Telch CF et al (1993) Group cognitive-behavioral therapy and group interpersonal psychotherapy for the nonpurging bulimic individual: a controlled comparison. J Consult Clin Psychol 61:296–305. https://doi.org/10.1037/0022-006X.61.2.296

Wilfley DE, Frank MA, Welch R et al (1998) Adapting interpersonal psychotherapy to a group format (IPT-G) for binge eating disorder: toward a model for adapting empirically supported treatments. Psychother Res 8:379–391. https://doi.org/10.1093/ptr/8.4.379

Wilfley DE, Welch RR, Stein RI et al (2002) A randomized comparison of group cognitive-behavioral therapy and group interpersonal psychotherapy for the treatment of overweight individuals with binge-eating disorder. Arch Gen Psychiatry 59:713–721. https://doi.org/10.1001/archpsyc.59.8.713

Wilson GT, Wilfley DE, Agras WS, Bryson SW (2010) Psychological treatments of binge eating disorder. Arch Gen Psychiatry 67:94–101. https://doi.org/10.1001/archgenpsychiatry.2009.170

Self-Help Interventions for the Treatment of Binge Eating

Anja Hilbert and Hans-Christian Puls

Abstract

Self-help interventions (SHIs) represent established psychological treatments for binge-eating disorder (BED) and bulimia nervosa (BN) with demonstrated short- and long-term efficacy. Implemented via self-help books, the internet, or smartphone applications, SHIs are highly accessible and may thus be suited to overcome patient- and provider-related barriers in traditional face-to-face psychological treatments. SHIs can be offered with or without professional guidance, with some evidence on more favorable outcomes with higher degrees of guidance. Key limitations of SHIs include low acceptability and participation, depicted through insufficient treatment engagement and completion as well as patient adherence, and should, therefore, be further investigated and enhanced. Predictors of outcome, including treatment-specific moderators and mediators, are largely unclear. The individual tailoring of interventions and their components to individual patients is part of the high potential of technology-based SHIs, although these remain underutilized and understudied. Cost-effectiveness compared to minimal treatment was suggested for SHIs and might be increased using complex models of care including SHIs. In sum, more research is needed to understand and further establish SHIs as psychological approaches to the treatment of BED and BN.

Keywords

Self-help · Treatment · Binge-eating disorder · Bulimia nervosa · Eating disorders · Guidance · Acceptability · Predictors of outcome · Cost-effectiveness

Learning Objectives

In this chapter, you will:

- Get familiarized with the need for alternative treatment options for a substantial number of patients with eating disorders who experience barriers to "traditional" face-to-face treatments.
- Get to know the different formats of self-help interventions (SHIs) and their respective efficacy and effectiveness for binge-eating disorder and bulimia nervosa.
- Understand the role of guidance and major limitations of SHIs, including low acceptability and participation.

(continued)

A. Hilbert (✉) · H.-C. Puls
Integrated Research and Treatment Center Adiposity Diseases, Behavioral Medicine, Department of Psychosomatic Medicine and Psychotherapy, Leipzig, Germany
e-mail: anja.hilbert@medizin.uni-leipzig.de; hans-christian.puls@medizin.uni-leipzig.de

G. K.W. Frank, L. A. Berner (eds.), *Binge Eating*, https://doi.org/10.1007/978-3-030-43562-2_19

- Learn about predictors of outcome, including treatment-specific moderators and mediators, cost-effectiveness, and the role of SHIs within complex models of care.
- Understand important questions for future research and clinical implications based on the available evidence on SHIs.

1 Introduction

Although a vast body of research has documented face-to-face cognitive-behavioral therapy (CBT) to be the most well-established treatment option for patients with binge-eating disorder (BED; Hilbert et al. 2019) and bulimia nervosa (BN; Svaldi et al. 2019), only a minority of patients receive an evidence-based psychological treatment such as CBT (Kazdin et al. 2017). Multiple reasons on both a patient and provider level may account for this "treatment gap." Patient-related reasons for not seeking or not completing face-to-face therapies include practical barriers (e.g., costs of treatment, reduced mobility, low availability of face-to-face treatments in rural areas; Ali et al. 2017; Vall and Wade 2015), the fear of stigma or shame (Corrigan et al. 2014), or patients with milder problems feeling that face-to-face therapies might be overly intense (Traviss-Turner et al. 2017). Provider-related barriers include a shortage of professionals with expertise in evidence-based treatments for eating disorders (EDs; Agras et al. 2017; Cooper and Bailey-Straebler 2015). To overcome this shortage, specialized and intensive training for mental health professionals in the field of eating disorders (EDs) is required, but not yet readily available (Wilson and Zandberg 2012). Indeed, Kazdin and Blase (2011) concluded from their analysis on the treatment gap in mental health care, that even a doubling of today's numbers of mental health providers would bring little benefit to patients with EDs.

To bypass both patient- and provider-related barriers to treatment for BED and BN, various forms of CBT-based self-help interventions

(SHIs) have been designed, conducted, and reviewed. What they all have in common is the independent implementation of CBT-based treatment principles with the patient following step-by-step instructions contained in a book or via technology-assisted systems. Within SHIs, patients are provided with information on their ED and specific therapeutic skills, which they are encouraged to use in order to achieve their individual treatment goals (e.g., establish a healthy and regular eating pattern, identify triggers and overcome maintaining factors of binge eating). While some SHIs are accompanied by varying degrees of contact to a professional ("guided self-help"), others are purely self-directed ("pure self-help") and therefore do not necessarily involve a professional provider or even an institution in which it takes place. As a general advantage of SHIs, patients can work through the manual at their own time and pace, easily embedded in their everyday lives. Overall, SHIs have the clear potential to be widely accessible and, at the same time, highly cost-effective for patients with BED and BN. A further reason for integrating existing CBT manuals into SHIs was seen in the existence of a subgroup of patients, who report rather moderate, but still clinically relevant eating problems and might, therefore, benefit from a less intensive treatment approach than face-to-face CBT (Fairburn 2013; Traviss-Turner et al. 2017). Because SHIs have been shown to be inherently empowering by nature (Sánchez-Ortiz et al. 2011), they might represent starting points into more intensive treatment (e.g., face-to-face CBT), especially for patients with more severe symptoms. On the contrary, however, patients who do not feel their symptoms to be sufficiently addressed and treated by SHIs might be discouraged from seeking professional face-to-face treatment even when it is available and indicated (Beintner et al. 2014).

The purpose of this chapter is to define and to describe the main formats of CBT-based SHIs and to summarize current evidence on their efficacy and effectiveness, in order to contextualize them within the range of available treatment options for patients with BED and BN. We aim to describe the role of guidance, acceptability, predictors of outcome, including moderators and mediators, and cost-effectiveness within SHIs. Further, the role of SHIs in complex models of care will be discussed. Finally, we will address directions for future research as well as clinical implications.

2 Formats

In light of the aforementioned shortcomings of traditional face-to-face treatments for patients with BED and BN, book-based formats of established CBT manuals represented the first and forerunning format of SHIs. Researchers and clinicians with expertise in the treatment of EDs integrated the main contents of CBT into structured text-based manuals, which today, after constant review and revision, are among the most well-established and thoroughly described SHIs for the treatment of BED and BN. The two most widely used self-help manuals "Overcoming Binge Eating" by Fairburn (1995, 2013) and "Getting Better Bite by Bite" by Schmidt and Treasure (1993; Schmidt et al. 2015) stand alongside a small number of other book-based manuals (e.g., "Working to Overcome Eating Difficulties" by Traviss et al. 2011, "Bulimia Nervosa and Binge-Eating: A Guide to Recovery" by Cooper 1996). To present an example of CBT-based SHIs, "Overcoming Binge Eating" in its second edition contains a set of theoretical chapters with information on the prevalence, epidemiology, maintenance, and treatment of binge-eating behaviors as well as the actual self-help manual including the key aspects of the treatment of binge eating (e.g., treatment motivation, healthy and regular eating patterns, triggers of binge eating, problem solving, and body image). Using psychoeducation, self-monitoring, and cognitive-behavioral exercises, the reader is provided with comprehensible and applicable therapeutic content for his/her eating problem.

With the rapidly rising importance of the Internet in daily life, its relevance in mental health care is also increasing (e.g., Hilbert et al. 2018). This "E-mental health" refers to the use of information and communication technology to support and improve mental health conditions and the care of which (Riper et al. 2012), and includes the internet-based delivery of SHIs in the treatment of patients with BED and BN. Internet-based SHIs might even better overcome the obstacles in reaching patients with BED and BN compared to book-based formats. The relative anonymity of the internet and its widespread and unlimited access may result in a low threshold for patients to consider and start an internet-based SHI. These benefits apply especially to patients with BED and BN who would not otherwise seek face-to-face treatment, for example, those with experiences of being stigmatized, those living in remote or psychotherapeutically underserved areas, and young individuals in the early stages of the ED development (Corrigan et al. 2014; Fairburn and Murphy 2015; Fairburn and Patel 2014). Further, similar to other technology-enhanced psychological interventions, internet-based SHIs carry the potential to be individually customized and tailored according to the patients' needs and actions (Bauer and Moessner 2013; Loucas et al. 2014), and might be designed to exchange real-time information (e.g., using compatible smartphone applications), with expected favorable effects on participation, adherence, and outcome (Beintner et al. 2014; Hildebrandt et al. 2017; Fairburn and Patel 2014). Examples for internet-based SHIs for BED and BN include the CBT-based guided self-help program "SalutBED/SalutBN" (Carrard et al. 2011b; de Zwaan et al. 2017) and the interactive and multimedia-based guided self-help program "Overcoming Bulimia Online" (Sánchez-Ortiz et al. 2011).

Within the last few years, researchers in the ED field have become increasingly intrigued by mobile and wireless technologies, including smartphone "apps," for the treatment of EDs (Anastasiadou et al. 2018; Juarascio et al. 2015b). By design, smartphone apps benefit from the rapidly progressing usage of and access to mobile phones (Kelly and Minges 2012), while they share the advantages of book- and internet-based formats of SHIs in enhancing reach and dissemination relative to face-to-face therapy (e.g., accessibility, anonymity). Further advantages of app-based SHIs for the treatment of EDs include their potential to customize psychological interventions to the individual patient (Agras et al. 2017), thus empowering the patient to address his/her individual problem behavior (Anastasiadou et al. 2018). App-based SHIs have been used as a sole means of support (e.g., for BED: Juarascio et al. 2015a), for relapse prevention, and as an adjunct to standard face-to-face treatment (Anastasiadou et al. 2018). Within the latter, communication between patients and clinicians, the assessment of therapeutic progress, and subsequent clinical decision-making may be improved (Juarascio et al. 2015b). App-based SHIs, in general, might be further enhanced by modern machine learning algorithms including Just-in-Time Adaptive Interventions (i.e., real-time interventions during app-identified moments of need; Juarascio et al. 2018). However, challenges and risks of app-based SHIs must be considered. An important weakness of mobile psychological interventions is that only a few of the vast number of existing apps are grounded in evidence-based treatment principles, and studies exploring their efficacy, validity, and clinical utility are scarce (Anastasiadou et al. 2018). Further, if patients use app-based SHIs in addition to face-to-face treatment, conflict between therapeutic recommendations from both treatments might emerge, possibly resulting in confusion and nonadherence by the patient (Juarascio et al. 2015b).

3 Efficacy

Beginning with the publication of "Overcoming Binge-Eating" as the first book-based SHI (Fairburn 1995), a mounting body of research has since documented the efficacy of SHIs in the treatment of EDs (e.g., Barakat et al. 2017; de Zwaan et al. 2017; Grilo et al. 2013; Kelly and Carter 2015; Wagner et al. 2015; Wilson et al. 2010) and was summarized in systematic reviews and meta-analyses (e.g., Beintner et al. 2014; Hilbert et al. 2019; Linardon and Wade 2018; Svaldi et al. 2019; Traviss-Turner et al. 2017). While SHIs are contraindicated for patients with

anorexia nervosa with their greater medical needs compared to patients with other EDs (Wilson and Zandberg 2012; Yim and Schmidt 2019), studies focused on the efficacy of SHIs in the treatment of BED and BN. In the following, we will provide a chronological overview on the efficacy of SHIs, moving from review papers to meta-analyses, first for BED and second for BN.

3.1 Binge-Eating Disorder

The National Institute for Health and Care Excellence (NICE) guideline recommendations (2004), in which SHIs were listed as recommended treatment options for BED, represented a starting point for the pertinent review paper on CBT-based SHIs for the treatment of EDs by Wilson and Zandberg (2012). A previous review by Sysko and Walsh (2008) including 26 controlled and uncontrolled studies on SHIs for BED and BN (6 of which related to BED alone) had concluded that CBT-based SHIs are significantly superior to wait-list control condition in reducing binge-eating frequency and producing abstinence from binge eating, especially in patients with BED. However, Sysko and Walsh (2008) noted little evidence for the efficacy of SHIs in comparison to active control conditions (i.e., other treatments). Thus, Wilson and Zandberg (2012) reviewed active treatment comparisons in a total of 10 controlled studies that compared book- and internet-based guided SHIs to other treatments for BED, such as interpersonal psychotherapy, pure SHIs, guided SHIs in combination with anti-obesity medication, or behavioral weight-loss treatment. Despite the limited evidence, the authors concluded that SHIs have specific effects in treating BED, consistently reducing binge-eating episodes, ED psychopathology, and mental comorbidity at post treatment and follow-up assessments (Wilson and Zandberg 2012). However, the examined SHIs did not yield significant weight loss in patients with BED and overweight or obesity, which is consistent with results of face-to-face CBT (e.g., Grilo et al. 2011). Body weight usually is considered a secondary outcome criterion in psychological intervention research in BED, while binge-eating frequency or abstinence commonly represents the primary outcome criterion (Hilbert et al. 2019).

In their comprehensive systematic review and meta-regression analysis on the participation and outcome in SHIs for BED and BN, Beintner et al. (2014) examined 50 SHI trials published through 2012 ($N = 2586$), 33 of which were randomized controlled trials (RCTs). Of the 62 analyzed SHI conditions, 29 conditions included patients with BED, and 43 conditions were supported by guidance. Book-, CD-ROM-, and internet-based SHIs were all administered. Overall, considerable effects of SHIs on the reduction of binge-eating frequency and ED psychopathology were found, whereas rates of abstinence from binge eating ranged widely at post treatment. Specifically, studies focusing on patients with BED compared to BN or mixed samples were likely to produce greater effects in all outcome measures (including higher participation and lower dropout rates), leading to the conclusion that SHIs may be especially well-suited for the treatment of BED. These results were expanded in the systematic review and meta-analysis by Traviss-Turner et al. (2017), who aggregated 30 RCTs on book- and internet-based guided SHIs for the treatment of EDs ($N = 2601$), 17 of which related to BED. In comparison to active and inactive control conditions (e.g., wait-list), guided SHIs showed significant effects of small-to-medium size in achieving abstinence from binge eating (effect size: 0.20) and reducing ED psychopathology (effect size: 0.46), respectively. As in Beintner et al. (2014), a meta-regression analysis suggested patients with BED to be more likely to achieve abstinence from binge eating in guided SHIs than those with BN.

More recently, Ghaderi et al. (2018) conducted a meta-analysis on 45 RCTs for the treatment of BED, 8 of which related to guided SHIs ($N = 282–384$, depending on the outcome variable). In comparison to wait-list, SHIs produced significantly higher abstinence from binge eating (small effect) and reduced binge-eating episodes, ED psychopathology, and depressive symptoms (medium effects), while there was no significant effect on body weight. Further, a meta-analysis

on various psychological and medical treatments for BED included 81 RCTs with 138 active intervention conditions, 14 conditions of which addressed guided or pure SHIs ($N = 498$; Hilbert et al. 2019). Compared to inactive control conditions, SHIs promoted abstinence from binge eating with a large effect (pooled abstinence rate: 46%) and reduced binge-eating frequency and ED psychopathology at post treatment with medium effects, while no significant effects on depressive symptoms or body weight were found, with the latter results being consistent with previous meta-analytic evidence (e.g., Ghaderi et al. 2018; Vocks et al. 2010). Interestingly, compared to active control conditions, SHIs involved significantly lower odds for abstinence from binge eating at 3- to 6-month follow-up and higher odds for dropout than face-to-face CBT, but no other short- and long-term differences were found for ED or general psychopathology. Further, a meta-regression analysis on a small number of studies did not suggest any differences between guided and pure SHIs in bringing about abstinence and reducing the number of binge-eating episodes.

In summary, SHIs, especially if based on CBT, have consistently documented efficacy for improving binge eating and ED psychopathology in patients with BED. In light of some evidence suggesting inferiority in relation to face-to-face CBT, SHIs may be considered as a treatment option when face-to-face CBT is not available or not acceptable (Hilbert et al. 2019). The importance of SHIs in the treatment of BED is also reflected in the updated NICE guidelines (NICE 2017), even recommending CBT-based guided SHIs as a first-line treatment for BED, given their established efficacy and low costs.

3.2 Bulimia Nervosa

The efficacy of SHIs for the treatment of BN is well-established when compared to inactive control conditions (Linardon and Wade 2018; Svaldi et al. 2019). As in BED, evidence on efficacy and consideration of costs led the NICE guidelines

(2017) to recommend CBT-based guided SHIs as the first-line treatment for BN (NICE 2017). The meta-regression analysis by Beintner et al. (2014) found CBT-based SHIs to be associated with smaller but significant improvements in BN compared to BED regarding abstinence from binge eating, frequency of binge-eating episodes, and ED psychopathology. Likewise, treatment completion and participation in SHIs were lower in BN compared to BED. The authors concluded that patients with BED are more likely to complete and thus benefit from SHIs than patients with BN, who might feel that their therapeutic goals (i.e., normalization of eating behavior through a reduction of restrictive eating) are harder to accomplish and might thus not be sufficiently addressed by SHIs. The meta-analysis by Traviss-Turner et al. (2017) supported this conclusion, as described above.

Further meta-analytic evidence on the efficacy of SHIs in the treatment of BN was presented by Linardon et al. (2017b), who included 37 RCTs on CBT for BN, four of which consisted of CBT-based SHIs alone, and found SHIs to be significantly more efficacious in producing abstinence from binge eating and compensatory behaviors, and reducing ED psychopathology, than inactive control conditions (medium effects). In a subsequent meta-analysis, Linardon and Wade (2018) examined 54 RCTs with 78 active intervention conditions for the treatment of BN, nine of which were focused on CBT-based guided SHIs. The authors found post treatment abstinence from binge eating and compensatory behaviors to be lower in guided SHIs than in treatments with greater involvement of a clinician (e.g., individual or group-based face-to-face CBT), while this effect was not observed at follow-up. Thus, guided SHIs were discussed to take longer to achieve therapeutic effects than, for example, face-to-face CBT. In contrast, the authors found no association between the number of treatment sessions and abstinence rates; thus, they assumed no effect of the amount of therapist contact. These results suggest that patients with BN benefit more, and more quickly, from intensive treatments in which the course and content of therapy are directed by a therapist, rather than by the patients themselves.

Most recently, Svaldi et al. (2019) conducted a comprehensive meta-analysis on the efficacy of a range of psychological and medical treatments for BN, including 79 RCTs with 127 active intervention conditions, with 10 conditions specifically focused on guided and pure SHIs ($N = 562$), mostly CBT-based. Compared to inactive control conditions, SHIs produced abstinence from binge eating with a medium-to-large effect size ($OR = 5.60$) and abstinence from compensatory behaviors with a medium effect ($OR = 4.23$). Further, SHIs were efficacious in reducing binge-eating frequency (medium-to-large effect), compensatory behaviors (medium-to-large effect), ED psychopathology (large effect), and depressive symptoms (small effect). Compared to active control conditions (e.g., CBT or combined treatment), SHIs showed medium-to-large effects regarding all aforementioned outcome measures. Of note, SHIs were associated with the greatest dropout rates (33%) from all treatments examined. Evidence from one included study supported the maintenance of effects at 3- and 6-month follow-up (Banasiak et al. 2005). The authors concluded that guided CBT-based SHIs are an alternative treatment option when face-to-face CBT, the method of choice, is not available. Notwithstanding, as with BED, more high-quality clinical studies comparing SHIs to other evidence-based interventions in a long-term perspective are still warranted (e.g., Linardon and Wade 2018; Svaldi et al. 2019).

4 Effectiveness

Moving from efficacy to effectiveness research, a number of studies focused on the transportability of efficacious CBT-based SHIs "from the lab to real world settings" (i.e., aiming to maximize their external validity by including any design, RCTs, nonrandomized controlled, and uncontrolled trials, Hans and Hiller 2013; e.g., DeBar et al. 2011; Grilo et al. 2013; Hildebrandt et al. 2017; Högdahl et al. 2013; Kazdin et al. 2017; Lynch et al. 2010; Striegel-Moore et al.

2010). Wilson and Zandberg (2012) evaluated their reviewed evidence with regard to effectiveness and concluded that the evidence base remains methodologically limited, but also that CBT-based SHIs were effective for the treatment of BED and BN in a number of different settings. The authors also underlined the point made by Kazdin and Blase (2011) that a treatment with a smaller effect on therapeutic outcome but with greater reach and scalability would nevertheless decrease the overall health burden of patients with EDs, thus decisively "addressing the unmet needs of countless individuals suffering from eating disorders" (Wilson and Zandberg 2012, p. 351).

To illustrate effectiveness approaches, Grilo et al. (2013) randomly assigned 48 patients with BED and obesity to either a CBT-related, book-based pure SHI (i.e., without any guidance, see Sect. 5) or treatment as usual, within a university-based medical health care center. To enhance the generalizability and applicability, only a few exclusion criteria were applied, and the initial visit for handing out and explaining the SHI book was performed by primary care physicians, who were not specifically trained in the treatment of EDs. While abstinence rates from binge eating at post treatment did not differ significantly between pure SHI (25%) and usual care (8%), monthly assessments of self-reported binge-eating frequency showed a significant reduction in the SHI condition, but not in usual care. Thus, pure SHIs may not have specific effectiveness when compared to usual care for patients with BED and obesity. In another example, a CBT-derived, book-based SHI with internet-based guidance was offered to 48 patients with BN from a specialized ED clinic, and its effects were compared to a nonrandomized control group ($N = 48$) undergoing a 16-week psychodynamic day patient program (Högdahl et al. 2013). The approximate clinician involvement was 11 h per patient in the SHI condition and more than 200 h in the day patient program. As anticipated, effects on binge eating, compensatory behaviors, and ED psychopathology tended to be larger in the day patient program than in the SHI condition, but there were no significant differences at post

treatment between conditions. The results suggested book-based SHIs with internet-based guidance to be an option for offering effective treatment with scarce CBT resources in clinical environments, which should be bolstered by future adequately powered effectiveness research.

5   The Role of Guidance

All formats of SHIs might be implemented with or without professional guidance and support at varying levels, with the common feature that possible guidance is primarily supportive and facilitating, thus being significantly less intensive than, for example, the therapeutic relationship within face-to-face CBT. The vast amount of different forms of professional guidance within SHIs can be classified by whom it is provided, and via its intensity and modality. The professional background of guides within SHIs for BED and BN have ranged from nurses without experience in the treatment of EDs (Walsh et al. 2004) to graduate students in clinical psychology with little or no experience in CBT (Wilson et al. 2010), to Master's- or Doctoral-level therapists (e.g., Peterson et al. 2009; Striegel-Moore et al. 2010). Varying in intensity, guidance can be administered by telephone, via the internet, or through face-to-face sessions (Traviss-Turner et al. 2017; Wilson and Zandberg 2012). Personally guided sessions were conducted in primary care (Carter and Fairburn 1998; Walsh et al. 2004) or in university-based, specialty ED clinics (e.g., Wilson et al. 2010), in individual or group format. For example, in an RCT pointing to the efficacy of a book-based SHI following CBT principles for the treatment of BED and BN (Hildebrandt et al. 2017), guidance was administered face-to-face, via one initial 60-min meeting with a therapist (i.e., psychologist or graduate student), followed by eight 25-min sessions. Another example can be found in a RCT proving the efficacy of internet-based SHI using CBT principles for the treatment of BED (INTERBED study; de Zwaan et al. 2017), in which guidance was implemented through two 90-min face-to-face sessions before and after treatment as well as weekly feedback via e-mail, provided by trained psychologists and physicians.

The question of whether guidance is necessary for a favorable treatment outcome in SHIs has been intensively discussed and must be answered regarding the ED to be treated. For BED, in the meta-analysis by Hilbert et al. (2019), a direct comparison between guided and pure SHIs for BED did not show significant differences regarding any of the outcome measures. Thus, in the treatment of BED, guidance might be of minor importance for a favorable outcome. Likely related to methodological differences, meta-regression analyses based on indirect comparisons estimated greater treatment effects and higher intervention completion rates in guided than unguided SHIs for both BED and BN (Beintner et al. 2014), but effects were stronger when focusing on BN alone. Thus, patients with BN compared to those with BED may need greater support to successfully work through CBT-based SHIs.

Further discussion topics represent the optimal form and pattern of guidance regarding its provider, intensity, and modality. In their review on CBT-based guided SHIs for EDs, Wilson and Zandberg (2012) pointed out that, based on some of their reviewed SHI studies, providers with lesser professional experience in the treatment of BED (e.g., Master's level psychologists with no prior experience in treating EDs; Striegel-Moore et al. 2010) achieved comparable outcomes to those with more experience. However, in their meta-regression analysis, Beintner et al. (2014) found that, with regard to the treatment of both BED and BN, guidance by an ED or mental health specialist or CBT therapist within CBT-based SHIs may yield a greater reduction of ED symptoms and higher intervention completion than guidance by nurses or general practitioners. Further, higher intervention completion rates were found to be achieved in CBT-based SHIs with face-to-face guidance compared to e-mail guidance. Overall, future research is warranted to identify the optimal type and dosage of guidance.

6 Acceptability

In contrast to the promising results demonstrating the efficacy of SHIs in the treatment of BED and BN, low levels of participation represent a major area of concern. High acceptability to patients and thus a sufficient level of participation within SHIs are essential for their scalability, dissemination, and implementation. Here, it is central to consider not only the engagement in an SHI and treatment or assessment completion, but also patients' adherence to the intervention.

Regarding treatment engagement, in a systematic review of 4 RCTs of internet-based CBT for EDs, which were guided SHIs (Fairburn and Murphy 2015), 16% to 24% of patients did not take up the intervention and, while treatment completion definitions differed between studies, completion rates were unsatisfactory. Further, a large variability of treatment dropout, ranging from 1% to 88% in RCTs and non-RCTs on guided and pure SHIs for BED and BN, was documented in the meta-regression analysis by Beintner et al. (2014). When comparing the different formats of SHIs, highest participation as defined by the completion of at least half of the intervention was suggested for book-based SHIs (65%), followed by CD-ROM- (38%) and internet-based SHIs (37%; Beintner et al. 2014). In their systematic review on app-based SHIs for AN, BN, and BED, Anastasiadou et al. (2018) documented satisfactory values regarding patient-reported acceptability and participation, but only one-third of the included RCTs, non-RCTs, and uncontrolled studies reported participation, which thus needs to be systematically reported in SHIs, including app-based formats.

Regarding reasons for low acceptability and thus insufficient participation within the various forms of SHIs, Walsh et al. (2004) pointed out that the low-intensity character of SHIs might yield in patients' low confidence that the treatment will be successful. Further reasons include the lack of personal contact (Robinson et al. 2006), motivation, time, and technical issues (Leung et al. 2013). In contrast, baseline treatment motivation might be positively associated with participation within SHIs (Leung et al. 2013). As noted above, participation can be depicted through patients' adherence to the SHI, which was positively associated with favorable treatment outcomes within internet-based SHIs (Carrard et al. 2011a; Manwaring et al. 2008). Importantly, adherence to internet- or app-based SHIs, such as the frequency of logins, can be easily assessed via automatic systems (Bauer and Moessner 2012). Therefore, adherence might be regularly monitored and should be fostered by suitable additional intervention components (e.g., more professional guidance) in case of decreasing levels of adherence in individual patients (Puls et al. 2019). Among others, this could be an important step in the direction of more personalized treatments, with likely benefits for participation and outcome within SHIs for BED and BN (Loucas et al. 2014).

7 Predictors, Moderators, and Mediators

In order to increase the efficacy of SHIs for patients with BED and BN, it is important to understand how, why, and for whom treatments work (Kraemer 2015). To this end, predictors of outcome, treatment-specific moderators (i.e., baseline variables interacting with the treatment type and associated with outcome) and mediators (i.e., variables changing due to treatment and thus interacting with outcome) need to be investigated. In their systematic review, Linardon et al. (2017a) summarized 65 RCTs and non-RCTs on CBT-based treatments for BED, BN, AN, and mixed samples, 14 of which were related to guided SHIs. Regarding baseline predictors, a history of AN in patients with BN as well as a higher frequency of compensatory behaviors and higher body mass index (BMI, kg/m^2) in mixed samples were associated with less favorable outcomes after guided SHIs. Moderators of guided SHIs were only studied for BED samples, showing that, among a range of clinical and sociodemographic variables, only higher binge-eating frequency was associated with less favorable outcomes within CBT-based guided SHIs

but with better outcomes within individual CBT. Similarly, there was no evidence for mediators in guided SHIs from BN samples. However, within BED, rapid response in binge eating (defined as a 65–70% reduction in binge eating by treatment week 4) predicted greater abstinence from binge eating at post treatment and 6-, 12-, and 18-month follow-up in CBT-based guided SHI, but not in behavioral weight loss treatment or interpersonal psychotherapy (Hilbert et al. 2015). Overall, the evidence on predictors, moderators, and mediators of SHI remains limited, so more research is needed to support the matching and tailoring of specific interventions to individual patients.

8 Cost-Effectiveness

Cost-effectiveness describes the degree to which an intervention's effectiveness is justified by its societal costs (Gray 2011; Ramsey et al. 2015), which comprises its direct costs (i.e., resulting from utilization of medical, psychological, or social services, nursing or informal care, or treatment-related travel) and indirect costs (i.e., resulting from loss of productivity, time spent for treatment, or sick leave). Thus, an intervention can be considered cost-effective if "it either is equally effective and associated with lower costs compared to the control intervention or if it is superior and the (additional) costs are considered reasonable given the additional health gain" (König et al. 2018, p. 156). As noted above, higher cost-effectiveness and thus enhanced scalability were discussed as major advantages of SHIs compared to traditional face-to-face treatments (e.g., Beintner et al. 2014; Wilson and Zandberg 2012; Yim and Schmidt 2019). In addition to their assumed lower societal costs and lower therapist involvement, guided SHIs may be conducted by a broader range of health care providers compared to traditional psychological face-to-face treatments (Fairburn and Patel 2014; see Sect. 5), which also might contribute to the cost-effectiveness of SHIs.

Focusing on the few studies providing specific evidence regarding CBT-based SHIs, König et al. (2018) conducted a comparative analysis on the cost-effectiveness of internet-based guided SHI and face-to-face CBT for BED ($N = 147$) and found no clear evidence for one treatment being more cost-effective. In the entire sample (i.e., including dropouts), the SHI condition was associated with lower costs but also with lower intervention effects than face-to-face CBT. Further, Aardoom et al. (2017) evaluated the cost-effectiveness of an internet-based SHI following CBT principles—psychoeducation and a fully-automated monitoring and feedback system, with three varying levels of therapist support—for patients with self-reported ED symptoms in comparison to wait-list ($N = 354$). The authors found no significant differences between the four conditions regarding societal costs and intervention effects (i.e., quality-adjusted life-years). While the mean societal costs per patient increased with higher therapist support within the SHI conditions, they were highest in the wait-list condition, due to an increased uptake of other medical treatments (e.g., inpatient mental health care). Thus, the authors suggested cost-effectiveness of SHIs compared to wait-list (Aardoom et al. 2017). This suggestion found support in the results from a study by Lynch et al. (2010), who analyzed the societal costs of a CBT-related, book-based guided SHI plus treatment as usual compared to treatment as usual alone for BN, BED, and recurrent binge eating ($N = 123$). Herein, greater intervention effects in the SHI condition and lower societal costs due to reduced use of other medical services in this condition were revealed. In sum, while these few studies are suggestive of greater cost-effectiveness of SHIs for patients with BED and BN compared to minimal treatment, but not compared to face-to-face CBT, the evidence remains scarce and inconclusive. Further research is therefore urgently needed to investigate the cost-effectiveness of SHIs in the treatment of BED and BN.

9 Self-Help Interventions in Complex Models of Care

In addition to the evaluation of CBT-based SHIs as a stand-alone approach to the treatment of BED and BN, SHIs have been tested within complex models of care conceptualizing the combination of different treatments on multiple steps of the care pathway (e.g., Tasca et al. 2019). Especially within such models, feasible and cost-effective ways to direct the adequate amount of treatment to subgroups of patients with specific symptom levels or treatment response patterns (e.g., patients with chronic EDs, at risk for dropout from treatment, with high levels of health care use, rapid responders) were demanded (Kazdin et al. 2017). Such tailored treatment allocation procedures were discussed to improve effectiveness and cost-effectiveness of psychological treatments (Agras et al. 2017; van Furth et al. 2016), however, the evidence remains very limited.

An example for a complex model of care with tailored treatment allocation is the three-level stepped care model (Jones et al. 2014; Wilfley et al. 2013), according to which 1551 college students were screened and allocated to one of four groups: (1) individuals at low ED risk were offered an internet- and CBT-based universal preventive pure SHI (entitled "StayingFit"; Taylor et al. 2012); (2) individuals at high ED risk were offered an internet-based selective preventive guided SHI ("StudentBodies—Targeted"; Beintner et al. 2012); (3) individuals with a clinical or subclinical ED other than full-syndrome AN were offered an internet-based guided SHI ("StudentBodies—Eating Disorders"; Jacobi et al. 2012), and individuals with full-syndrome AN or other medical concerns were referred to face-to-face treatment (e.g., CBT, interpersonal psychotherapy). Additionally, if individuals showed no considerable symptom reduction in their intervention category, they were directed to a more intensive intervention. Cost-effectiveness estimates suggested that the proposed model was less costly and resulted in fewer college students in need of face-to-face treatment compared to standard care (Kass et al. 2017). Another example of tailored treatment allocation within a complex model of care can be found in a multisite RCT comparing individual CBT with CBT-oriented, book-based guided SHI, in which 293 patients with BN were identified as responders or nonresponders, depending on the level of symptom reduction at treatment week 6 out of 18 (Mitchell et al. 2011). Nonresponders were offered antidepressant treatment with fluoxetine in addition to their ongoing treatment. While abstinence from binge eating did not differ between the two conditions at post treatment, medication use was lower in the SHI condition, and, at follow-up, the SHI condition led to a greater reduction of binge eating and compensatory behaviors than CBT.

An example for SHIs within a complex model of care for BED without tailored treatment allocation was provided by Tasca et al. (2019): A total of 135 patients received a book-based pure SHI, using CBT principles, followed by either 16 weeks of group psychodynamic-interpersonal psychotherapy or a no-treatment control condition. The pure SHI produced significant reductions in binge-eating frequency and ED psychopathology, but the effect of the subsequently added psychotherapy on ED symptoms was not significantly different from the control condition. However, compared to the control condition, subsequent psychotherapy yielded significantly greater improvements in interpersonal problems and attachment avoidance, which are relevant maintenance factors of binge eating (Ivanova et al. 2015). In sum, complex models of care that include pure or guided SHIs as a first-step intervention might represent cost-effective approaches for the treatment of patients with BED or BN, but further research in this area is clearly needed, especially to elucidate the utility of tailored treatment allocation procedures.

An example for SHIs in relapse prevention after ED treatment was provided by Jacobi et al. (2017), who conducted an RCT comparing a 9-month internet-based guided SHI using CBT principles with treatment as usual following inpatient treatment for BN ($N = 253$). In general, the two conditions did not significantly differ in abstinence from binge eating, compensatory

behaviors, and binge-eating frequency at post treatment and follow-up. The SHI significantly reduced the frequency of vomiting episodes at post treatment compared to treatment as usual, but this difference was no longer significant at follow-up. Interestingly, in patients who still reported binge-eating and compensatory behaviors at the end of inpatient treatment, differences between conditions in favor of the SHI were larger at post treatment and follow-up. These results suggest that SHIs for relapse prevention might be especially suited for nonresponders in inpatient treatment of BN.

10 Future Directions

Despite the potential of CBT-based SHIs to overcome barriers to "traditional" care for the treatment of BED and BN, more research is needed to address remaining questions: Further studies should compare CBT-based SHIs with established face-to-face treatments regarding efficacy, acceptability, participation, and cost-effectiveness, (Beintner et al. 2014; Wilson and Zandberg 2012; Yim and Schmidt 2019). Identifying reliable predictors of outcome within SHIs may allow to answer the question "what works best for whom?" and, subsequently, to tailor specific intervention components to individual patients (Loucas et al. 2014). The benefits of the different formats of SHIs should be further investigated, especially with regard to specific patient groups (e.g., app-based SHIs for the treatment of younger patients), and the role of guidance needs to be further elucidated, focusing on amount, providers, and settings of guidance (Hilbert et al. 2019). Participation and adverse events should routinely be assessed and reported in studies on SHIs for BED and BN. The latter is particularly important in SHIs without any form of guidance and monitoring of clinical risk. Finally, of specific interest is the use of SHIs for universal or selective prevention, treatment, and relapse prevention based on a conceptualization of complex models of care and using tailored allocation (Yim and Schmidt 2019), in order to offer care to "the vast majority of individuals in

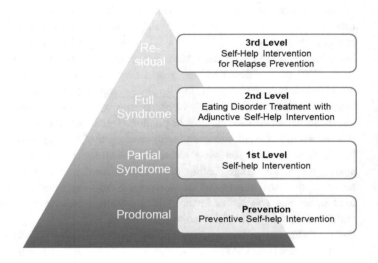

need of services for EDs who are not otherwise
served" (Kazdin et al. 2017, p. 21).

As noted above, growing evidence and clinical
guideline recommendations (NICE 2017) justify
the use of SHIs in the treatment of BED and BN,
especially if CBT-based. However, depending on
national health care policies, SHIs remain largely
underutilized (Fitzsimmons-Craft et al. 2019) and
their potential to reduce the "treatment gap" (see
Sect. 1) is not yet exhausted. Thus, systematic
research on the dissemination and implementation of psychological treatments, including
CBT-based SHIs, to EDs is therefore highly
warranted (Agras et al. 2017; Kazdin et al.
2017). Importantly, to inform health care policies
with evidence from high-quality research, a
mutual exchange between researchers and policy
makers was demanded (Brownell and Roberto
2015).

Practically, if clinicians wish to use CBT-based
SHIs, they may consider symptom severity of an
individual patient to guide their treatment-related
decision-making, for example, offering a higher
level of care corresponding to the patient's need.
Figure 1 illustrates a possible use of SHIs
according to the eating disorder symptom severity,
ranging from a preventive or early interventive use
in case of eating disorder risk or partial syndrome
to an adjunctive or relapse preventive use in case
of full syndrome or residual symptoms. As noted

above, further clinical decision points include
other baseline characteristics (e.g., patient
expectations and practical aspects) and process-
related aspects (e.g., use of guidance, early symptom change, and patients' adherence). For example, for some patients with full-syndrome BED or
BN, those with lower symptom severity, high
skills for SHI use, good adherence, and early
symptom reduction, SHIs, particularly if guided,
may be sufficient for achieving symptom remission. In general, however, given the current evidence, SHIs may be considered for the treatment of
BED and BN if face-to-face psychological treatment such as CBT is not available. In any use of
SHIs, clinicians should transparently communicate
goals, utilization, and outcomes of SHIs to their
patients (Yim and Schmidt 2019). Importantly,
patients should be thoroughly advised how to use
the SHI in the best possible way, especially if it is
technology based. A general requirement is that
SHIs should be free of any disturbances, such as
technical problems, inconvenient handling, or
non-fitting expectations of the patient.

11 Summary

After more than two decades of research,
CBT-based SHIs today represent a well-
supported treatment option for patients with

BED and BN. A growing body of research documented not only their efficacy, but also, to a lesser extent, their comparative efficacy and their effectiveness. Especially in the treatment of BN, professional guidance seems to be an important component of CBT-based SHIs. Guidance was in some studies associated with better treatment outcomes and higher levels of participation. Acceptability and participation might be linked; regardless, they represent key aspects for the design and conduct of SHIs, and should, therefore, be systematically monitored, consistently described, and further enhanced. In this context, individual tailoring of interventions and their components could be a critical line of action for the further development of SHIs. Cost-effectiveness of SHIs compared to minimal treatment was suggested by some studies. Placing SHIs in complex models of care was discussed to increase effectiveness and cost-effectiveness of SHIs. Finally, despite the promising evidence on the efficacy of CBT-based SHIs, more efforts leading to optimal dissemination and implementation of evidence-based SHIs in the treatment of patients with BED and BN are urgently needed.

References

Aardoom JJ, Dingemans AE, van Ginkel JR et al (2017) Cost-utility of an internet-based intervention with or without therapist support in comparison with a waiting list for individuals with eating disorder symptoms: a randomized controlled trial. Int J Eat Disord 49:1068–1076

Agras WS, Fitzsimmons-Craft EE, Wilfley DE (2017) Evolution of cognitive-behavioral therapy for eating disorders. Behav Res Ther 88:26–36

Ali K, Farrer L, Fassnacht DB et al (2017) Perceived barriers and facilitators towards help-seeking for eating disorders: a systematic review. Int J Eat Disord 50:9–21

Anastasiadou D, Folkvord F, Lupiañez-Villanueva F (2018) A systematic review of mHealth interventions for the support of eating disorders. Eur Eat Disord Rev 26:394–416

Banasiak SJ, Paxton SJ, Hay P (2005) Guided self-help for bulimia nervosa in primary care: a randomized controlled trial. Psychol Med 35:1283–1294

Barakat S, Maguire S, Surgenor L et al (2017) The role of regular eating and self-monitoring in the treatment of

bulimia nervosa: a pilot study of an online guided self-help CBT program. Behav Sci 7:39

Bauer S, Moessner M (2012) Technology-enhanced monitoring in psychotherapy and e-mental health. J Ment Health 21:355–363

Bauer S, Moessner M (2013) Harnessing the power of technology for the treatment and prevention of eating disorders. Int J Eat Disord 46:508–515

Beintner I, Jacobi C, Taylor CB (2012) Effects of an internet-based prevention programme for eating disorders in the USA and Germany - a meta-analytic review. Eur Eat Disord Rev 20:1–8

Beintner I, Jacobi C, Schmidt UH (2014) Participation and outcome in manualized self-help for bulimia nervosa and binge eating disorder - a systematic review and metaregression analysis. Clin Psychol Rev 34:158–176

Brownell KD, Roberto CA (2015) Strategic science with policy impact. Lancet 385:2445–2446

Carrard I, Crépin C, Rouget P, Lam T, van der Linden M et al (2011a) Acceptance and efficacy of a guided internet self-help treatment program for obese patients with binge eating disorder. Clin Pract Epidemiol Ment Health 7:8–18

Carrard I, Crépin C, Rouget P, Lam T, Golay A et al (2011b) Randomised controlled trial of a guided self-help treatment on the internet for binge eating disorder. Behav Res Ther 49:482–491

Carter JC, Fairburn CG (1998) Cognitive-behavioral self-help for binge eating disorder: a controlled effectiveness study. J Consult Clin Psychol 66:616–623

Cooper PJ (1996) Bulimia nervosa & binge-eating: a guide to recovery, Repr. New York University Press, New York

Cooper Z, Bailey-Straebler S (2015) Disseminating evidence-based psychological treatments for eating disorders. Curr Psychiatry Rep 17:551

Corrigan PW, Druss BG, Perlick DA (2014) The impact of mental illness stigma on seeking and participating in mental health care. Psychol Sci Public Interest 15:37–70

de Zwaan M, Herpertz S, Zipfel S et al (2017) Effect of internet-based guided self-help vs individual face-to-face treatment on full or subsyndromal binge eating disorder in overweight or obese patients: the INTER-BED randomized clinical trial. JAMA Psychiat 74:987–995

DeBar LL, Striegel-Moore RH, Wilson GT et al (2011) Guided self-help treatment for recurrent binge eating: replication and extension. Psychiatr Serv 62:367–373

Fairburn CG (1995) Overcoming binge eating. Guilford Press, New York

Fairburn CG (2013) Overcoming binge eating: the proven program to learn why you binge and how you can stop, 2nd edn. Guilford Press, New York

Fairburn CG, Murphy R (2015) Treating eating disorders using the internet. Curr Opin Psychiatry 28:461–467

Fairburn CG, Patel V (2014) The global dissemination of psychological treatments: a road map for research and practice. Am J Psychiatry 171:495–498

Fitzsimmons-Craft EE, Balantekin KN, Graham AK et al (2019) Results of disseminating an online screen for eating disorders across the U.S.: reach, respondent characteristics, and unmet treatment need. Int J Eat Disord 52:721–729

Ghaderi A, Odeberg J, Gustafsson S et al (2018) Psychological, pharmacological, and combined treatments for binge eating disorder: a systematic review and meta-analysis. PeerJ 21:e5113

Gray A (2011) Applied methods of cost-effectiveness analysis in health care. Handbooks in health economic evaluation series. Oxford University Press, Oxford

Grilo CM, Masheb RM, Wilson GT et al (2011) Cognitive-behavioral therapy, behavioral weight loss, and sequential treatment for obese patients with binge-eating disorder: a randomized controlled trial. J Consult Clin Psychol 79:675–685

Grilo CM, White MA, Gueorguieva R et al (2013) Self-help for binge eating disorder in primary care: a randomized controlled trial with ethnically and racially diverse obese patients. Behav Res Ther 51:855–861

Hans E, Hiller W (2013) Effectiveness of and dropout from outpatient cognitive behavioral therapy for adult unipolar depression: a meta-analysis of nonrandomized effectiveness studies. J Consult Clin Psychol 81:75–88

Hilbert A, Hildebrandt T, Agras WS et al (2015) Rapid response in psychological treatments for binge eating disorder. J Consult Clin Psychol 83:649–654

Hilbert A, Opitz L, de Zwaan M (2018) Internet-based interventions for eating disorders. In: Agras WS, Robinson A (eds) The Oxford handbook of eating disorders, vol 1, 2nd edn. Oxford University Press, New York

Hilbert A, Petroff D, Herpertz S et al (2019) Meta-analysis of the efficacy of psychological and medical treatments for binge-eating disorder. J Consult Clin Psychol 87:91–105

Hildebrandt T, Michaelides A, Mackinnon D et al (2017) Randomized controlled trial comparing smartphone assisted versus traditional guided self-help for adults with binge eating. Int J Eat Disord 50:1313–1322

Högdahl L, Birgegård A, Björck C (2013) How effective is bibliotherapy-based self-help cognitive behavioral therapy with internet support in clinical settings? Results from a pilot study. Eat Weight Disord 18:37–44

Ivanova IV, Tasca GA, Hammond N et al (2015) Negative affect mediates the relationship between interpersonal problems and binge-eating disorder symptoms and psychopathology in a clinical sample: a test of the interpersonal model. Eur Eat Disord Rev 23:133–138

Jacobi C, Völker U, Trockel MT et al (2012) Effects of an internet-based intervention for subthreshold eating disorders: a randomized controlled trial. Behav Res Ther 50:93–99

Jacobi C, Beintner I, Fittig E et al (2017) Web-based aftercare for women with bulimia nervosa following inpatient treatment: randomized controlled efficacy trial. J Med Internet Res 19(9):e321

Jones M, Kass AE, Trockel M et al (2014) A population-wide screening and tailored intervention platform for eating disorders on college campuses: the healthy body image program. J Am Coll Heal 62:351–356

Juarascio AS, Goldstein SP, Manasse SM et al (2015a) Perceptions of the feasibility and acceptability of a smartphone application for the treatment of binge eating disorders: qualitative feedback from a user population and clinicians. Int J Med Inform 84:808–816

Juarascio AS, Manasse SM, Goldstein SP et al (2015b) Review of smartphone applications for the treatment of eating disorders. Eur Eat Disord Rev 23:1–11

Juarascio AS, Parker MN, Lagacey MA et al (2018) Just-in-time adaptive interventions: a novel approach for enhancing skill utilization and acquisition in cognitive behavioral therapy for eating disorders. Int J Eat Disord 51:826–830

Kass AE, Balantekin KN, Fitzsimmons-Craft EE et al (2017) The economic case for digital interventions for eating disorders among United States college students. Int J Eat Disord 50:250–258

Kazdin AE, Blase SL (2011) Rebooting psychotherapy research and practice to reduce the burden of mental illness. Perspect Psychol Sci 6:21–37

Kazdin AE, Fitzsimmons-Craft EE, Wilfley DE (2017) Addressing critical gaps in the treatment of eating disorders. Int J Eat Disord 50:170–189

Kelly AC, Carter JC (2015) Self-compassion training for binge eating disorder: a pilot randomized controlled trial. Psychol Psychother 88:285–303

Kelly T, Minges M (2012) Information and communications for development 2012: maximizing mobile. The World Bank, Washington, DC

König H, Bleibler F, Friederich H et al (2018) Economic evaluation of cognitive behavioral therapy and internet-based guided self-help for binge-eating disorder. Int J Eat Disord 51:155–164

Kraemer HC (2015) Messages for clinicians: moderators and mediators of treatment outcome in randomized clinical trials. Am J Psychiatry 173:672–679

Leung SF, Ma LCJ, Russell J (2013) An open trial of self-help behaviours of clients with eating disorders in an online programme. J Adv Nurs 69:66–76

Linardon J, Wade TD (2018) How many individuals achieve symptom abstinence following psychological treatments for bulimia nervosa? A meta-analytic review. Int J Eat Disord 51:287–294

Linardon J, de la Piedad GX, Brennan L (2017a) Predictors, moderators, and mediators of treatment outcome following manualised cognitive-behavioural therapy for eating disorders: a systematic review. Eur Eat Disord Rev 25:3–12

Linardon J, Wade TD, de la Piedad GX et al (2017b) The efficacy of cognitive-behavioral therapy for eating disorders: a systematic review and meta-analysis. J Consult Clin Psychol 85:1080–1094

Loucas CE, Fairburn CG, Whittington C et al (2014) E-therapy in the treatment and prevention of eating disorders: a systematic review and meta-analysis. Behav Res Ther 63:122–131

Lynch FL, Striegel-Moore RH, Dickerson JF et al (2010) Cost-effectiveness of guided self-help treatment for recurrent binge eating. J Consult Clin Psychol 78:322–333

Manwaring JL, Bryson SW, Goldschmidt AB et al (2008) Do adherence variables predict outcome in an online program for the prevention of eating disorders? J Consult Clin Psychol 76:341–346

Mitchell JE, Agras S, Crow S et al (2011) Stepped care and cognitive-behavioural therapy for bulimia nervosa: randomised trial. Br J Psychiatry 198:391–397

National Institute for Health and Care Excellence (2004) Eating disorders - core interventions in the treatment and management of anorexia nervosa, bulimia nervosa and related eating disorders: NICE clinical guideline no 9. NICE, London

National Institute for Health and Care Excellence (2017) Eating disorders: recognition and treatment. NICE, London

Peterson CB, Mitchell JE, Crow SJ et al (2009) The efficacy of self-help group treatment and therapist-led group treatment for binge eating disorder. Am J Psychiatry 166:1347–1354

Puls H-C, Schmidt R, Herpertz S et al (2019) Adherence as a predictor of dropout in internet-based guided self-help for adults with binge-eating disorder and overweight or obesity. Int J Eat Disord: Epub ahead of print.

Ramsey SD, Willke RJ, Glick H et al (2015) Cost-effectiveness analysis alongside clinical trials II-an ISPOR good research practices task force report. Value Health 18:161–172

Riper H, Smit F, Van der Zanden R et al (2012) E-mental health: high tech, high touch, high trust. Trimbos Instituut, Utrecht

Robinson S, Perkins S, Bauer S et al (2006) Aftercare intervention through text messaging in the treatment of bulimia nervosa--feasibility pilot. Int J Eat Disord 39:633–638

Sánchez-Ortiz VC, House J, Munro C et al (2011) "A computer isn't gonna judge you": a qualitative study of users' views of an internet-based cognitive behavioural guided self-care treatment package for bulimia nervosa and related disorders. Eat Weight Disord 16:e93–e101

Schmidt U, Treasure J (1993) Eating bit(e) by bit(e) - a survival kit for suffers of bulimia nervosa and binge eating disorders. Lawrence Erlbaum Associates, Hove

Schmidt U, Treasure J, Alexander J (2015) Getting better bite by bite: a survival kit for sufferers of bulimia nervosa and binge eating disorders, 2nd edn. Taylor and Francis, London

Striegel-Moore RH, Wilson GT, DeBar L et al (2010) Cognitive behavioral guided self-help for the treatment of recurrent binge eating. J Consult Clin Psychol 78:312–321

Svaldi J, Schmitz F, Baur J et al (2019) Efficacy of psychotherapies and pharmacotherapies for bulimia nervosa. Psychol Med 49:898–910

Sysko R, Walsh BT (2008) A critical evaluation of the efficacy of self-help interventions for the treatment of bulimia nervosa and binge-eating disorder. Int J Eat Disord 41:97–112

Tasca GA, Koszycki D, Brugnera A et al (2019) Testing a stepped care model for binge-eating disorder: a two-step randomized controlled trial. Psychol Med 49:598–606

Taylor CB, Taylor K, Jones M et al (2012) Obesity prevention in defined (high school) populations. Int J Obes 2:30–32

Traviss GD, Heywood-Everett S, Hill AJ (2011) Guided self-help for disordered eating: a randomised control trial. Behav Res Ther 49:25–31

Traviss-Turner GD, West RM, Hill AJ (2017) Guided self-help for eating disorders: a systematic review and Metaregression. Eur Eat Disord Rev 25:148–164

Vall E, Wade TD (2015) Predictors of treatment outcome in individuals with eating disorders: a systematic review and meta-analysis. Int J Eat Disord 48:946–971

van Furth EF, van der Meer A, Cowan K (2016) Top 10 research priorities for eating disorders. Lancet Psychiatry 3:706–707

Vocks S, Tuschen-Caffier B, Pietrowsky R et al (2010) Meta-analysis of the effectiveness of psychological and pharmacological treatments for binge eating disorder. Int J Eat Disord 43:205–217

Wagner G, Penelo E, Nobis G et al (2015) Predictors for good therapeutic outcome and drop-out in technology assisted guided self-help in the treatment of bulimia nervosa and bulimia like phenotype. Eur Eat Disord Rev 23:163–169

Walsh BT, Fairburn CG, Mickley D et al (2004) Treatment of bulimia nervosa in a primary care setting. Am J Psychiatry 161:556–561

Wilfley DE, Agras WS, Taylor CB (2013) Reducing the burden of eating disorders: a model for population-based prevention and treatment for university and college campuses. Int J Eat Disord 46:529–532

Wilson GT, Zandberg LJ (2012) Cognitive-behavioral guided self-help for eating disorders: effectiveness and scalability. Clin Psychol Rev 32:343–357

Wilson GT, Wilfley DE, Agras WS et al (2010) Psychological treatments of binge eating disorder. Arch Gen Psychiatry 67:94–101

Yim SH, Schmidt U (2019) Self-help treatment of eating disorders. Psychiatr Clin North Am 42:231–241

New Avenues for the Treatment of Binge Eating Based on Implicit Processes

Kerri N. Boutelle, Dawn M. Eichen, and Carol B. Peterson

Abstract

Current treatments for individuals with binge eating, including binge-eating disorder (BED) and bulimia nervosa (BN), are successful for about 50 and 30%, respectively. Thus, it is essential to explore novel treatments that may be successful for more individuals with binge eating including those resistant to current treatments. As our understanding of the neurobiology of binge eating has improved, additional mechanisms to target in treatments have been identified. This chapter summarizes the research to date on several newer models of treatment, largely based on mechanisms identified from the current research on the neurobiology of binge-eating disorder and overeating. Models explored include neuromodulation, neurocognitive training, a model based on the behavioral susceptibility model, and ecological momentary intervention. Taken together, these novel treatments show initial promise, but most require significantly more research before they are widely adapted as evidence-based treatments for binge eating.

Keywords

Neuromodulation · Neurocognitive training · Regulation of cues · Cue-exposure treatment · Appetite awareness training and ecological momentary intervention

Learning Objectives

In this chapter, you will:

1. Understand the rationale for why novel methods of treatment for individuals with binge eating are needed.
2. Explain mechanisms that are targeted in novel methods of treatment for individuals with binge eating.
3. Summarize the research to date on novel methods of treatment for individuals with binge eating.

K. N. Boutelle (✉)
Department of Pediatrics, Family Medicine and Public Health, and Psychiatry, University of California San Diego, La Jolla, CA, USA
e-mail: kboutelle@ucsd.edu

D. M. Eichen
Department of Pediatrics, University of California San Diego, La Jolla, CA, USA

C. B. Peterson
Department of Psychiatry and Behavioral Sciences, University of Minnesota, Minneapolis, MN, USA

1 Why New Treatment Models Are Needed for Binge Eating?

Binge eating is considered a core symptom among several eating disorders, including binge-eating disorder (BED), bulimia nervosa (BN), and

© Springer Nature Switzerland AG 2020
G. K.W. Frank, L. A. Berner (eds.), *Binge Eating*, https://doi.org/10.1007/978-3-030-43562-2_20

anorexia nervosa, binge-eating/purging subtype (AN-BP). All of these disorders are associated with impairments due to elevated psychiatric comorbidity, reduced quality of life, and medical complications (Herpertz-Dahlmann 2014; Hudson et al. 2007; Treasure et al. 2010; Kessler et al. 2013; Mitchell 2016; Rieger et al. 2005). Binge eating is also highly comorbid with obesity (OB), putting many individuals at risk for additional medical complications, including cardiovascular disease, type 2 diabetes, cancer, osteoarthritis, and all-cause mortality (Dixon 2010; Flegal et al. 2013). Meta-analyses show that after treatment, abstinence rates for individuals with BED are approximately 45–54% (Hilbert et al. 2019), while abstinence rates for individuals with BN are approximately 30% (Linardon and Wade 2018). Furthermore, AN-BP is associated with poorer long-term outcomes than AN-R (Zipfel et al. 2000). Thus, there is a need for additional treatments that address binge eating which can improve outcomes. The recent growth of research on the neurocognitive and neurobiological underpinnings of binge eating and obesity can provide potential mechanisms for treatment targets. Although OB is not an eating disorder per se, uncontrolled eating is a contributor to higher body weight and can be considered on the same behavioral continuum with binge eating (Vainik et al. 2019).

Current guidelines from the American Psychiatric Association (Yager et al. 2012, 2006) and the National Institute for Health and Care Excellence (National Institute for Health and Care Excellence 2004) support the use of cognitive behavioral therapy (CBT) as the primary treatment for individuals with BED and for BN (Wilfley et al. 2011). CBT focuses on disrupting the restraint/binge-eating cycle by targeting behaviors as well as thoughts related to eating, weight, and shape. CBT results with significant reductions in binge eating. However, CBT does not consistently produce significant weight loss and weight regain is common (Brownley et al. 2016; Vocks et al. 2010; Bulik et al. 2012; Grilo et al. 2011), leaving many individuals with binge eating at risk for the comorbidities associated

with OB. Additionally, although some do respond to CBT, not all individuals have favorable outcomes. Randomized controlled trials have also supported the use of interpersonal therapy (IPT) for the treatment of BED and BN, with emerging data providing preliminary support for dialectical behavioral therapy (DBT) as well (Wilfley et al. 2011; Kass et al. 2013; Linardon et al. 2017). However, consistent with CBT findings a sizeable percentage do not respond to treatment or experience a subsequent relapse. Clearly, novel interventions are needed as primary and ancillary treatments.

Talk therapy (including CBT, DBT, and IPT) assumes that individuals can access and consciously control cognitions that maintain binge eating. However, dual-process theories (Strack and Deutsch 2004) suggest that implicit processes may also exert control over behavior, even when the behavior is not consistent with long-term goals. Addressing more implicit mechanisms could enhance treatments for disorders characterized by binge eating and result in reductions in binge eating and, in those who have overweight or obesity, enhanced weight loss.

2 Neurobiology of Binge Eating and Obesity

Several reviews detail the neurobiological basis of BED and BN (Balodis et al. 2015; Kessler et al. 2016) and other chapters in this book detail reward-related and executive control-related alterations documented in individuals with binge eating. Overall, research indicates that individuals with binge eating have altered reward responsivity, particularly as it relates to palatable food cues. Additionally, individuals with obesity show similar reward-related alterations. For example, participants with obesity demonstrated greater activation in the corticolimbic regions (lateral OFC, caudate, anterior cingulate) over time, suggesting that people with OB have sustained responses in brain regions implicated in reward, even after eating (Dimitropoulos et al. 2012).

Of note, reward processing regions including the ventral-striatal and insular cortex discriminated between individuals with BED, obesity, healthy weight, and BN, further supporting the neurobiological basis of these disorders (Weygandt et al. 2012). Converging evidence also suggests that individuals with binge eating have difficulties activating the areas of the brain essential for inhibitory control, even on tasks that do not include food-specific stimuli (Balodis et al. 2013; Marsh et al. 2009). In summary, neurobiological and neurocognitive research suggests strong underlying mechanisms related to reward sensitivity and inhibitory control.

3 Neurocognition of Binge Eating and Obesity

Research on the neurocognitive underpinnings of binge eating and OB focus on cognitive processes with strong implications for treatment, including reward sensitivity, and overall executive functioning, particularly inhibitory control (Kessler et al. 2016; Eichen et al. 2017). More in-depth discussion of these factors is reviewed in other chapters. Considering the comorbid nature of binge eating and OB, it is possible that having OB could compound the neurocognitive changes seen in individuals with disorders characterized by binge eating.

3.1 Reward Sensitivity

There is a growing body of research that suggests reward processing dictates attention (Anderson 2016). Attention bias models of binge-eating behaviors focus on the role of appetitive motivation and incentive salience (Berridge 2009). Thus, attention biases to food would be indicative of impaired reward sensitivity toward food. These models suggest that conditioned palatable food cues elicit attention bias toward those stimuli, exacerbating cravings and a compulsive motivation to eat, resulting in binge-eating episodes as individuals seek the anticipated food reward and

alleviation of cravings (Stojek et al. 2018). In general, data suggest that an attention bias to food cues predisposes and individual to overconsumption. In particular, a large body of research suggests that increased attention bias toward palatable food cues predisposes individuals to consume those foods (Braet and Crombez 2003; Castellanos et al. 2009; Graham et al. 2011; Long et al. 1994; Nijs et al. 2010a, b). Several studies have demonstrated an attentional bias toward food cues among individuals who binge eat (Stojek et al. 2018; Albery et al. 2016; Schag et al. 2013; Schmitz et al. 2014; Shank et al. 2015; Schmidt et al. 2016). Taken together, these data suggest that individuals who binge eat have altered reward processing, as evidenced by increased reactivity to food stimuli. Impaired reward sensitivity, specifically manifested as attention bias to food cues, may be a promising mechanism to target in the development of new treatments.

3.2 Executive Function

The other major focus of research examining the neurocognitive underpinnings of binge eating and OB has focused on executive functioning, in particular, inhibitory control. Executive function or cognitive control broadly refers to top-down mental processes that enable effortful control over behavior. Executive function plays a strong role in regulating eating behavior and related thoughts and impairments in these areas are associated with dysfunction found in eating and weight disorders. Self-regulatory goals (e.g., to maintain a healthy diet) are also compromised by poor working memory [i.e., the ability to keep goal-relevant information in mind in the face of distractors (Hofmann et al. 2012)].

In some studies, individuals with BED and OB have greater impairments than those with OB without BED. For example, individuals with BED and OB performed worse on tasks related to self-regulation, planning, working memory, and decision-making compared to individuals with OB without BED (Reiter et al. 2017; Boeka and Lokken 2011; Duchesne et al. 2010;

Manasse et al. 2014; Manwaring et al. 2011). A recent meta-analysis suggested individuals with BN have moderate to large impairments in executive function (Hirsta et al. 2017).

Interestingly, individuals with OB in the absence of BED also show neurocognitive deficits (Lang et al. 2014; Liang et al. 2014; Smith et al. 1992). Evidence for impaired cognitive flexibility is inconsistent (Prickett et al. 2015). It should be noted that studies that have demonstrated a relationship between OB, impaired decision-making, and central coherence have not consistently controlled for OB comorbidities, such as cardiovascular disease or depression, that could potentially explain the relationship (Prickett et al. 2015). To date, there does not appear to be a consistent relationship between the level of executive function impairment and the degree of obesity (Sargenius et al. 2017).

With regard to inhibitory control, there is strong evidence that impairments in inhibitory control persist across diagnoses with binge eating (Wu et al. 2013). Individuals with BED and OB and individuals with OB without BED both showed deficits in general inhibitory control when compared to healthy controls, but at times do not differ from each other (Lavagnino et al. 2016), suggesting that the cognitive impairments may not solely be related to binge eating. However, some research suggests individuals with BED have greater impulsivity specifically toward food (Giel et al. 2017a). Increased BMI is associated with higher levels of impulsivity regardless of binge-eating status, although there is inconclusive evidence as to whether this impulsivity is related to food cues or is generalized (Schiff et al. 2016). Individuals with BN also demonstrated greater impulsivity than healthy controls (Kemps and Wilsdon 2010).

4 Novel Treatments for Binge Eating

4.1 Neurocognitive Training

Neurocognitive training refers to programs that can enhance cognitive processes to facilitate changes in behavior. Neurocognitive training can be delivered in several forms including via computer or smartphone. Based on the literature reviewed above, research on neurocognitive training for binge eating and OB have focused on attention bias and executive function (inhibitory control, working memory, and cognitive flexibility) that contribute to higher-level functions like reasoning and problem solving.

Several meta-analyses support the preliminary potential of attention bias and executive function training for changing appetitive behaviors; however, effects tend to be short term and small in size (Allom et al. 2016; Jones et al. 2016; Turton et al. 2016). The typical protocol for attention bias modification is a visual dot-probe task, in which participants are continually asked to respond to one of two stimuli; in the training version of the task, participants are repeatedly asked to respond to (and thus attend to) nonfood stimuli being presented instead of food stimuli.

A recent study on modifying attention bias in patients with BED showed that bias could be modified in the expected direction and cravings decreased, however, the effect was relatively brief (Schmitz and Svaldi 2017). An open-label case series among individuals with OB and binge eating demonstrated that the bias modification program was a feasible and acceptable treatment (Boutelle et al. 2016). Further, initial efficacy data showed decreases in weight, eating disorder symptoms, binge eating, loss of control, and responsivity to food in the environment, as well as changes in attention bias. The majority of these effects remained after 3 months. Another study comparing an attention bias modification program with a sham condition in individuals with BED and bulimia nervosa showed significant reductions in binge eating, eating disorder symptoms, trait food craving, and food cue reactivity in both arms. However, food intake, approach bias, and attention bias toward food did not change (Brockmeyer et al. 2019). It is possible that some methodological issues may have contributed to these findings. Participants in the approach bias assessment and training were instructed to determine their response by looking at the shape outside the picture vs. the content of the picture which may have lessened the association of food with the approach/avoidance cue. Further, the dot-probe task only assesses a snapshot of attention whereas eye-tracking data could have provided a more complete picture of attentional processes.

Attention bias modification programs have also been piloted in individuals with OB. Attention bias modification programs have been shown to change attention bias for chocolate and have been associated with reduced chocolate intake (Schumacher et al. 2016). Another study demonstrated the ability to modify attentional bias, and that the retraining effects were maintained at 24 h and 1 week follow-up as well as extending to new food pictures (Kemps et al. 2016). In children with OB, a single-session attention bias modification program, compared to a control program, showed changes in bias scores and influenced how much children ate in the absence of hunger immediately following training in the laboratory setting (Boutelle et al. 2014a). In summary, studies investigating attention bias modification have shown initially promising results, although given some mixed findings, it is not yet clear to what extent attention bias modification may be helpful in specifically targeting binge eating versus overeating in general. Further, it is possible that attention bias modification may be more efficacious as an adjunct treatment rather than a stand-alone treatment.

The other area of focus for neurocognitive training is executive functioning, in particular inhibitory control training. The typical protocol for inhibitory control training is a stop-start signal task or go/no-go task in which participants are repeatedly asked to inhibit responses to stimuli when presented with a "stop" cue. A study evaluating a three-session inhibition training program to suppress gaze toward food pictures showed that the training program was feasible and acceptable, and that patients in the training group significantly improved inhibitory control toward high-caloric food stimuli. Although both the intervention and the control groups decreased binge eating, no changes were found in food craving, food addiction, liking, and wanting

ratings (Giel et al. 2017b). A proof-of-concept study using a within-subjects design evaluated the impact of a single session food-specific or general go/no-go training in women with bulimia nervosa or BED on subsequent food intake in the lab (Turton et al. 2018). Results suggested that the training was feasible and well accepted by patients; however, no significant differences in eating behavior or eating disorder symptoms 24 h following training were detected. The authors of this study proposed that given the significant and chronic nature of eating disorder symptoms in this sample, greater duration and frequency of training sessions may have been needed to see an impact on behavior.

Several additional studies have focused on inhibitory control in the treatment of individuals with OB. One small randomized control trial evaluated the efficacy of ImpulsE (an eclectic treatment that includes elements of classic CBT, DBT, and acceptance and commitment therapy, with a stop-signal inhibitory control training) compared to treatment as usual (CBT-based approach) in 69 adults with obesity and impulsive/disinhibited eating behaviors (Preuss et al. 2017). The majority of the sample met diagnostic criteria for an eating disorder (33% BED; 40.6% Other Specified Feeding and Eating Disorder, including atypical BED). Both treatments reduced inhibition and eating disorder pathology; however, ImpulsE was more efficacious at reducing number of days with binge eating among individuals with BED.

A recent pilot study comparing gamified and non-gamified, personalized, and adaptive inhibitory control training for weight loss in adults with obesity showed that the inhibitory control training was feasible and acceptable, but that there was no difference among the conditions and no significant impact on weight (Forman et al. 2019). Another pilot study combined attention bias and inhibitory control trainings to compare food-specific trainings to generic stimuli (flowers, animals) (Stice et al. 2017). Results showed that individuals who had food-specific training demonstrated reduced activation in reward and attention regions of the brain to high-calorie food stimuli and had greater percent body fat loss at 4 weeks following the training compared to those in the general training. However, the effects on body fat were not persistent at the 6-month follow-up. To date, research has focused on targeting weight change so it is not yet clear the extent to which inhibitory control training may target binge eating specifically vs. overeating in general.

Another form of cognitive training is approach bias modification (ABM). ABM aims to target implicit cognitive biases that increase attention to and approach toward specific cues. The task involves either pushing a joystick away (avoid) or pulling a joystick towards (approach) in response to an image. These programs have successfully been deployed to treat addictions, particularly in regard to alcohol (Eberl et al. 2013; Wiers et al. 2011) and have recently been applied toward food cues (Schumacher et al. 2016; Dickson et al. 2016; Brockmeyer et al. 2015). One clinical trial investigated the efficacy of real ABM (in which all unhealthy food cues require the avoid response) versus a sham ABM (in which food cues require an equal number of approach and avoid responses) in individuals with BED or bulimia nervosa (Brockmeyer et al. 2019). Results showed that both treatments were effective at reducing binge eating, eating disorder symptomatology, food craving, and food cue reactivity. There were limited differences between the two groups except that the real ABM training resulted in greater reduction of cognitive eating disorder symptomatology, as measured by the global score on the Eating Disorder Examination, than the sham ABM. Authors concluded that ABM may be better suited as an adjunct treatment for eating disorders and future research in this domain is necessary.

In summary, emerging research suggests that cognitive trainings, including attention bias, inhibitory control, and approach bias modification, are feasible and acceptable among participants with binge eating and OB. However, the mixture of methods, stimuli, and efficacy data suggest that further research is needed.

4.2 Neuromodulation

Neuromodulation involves the use of magnetic stimulation or electrical current to decrease or increase nerve cell activity in brain regions and neural circuits related to eating disorder psychopathology (Dalton et al. 2017) including binge eating. Invasive neuromodulation, such as deep brain stimulation, has not been examined in BN or BED to date. Emerging noninvasive neuromodulation treatments including transcranial magnetic stimulation (TMS) and transcranial direct current stimulation (tDCS) are designed to target specific brain regions hypothesized to contribute to binge-eating maintenance including reward and behavioral inhibition, depending on symptoms presentation (Pisetsky et al. 2019). TMS relies on the use of electric current through a magnetic field to suppress or increase brain activity in designated regions and is typically administered in a repetitive series of treatments (rTMS). For tDCS, electrodes are placed on specific regions of the scalp to deliver electrical current, typically during the completion of a task.

Emerging data support the potential efficacy of tDCS and rTMS in the treatment of binge eating and OB, as well as anorexia nervosa (Dalton et al. 2018; McClelland et al. 2016; Kekic et al. 2016). Burgess and colleagues (Burgess et al. 2016) used tDCS targeting the right dorsolateral prefrontal cortex (DLPFC, a region of the brain associated with reward, inhibition, and emotion regulation) compared to a sham condition. Compared to the sham condition, tDCS was associated with reductions in food intake, food cravings, and the desire to binge eat among adults with full and subthreshold BED (particularly among males). One study that examined rTMS in BED using a case study design found that 20 sessions targeting the left DLPFC resulted in reductions in binge eating and associated psychopathology (Bazcynski et al. 2014). A recent rTMS study among adults with OB using deep TMS to target the insula observed reductions in food craving and body weight (Ferrulli et al. 2019). The most recently emerging noninvasive approach, real-time fMRI neurofeedback, is promising but has not been examined in BED (Val-Laillet et al. 2015). An important consideration in neuromodulation treatment studies is potential placebo effects, given the negative findings of a recent tDCS study examining OB and food cravings (Ray et al. 2019). Studies investigating rTMS for BN have yielded inconsistent data, with one study showing a reduction in food cravings (Van den Eynde et al. 2010) and a more recent double-blind study finding no differences in binge-eating outcome measures between rTMS and sham conditions (Gay et al. 2016).

In summary, recent treatment studies using noninvasive neuromodulation approaches including rTMS and tDCS have shown promise in reducing behavioral and cognitive symptoms of binge eating and OB. The robustness of these findings requires replication in randomized controlled trials as well as a broadening of clinical targets (e.g., emotion regulation). In addition, the potential efficacy of real-time fMRI interventions for the treatment of BED and OB warrants empirical examination.

4.3 Ecological Momentary Intervention

Ecological momentary assessment (EMA), which relies on the use of experience sampling in the natural environment using ambulatory technology including cell phones, has informed the identification of eating disorder symptom precipitants and maintenance mechanisms [see review by Engel and colleagues (2016)]. In BED, for example, EMA has highlighted the importance of negative reinforcement associated with binge eating to reduce momentary negative emotions including guilt (Berg et al. 2015). More recent eating disorder studies have used passive assessments to examine variables associated with binge eating including sleep patterns (Mason et al. 2019), and recent EMA investigations have demonstrated the important potential influence of neurocognition

on the momentary relationship between emotion and binge eating (Smith et al. 2019). Data also suggest that the use of technology delivery of traditional CBT treatment may improve outcome, potentially targeting momentary variables more effectively. Hildebrandt and colleagues (2017) conducted a randomized trial administering guided self-help cognitive-behavioral therapy using a smartphone application to a mixed sample of adults with BED or bulimia nervosa and found comparable outcomes between the technology-delivered treatment condition and the traditional guided self-help treatment (with greater meal adherence among the technology-based treatment).

Evolving from EMA, emerging treatments using ecological momentary intervention (EMI) employ interactive technology to deliver "just-in-time" treatments in response to "real time" data recorded in the natural environment (Smith and Juarascio 2019). This approach to treatment typically involves mobile health technology (mHealth), which has been examined in heterogeneous eating disorder populations with mixed findings (Smith and Juarascio 2019; Anastasiadou et al. 2018). As noted by Smith and Juarascio (2019), mHealth interventions are becoming increasingly more, particularly ambulatory assessments (e.g., heart rate variability) along with self-reported mood and eating behaviors, that can provide data to facilitate tailored interventions. For example, the detection of increased negative emotion, meal skipping, and other momentary precipitants can prompt alerts that the individual is "at risk" of binge eating. These signals can be accompanied by messages specifying strategies (e.g., mindfulness techniques, reminders to eat a planned meal/snack, and behavioral approaches) to prevent the occurrence of binge eating.

In summary, EMI and "just in time" treatments show promise as an emerging treatment for binge eating as well as other types of eating disorders but require more extensive investigations in randomized controlled trials.

4.4 Other New Models for the Treatment of Binge Eating and Uncontrolled Eating

According to the Behavioral Susceptibility Theory (BST), individual characteristics in appetitive traits are genetically determined and impact how an individual interacts with the current food environment (Carnell et al. 2013; Carnell and Wardle 2008; Llewellyn and Wardle 2015). The BST highlights dual processes: eating onset (responsiveness to signals to start eating, i.e., food responsiveness) and eating offset (responsiveness to signals to stop eating, i.e., satiety responsiveness). This dual susceptibility was initially described by Stanley Schachter in his Externality Theory, which hypothesized that individuals with OB have more difficulty sensing internal satiety signals and are more reactive to external cues to eat than their lean counterparts (Schachter 1971; Schachter and Rodin 1974). Based on these theories, there is an opportunity for the treatment of binge eating and uncontrolled eating, and the OB that results in some cases, by targeting both food responsiveness and satiety responsiveness. This model, called Regulation of Cues (ROC), uses psychoeducation and experiential learning exercises to enhance sensitivity to internal cues to promote eating offset and encourage proactive management of external cues to inhibit eating onset. Data suggest that targeting these appetitive behaviors offers a promising approach for weight-loss among adults who binge eat (83) as well as for children with overweight and obesity (Boutelle et al. 2011, 2014b). Currently, the ROC model is being tested in a study of adults with OB (Boutelle et al. 2019), which will evaluate the impact on weight and binge eating. Considering that binge eating is an extreme form of overeating, which includes feelings of loss of control, gaining control over eating by improving responsiveness to hunger/satiety cues and decreasing responsiveness to external cues could result in a significant advancement in clinical care for individuals with binge eating and OB.

ROC includes four main components: psychoeducation, coping skills to address sensitivity to food cues, experiential learning, and self-monitoring of both hunger/satiety and cravings, and physical activity. In line with the theory, sessions focus on improving satiety responses or decreasing sensitivity to food cues in the external environment.

The psychoeducation components in ROC describe "Tricky Hungers," which detail ways the body is "tricked" into feeling hungry when it is not. The overall goal of psychoeducation is to increase participant's awareness of the situations, thoughts, and moods that lead to overeating and improved awareness of true physical hunger. Physiological and neurobiological models of overeating past nutritional needs are presented so that participants understand how these vulnerabilities lead to overeating. Participants are educated on these factors to reduce guilt regarding overeating by helping participants understand the biological and psychological processes that contribute to overeating and binge eating. The initial sessions in ROC focus on appetite awareness. Following psychoeducation, to improve sensitivity to satiety, participants are taught to self-monitor their hunger, before, during, and after each meal, as well as 20 min after eating, on a 1–5 scale, with 1 = "starving" and 5 = "stuffed." For the experiential exercise, participants bring a meal to treatment and eat and monitor their hunger as described at the beginning of each session. Conditions are manipulated to simulate eating under different conditions (boredom, sadness, when full, when hungry). Coping skills are taught to identify and manage any instances of Tricky Hunger, and include physiological skills (deep breathing, relaxation, mindfulness), behavioral skills (delay, activity substitution) and cognitive skills (cognitive restructuring, distraction).

In subsequent sessions, content focuses on decreasing sensitivity to food cues in the environment. Since food cue responsiveness is learned, participants are provided information about basic learning theory and how physiological responses to food cues develop and can be managed (e.g., how cravings are formed). Coping skills are presented to assist in mastery and tolerance of food cue sensitivity to reduce responding to cravings. Participants learn to self-monitor their cravings (defined as urges to eat when not physically hungry). Craving is monitored with a 5-point scale, 1 = "not craving it at all" and 5 = "craving is overwhelming" and participants rate cravings during the day (ideally one craving a day at minimum). Participants create a craving hierarchy and bring their own highly craved foods to sessions. Using their highly craved foods, they complete cue-exposure treatment (CET-Food) exposures at each session. During the exposure, participants rate their cravings while looking at the food, holding the food, smelling the food, after taking two small bites of the food, and rate their cravings at 30 s intervals for the duration of the exposure. After 5 min, the participants dispose the food without eating it. Exposures are only conducted when participants are physically sated and utilize their preferred foods.

Throughout the treatment, physical activity is recommended according to national guidelines (150–250 min of at least moderate-intensity activity). Psychoeducation focuses on how physical activity could improve sensitivity to body sensations, inhibitory control, and aid in mastering and tolerating physiological and psychological arousal, resisting cravings, and overeating. Participants are instructed to self-monitor their physical activity each day.

Models such as ROC could potentially provide additional treatments to those with binge eating and OB, however, randomized trials are underway with individuals with OB (86) and more studies are needed to evaluate the effectiveness with both BED and OB, and possibly BN.

5 Summary

Current treatments are only successful at eliminating binge eating for about 30–50% of individuals with BN or BED, and, in the case of BED, few are successful at encouraging weight loss that can help reduce significant comorbidities. Thus, new primary and ancillary treatments are needed to improve outcomes. Neurobiological and neurocognitive underpinnings are promising targets for treatment. Treatments targeting these aspects that show preliminary promise include neurocognitive training and neuromodulation. EMI and mHealth approaches use the power of technology to deploy interventions during moments when patients may need it the most. Lastly, alternative behavioral models such as the ROC model that targets overeating have the potential to impact both binge eating and weight loss.

References

Albery I, Wilcockson T, Frings D, Moss A, Caselli G, Spada M (2016) Examining the relationship between selective attentional bias for food- and body-related stimuli and purging behaviour in bulimia nervosa. Appetite 107:208–212

Allom V, Mullan B, Hagger M (2016) Does inhibitory control training improve health behaviour? A meta-analysis. Health Psychol Rev 10:168–186

Anastasiadou D, Folkvord F, Lupianez-Villanueva F (2018) A systematic review of mHealth interventions for the support of eating disorders. Eur Eat Disord Rev 26:394–416

Anderson B (2016) The attention habit: how reward learning shapes attentional selection. Ann N Y Acad Sci 1369:24–39

Balodis I, Kober H, Worhunsky P, White M, Stevens M, Pearlson G et al (2013) Monetary reward processing in obese individuals with and without binge eating disorder. Biol Psychiatry 73:877–886

Balodis I, Grilo C, Potenza M (2015) Neurobiological features of binge eating disorder. CNS Spectr 20:557–565

Bazcynski T, de Aquino CC, Nazar B, Carta M, Arias-Carrion O, Silva A et al (2014) High-frequency rTMS to treat refractory binge eating disorder and comorbid depression: a case report. CNS Neurol Disord Drug Targets 13:771–775

Berg K, Crosby RD, Cao L, Crow S, Engel SG, Wonderlich SA et al (2015) Negative affect prior to and following overeating-only, loss of control eating-only, and binge eating episodes in obese adults. Int J Eat Disord 48:641–653

Berridge K (2009) 'Liking' and 'wanting' food rewards: brain substrates and roles in eating disorders. Physiol Behav 97:537–550

Boeka A, Lokken K (2011) Prefrontal systems involvement in binge eating. Eat Weight Disord 16:e121–e126

Boutelle K, Zucker N, Peterson C, Rydell S, Cafri G, Harnack L (2011) Two novel treatments to reduce overeating in overweight children: a randomized controlled trial. J Consult Clin Psychol 79:759–771

Boutelle K, Kuckertz J, Carlson J, Amir N (2014a) A pilot study evaluating a one-session attention modification training to decrease overeating in obese children. Appetite 76:180–185

Boutelle K, Zucker N, Peterson C, Rydell S, Carlson J, Harnack L (2014b) An intervention based on Schachter's externality theory for overweight children: the regulation of cues pilot. J Pediatr Psychol 39:405–417

Boutelle K, Monreal T, Strong D, Amir N (2016) An open trial evaluating an attention bias modification program for overweight adults who binge eat. J Behav Ther Exp Psychiatry 52:138–146

Boutelle K, Eichen D, Peterson C, Strong D, Rock C, Marcus B (2019) Design of the PACIFIC study: a randomized controlled trial evaluating a novel treatment for adults with overweight and obesity. Contemp Clin Trials 84:105824

Braet C, Crombez G (2003) Cognitive interference due to food cues in childhood obesity. J Clin Child Adolesc Psychol 32:32–39

Brockmeyer T, Hahn C, Reetz C, Schmidt U, Friederich H (2015) Approach bias modification in good craving-a proof-of-concept study. Eur Eat Disord Rev 23:352–360

Brockmeyer T, Friederich H, Kuppers C, Chowdhury S, Harms L, Simmonds J et al (2019) Approach bias modification training in bulimia nervosa and binge-eating disorder: a pilot randomized controlled trial. Int J Eat Disord 52:520–529

Brownley K, Berkman N, Peat C, Lohr K, Cullen K, Bann C et al (2016) Binge-eating disorder in adults: a systematic review and meta-analysis. Ann Intern Med 165:409–420

Bulik C, Marcus M, Zerwas S, Levine M, La Via M (2012) The changing "weightscape" of bulimia nervosa. Am J Psychiatry 169:1031–1036

Burgess EE, Sylvester MD, Morse K, Amthor F, Mrug S, Lokken K et al (2016) Effects of transcranial direct current stimulation (tDCS) on binge eating disorder. Int J Eat Disord 49:930–936

Carnell S, Wardle J (2008) Appetite and adiposity in children: evidence for a behavioral susceptibility theory of obesity. Am J Clin Nutr 88:22–29

Carnell S, Benson L, Pryor K, Driggin E (2013) Appetitive traits from infancy to adolescence: using behavioral and neural measures to investigate obesity risk. Physiol Behav 121:79–88

Castellanos E, Charboneau E, Dietrich M, Park S, Bradley B, Mogg K et al (2009) Obese adults have visual attention bias for food cue images: evidence for altered reward system function. Int J Obes 33:1063–1073

Dalton B, Campbell I, Schmidt U (2017) Neuromodulation and neurofeedback treatments in eating disorders and obesity. Curr Opin Psychiatry 30:458–473

Dalton B, Bartholdy S, McClelland J, Kekic M, Rennalls S, Werthmann J et al (2018) Randomised controlled feasibility trial of real versus sham repetitive transcranial magnetic stimulation treatment in adults with severe and enduring anorexia nervosa: the TIARA study. BMJ Open 8:e021531

Dickson H, Kavanagh D, MacLeod C (2016) The pulling power of chocolate: effects of approach-avoidance training on approach bias and consumption. Appetite 99:46–51

Dimitropoulos A, Tkach J, Ho A, Kennedy J (2012) Greater corticolimbic activation to high-calorie food cues after eating in obese vs. normal-weight adults. Appetite 58:303–312

Dixon J (2010) The effect of obesity on health outcomes. Mol Cell Endocrinol 316:104–108

Duchesne M, Mattos P, Appolinario J, de Freitas S, Coutinho G, Santos C et al (2010) Assessment of executive functions in obese individuals with binge eating disorder. Rev Bras Psiquiatr 32:381–388

Eberl C, Wiers R, Pawelczack S, Rinck M, Becker E, Lindenmeyer J (2013) Approach bias modification in alcohol dependence: do clinical effects replicate and for whom does it work best? Dev Cogn Neurosci 4:38–51

Eichen D, Matheson B, Appleton-Knapp S, Boutelle K (2017) Neurocognitive treatments for eating disorders and obesity. Curr Psychiatry Rep 19:62

Engel SG, Crosby RD, Thomas G, Bond D, Lavender J, Mason T et al (2016) Ecological momentary assessment in eating disorder and obesity research: a review of the recent literature. Curr Psychiatry Rep 18:37

Ferrulli A, Macri C, Terruzzi I, Massarini S, Ambrogi F, Adamo M et al (2019) Weight loss induced by deep transcranial magnetic stimulation in obesity: a randomized, double-blind, sham-controlled study. Diabetes Obes Metab 21:1849–1860

Flegal K, Kit B, Orpana H, Graubard B (2013) Association of all-cause mortality with overweight and obesity using standard body mass index categories: a systematic review and meta-analysis. JAMA 309:71–82

Forman EM, Manasse SM, Dallal DH, Crochiere R, Loyka C, Butryn M et al (2019) Computerized neurocognitive training for improving dietary health and facilitating weight loss. J Behav Med 42:1029–1040

Gay A, Jaussent I, Sigaud T, Billard S, Attal J, Seneque M et al (2016) A lack of clinical effect of high-frequency rTMS to dorsolateral prefrontal cortex on bulimic symptoms: a randomised, double-blind trial. Eur Eat Disord Rev 24:474–481

Giel K, Teufel M, Junne F, Zipfel S, Schag K (2017a) Food-related impulsivity in obesity and binge eating disorder-a systematic update of the evidence. Nutrients 9:E1170

Giel K, Speer E, Schag K, Leehr E, Zipfel S (2017b) Effects of a food-specific inhibition training in individuals with binge eating disorder-findings from a randomized controlled proof-of-concept study. Eat Weight Disord 22:345–351

Graham R, Hoover A, Ceballos N, Komogortsev O (2011) Body mass index moderates gaze orienting biases and pupil diameter to high and low calorie food images. Appetite 56:577–586

Grilo C, Masheb R, Wilson G, Gueorguieva R, White M (2011) Cognitive-behavioral therapy, behavioral weight loss, and sequential treatment for obese patients with binge-eating disorder: a randomized controlled trial. J Consult Clin Psychol 79:675–685

Herpertz-Dahlmann B (2014) Adolescent eating disorders: update on definitions, symptomatology, epidemiology, and comorbidity. Child Adolesc Psychiatr Clin N Am 24:177–196

Hilbert A, Petroff D, Herpertz S, Pietrowsky R, Tuschen-Caffier B, Vocks S et al (2019) Meta-analysis of the efficacy of psychological and medical treatments for binge-eating disorder. J Consult Clin Psychol 87:91–105

Hildebrandt T, Michaelides A, MacKinnon D, Greif R, DeBar L, Sysko R (2017) Randomized controlled trial comparing smartphone assisted versus traditional guided self-help for adults with binge eating. Int J Eat Disord 50:1313–1322

Hirsta R, Bearda C, Colaby K, Quittnera Z, Milsa B, Lavender J (2017) Anorexia nervosa and bulimia nervosa: a meta-analysis of executive functioning. Neurosci Biobehav Rev 83:678–690

Hofmann W, Schmeichel B, Baddeley A (2012) Executive functions and self-regulation. Trends Cogn Sci 16:174–180

Hudson J, Hiripi E, Pope H, Kessler R (2007) The prevalence and correlates of eating disorders in the National Comorbidity Survey Replication. Biol Psychiatry 61:348–358

Jones A, Di Lemma L, Robinson E, Christiansen P, Nolan S, Tudur-Smith C et al (2016) Inhibitory control training for appetitive behaviour change: a meta-analytic investigation of mechanisms of action and moderators of effectiveness. Appetite 97:16–28

Kass A, Kolko R, Wilfley D (2013) Psychological treatments for eating disorders. Curr Opin Psychiatry 26:549–555

Kekic M, Boysen E, Campbell I, Schmidt U (2016) A systematic review of the clinical efficacy of transcranial direct current stimulation (tDCS) in psychiatric disorders. J Psychiatr Res 74:70–86

Kemps E, Wilsdon A (2010) Preliminary evidence for a role for impulsivity in cognitive disinhibition in bulimia nervosa. J Clin Exp Neuropsychol 32:515–521

Kemps E, Tiggemann M, Hollitt S (2016) Longevity of attentional bias modification effects for food cues in overweight and obese individuals. Psychol Health 31:115–129

Kessler R, Berglund PA, Chiu W, Deitz A, Hudson J, Shahly V et al (2013) The prevalence and correlates of binge eating disorder in the World Health Organization World Mental Health Surveys. Biol Psychiatry 73:904–914

Kessler R, Hutson P, Herman B, Potenza M (2016) The neurobiological basis of binge-eating disorder. Neurosci Biobehav Rev 63:223–238

Lang K, Lopez C, Stahl D, Tchanturia K, Treasure J (2014) Central coherence in eating disorders: an updated systematic review and meta-analysis. World J Biol Psychiatry 15:586–598

Lavagnino L, Arnone D, Cao B, Soares J, Selvaraj S (2016) Inhibitory control in obesity and binge eating disorder: a systematic review and meta-analysis of neurocognitive and neuroimaging studies. Neurosci Biobehav Rev 68:714–726

Liang J, Matheson B, Kaye W, Boutelle K (2014) Neurocognitive correlates of obesity and obesity-related behaviors in children and adolescents. Int J Obes 38:494–506

Linardon J, Wade T (2018) How many individuals achieve symptom abstinence following psychological treatments for bulimia nervosa? A meta-analytic review. Int J Eat Disord 51:287–294

Linardon J, Fairburn C, Fitzsimmons-Craft E, Wilfley D, Brennan L (2017) The empirical status of the third-wave behaviour therapies for the treatment of eating disorders: a systematic review. Clin Psychol Rev 58:125–140

Llewellyn C, Wardle J (2015) Behavioral susceptibility to obesity: gene-environment interplay in the development of weight. Physiol Behav 152:494–501

Long C, Hinton C, Gillespie N (1994) Selective processing of food and body size words: application of the Stroop test with obese restrained eaters, anorexics, and normals. Int J Eat Disord 15:279–283

Manasse S, Juarascio A, Forman E, Berner L, Butryn M, Ruocco A (2014) Executive functioning in overweight individuals with and without loss-of-control eating. Eur Eat Disord Rev 22:373–377

Manwaring JL, Green L, Myerson J, Strube M, Wilfley D (2011) Discounting of various types of rewards by women with and without binge eating disorder: evidence for general rather than specific differences. Psychol Rec 61:561–582

Marsh R, Steinglass J, Gerber A, Graziano O'Leary K, Wang Z, Murphy D et al (2009) Deficient activity in the neural systems that mediate self-regulatory control in bulimia nervosa. Arch Gen Psychiatry 66:51–63

Mason T, Engwall A, Mead M, Irish L (2019) Sleep and eating disorders among adults enrolled in a commercial weight loss program: associations with self-report and objective sleep measures. Eat Weight Disord 24:307–312

McClelland J, Kekic M, Bozhilova N, Nestler S, Dew T, Van den Eynde F et al (2016) A randomised controlled trial of neuronavigated repetitive transcranial magnetic stimulation (rTMS) in anorexia nervosa. PLoS One 11: e0148606

Mitchell J (2016) Medical comorbidity and medical complications associated with binge-eating disorder. Int J Eat Disord 49:319–323

National Institute for Health and Care Excellence (2004) Core interventions in the treatment and management of anorexia nervosa, bulimia nervosa and related eating disorders (Clinical guideline 9). National Collaborating Centre for Medical Health

Nijs I, Franken I, Muris P (2010a) Food-related Stroop interference in obese and normal-weight individuals: behavioral and electrophysiological indices. Eat Behav 11:258–265

Nijs IM, Muris P, Euras AS, Franken IH (2010b) Differences in attention to food and food intake between overweight/obese and normal-weight females under conditions of hunger and satiety. Appetite 54:243–254

Pisetsky E, Schaefer L, Wonderlich S, Peterson C (2019) Emerging psychological treatments in eating disorders. Psychiatr Clin North Am 42:219–229

Preuss H, Pinnow M, Schnicker K, Legenbauer T (2017) Improving inhibitory control abilities (ImpulsE)-A promising approach to treat impulsive eating? Eur Eat Disord Rev 25:533–543

Prickett C, Brennan L, Stolwyk R (2015) Examining the relationship between obesity and cognitive function: a systematic literature review. Obes Res Clin Pract 9:93–113

Ray M, Sylvester M, Helton A, Pittman B, Wagstaff L, McRae TR 3rd et al (2019) The effect of expectation on transcranial direct current stimulation (tDCS) to suppress food craving and eating in individuals with overweight and obesity. Appetite 136:1–7

Reiter A, Heinze H, Schlagenhauf F, Deserno L (2017) Impaired flexible reward-based decision-making in binge eating disorder: evidence from computational modeling and functional neuroimaging. Neuropsychopharmacology 42:628–637

Rieger E, Wilfley D, Stein R, Marino V, Crow S (2005) A comparison of quality of life in obese individuals with and without binge eating disorder. Int J Eat Disord 37:234–240

Sargenius H, Lydersen S, Hestad K (2017) Neuropsychological function in individuals with morbid obesity: a cross-sectional study. BMC Obes 4:6

Schachter S (1971) Some extraordinary facts about obese humans and rats. Am Psychol 26:129–144

Schachter S, Rodin J (1974) Obese humans and rats. Erlbaum, Hillsdale, NJ

Schag K, Teufel M, Junne F, Preissl H, Hautzinger M, Zipfel S et al (2013) Impulsivity in binge eating disorder: food cues elicit increased reward responses and disinhibition. PLoS One 16:e76542

Schiff S, Amodio P, Testa G, Nardi M, Montagnese S, Caregaro L et al (2016) Impulsivity toward food reward is related to BMI: evidence from intertemporal choice in obese and normal-weight individuals. Brain Cogn 110:112–119

Schmidt R, Luthold P, Kittel R, Tetzlaff A, Hilbert A (2016) Visual attentional bias for food in adolescents with binge-eating disorder. J Psychiatr Res 80:20–29

Schmitz F, Svaldi J (2017) Effects of bias modification training in binge eating disorder. Behav Ther 48:707–717

Schmitz F, Naumann E, Trentowska M, Svaldi J (2014) Attentional bias for food cues in binge eating disorder. Appetite 80:70–80

Schumacher S, Kemps E, Tiggemann M (2016) Bias modification training can alter approach bias and chocolate consumption. Appetite 96:219–224

Shank L, Tanofsky-Kraff M, Nelson E, Shomaker L, Ranzenhofer L, Hannallah L et al (2015) Attentional bias to food cues in youth with loss of control eating. Appetite 87:68–75

Smith K, Juarascio A (2019) From ecological momentary assessment (EMA) to ecological momentary intervention (EMI): past and future directions for ambulatory assessment and interventions in eating disorders. Curr Psychiatry Rep 21:53

Smith D, Marcus M, Kaye W (1992) Cognitive-behavioral treatment of obese binge eaters. Int J Eat Disord 12:257–262

Smith K, Mason T, Crosby R, Engel S, Wonderlich S (2019) A multimodal, naturalistic investigation of relationships between behavioral impulsivity, affect, and binge eating. Appetite 136:59057

Stice E, Yokum S, Veling H, Kemps E, Lawrence N (2017) Pilot test of a novel food response and attention training treatment for obesity: brain imaging data suggest actions shape valuation. Behav Res Ther 94:60–70

Stojek M, Shank L, Vannucci A, Bongiorno D, Nelson E, Waters A et al (2018) A systematic review of attentional biases in disorders involving binge eating. Appetite 123:367–389

Strack F, Deutsch R (2004) Reflective and impulsive determinants of social behavior. Personal Soc Psychol Rev 8:220–247

Treasure J, Claudino A, Zucker N (2010) Eating disorders. Lancet 375:583–593

Turton R, Bruidegom K, Cardi V, Hirsch C, Treasure J (2016) Novel methods to help develop healthier eating habits for eating and weight disorders: a systematic review and meta-analysis. Neurosci Biobehav Rev 61:132–155

Turton R, Nazar B, Burgess E, Lawrence N, Cardi V, Treasure J et al (2018) To go or not to go: a proof of concept study testing food-specific inhibition training for women with eating and weight disorders. Eur Eat Disord Rev 26:11–21

Vainik U, Garcia-Garcia I, Dagher A (2019) Uncontrolled eating: a unifying heritable trait linked with obesity, overeating, personality and the brain. Eur J Neurosci 50:2430–2445

Val-Laillet D, Aarts E, Weber B, Ferrari M, Quaresima V, Stoeckel L et al (2015) Neuroimaging and neuromodulation approaches to study eating behavior and prevent and treat eating disorders and obesity. Neuroimage Clin 24:1–31

Van den Eynde F, Claudino A, Mogg A, Horrell L, Stahl D, Ribeiro W et al (2010) Repetitive transcranial magnetic stimulation reduces cue-induced food craving in bulimic disorders. Biol Pschiatry 67:793–795

Vocks S, Tuschen-Caffier B, Pietrowsky R, Rustenbach S, Kersting A, Herpertz S (2010) Meta-analysis of the effectiveness of psychological and pharmacological treatments for binge eating disorder. Int J Eat Disord 43:205–217

Weygandt M, Schaefer A, Schienle A, Haynes J (2012) Diagnosing different binge-eating disorders based on reward-related brain activation patterns. Hum Brain Mapp 33:2135–2146

Wiers RW, Eberl C, Rinck M, Becker E, Lindenmeyer J (2011) Retraining automatic action tendencies changes alcoholic patients' approach bias for alcohol and improves treatment outcome. Psychol Sci 22:490–497

Wilfley D, Kolko R, Kass A (2011) Cognitive-behavioral therapy for weight management and eating disorders in children and adolescents. Child Adolesc Psychiatr Clin N Am 20:271–285

Wu M, Hartmann M, Skunde M, Herzog W, Friederich H (2013) Inhibitory control in bulimic-type eating disorders: a systematic review and meta-analysis. PLoS One 8:e83412

Yager J, Devlin M, Halmi KA, Herzog D, Mitchelolo J, Powers P et al (2006) Practice guidelines for the treatment of patients with eating disorders, 3rd edn. American Psychiatric Association, Arlington, VA, pp 1–128

Yager J, Devlin M, Halmi K, Herzog D, Mitchell J, Powers D et al (2012) Guideline watch (August 2012): practice guideline for the treatment of patients with eating disorders, 3rd edn. American Psychiatric Association, Arlington, VA, pp 1–18

Zipfel S, Lowe B, Reas D, Deter H, Herzog W (2000) Long-term prognosis in anorexia nervosa: lessons from a 21-year follow-up study. Lancet 355:721–722

Part V

Research Agenda

Binge-Eating Disorder: Unanswered Questions

B. Timothy Walsh and Michael J. Devlin

Abstract

Binge-eating disorder (BED) was officially recognized in *DSM-5* in 2013 and remains a focus of significant clinical and research interest. This chapter describes several important but unanswered questions about the nature of BED, including how best to understand the disturbances in eating behavior. Laboratory studies during which individuals with BED are asked to engage in binge eating permit objective measurement of eating behavior and have convincingly documented that the binge-eating episodes of individuals with BED are abnormal. Surprisingly, non-binge eating behaviors during laboratory studies also differ from those of comparable controls. In contrast, several studies of eating behavior in nonlaboratory settings using ecological momentary assessment (EMA) suggest that the eating behavior of individuals with BED closely resembles that of individuals without BED. In addition, changes in binge eating are not tightly linked to changes in body weight and the presence of BED is not a robust predictor of response to treatments aimed at weight reduction. This chapter briefly reviews these findings and underscores the importance of addressing the unanswered questions that emerge.

Keywords

Eating behavior · Binge eating · DSM-5 · Binge-eating disorder (BED) · Bulimia nervosa

Learning Objectives

After reading this chapter, you will be able to:

1. Describe what is known, and what is not, about the eating behavior of individuals with binge-eating disorder.
2. Describe what is known, and what is not, about the psychological characteristics of individuals with binge-eating disorder.
3. Describe the characteristics and questions about the treatment response of individuals with binge-eating disorder.

1 Introduction

The syndrome of binge eating was described in 1959 by Albert J. Stunkard, one of the pioneering investigators examining the interactions between mental health and obesity (Stunkard 1959). However, it was not until 1993, during the development of DSM-IV, that explicit criteria for what

B. T. Walsh (✉) · M. J. Devlin
New York State Psychiatric Institute, Columbia University Irving Medical Center, New York, NY, USA
e-mail: btw1@cumc.columbia.edu; mjd5@cumc.columbia.edu

© Springer Nature Switzerland AG 2020
G. K.W. Frank, L. A. Berner (eds.), *Binge Eating*, https://doi.org/10.1007/978-3-030-43562-2_21

was termed binge-eating disorder (BED) were proposed by Robert Spitzer who, in 1980, had led the nosological revolution in psychiatry represented by DSM-III (Spitzer et al. 1993). Although Spitzer vigorously advocated for the formal inclusion of BED in DSM-IV, the committees overseeing the development of DSM-IV believed that insufficient data were available about the clinical characteristics of individuals with BED and therefore included the criteria for BED only in an appendix of DSM-IV containing criteria sets for further study.

The availability of those criteria sparked a remarkable surge in research on BED so that, when DSM-5 was being developed, there were almost 1000 peer-reviewed publications on the nature, course, complications, and treatment of BED. A comprehensive review prepared under the auspices of the DSM-5 Work Group on Eating Disorders summarized these data, and found solid evidence supporting the potential clinical utility of BED (Wonderlich et al. 2009). In empirical taxometric studies, BED was distinct from anorexia nervosa and bulimia nervosa. In laboratory studies that permitted the objective measurement of eating behavior, individuals with BED consumed more food than individuals of similar body weight without BED. Individuals with BED experienced more symptoms of mood and anxiety than comparable individuals without BED, and, although the evidence was mixed, there were preliminary indications that individuals with BED tended to gain more weight over time and suggestions that they fared worse in standard weight-loss treatment programs. It was this extensive literature that provided the foundation for the official recognition of BED in DSM-5, published in 2013.

Since the arrival of DSM-5, BED has continued to draw substantial interest. In 2015, the FDA approved the use of the stimulant lisdexamfetamine (Vyvanse®) for individuals with BED, the first medication approved for this indication, and is currently considering whether to approve dasotraline, a serotonergic and dopaminergic reuptake blocker. At the time of this writing, a PubMed search found almost 400 articles with "binge-eating disorder" in the title published since 2013.

Such information provides considerable evidence of the utility of this diagnostic category for patients, clinicians, and researchers. However, in this chapter, we hope to highlight several important and fundamental questions about the nature of BED that have not been successfully answered.

2 The Definition of Binge Eating

The definition of an episode of binge eating, as articulated in DSM-IV and -5, requires two elements: the consumption of an objectively large amount of food and a sense of loss of control over eating during the episode. Although there is no explicit definition of what constitutes a large amount of food, and the sense of loss of control is subjective, the use of structured diagnostic instruments to assess binge eating, such as the Eating Disorder Examination (EDE) and the Eating Disorder Assessment for DSM-5 (EDA-5), allow clinical investigators to achieve reasonable reliability in the diagnosis of BED and bulimia nervosa, for both of which binge eating is a *sine qua non* (Grilo et al. 2004; Peterson et al. 2007; Sysko et al. 2015).

A number of studies have examined the relative significance of objective overeating versus the sense of loss of control during episodes. The data emerging from these studies are complex, but suggest that the two elements in the definition of binge eating are linked to somewhat different clinical phenomena. The sense of loss of control is more closely associated with emotional distress and psychopathology, whereas overeating is associated with being overweight, and, in at least some samples, a greater tendency to gain weight (Sonneville et al. 2013; Goldschmidt 2017).

However, a number of studies, especially those focusing on youth, have raised questions about the importance of the consumption of an objectively large amount of food in the definition of binge eating. Particularly during growth and development, it is challenging to judge what is an "unusually large" amount of food; for example, Shomaker et al. (2010) found that, late in puberty, healthy adolescent males consumed approximately 2000 kcals at a single lunch meal. And, as noted above, the sense of loss of control is clearly linked to measures of psychological distress. Therefore, especially among youth, it is possible that frequent loss of control eating is more clinically meaningful than the diagnosis of BED (Kelly et al. 2014). This perspective regarding the greater salience of the sense of loss of control versus binge size has also been explored in relation to bulimia nervosa and would also be important to examine in relation to the binge-eating/purging subtype of anorexia nervosa (Wolfe et al. 2009; Brownstone et al. 2013).

In this context, it is important to note that the recently adopted ICD-11 defines binge eating solely on the basis of recurrent episodes of out of control eating, without the requirement that they be unusually large. Therefore, individuals whose behavior does not meet the DSM-5 criteria for bulimia nervosa or BED may nonetheless meet the ICD-11 criteria. An important unanswered question is to what degree the accumulated information regarding the characteristics of individuals meeting the DSM-5 criteria for BED (and for bulimia nervosa) can be extended to individuals meeting the ICD-11 criteria. An additional complicating factor is the heterogeneity in how the loss of control is experienced by patients, leading to challenges in how to operationalize its definition.

3 Laboratory Studies of Eating Behavior

As noted in the review of Wonderlich et al. (2009), laboratory studies of eating behavior provide some of the most convincing evidence of the validity of BED. In these studies, individuals were presented with a single food item or an array of foods from which to choose and were either asked to binge eat or were exposed to conditions aimed to induce overeating. With impressive consistency, in these studies, individuals with BED consumed significantly more food than did individuals without BED of comparable body weight, providing convincing evidence that the binge eating of individuals

with BED is objectively abnormal (Walsh and Boudreau 2003).

However, other results from laboratory studies are puzzling. Two of the laboratory studies reported the duration of meals, and found that the average duration of the binge meals of individuals with BED was significantly longer than that of individuals without BED (Walsh and Boudreau 2003). As a result, there was no evidence that the rate of calorie intake among individuals with BED was faster, despite the fact that "eating much more rapidly than normal" is included in criterion B in DSM-5. These laboratory data, although only from two studies, suggests that the disturbance might be better characterized as "eating much longer than normal" rather than "much more rapidly."

Even more puzzling are observations from the several studies in which individuals with BED were asked to eat normally, without binge eating. Consistently, individuals with BED consumed more calories than did individuals without BED (Walsh and Boudreau 2003) in non-binge laboratory meals, seemingly implying that eating behavior is abnormal regardless of whether individuals are binge eating. It is unclear how to understand these results, but one possibility relates to disinhibition. Individuals with BED consistently score higher on the disinhibition scale of Stunkard's three-factor eating questionnaire (TFEQ) (Stunkard and Messick 1985). It is possible that the conditions of a laboratory study, especially one in which subjects are presented with a large, buffet-style array of foods, provoke greater disinhibition among subjects with BED, leading to greater consumption whether or not they are asked to binge eat (Guss et al. 1994). Unfortunately, this raises questions about the degree to which behavior observed during such laboratory studies is representative of eating behavior outside the laboratory. To definitively resolve such questions, techniques are needed to allow the objective assessment of eating behavior among free-living individuals, a technology that is currently under development (e.g., Farooq et al. 2019).

In contrast, interpretation of data from laboratory studies of individuals with bulimia nervosa appears more straightforward. When asked to binge eat, individuals with bulimia nervosa who are able to comply with the request consume far more calories than do comparable individuals without bulimia nervosa. But, when asked to eat a normal, non-binge meal, individuals with bulimia nervosa consume fewer calories than controls (Walsh and Boudreau 2003). These results are generally consistent with clinical experience. For example, an important element of the model on which Cognitive Behavioral Therapy (CBT) for bulimia nervosa is based is that episodes of binge eating often follow periods of restriction, and an important early step in CBT is to encourage patients to eat several meals spaced over the day. In addition, the available laboratory studies suggest that both the caloric content of binge episodes of individuals with bulimia nervosa and the rate of caloric consumption during binge episodes are substantially greater than those of individuals with BED. Such results suggest that the eating behavior of individuals with BED more closely resembles that of comparable peers than does the eating behavior of individuals with bulimia nervosa. We would also note that, to our knowledge, there have been no objective studies of the binge eating of individuals with binge-eating/purging subtype of anorexia nervosa.

4 Eating Behavior Outside the Laboratory

A number of studies have asked individuals with and without BED to record their eating behavior and emotional state in real time by using hand-held devices, a method termed Ecological Momentary Assessment (EMA). Although this method relies on participant self-report and is therefore not an objective measure of eating behavior, several of studies using EMA have yielded results that raise provocative questions about the nature of binge eating as defined in DSM.

Greeno et al. (2000) obtained data from 40 women with BED and 38 women without BED, as assessed by the Eating Disorder Examination. All of the women with BED reported binge episodes, but, unexpectedly, so did 66% of the women who had not received a diagnosis of BED, and most of these individuals reported binge eating more than twice a week, the DSM-IV standard for the diagnosis. The study confirmed that individuals with BED experienced considerably more negative affect than individuals without BED and also found that the mean ratings of negative affect and loss of control over eating of the individuals without BED who reported binge eating were intermediate between those of individuals with BED and individuals without BED who did not report binge eating.

Le Grange et al. (2001) examined information from EMA monitoring for 2 weeks of 18 women with BED and 17 women without BED. The frequency of binge eating was similar in both groups. However, the individuals with BED did report higher levels of emotional distress on a number of psychological measures.

Goldschmidt et al. (2012) used EMA to assess nine adults with BED and 13 without BED over 7 days. The average number of calories consumed per eating episode was slightly over 550 kcals, and was virtually identical for the BED and non-BED groups. Overall, individuals reported eating 3.4 times per day; individuals with BED described 11.9% of episodes as binge eating versus 7.7% of episodes for individuals without BED. The authors concluded that a sense of loss of control, rather than the amount of food consumed, was associated with emotional distress among individuals with BED.

These studies are impressively consistent in suggesting that what distinguishes individuals with BED from comparable individuals without BED is not the amount of food consumed during episodes of eating as much as the level of distress, generally, and a sense of loss of control over eating, in particular.

5 Weight Change During Treatment

A wide range of treatments have been found to be effective in reducing the frequency of binge eating in BED, including CBT, Interpersonal

Psychotherapy (IPT), behavioral weight management, antidepressants, the stimulant lisdexamfetamine (Vyvanse®) and the anticonvulsant topiramate (Topamax®) (Brownley et al. 2016). In many trials, the reduction in the frequency of binge eating is dramatic, approaching complete cessation. However, weight loss does not follow automatically. In fact, most psychotherapeutic interventions, although clearly effective as assessed by reductions in binge eating, result in no reduction in weight whatsoever. The only interventions that routinely lead to reductions in both binge eating and weight are those that are known to reduce appetite and lead to weight loss, such as lisdexamfetamine and topiramate.

The absence of a change in weight must indicate that the average caloric intake has not changed, despite the reduction or elimination of binge eating. There are at least three reasonable explanations for the lack of a consistent association between a reduction in binge eating and a loss of weight. One is that, after cessation of binge eating, the calories previously consumed during binge-eating episodes are redistributed over the day vs being concentrated in binge episodes, which typically occur during the evening or at night. Another possibility is that eating behavior does not actually change substantially, but the individual's perspective on the behavior does. The individual may feel better emotionally and no longer view their eating behavior, even episodes of overeating, as being as out of control. A final possibility is that the reduction or cessation of binge eating with treatment is indeed associated with an actual reduction in daily caloric intake, but that rather than resulting in weight loss, this caloric reduction stabilizes weight following an upward trend prior to treatment. Consistent with this is evidence that binge eating is associated with weight gain over time and with weight fluctuation (Spitzer et al. 1992; Fairburn et al. 2000). It is quite plausible that individuals might be more likely to enter treatment while in a weight gain phase, and that treatment interrupts this weight gain. These explanations are not mutually exclusive, and, as noted above, the only way to determine to what degree each is true is to find a way to measure eating behavior

objectively during treatment of BED outside of the laboratory.

Dissecting the factors associated with weight change during the treatment of individuals with binge-eating/purging subtype of anorexia nervosa or with bulimia nervosa is more complex given the caloric restriction and purging that also may affect energy balance.

6 Prediction of Response to Treatment

The question of the prognostic significance of binge eating among individuals who are overweight or obese seeking weight loss treatment is one for which a relatively clear answer has emerged. Despite early suggestions that BED negatively impacted the response to treatment, reviews addressing this question (Mitchell et al. 2008; De Zwaan 2010; Tronieri and Wadden 2018) have concluded that, with relatively few exceptions, binge eating or BED prior to treatment does not robustly predict response to weight loss treatment. One key study, the large-scale multicenter Look AHEAD trial examining weight loss in obese individuals with type 2 diabetes mellitus following an intensive lifestyle intervention vs. enhanced usual care, found that, while continued and de novo binge eating following treatment were associated with less favorable weight outcome, baseline binge eating, which had a rather low prevalence of 6.9%, was not a negative predictor (Gorin et al. 2008; Chao et al. 2017). It is important to note that overweight individuals with BED presenting for eating disorder rather than weight toss treatment are a distinct group, and findings regarding weight loss outcomes may not generalize to them. In this group, even in treatment focused on weight reduction, such as behavioral weight loss (Wilson et al. 2010), weight reduction is often minimal. Moreover, the relationships between baseline features, including binge-eating frequency and BMI, treatment assignment, and outcome may be complex (Sysko et al. 2010).

A similar set of questions has arisen regarding patients presenting for bariatric surgery.

Consistent with findings from behavioral weight loss programs, most available evidence suggests that preoperative binge eating is not associated with less favorable weight outcomes (Livhits et al. 2012), but that continued, remerging, or de novo binge eating following surgery does have a negative impact on weight outcomes (Meany et al. 2014; Wimmelmann et al. 2014). Most recently, Devlin et al. (2018) reported on 7-year outcomes from the multicenter Longitudinal Assessment of Bariatric Surgery (LABS) psychosocial study. They found that loss of control eating following Roux-en-Y gastric bypass was associated with less long-term weight loss and greater regain from weight nadir. They also found that a higher postoperative Eating Disorder Examination-Bariatric Surgery Version global score, a widely accepted measure of core eating disorder psychopathology that often accompanies loss of control eating, was associated with poorer weight outcomes. Thus, although binge eating should not be viewed as a contraindication to treatment for weight disorders, via either behavioral weight loss or bariatric surgery, it is a clinically relevant behavior to monitor and perhaps to modify over the course of and following treatment.

7   BED and Shape/Weight Overvaluation

There is some indication that body dissatisfaction in youth may be associated with LOC eating or influence its course, and thus may have important implications for prevention and/or treatment of disorders that include binge eating. An important unanswered question regarding the definition of BED relates to the role of overvaluation of shape and weight (i.e., undue influence of shape and/or weight on self-evaluation) in the overall diagnostic construct. This is particularly notable since the other two major adult eating disorders that include binge eating, bulimia nervosa and the binge-eating/purging subtype of anorexia nervosa, both include overvaluation of shape and weight as a diagnostic criterion. Studies in clinical (Grilo et al. 2015a) and community (Grilo

et al. 2015b) samples conducted to inform the development of DSM-5 eating disorder criteria indicated that a substantial proportion of individuals with clinically significant BED (i.e., showing a degree of distress and/or impairment commensurate with categorization as illness) did not exhibit significant distress related to shape and/or weight, arguing against its inclusion as a diagnostic criterion. However, the argument for overvaluation of shape/weight as a diagnostic *specifier* for BED is cogently summarized by Grilo (2013). Although shape/weight overvaluation is not associated with binge frequency or BMI, it is reliably associated with specific eating-related psychopathology and greater associated symptoms like depression and low self-esteem (Grilo 2013). More recently, a network analysis of BED conducted by Wang and colleagues (2019) identified overvaluation of shape as the symptom with highest centrality, with behavioral features appearing to be somewhat less central. Coffino and colleagues (2019) using data from the National Epidemiologic Survey on Alcohol and Related Conditions (NESARC), reported that shape or weight overvaluation in individuals with BED was associated with greater functional impairment on certain indices, such as interference with normal daily activities, problems getting along with others, and problems fulfilling responsibilities, but not with differences on the Short-Form Health Survey (SF-12) physical or mental disability scales. Therefore, research to date has provided limited support for shape/weight overvaluation as a diagnostic specifier in BED, but more studies are needed.

Another key remaining question with regard to overvaluation of shape and weight is whether it plays a different role in promoting or maintaining binge eating when it occurs in the absence of any compensatory behaviors like dietary restriction or purging, and how that role may depend on development. Emerging evidence suggests that, while there is little support for a perceptual abnormality in BED, there may be cognitive biases similar to those seen in bulimia and anorexia nervosa (Sebastian et al. 1996; Dobson and Dozois 2004), such as attention bias (increased attention to unattractive aspects of one's own body) and body-related memory bias, in individuals with BED. While the evidence on behavioral features of body image disturbance in BED is mixed, some studies suggest that obese individuals with BED, compared to individuals without BED of similar weight, display more body avoidance and/or checking (Reas et al. 2005). However, individuals with BED do not, by definition, engage in behaviors to compensate for binge eating and control their shape and weight. As such, the causal role that overvaluation of shape or weight may play in maintaining disordered eating in BED is not known, and future study is needed to determine whether targeting overvaluation of shape/weight is helpful in reducing binge eating in BED.

8 Conclusions

In this chapter, we have briefly described the results of a range of studies that underscore major limitations in our knowledge of binge eating and, in particular, BED. It is clear that individuals with BED experience substantially much more emotional distress, generally, and about their eating, shape, and weight, specifically, than do comparable individuals without BED. This well-replicated finding supports the clinical utility of BED, as, in the care of individuals with BED, it is important to assess these psychological features and to address them in treatment. What is surprisingly unclear is to what degree the eating behavior of individuals with BED differs substantially from individuals without BED. Laboratory studies convincingly demonstrate major differences in eating behavior, but these have not been borne out by studies using EMA. Furthermore, cessation of binge eating does not reliably lead to weight loss, and binge eating is not a robust predictor of response to behavioral weight loss treatment or to bariatric surgery. These findings indicate that our understanding of the fundamental nature of binge eating unaccompanied by compensatory behaviors is inadequate. We suggest that future studies examine questions such as those raised in this chapter using objective measures of eating behavior from ambulatory individuals as such techniques become available.

References

Brownley KA, Berkman ND, Peat CM, Lohr KN, Cullen KE, Bann CM, Bulik CM (2016) Binge-eating disorder in adults: a systematic review and meta-analysis. Ann Intern Med 165:409–420

Brownstone LM, Bardone-Cone AM, Fitzsimmons-Craft EE, Printz KS, Le Grange D, Mitchell JE, Crow SJ, Peterson CB, Crosby RD, Klein MH, Wonderlich SA, Joiner TE (2013) Subjective and objective binge eating in relation to eating disorder symptomatology, negative affect, and personality dimensions. Int J Eat Disord 46:66–76

Chao AM, Wadden TA, Gorin AA, Shaw Tronieri J, Pearl RL, Bakizada ZM, Yanovski SZ, Berkowitz RI (2017) Binge eating and weight loss outcomes in individuals with type 2 diabetes: 4-year results from the Look AHEAD Study. Obesity (Silver Spring) 25:1830–1837

Coffino JA, Udo T, Grilo CM (2019) The significance of overvaluation of shape or weight in binge-eating disorder: results from a national sample of U.S. adults. Obesity (Silver Spring) 27:1367–1371

De Zwaan M (2010) Obesity treatment for binge-eating disorder in the obese. In: Grilo CM, Mitchell JE (eds) The treatment of eating disorders. Guilford, New York

Devlin MJ, King WC, Kalarchian MA, Hinerman A, Marcus MD, Yanovski SZ, Mitchell JE (2018) Eating pathology and associations with long-term changes in weight and quality of life in the longitudinal assessment of bariatric surgery study. Int J Eat Disord 51:1322–1330

Dobson KS, Dozois DJA (2004) Attentional biases in eating disorders: a meta-analytic review of Stroop performance. Clin Psychol Rev 23:1001–1022

Fairburn CG, Cooper Z, Doll HA, Norman P, O'connor M (2000) The natural course of bulimia nervosa and binge eating disorder in young women. Arch Gen Psychiatry 57:659–665

Farooq M, Doulah A, Parton J, Mccrory MA, Higgins JA, Sazonov E (2019) Validation of sensor-based food intake detection by multicamera video observation in an unconstrained environment. Nutrients 11:609

Goldschmidt AB (2017) Are loss of control while eating and overeating valid constructs? A critical review of the literature. Obes Rev 18:412–449

Goldschmidt AB, Engel SG, Wonderlich SA, Crosby RD, Peterson CB, Le Grange D, Tanofsky-Kraff M, Cao L, Mitchell JE (2012) Momentary affect surrounding loss of control and overeating in obese adults with and without binge eating disorder. Obesity (Silver Spring) 20:1206–1211

Gorin AA, Niemeier HM, Hogan P, Coday M, Davis C, Dilillo VG, Gluck ME, Wadden TA, West DS, Williamson D, Yanovski SZ, Look ARG (2008)

Binge eating and weight loss outcomes in overweight and obese individuals with type 2 diabetes: results from the Look AHEAD trial. Arch Gen Psychiatry 65:1447–1455

Greeno CG, Wing RR, Shiffman S (2000) Binge antecedents in obese women with and without binge eating disorder. J Consult Clin Psychol 68:95–102

Grilo CM (2013) Why no cognitive body image feature such as overvaluation of shape/weight in the binge eating disorder diagnosis? Int J Eat Disord 46:208–211

Grilo CM, Masheb RM, Lozano-Blanco C, Barry DT (2004) Reliability of the eating disorder examination in patients with binge eating disorder. Int J Eat Disord 35:80–85

Grilo CM, Ivezaj V, White MA (2015a) Evaluation of the DSM-5 severity indicator for binge eating disorder in a clinical sample. Behav Res Ther 71:110–114

Grilo CM, Ivezaj V, White MA (2015b) Evaluation of the DSM-5 severity indicator for bulimia nervosa in a community sample. Behav Res Ther 66:72–76

Guss JL, Kissileff HR, Walsh BT, Devlin MJ (1994) Binge eating behavior in patients with eating disorders. Obes Res 2:355–363

Kelly NR, Shank LM, Bakalar JL, Tanofsky-Kraff M (2014) Pediatric feeding and eating disorders: current state of diagnosis and treatment. Curr Psychiatry Rep 16:446

Le Grange D, Gorin A, Catley D, Stone AA (2001) Does momentary assessment detect binge eating in overweight women that is denied at interview? Eur Eat Disord Rev 9:309–324

Livhits M, Mercado C, Yermilov I, Parikh JA, Dutson E, Mehran A, Ko CY, Gibbons MM (2012) Preoperative predictors of weight loss following bariatric surgery: systematic review. Obes Surg 22:70–89

Meany G, Conceicao E, Mitchell JE (2014) Binge eating, binge eating disorder and loss of control eating: effects on weight outcomes after bariatric surgery. Eur Eat Disord Rev 22:87–91

Mitchell JE, Devlin MJ, De Zwaan M, Crow SJ, Peterson CB (2008) Binge-eating disorder. Clinical foundations and treatment. Guilford, New York

Peterson CB, Miller KB, Johnson-Lind J, Crow SJ, Thuras P (2007) The accuracy of symptom recall in eating disorders. Compr Psychiatry 48:51–56

Reas DL, Grilo CM, Masheb RM, Wilson GT (2005) Body checking and avoidance in overweight patients with binge eating disorder. Int J Eat Disord 37:342–346

Sebastian SB, Williamson DA, Blouin DC (1996) Memory bias for fatness stimuli in the eating disorders. Cogn Ther Res 20:275–286

Shomaker LB, Tanofsky-Kraff M, Savastano DM, Kozlosky M, Columbo KM, Wolkoff LE, Zocca JM, Brady SM, Yanovski SZ, Crocker MK, Ali A,

Yanovski JA (2010) Puberty and observed energy intake: boy, can they eat! Am J Clin Nutr 92:123–129

Sonneville KR, Horton NJ, Micali N, Crosby RD, Swanson SA, Solmi F, Field AE (2013) Longitudinal associations between binge eating and overeating and adverse outcomes among adolescents and young adults: does loss of control matter? JAMA Pediatr 167:149–155

Spitzer RL, Devlin M, Walsh BT, Hasin D, Wing R, Marcus M, Stunkard A, Wadden T, Yanovski S, Agras S, Mitchell J, Nonas C (1992) Binge eating disorder—a multisite field trial of the diagnostic-criteria. Int J Eat Disord 11:191–203

Spitzer RL, Yanovski S, Wadden T, Wing R, Marcus MD, Stunkard A, Devlin M, Mitchell J, Hasin D, Horne RL (1993) Binge eating disorder: its further validation in a multisite study. Int J Eat Disord 13:137–153

Stunkard AJ (1959) Eating patterns and obesity. Psychiatry Q 33:284–295

Stunkard AJ, Messick S (1985) The three-factor eating questionnaire to measure dietary restraint, disinhibition and hunger. J Psychosom Res 29:71–83

Sysko R, Hildebrandt T, Wilson GT, Wilfley DE, Agras WS (2010) Heterogeneity moderates treatment response among patients with binge eating disorder. J Consult Clin Psychol 78:681–690

Sysko R, Glasofer DR, Hildebrandt T, Klimek P, Mitchell JE, Berg KC, Peterson CB, Wonderlich SA, Walsh BT (2015) The eating disorder assessment for DSM-5 (EDA-5): development and validation of a structured interview for feeding and eating disorders. Int J Eat Disord 48:452–463

Tronieri JS, Wadden TA (2018) Behavioral assessment of patients with obesity. In: Wadden TA, Bray GA (eds) Handbook of obesity treatment. Guilford, New York

Walsh BT, Boudreau G (2003) Laboratory studies of binge eating disorder. Int J Eat Disord 34:S30–S38

Wang SB, Jones PJ, Dreier M, Elliott H, Grilo CM (2019) Core psychopathology of treatment-seeking patients with binge-eating disorder: a network analysis investigation. Psychol Med 49:1923–1928

Wilson GT, Wilfley DE, Agras WS, Bryson SW (2010) Psychological treatments of binge eating disorder. Arch Gen Psychiatry 67:94–101

Wimmelmann CL, Dela F, Mortensen EL (2014) Psychological predictors of mental health and health-related quality of life after bariatric surgery: a review of the recent research. Obes Res Clin Pract 8:e314–e324

Wolfe BE, Baker CW, Smith AT, Kelly-Weeder S (2009) Validity and utility of the current definition of binge eating. Int J Eat Disord 42:674–686

Wonderlich SA, Gordon KH, Mitchell JE, Crosby RD, Engel SG (2009) The validity and clinical utility of binge eating disorder. Int J Eat Disord 42:687–705

Overcoming Barriers to the Treatment of Binge Eating

Sally Bilić, Johanna Sander, and Stephanie Bauer

Abstract

Despite the significant burden of illness associated with binge-eating disorder, only a minority of affected individuals seek and receive adequate professional treatment. In this chapter, we provide an overview on the various individual, structural, and practical barriers that hinder help-seeking and uptake of treatment for eating disorders in general and binge-eating disorder in particular. Individual factors may include limited self-recognition, a lack of mental health literacy, shame, and stigmatization as well as high ambivalence to change. Socioenvironmental factors may include a lack of social support, stereotypes, and misconceptions. In addition, the limited availability, accessibility, and affordability of treatment play a major role. Finally, a lack of knowledge and experience among health professionals from outside the eating disorders field may contribute to delayed diagnosis and suboptimal treatment. We will describe strategies to increase the reach of preventive and therapeutic interventions. These include, for instance, advocacy efforts and initiatives on the health policy level, screening tools for the early detection of eating disorders, and technology-enhanced interventions to provide support to currently underserved populations. Finally, priorities for future research on barriers and facilitators to care in the eating disorders field are discussed.

Keywords

Eating disorders · Binge-eating disorder · Help-seeking · Treatment · Barriers · Dissemination

Learning Objectives

In this chapter, you will:

- Learn that although many individuals with eating disorders are in need of professional help, only a small proportion seek and receive adequate treatment.
- Get an overview about barriers that impede help-seeking and treatment uptake for eating disorders in general and binge-eating disorder in particular.
- Learn about opportunities how to overcome barriers to care for eating disorders.

S. Bilić · J. Sander · S. Bauer (✉)
Center for Psychotherapy Research, University Hospital Heidelberg, Heidelberg, Germany
e-mail: stephanie.bauer@med.uni-heidelberg.de

1 Introduction

Even though eating disorders are associated with severe impairment, only a small proportion of

© Springer Nature Switzerland AG 2020
G. K.W. Frank, L. A. Berner (eds.), *Binge Eating*, https://doi.org/10.1007/978-3-030-43562-2_22

those affected actually seek help and ultimately utilize treatment. Among those who do, their first contact with treatment providers is often delayed and only about one-third of those seeking help actually receive eating disorder-specific treatment (Hart et al. 2011). These facts illustrate the enormous treatment gap, i.e., the discrepancy between those with an eating disorder who are in need of treatment and those who actually receive specialist treatment (Kazdin et al. 2017). This treatment gap is not only present in the field of eating disorders: It is estimated that between 50 and 70% of US citizens who experience severe mental health problems do not receive any form of treatment, with even lower treatment rates among ethnic minorities (Kessler et al. 2005; Kohn et al. 2018). According to the World Health Organization, this is true for the vast majority of people meeting criteria for a psychiatric disorder worldwide. For Central and South America, the average treatment gap is approximately 75% and even higher in the indigenous population (Kohn et al. 2018). Extremely low treatment rates have also been reported for low-income countries in Eastern Europe and Central Asia. In some countries, only 3% or less of those affected by severe mental illness actually receive treatment (Betancourt and Chambers 2016; World Health Organization 2019).

The treatment gap with respect to eating disorders originates on the one hand from the fact that eating disorders often remain unrecognized and thus undiagnosed for quite a long time. This results in delayed referral to and receipt of eating disorder-specific treatment, which, in turn, leads to prolonged suffering and potentially a chronic disease progression. On the other hand, there is a lack of specialized services for eating disorders, and due to their limited availability, many individuals will receive either no treatment or receive inadequate care. Therefore, there is increasing consensus in the mental health and eating disorders field, that new ways of delivering effective interventions to individuals in need of services are required to reduce the treatment gap (Fairburn and Wilson 2013; Kazdin 2017). In addition, there are other individual, socioenvironmental and treatment-related factors that impede help-seeking and access to services for eating disorders.

In this chapter, we give an overview of the current evidence based on barriers preventing individuals with eating disorders from receiving treatment, with specific regard to barriers relevant for individuals with binge-eating disorder (BED). Furthermore, we describe strategies how to overcome these barriers that may ultimately contribute to a reduction of the burden of illness associated with binge-eating-related disorders on a population level.

2 Individual Barriers to Treatment

A characteristic feature of eating disorders that prevents affected individuals from seeking help and consulting a healthcare professional is the *missing capacity to understand their symptoms as representing a mental health problem* that should be treated (e.g., Gulliksen et al. 2015). In more than half of the cases, eating disorder symptoms are first experienced under the age of 16 years, and nearly half of those affected wait more than a year after symptom identification before seeking help. Due to long waiting times for treatment in many healthcare systems, the beginning of treatment in those who actually seek help is further delayed (Beat 2015).

In the early stages of an eating disorder, *denial of the problematic aspects* of eating disorder-related behaviors also contributes to late or no help-seeking (Roehrig and McLean 2010; Schoen et al. 2012). These are barriers of utmost concern, as we know that early diagnosis and treatment enhance the chances for full recovery, and may avoid long-term suffering and high direct and indirect costs.

Compared to individuals with anorexia nervosa or bulimia nervosa, the delay in help-seeking is even more pronounced in individuals with BED because their symptoms may not be self-perceived as

(continued)

being related to a serious mental illness that requires treatment (Beat 2015).

Individuals with BED may be less likely to recognize their symptoms as an eating disorder because they do not match the stereotype of eating disorders being mostly a phenomenon in underweight young and white females. Limited *eating disorder self-recognition* has been found to be an important factor contributing to delayed help-seeking (Grillot and Keel 2018). Studies showed that being female increases the likelihood of help-seeking for eating or weight-related problems, suggesting that males may be less likely to seek help because of being outside the stereotype (Forrest et al. 2017). This may also apply for individuals from ethnically diverse backgrounds, who may be less likely to seek help than white women (Cachelin et al. 2006).

Poor eating disorder self-recognition may also be related to *a lack of mental health literacy* regarding eating disorders, i.e., lacking awareness and understanding about eating disorders and limited knowledge about available treatment services, which also impedes help-seeking (Gratwick-Sarll et al. 2013).

The literature indicates that individuals are generally more likely to seek help for physical complaints than for mental health problems. This may be particularly the case for individuals with BED who may be rather willing to seek treatment for presumably less tabooed physical comorbidities of their eating disorder. This is reflected in an elevated use of healthcare services among those with BED (Striegel-Moore et al. 2005). However, eating disorders in this population often remain undetected and undiagnosed. Accordingly, not only limited self-recognition and poor mental health literacy among individuals with eating disorders, but also insufficient knowledge and understanding of the diverse and heterogeneous symptomatology of the different forms of eating disorders on the side of healthcare providers may contribute to unmet treatment needs (see below).

Other barriers relevant in terms of reluctance to seek treatment are the individual's *internal perceptions and beliefs* about eating disorders and about what others might think of them or how others may judge them. Many individuals with eating disorders *fear to be stigmatized* when disclosing eating disorder-related problems and thus try to conceal their condition (Mond et al. 2010). A recent literature review found *stigma* to be the most frequently identified barrier to help-seeking for eating disorders (Ali et al. 2017). Individuals with eating disorders also often report *feelings of guilt, shame, and embarrassment* or do not dare to talk about their thoughts and feelings based on the inner belief that they should be able to handle the eating disorder by themselves (Becker et al. 2010).

Another factor that negatively impacts the process of seeking and receiving professional help is *weight stigma*. Weight-based stigma and weight-related social discrimination have been found to be highly prevalent in the western culture (Puhl and Suh 2015). Recently, weight-based stigma and discrimination have been described as a global health problem that does not occur only in high-income countries, but is also increasingly common in low- and middle-income countries (Brewis et al. 2018). Weight stigma has been found to be a risk factor associated with eating disorder symptoms and particularly with BED (Almeida et al. 2011). Research has shown that individuals with BED internalize negative societal weight biases and also tend to have stigmatizing thoughts and attitudes toward themselves (Barnes et al. 2014).

The perception of being the target of weight stigma among individuals with BED, i.e., concerns of being negatively judged, avoided, or rejected because of a larger body size contributes to poor eating disorder related help-seeking in these individuals.

Research also revealed that individuals with eating disorder symptoms consider weight loss and weight control strategies desirable (e.g., Mond et al. 2010), contributing to low motivation and high ambivalence to change which are also important factors hindering individuals from seeking help. In case of long illness duration, sufferers may also have low confidence to be able to change entrenched behavior patterns, and may fear to lose the sense of control provided by the eating disorder, e.g., overeating habits, body weight, and life choices in general (Gulliksen et al. 2015; Leavey et al. 2011). Taken together, these intraindividual beliefs and perceptions may lead to passive and ambivalent attitudes toward help-seeking, treatment initiation, and recovery.

3 Socioenvironmental Barriers to Treatment

In addition to barriers arising from within a person, there are also factors in the individual's social environment and in the service landscape that prevent people from receiving adequate treatment for their eating disorder. Research revealed that stereotypical misconceptions and stigmatizing attitudes related to people with eating disorders are widespread among the public and also among families and friends of individuals with eating disorders. This fact is assumed to be a major obstacle in preventing those affected from seeking help.

Empirical studies indicate that eating disorders are more stigmatized than other mental or physical health conditions and that the public tends to perceive eating disorders to be rather a lifestyle choice and not a "real" illness that requires professional help (Ebneter and Latner 2013; Roehrig and McLean 2010). Individuals with anorexia or bulimia nervosa are frequently viewed as attention-seeking, personally responsible and blameworthy for their condition and should just "pull themselves together" to overcome their problem (Crisp 2005). Accordingly, lack of social support and encouragement by others including missing family support and poor understanding from peers, often prevent individuals with eating

disorders from disclosing their problem and seeking help (Akey et al. 2013; Becker et al. 2010).

Stigmatizing attributions reflecting a lack of self-control, personal responsibility, and blameworthiness are even more pronounced toward individuals with BED compared to those with anorexia or bulimia (Puhl and Suh 2015), while at the same time BED is perceived to be less severe, less impairing, and more easily treatable (Reas 2017).

Stereotypical ideas and stigmatizing beliefs of eating disorder patients are also present among healthcare professionals, and are related to *negative reactions of clinicians* toward individuals with eating disorders reflecting frustration, anger, worries, or hopelessness, accompanied by lack of improvement in patients. Findings of a literature review showed that this particularly applies to inexperienced clinicians, and that medical practitioners frequently report feelings of lacking competence concerning the treatment of eating disorders (Thompson-Brenner et al. 2012).

Another challenge concerns the fact that healthcare professionals frequently fail to identify patients' symptoms as being related to an eating disorder. This is considered to be particularly due to a lack of knowledge and experience regarding eating disorders and treatment options among primary care physicians who, in many cases, are the first point of contact if an individual decides to seek help (Dickerson et al. 2011). According to a study by Linville et al. (2010), 92% of physicians and nurse practitioners believe they have missed an eating disorder diagnosis in the past.

Especially for BED, although much more common than anorexia and bulimia nervosa, diagnosis and treatment rates are alarmingly low (Kornstein 2017). Hudson et al. (2007) reported that only 3.2% of those fulfilling the diagnostic criteria for BED have ever received a formal diagnosis. Findings were confirmed by the BEST study ("Binge-Eating Self-Help

Treatment") that analyzed health insurance data of 100 women who had met diagnostic criteria for BED within the 12 months prior to study enrollment. The results showed that only 4% of this sample received an eating disorder diagnosis within that period, even though 99% had at least one contact to a healthcare provider (Dickerson et al. 2011).

> The risk of delayed detection or not being diagnosed with an eating disorder at all is even higher in individuals with BED compared to other eating disorders, because the BED symptoms do not match the widespread stereotypical view of eating disorders. Knowledge concerning the diagnosis and treatment among nonspecialist healthcare providers is often scarce.

A recent study including 405 healthcare providers from the United States revealed that 93.0% of the general healthcare providers and 88.6% of psychiatrists among the survey respondents were not able to correctly identify the diagnostic criteria for BED (Chao et al. 2019). However, access to treatment is not only hindered by inadequate detection and the absence or delay of diagnosis: Even if an eating disorder is diagnosed and the individual is referred to mental health treatment, the limited availability and affordability of appropriate services are major challenges. Several reviews came to similar conclusions with respect to eating disorders and other mental health problems and identified practical barriers such as lack of time, costs, and service restrictions as factors why affected individuals do not receive treatment (Brown et al. 2016; Gulliver et al. 2010; Innes et al. 2017).

4 How to Overcome Barriers to Treatment?

There is broad consensus among various stakeholders that facilitating access to prevention and treatment is of utmost importance in order to reduce the burden of illness related to eating disorders. But how can we achieve this aim and, more specifically, who can do what in order to reduce barriers to adequate care? Given the manifold factors that contribute to the treatment gap, there is a need for strategies and action on different levels.

On the individual level, interventions to improve mental health literacy, i.e., to improve knowledge around eating disorders and their treatment, as well as interventions to reduce stigma related to eating disorders are assumed to facilitate self-recognition, disclosure, and help-seeking among affected individuals. However, changing individuals' attitudes and self-perception is challenging. Different specific approaches such as educating about eating disorders in general or explaining the etiology of eating disorders have been discussed in the literature, but it is currently unknown which procedures are most effective in reducing stigma and there is a clear need for more systematic evaluation research in this area (Doley et al. 2017).

In order to counteract the public stigma and raise awareness around eating disorders in general by educating the broader public, large-scale campaigns and coordinated advocacy efforts are required. One example of such an effort is an initiative called the "Nine truths about eating disorders" which was developed by an international group of eating disorder experts, professional organizations, and advocacy groups in 2015. A set of nine facts related to the etiology, severity, consequences, and treatment of eating disorders was compiled, translated into over 30 languages, and disseminated in different formats (e.g., print, online, and video) in various countries (Schaumberg et al. 2017). However, the actual reach of this resource (i.e., the size of the population that has accessed the document) is currently unknown and challenging to determine. An example from the field of eating disorder prevention, i.e., the Body Project, shows that it is indeed possible to implement and disseminate an intervention to large populations in many different countries if researchers, clinicians, and community stakeholder work together (Becker and Stice 2017). Such initiatives promise to improve knowledge and change attitudes toward eating disorders not only on the individual level (which should result in improved help-seeking attitudes and behaviors) but on a societal level. Advocacy ultimately aims at a shift on the health political level by convincing policy makers that eating disorders are serious mental illnesses that may be prevented and treated in a cost-effective manner if adequate structural and financial support is available. As long as care for eating disorders is not fully covered by health insurance, cost of treatment will remain a major barrier to treatment and prevent affected individuals from accessing specialist treatment even if they may be aware that they are in need of it. The field is challenged to engage various stakeholder groups in the research process to enhance the evidence base of advocacy work and to ultimately quantify its impact. This is in line with the recent call for more strategic science, i.e., research that may answer policy-related questions relevant for care around eating disorders (Thomas et al. 2017).

On the level of healthcare providers, it is not only important to improve the availability of expert treatment and increase the number of eating disorder specialists, but also to work toward an improved detection of eating disorders in primary care and among professionals from outside the mental health field. As described above, many individuals with BED get in contact with health care providers, but only in a minority the illness is recognized and diagnosed, and an even smaller proportion actually receives timely specialized treatment. Thus, the provision of guidelines for standardized eating disorder diagnosis, the training of professionals from outside of our own field and their sensitization for eating disorders in general and BED in particular should result in higher numbers of affected individuals that get timely access to treatment. This also highlights the importance of networking between different groups of professionals (e.g., primary care physicians, mental health professionals, and eating disorder experts) and a close connection of different health care sectors which is essential for an improved continuity of care.

Another approach that is considered promising, in terms of reducing barriers to care

and facilitating early eating disorder detection and timely treatment allocation, is the implementation of online and mobile screening tools. While such tools do not intend to replace standard diagnostic procedures in the face-to-face setting, they may efficiently assess eating disorder-related risk factors, attitudes, and behaviors and provide immediate feedback to users. Depending on their entries, this feedback may recommend users to see a professional in order to clarify their need for treatment. Currently, it is still largely unknown to which extent such screening and feedback may improve actual help-seeking behavior and reduce the time until uptake of eating disorder treatment. Research indicates that a large proportion of the individuals who decide to complete screening assessments for eating disorders on the internet actually report substantial eating disorder-related impairment and do not currently receive treatment (e.g., Fitzsimmons-Craft et al. 2019a, b; McLean et al. 2019). This means that such tools may reach currently underserved groups, but their impact still needs to be evaluated in rigorous research.

A related topic concerns the question whether technology-enhanced interventions (beyond screening tools) may reduce barriers to care. This could be the case in two different ways: First, e-mental health interventions could be used to facilitate access to conventional face-to-face care and second, technology could be used to deliver the actual intervention without an in-person contact between client and provider. Only few studies in the eating disorder field have focused on the first scenario: Two studies from Germany provided preliminary evidence that an internet-based program for prevention and early intervention may facilitate access to conventional care in individuals who experience substantial eating disorder-related impairment during their participation and reported that they would not have sought face-to-face treatment without the support they received from the online program (Kindermann et al. 2016; Moessner et al. 2016a). Recently, a brief online program that specifically aims at addressing sufferers' ambivalence to change and their reluctance to enter treatment was evaluated in the United Kingdom. The results showed that only a third of the participants actually used the intervention and that participation in the program was not associated with a higher probability of participants accessing specialist eating disorder services (Denison-Day et al. 2019).

Concerning the second scenario of harnessing the potential of technology, i.e., delivering preventive or therapeutic interventions via technology, a much larger body of research is available. Several reviews and meta-analyses indicate that such interventions may reduce eating disorder-related risk factors and impairment (e.g., Aardoom et al. 2013; Melioli et al. 2016). While many studies have included mixed samples, some have specifically explored the effects of online interventions for BED. For example, one program was developed based on the self-help book "Overcoming Binge Eating" using principles of cognitive behavioral therapy (CBT; Fairburn 1995). The internet-based intervention includes structured modules and exercises as well as a self-monitoring diary and continuous e-mail support by a psychologist. In a randomized trial conducted in Europe, participants in the intervention group showed significantly larger improvements in binge-eating behavior, drive for thinness, body dissatisfaction, and interoceptive awareness compared to a waitlist control group (Carrard et al. 2011).

Wagner et al. (2016) reported similar results for another CBT-oriented guided self-help intervention that consists of structured writing exercises (on topics such as body image and problem solving), psychoeducation, an online diary, week schedules', and regular electronic feedback by a therapist. It was shown that the intervention group experienced significantly fewer objective binge-eating episodes, and more improvement in dietary restraint, and weight and shape concerns compared to a control group that did not have access to the online program (Wagner et al. 2016).

Of particular relevance for service research is of course the direct comparison between online treatment and conventional evidence-based psychotherapy in the face-to-face setting. To this end, De Zwaan and colleagues studied the efficacy of a therapist-guided online intervention in comparison to outpatient CBT for full-syndrome or subthreshold BED in Germany (De Zwaan et al. 2017). The findings indicate that both groups improved and did not differ in general psychopathology, BMI, and quality of life. However, the results showed that face-to-face CBT was superior to the online program in terms of key BED symptoms, i.e., the conventional intervention was more effective in reducing the number of objective binge-eating days and eating disorder psychopathology until the 6-month follow-up. However, these group differences were not present anymore at the 18-month follow-up, so that the authors conclude that internet-based guided self-help interventions probably act with a slower trajectory of improvement compared to face-to-face treatment. In subsequent health–economic analyses no significant differences between the interventions were found, i.e., there was only a trend of CBT being more effective but at the same time more costly (Konig et al. 2018).

It makes intuitive sense to assume that e-health interventions may help us to overcome some of the major barriers to eating disorder treatment: Due to technological advances and today's broad and permanent availability of internet and mobile technology, the theoretical reach of such interventions is huge. We can offer timely, easy, and low-threshold access to large groups of individuals who are unable to utilize conventional care (e.g., due to a lack of providers in the region or country they live) or who are not willing to utilize such care (e.g., due to shame and embarrassment). However, there is a clear need for more studies on the actual reach of technology-enhanced interventions and on effective dissemination strategies (Bauer and Moessner 2013; Lipson et al. 2017; Nacke et al. 2019). Most of the studies so far include rather small samples and it is unknown to which extent underserved populations may be actually reached by such interventions. Actually, recent research indicates that it is challenging to convince the target group of young people to register for and utilize online eating disorder prevention programs (Fitzsimmons-Craft et al. 2019a, b; Moessner et al. 2016b). There is also evidence that different access paths via which users arrive at an online intervention and specific strategies to approach the target group (e.g., via high schools vs. online/social media) result in participant groups with different sociodemographic characteristics and severity of eating disorder-related impairment and with different duration and intensity of intervention use (Bauer et al. 2019). In addition to identifying most promising and cost-effective dissemination strategies, it is also important to assess the cost of intervention delivery in order to lay the basis for a sustained implementation of technology-enhanced interventions into routine healthcare (Minarik et al. 2013; Moessner et al. 2016b).

> Future developments should focus on the sustained implementation of screening tools and improved diagnostic procedures in order to facilitate earlier symptom recognition. Also, there is a need for broadly available easy access interventions (e.g., delivered via technology) as a complement to conventional healthcare.

5   Conclusion

Even highly effective interventions may not alleviate the burden of disease associated with BED if they are not broadly available and/or not widely utilized by affected individuals. In terms of an intervention's public health impact, its effectiveness and reach are of equal importance. Past service research has focused primarily on developing and evaluating evidence-based treatments for BED which is of course a prerequisite for the provision of adequate professional care. However, future studies also need to more systematically investigate ways how to overcome barriers to care in order to ultimately increase the reach of evidence-based care.

As described in this chapter, the reasons why so many individuals affected by eating disorders do not seek or receive professional help are manifold. While some of these factors may be specific to certain healthcare contexts or specific groups of individuals, others may be relevant universally. There is a clear need for more systematic research to better understand the various barriers to care, their interplay, and their relative contribution to the treatment gap. Particularly, more international research is needed as well as studies in samples with low socioeconomic status, migration status, and ethnically diverse backgrounds.

The good news is that many of the barriers that have been identified are modifiable, i.e., the opportunities for policy makers, mental health professionals, advocacy groups, activists, and researchers to contribute to a reduction of barriers are also manifold. There is no single solution to this challenge and joint efforts on multiple levels are required in order to improve the availability and accessibility of effective preventive and therapeutic interventions for BED.

References

Aardoom JJ, Dingemans AE, Spinhoven P, Van Furth EF (2013) Treating eating disorders over the internet: a systematic review and future research directions. Int J Eat Disord 46(6):539–552. https://doi.org/10.1002/eat.22135

Akey JE, Rintamaki LS, Kane TL (2013) Health belief model deterrents of social support seeking among people coping with eating disorders. J Affect Disord 145(2):246–252. https://doi.org/10.1016/j.jad.2012.04.045

Ali K, Farrer L, Fassnacht DB, Gulliver A, Bauer S, Griffiths KM (2017) Perceived barriers and facilitators towards help-seeking for eating disorders: a systematic review. Int J Eat Disord 50(1):9–21. https://doi.org/10.1002/eat.22598

Almeida L, Savoy S, Boxer P (2011) The role of weight stigmatization in cumulative risk for binge eating. J Clin Psychol 67(3):278–292. https://doi.org/10.1002/jclp.20749

Barnes RD, Ivezaj V, Grilo CM (2014) An examination of weight bias among treatment-seeking obese patients with and without binge eating disorder. Gen Hosp Psychiatry 36(2):177–180. https://doi.org/10.1016/j.genhosppsych.2013.10.011

Bauer S, Moessner M (2013) Harnessing the power of technology for the treatment and prevention of eating disorders. Int J Eat Disord 46(5):508–515

Bauer S, Bilic S, Özer F, Moessner M (2019) Dissemination of an internet-based program for the prevention and early intervention in eating disorders: relationship between access paths, user characteristics, and program utilization. Z Kinder Jugendpsychiatr Psychother 48:25. https://doi.org/10.1024/1422-4917/a000662

Beat (2015) The costs of eating disorders: social, health and economic impacts. Beat, UK

Becker CB, Stice E (2017) From efficacy to effectiveness to broad implementation: evolution of the body project. J Consult Clin Psychol 85(8):767–782

Becker AE, Hadley Arrindell A, Perloe A, Fay K, Striegel-Moore RH (2010) A qualitative study of perceived social barriers to care for eating disorders: perspectives from ethnically diverse health care consumers. Int J Eat Disord 43(7):633–647. https://doi.org/10.1002/eat.20755

Betancourt TS, Chambers DA (2016) Optimizing an era of global mental health implementation science. JAMA Psychiat 73(2):99. https://doi.org/10.1001/jamapsychiatry.2015.2705

Brewis A, SturtzSreetharan C, Wutich A (2018) Obesity stigma as a globalizing health challenge. Glob Health 14(1):20. https://doi.org/10.1186/s12992-018-0337-x

Brown A, Rice SM, Rickwood DJ, Parker AG (2016) Systematic review of barriers and facilitators to accessing and engaging with mental health care among at-risk young people. Asia Pac Psychiatry 8(1):3–22. https://doi.org/10.1111/appy.12199

Cachelin FM, Striegel-Moore RH, Regan PC (2006) Factors associated with treatment seeking in a community sample of European American and Mexican American women with eating disorders. Eur Eat Disord Rev 14(6):422–429. https://doi.org/10.1002/erv.720

Carrard I, Crepin C, Rouget P, Lam T, Golay A, Van der Linden M (2011) Randomised controlled trial of a guided self-help treatment on the internet for binge eating disorder. Behav Res Ther 49(8):482–491. https://doi.org/10.1016/j.brat.2011.05.004

Chao AM, Rajagopalan AV, Tronieri JS, Walsh O, Wadden TA (2019) Identification of binge eating disorder criteria: results of a National Survey of Healthcare Providers. J Nurs Scholarsh 51(4):399–407. https://doi.org/10.1111/jnu.12468

Crisp A (2005) Stigmatization of and discrimination against people with eating disorders including a report of two nationwide surveys. Eur Eat Disord Rev 13(3):147–152. https://doi.org/10.1002/erv.648

De Zwaan M, Herpertz S, Zipfel S, Svaldi J, Friederich HC, Schmidt F et al (2017) Effect of internet-based guided self-help vs individual face-to-face treatment on full or subsyndromal binge eating disorder in overweight or obese patients: the INTERBED randomized clinical trial. JAMA Psychiat 74(10):987–995. https://doi.org/10.1001/jamapsychiatry.2017.2150

Denison-Day J, Muir S, Newell C, Appleton KM (2019) A Web-based intervention (MotivATE) to increase attendance at an eating disorder service assessment appointment: Zelen randomized controlled trial. JMIR 21(2):e11874. https://doi.org/10.2196/11874

Dickerson JF, DeBar L, Perrin NA, Lynch F, Wilson GT, Rosselli F, Kraemer HC, Striegel-Moore RH (2011)

Health-service use in women with binge eating disorders. Int J Eat Disord 44(6):524–530. https://doi.org/10.1002/eat.20842

Doley JR, Hart LM, Stukas AA, Petrovic K, Bouguettaya A, Paxton SJ (2017) Interventions to reduce the stigma of eating disorders: a systematic review and meta-analysis. Int J Eat Disord 50(3):210–230. https://doi.org/10.1002/eat.22691

Ebneter DS, Latner JD (2013) Stigmatizing attitudes differ across mental health disorders: a comparison of stigma across eating disorders, obesity, and major depressive disorder. J Nerv Ment Dis 201(4):281–285. https://doi.org/10.1097/NMD.0b013e318288e23f

Fairburn CG (1995) Overcoming binge eating. Guilford, New York, NY

Fairburn CG, Wilson GT (2013) The dissemination and implementation of psychological treatments: problems and solutions. Int J Eat Disord 46(5):516–521. https://doi.org/10.1002/eat.22110

Fitzsimmons-Craft EE, Balantekin KN, Graham AK, Smolar L, Park D, Mysko C et al (2019a) Results of disseminating an online screen for eating disorders across the US: reach, respondent characteristics, and unmet treatment need. Int J Eat Dis 52(6):721–729. https://doi.org/10.1002/eat.23043

Fitzsimmons-Craft EE, Firebaugh ML, Graham AK, Eichen DM, Monterubio GE, Balantekin KN et al (2019b) State-wide university implementation of an online platform for eating disorders screening and intervention. Psychol Serv 16(2):239. https://doi.org/10.1037/ser0000264

Forrest LN, Smith AR, Swanson SA (2017) Characteristics of seeking treatment among U.S. adolescents with eating disorders. Int J Eat Disord 50(7):826–833. https://doi.org/10.1002/eat.22702

Gratwick-Sarll K, Mond J, Hay P (2013) Self-recognition of eating-disordered behavior in college women: further evidence of poor eating disorders "mental health literacy"? Eat Disord 21(4):310–327. https://doi.org/10.1080/10640266.2013.797321

Grillot CL, Keel PK (2018) Barriers to seeking treatment for eating disorders: the role of self-recognition in understanding gender disparities in who seeks help. Int J Eat Disord 51(11):1285–1289. https://doi.org/10.1002/eat.22965

Gulliksen KS, Nordbo RH, Espeset EM, Skarderud F, Holte A (2015) The process of help-seeking in anorexia nervosa: patients' perspective of first contact with health services. Eat Disord 23(3):206–222. https://doi.org/10.1080/10640266.2014.981429

Gulliver A, Griffiths KM, Christensen H (2010) Perceived barriers and facilitators to mental health help-seeking in young people: a systematic review. BMC Psychiatry 10(1):113. https://doi.org/10.1186/1471-244X-10-113

Hart LM, Granillo MT, Jorm AF, Paxton SJ (2011) Unmet need for treatment in the eating disorders: a systematic review of eating disorder specific treatment seeking among community cases. Clin Psychol Rev 31

(5):727–735. https://doi.org/10.1016/j.cpr.2011.03.004

Hudson JI, Hiripi E, Pope HG Jr, Kessler RC (2007) The prevalence and correlates of eating disorders in the National Comorbidity Survey Replication. Biol Psychiatry 61(3):348–358. https://doi.org/10.1016/j.biopsych.2006.03.040

Innes NT, Clough BA, Casey LM (2017) Assessing treatment barriers in eating disorders: a systematic review. Eat Disord 25(1):1–21. https://doi.org/10.1080/10640266.2016.1207455

Kazdin AE (2017) Addressing the treatment gap: a key challenge for extending evidence-based psychosocial interventions. Behav Res Ther 88:7–18. https://doi.org/10.1016/j.brat.2016.06.004

Kazdin AE, Fitzsimmons-Craft EE, Wilfley DE (2017) Addressing critical gaps in the treatment of eating disorders. Int J Eat Disord 50(3):170–189. https://doi.org/10.1002/eat.22670

Kessler RC, Demler O, Frank RG, Olfson M, Pincus HA, Walters EE, Wang P, Wells KB, Zaslavsky AM (2005) Prevalence and treatment of mental disorders, 1990 to 2003. N Engl J Med 352(24):2515–2523. https://doi.org/10.1056/NEJMsa043266

Kindermann S, Ali K, Minarik C, Moessner M, Bauer S (2016) Enhancing help-seeking behavior in individuals with eating disorder symptoms via internet: a case report. Ment Health Prevention 4(2):69–74. https://doi.org/10.1016/j.mhp.2016.04.002

Kohn R, Ali AA, Puac-Polanco V, Figueroa C, López-Soto V et al (2018) Mental health in the Americas: an overview of the treatment gap. Rev Panam Salud Publica 42:e165. https://doi.org/10.26633/RPSP.2018.165

Konig HH, Bleibler F, Friederich HC, Herpertz S, Lam T, Mayr A et al (2018) Economic evaluation of cognitive behavioral therapy and internet-based guided self-help for binge eating disorder. Int J Eat Disord 51(2):155–164. https://doi.org/10.1002/eat.22822

Kornstein SG (2017) Epidemiology and recognition of binge-eating disorder in psychiatry and primary care. J Clin Psychiatry 78(Suppl 1):3–8. https://doi.org/10.4088/JCP.sh16003su1c.01

Leavey G, Vallianatou C, Johnson-Sabine E, Rae S, Gunputh V (2011) Psychosocial barriers to engagement with an eating disorder service: a qualitative analysis of failure to attend. Eat Disord 19(5):425–440. https://doi.org/10.1080/10640266.2011.609096

Linville D, Benton A, O'Neil M, Sturm K (2010) Medical providers' screening, training and intervention practices for eating disorders. Eat Disord 18(2):110–131. https://doi.org/10.1080/10640260903585532

Lipson SK, Jones JM, Taylor CB, Wilfley DE, Eichen DM, Fitzsimmons-Craft EE, Eisenberg D (2017) Understanding and promoting treatment-seeking for eating disorders and body image concerns on college campuses through online screening, prevention and intervention. Eat Behav 25:68–73. https://doi.org/10.1016/j.eatbeh.2016.03.020

McLean S, Caldwell B, Roberton M (2019) Reach out and recover: intentions to seek treatment in individuals using online support for eating disorders. Int J Eat Disord 52(10):1137–1149. https://doi.org/10.1002/eat.23133

Melioli T, Bauer S, Franko DL, Moessner M, Ozer F, Chabrol H, Rodgers RF (2016) Reducing eating disorder symptoms and risk factors using the internet: a meta-analytic review. Int J Eat Disord 49(1):19–31. https://doi.org/10.1002/eat.22477

Minarik C, Moessner M, Özer F, Bauer S (2013) Implementierung und Dissemination eines internetbasierten Programms zur Prävention und frühen Intervention bei Essstörungen. Psychiatr Prax 40:332–338. https://doi.org/10.1055/s-0033-1349488

Moessner M, Minarik C, Ozer F, Bauer S (2016a) Can an internet-based program for the prevention and early intervention in eating disorders facilitate access to conventional professional healthcare? J Ment Health 25(5):441–447. https://doi.org/10.3109/09638237.2016.1139064

Moessner M, Minarik C, Ozer F, Bauer S (2016b) Effectiveness and cost-effectiveness of school-based dissemination strategies of an internet-based program for the prevention and early intervention in eating disorders: a randomized trial. Prev Sci 17(3):306–313. https://doi.org/10.1007/s11121-015-0619-y

Mond JM, Hay PJ, Paxton SJ, Rodgers B, Darby A, Nillson J, Quirk F, Owen C (2010) Eating disorders "mental health literacy" in low risk, high risk and symptomatic women: implications for health promotion programs. Eat Disord 18(4):267–285. https://doi.org/10.1080/10640266.2010.490115

Nacke B, Beintner I, Gorlich D, Vollert B, Schmidt-Hantke J, Hutter K et al (2019) everyBody-tailored online health promotion and eating disorder prevention for women: study protocol of a dissemination trial. Internet Interv 16:20–25. https://doi.org/10.1016/j.invent.2018.02.008

Puhl R, Suh Y (2015) Stigma and eating and weight disorders. Curr Psychiatry Rep 17(3):552. https://doi.org/10.1007/s11920-015-0552-6

Reas DL (2017) Public and healthcare professionals' knowledge and attitudes toward binge eating disorder: a narrative review. Nutrients 9(11):E1267. https://doi.org/10.3390/nu9111267

Roehrig JP, McLean CP (2010) A comparison of stigma toward eating disorders versus depression. Int J Eat Disord 43(7):671–674. https://doi.org/10.1002/eat.20760

Schaumberg K, Welch E, Breithaupt L, Hübel C, Baker JH, Munn-Chernoff MA et al (2017) The science behind the academy for eating disorders' nine truths about eating disorders. Eur Eat Disord Rev 25(6):432–450

Schoen EG, Lee S, Skow C, Greenberg ST, Bell AS, Wiese JE, Martens JK (2012) A retrospective look at the internal help-seeking process in young women with eating disorders. Eat Disord 20(1):14–30. https://doi.org/10.1080/10640266.2012.635560

Striegel-Moore RH, Dohm FA, Kraemer HC, Schreiber GB, Crawford PB, Daniels SR (2005) Health services use in women with a history of bulimia nervosa or binge eating disorder. Int J Eat Disord 37(1):11–18. https://doi.org/10.1002/eat.20090

Thomas JJ, Klump KL, Weissman RS (2017) Introduction to the special issue on evidence-based advocacy and strategic science in eating disorders. Int J Eat Disord 50 (3):169–169. https://doi.org/10.1002/eat.22684

Thompson-Brenner H, Satir DA, Franko DL, Herzog DB (2012) Clinician reactions to patients with eating disorders: a review of the literature. Psychiatr Serv 63 (1):73–78. https://doi.org/10.1176/appi.ps.201100050

Wagner B, Nagl M, Dolemeyer R, Klinitzke G, Steinig J, Hilbert A, Kersting A (2016) Randomized controlled trial of an internet-based cognitive-behavioral treatment program for binge-eating disorder. Behav Ther 47(4):500–514. https://doi.org/10.1016/j.beth.2016.01.006

World Health Organization (2019) Closing the mental health treatment gap in Central Asia. World Health Organization Regional Office for Europe. Available http://www.euro.who.int/en/countries/kyrgyzstan/news/news/2019/6/closing-the-mental-health-treatment-gap-in-central-asia

Printed in the United States
by Baker & Taylor Publisher Services